Lecture Notes in Computer Science 11707

More information about this series at http://www.springer.com/series/7409

Sven Hartmann · Josef Küng ·
Sharma Chakravarthy · Gabriele Anderst-Kotsis ·
A Min Tjoa · Ismail Khalil (Eds.)

Database and Expert Systems Applications

30th International Conference, DEXA 2019
Linz, Austria, August 26–29, 2019
Proceedings, Part II

Springer

Editors
Sven Hartmann
Clausthal University of Technology
Clausthal-Zellerfeld, Germany

Josef Küng
Johannes Kepler University of Linz
Linz, Austria

Sharma Chakravarthy
The University of Texas at Arlington
Arlington, TX, USA

Gabriele Anderst-Kotsis
Johannes Kepler University of Linz
Linz, Austria

A Min Tjoa ⓘ
Software Competence Center Hagenberg
Hagenberg im Mühlkreis, Austria

Ismail Khalil
Johannes Kepler University of Linz
Linz, Austria

ISSN 0302-9743   ISSN 1611-3349 (electronic)
Lecture Notes in Computer Science
ISBN 978-3-030-27617-1   ISBN 978-3-030-27618-8 (eBook)
https://doi.org/10.1007/978-3-030-27618-8

LNCS Sublibrary: SL3 – Information Systems and Applications, incl. Internet/Web, and HCI

This Springer imprint is published by the registered company Springer Nature Switzerland AG
The registered company address is: Gewerbestrasse 11, 6330 Cham, Switzerland

Preface

This volume contains the papers presented at the 30th International Conference on Database and Expert Systems Applications (DEXA 2019), which was held in Linz, Austria, during August 26–29, 2019. On behalf of the Program Committee, we commend these papers to you and hope you find them useful.

Database, information, and knowledge systems have always been a core subject of computer science. The ever-increasing need to distribute, exchange, and integrate data, information, and knowledge has added further importance to this subject. Advances in the field will help facilitate new avenues of communication, to proliferate interdisciplinary discovery, and to drive innovation and commercial opportunity.

DEXA is an international conference series that showcases state-of-the-art research activities in database, information, and knowledge systems. The conference and its associated workshops provide a premier annual forum to present original research results and to examine advanced applications in the field. The goal is to bring together developers, scientists, and users to extensively discuss requirements, challenges, and solutions in database, information, and knowledge systems.

DEXA 2019 solicited original contributions dealing with any aspect of database, information, and knowledge systems. Suggested topics included, but were not limited to:

- Acquisition, Modeling, Management, and Processing of Knowledge
- Authenticity, Privacy, Security, and Trust
- Availability, Reliability, and Fault Tolerance
- Big Data Management and Analytics
- Consistency, Integrity, Quality of Data
- Constraint Modeling and Processing
- Cloud Computing and Database-as-a-Service
- Database Federation and Integration, Interoperability, Multi-Databases
- Data and Information Networks
- Data and Information Semantics
- Data Integration, Metadata Management, and Interoperability
- Data Structures and Data Management Algorithms
- Database and Information System Architecture and Performance
- Data Streams and Sensor Data
- Data Warehousing
- Decision Support Systems and Their Applications
- Dependability, Reliability, and Fault Tolerance
- Digital Libraries and Multimedia Databases
- Distributed, Parallel, P2P, Grid, and Cloud Databases
- Graph Databases
- Incomplete and Uncertain Data
- Information Retrieval

- Information and Database Systems and Their Applications
- Mobile, Pervasive, and Ubiquitous Data
- Modeling, Automation, and Optimization of Processes
- NoSQL and NewSQL Databases
- Object, Object-Relational, and Deductive Databases
- Provenance of Data and Information
- Semantic Web and Ontologies
- Social Networks, Social Web, Graph, and Personal Information Management
- Statistical and Scientific Databases
- Temporal, Spatial, and High-Dimensional Databases
- Query Processing and Transaction Management
- User Interfaces to Databases and Information Systems
- Visual Data Analytics, Data Mining, and Knowledge Discovery
- WWW and Databases, Web Services
- Workflow Management and Databases
- XML and Semi-Structured Data

Following the call for papers, which attracted 157 submissions, there was a rigorous review process that saw each submission refereed by 3 to 6 international experts. The 32 submissions judged best by the Program Committee were accepted as full research papers, yielding an acceptance rate of 20%. A further 34 submissions were accepted as special research papers.

As is the tradition of DEXA, all accepted papers are published by Springer. Authors of selected papers presented at the conference were invited to submit substantially extended versions of their conference papers for publication in special issues of international journals. The submitted extended versions underwent a further review process.

The success of DEXA 2019 was the result of collegial teamwork from many individuals. We wish to thank all authors who submitted papers and all conference participants for the fruitful discussions.

We are grateful to Dirk Draheim, (Technical University of Tallinn), Vladimir Marik (Technical University of Prague), Axel Polleres (Vienna Business School), and Stefanie Rinderle Ma (University of Vienna) for their keynote talks.

This edition of DEXA also featured four international workshops covering a variety of specialized topics:

- BIOKDD 2019: The 10th International Workshop on Biological Knowledge Discovery from Data
- IWCFS 2019: The Third International Workshop on Cyber-Security and Functional Safety in Cyber-Physical Systems
- MLKgraphs 2019: The First International Workshop on Machine Learning and Knowledge Graphs
- TIR 2019: The 16th International Workshop on Technologies for Information Retrieval

We would like to express our thanks to all institutions actively supporting this event, namely:

- Johannes Kepler University Linz (JKU)
- Software Competence Center Hagenberg (SCCH)
- International Organization for Information Integration and Web based applications and Services (@WAS)

Finally, we hope that all the participants of DEXA 2019 enjoyed the program that was put together.

August 2019

Sven Hartmann
Josef Küng
Sharma Chakravarthy

Organization

General Chair

A Min Tjoa   Technical University of Vienna, Austria

Program Committee Chairs

Sharma Chakravarthy   University of Texas at Arlington, USA
Sven Hartmann   Clausthal University of Technology, Germany
Josef Küng   Johannes Kepler University Linz, Austria

Steering Committee

Gabriele Anderst-Kotsis   Johannes Kepler University Linz, Austria
A Min Tjoa   Software Competence Center Hagenberg, Austria
Ismail Khalil   Johannes Kepler University Linz, Austria

Program Committee and Reviewers

Sonali Agarwal   Indian Institute of Information Technology Allahabad,
                                    India
Riccardo Albertoni   Institute of Applied Mathematics and Information
                                    Technologies, Italian National Council of Research,
                                    Italy
Idir Amine Amarouche   University Houari Boumediene, Algeria
Rachid Anane   Coventry University, UK
Mustafa Atay   Winston-Salem State University, USA
Faten Atigui   CNAM, France
Ladjel Bellatreche   ENSMA, France
Nadia Bennani   INSA Lyon, France
Karim Benouaret   Université Claude Bernard Lyon 1, France
Djamal Benslimane   Lyon 1 University, France
Morad Benyoucef   University of Ottawa, Canada
Mikael Berndtsson   University of Skovde, Sweden
Catherine Berrut   Grenoble University, France
Vasudha Bhatnagar   Delhi University, India
Athman Bouguettaya   University of Sydney, Australia
Omar Boussaid   University of Lyon/Lyon 2, France
Stephane Bressan   National University of Singapore, Singapore
Barbara Catania   DISI, University of Genoa, Italy
Sharma Chakravarthy   The University of Texas at Arlington, USA
Cindy Chen   University of Massachusetts Lowell, USA

Max Chevalier	IRIT - SIG, Université de Toulouse, France
Soon Ae Chun	City University of New York, USA
Alfredo Cuzzocrea	University of Trieste, Italy
Deborah Dahl	Conversational Technologies, USA
Jérôme Darmont	Université de Lyon (ERIC Lyon 2), France
Soumyava Das	Teradata, USA
Vincenzo Deufemia	Università degli Studi di Salerno, Italy
Juliette Dibie-Barthélemy	AgroParisTech, France
Dejing Dou	University of Oregon, USA
Cedric du Mouza	CNAM, France
Johann Eder	University of Klagenfurt, Austria
Suzanne Embury	The University of Manchester, UK
Markus Endres	University of Augsburg, Germany
Noura Faci	Lyon 1 University, France
Bettina Fazzinga	ICAR-CNR, Italy
Stefano Ferilli	University of Bari, Italy
Flavio Ferrarotti	Software Competence Center Hagenberg, Austria
Vladimir Fomichov	School of Business Informatics, National Research University Higher School of Economics, Russia
Flavius Frasincar	Erasmus University Rotterdam, The Netherlands
Bernhard Freudenthaler	Software Competence Center Hagenberg GmbH, Austria
Steven Furnell	Plymouth University, UK
Joy Garfield	University of Worcester, UK
Claudio Gennaro	ISTI-CNR, Italy
Manolis Gergatsoulis	Ionian University, Greece
Javad Ghofrani	HTW Dresden University of Applied Sciences, Germany
Vikram Goyal	IIIT Delhi, India
Carmine Gravino	University of Salerno, Italy
Sven Groppe	Lübeck University, Germany
William Grosky	University of Michigan, USA
Francesco Guerra	Università degli Studi Di Modena e Reggio Emilia, Italy
Giovanna Guerrini	University of Genova, Italy
Allel Hadjali	ENSMA, France
Abdelkader Hameurlain	Paul Sabatier University, France
Ibrahim Hamidah	Universiti Putra Malaysia, Malaysia
Takahiro Hara	Osaka University, Japan
Ionut Emil Iacob	Georgia Southern University, USA
Sergio Ilarri	University of Zaragoza, Spain
Abdessamad Imine	INRIA Grand Nancy, France
Yasunori Ishihara	Nanzan University, Japan
Peiquan Jin	University of Science and Technology of China, China
Anne Kao	Boeing, USA
Dimitris Karagiannis	University of Vienna, Austria

Stefan Katzenbeisser	University of Passau, Germany
Anne Kayem	Hasso-Plattner-Institute, Germany
Uday Kiran Rage	University of Tokyo, Japan
Carsten Kleiner	University of Applied Sciences and Arts Hannover, Germany
Henning Koehler	Massey University, New Zealand
Michal Krátký	Technical University of Ostrava, Czech Republic
Petr Kremen	Czech Technical University in Prague, Czech Republic
Anne Laurent	LIRMM, University of Montpellier 2, France
Lenka Lhotska	Czech Technical University, Czech Republic
Wenxin Liang	Dalian University of Technology, China
Chuan-Ming Liu	National Taipei University of Technology, Taiwan
Hong-Cheu Liu	University of South Australia, Australia
Jorge Lloret Gazo	University of Zaragoza, Spain
Hui Ma	Victoria University of Wellington, New Zealand
Qiang Ma	Kyoto University, Japan
Zakaria Maamar	Zayed University, UAE
Elio Masciari	ICAR-CNR, Università della Calabria, Italy
Brahim Medjahed	University of Michigan, USA
Jun Miyazaki	Tokyo Institute of Technology, Japan
Lars Moench	University of Hagen, Germany
Riad Mokadem	IRIT, Paul Sabatier University, France
Anirban Mondal	Ashoka University, India
Yang-Sae Moon	Kangwon National University, Republic of Korea
Franck Morvan	IRIT, Paul Sabatier University, France
Francesc Munoz-Escoi	Universitat Politècnica de València, Spain
Ismael Navas-Delgado	University of Málaga, Spain
Wilfred Ng	Hong Kong University of Science and Technology, SAR China
Javier Nieves Acedo	IK4-Azterlan, Spain
Marcin Paprzycki	Polish Academy of Sciences, Warsaw Management Academy, Poland
Oscar Pastor Lopez	Universitat Politècnica de València, Spain
Clara Pizzuti	Institute for High Performance Computing and Networking, National Research Council, Italy
Elaheh Pourabbas	National Research Council, Italy
Rodolfo Resende	Federal University of Minas Gerais, Brazil
Claudia Roncancio	Grenoble University, LIG, France
Viera Rozinajova	Slovak University of Technology in Bratislava, Slovakia
Massimo Ruffolo	ICAR-CNR, Italy
Marinette Savonnet	University of Burgundy, France
Florence Sedes	IRIT, Paul Sabatier University, France
Nazha Selmaoui	University of New Caledonia, New Caledonia
Michael Sheng	Macquarie University, Australia
Patrick Siarry	Université Paris 12, LiSSi, France

Hala Skaf-Molli	Nantes University, France
Srinivasa Srinath	IIIT Bangalore, India
Bala Srinivasan (Retried)	Monash University, Australia
Olivier Teste	IRIT, University of Toulouse, France
Stephanie Teufel	University of Fribourg, Switzerland
Jukka Teuhola	University of Turku, Finland
Jean-Marc Thevenin	University of Toulouse 1 Capitole, France
Vicenc Torra	Maynooth University, Ireland
Traian Marius Truta	Northern Kentucky University, USA
Lucia Vaira	University of Salento, Italy
Krishnamurthy Vidyasankar	Memorial University of Newfoundland, Canada
Marco Vieira	University of Coimbra, Portugal
Ming Hour Yang	Chung Yuan Christian University, Taiwan
Xiaochun Yang	Northeastern University, China
Haruo Yokota	Tokyo Institute of Technology, Japan
Qiang Zhu	The University of Michigan, USA
Yan Zhu	Southwest Jiaotong University, China
Marcin Zimniak	Leipzig University, Germany
Ester Zumpano	University of Calabria, Italy

Organizers

Institute for
Telecooperation

www.iiwas.org

Contents – Part II

Semantic Web and Ontologies

Information Processing

Temporal, Spatial, and High Dimensional Databases

Knowledge Discovery

Web Services

Contents – Part I

Authenticity, Privacy, Security and Trust

Consistency, Integrity, Quality of Data

Decision Support Systems

Data Mining and Warehousing

Distributed, Parallel, P2P, Grid and Cloud Databases

Looking into the Peak Memory Consumption of Epoch-Based Reclamation in Scalable in-Memory Database Systems

Hitoshi Mitake[1], Hiroshi Yamada[2], and Tatsuo Nakajima[1(✉)]

[1] Department of Computer Science and Engineering,
Waseda University, Shinjuku, Tokyo, Japan
{mitake, tatsuo}@dcl.cs.waseda.ac.jp
[2] Department of Computer and Information Sciences,
TUAT, Fuchu, Tokyo, Japan
hiroshiy@cc.tuat.ac.jp

Abstract. Deferred memory reclamation is an essential mechanism of scalable in-memory database management systems (DBMSs) that releases stale objects asynchronously to free operations. Modern scalable in-memory DBMSs commonly employ a deferred reclamation mechanism named epoch-based reclamation (EBR). However, no existing research has studied the EBR's trade-off between performance improvements and memory consumption; its peak memory consumption makes capacity planning difficult and sometimes causes disruptive performance degradation. We argue that gracefully controlling the peak memory usage is a key to achieving stable throughput and latency of scalable EBR-based in-memory DBMSs. This paper conducts a quantitative analysis and evaluation of a representative EBR-based DBMS, Silo, from the viewpoint of memory management. Our evaluation reveals that the integration of conventional solutions fails to achieve stable performance with lower memory utilization, and Glasstree-based Silo achieves a 20% higher throughput, latencies characterized by an 81% lower standard deviation, and 34% lower peak memory usage than Masstree-based Silo even under read-majority workloads.

Keywords: In-memory database · Epoch-based reclamation · Multicore scalability · Index tree structure

1 Introduction

In-memory database management systems (DBMSs) are promising components that achieve considerably higher performance than traditional disk-based DBMSs because modern commodity servers are equipped with multiple terabytes of DRAM [24, 25]. Exploring the design of in-memory DBMSs, such as key-value stores (KVSes) [16] and relational database management systems (RDBMSs) [9, 21], is a popular research topic. Some of these efforts have resulted in successful, commercially available systems

© Springer Nature Switzerland AG 2019
S. Hartmann et al. (Eds.): DEXA 2019, LNCS 11707, pp. 3–18, 2019.
https://doi.org/10.1007/978-3-030-27618-8_1

like VoltDB[1]. Hekaton [4] and Silo [22] are the latest in-memory RDBMSs in which transaction throughput is scalable on multicore platforms.

In scalable in-memory DBMSs, deferred memory reclamation is an essential mechanism for attaining multicore scalability [5]. Even if the highly concurrent data structures utilize synchronization techniques, such as lock-free approaches [5], naive memory management such as reference counts involve frequent writes to shared memory areas, thus resulting in cache line contentions that can produce scalability bottlenecks [17]. To minimize the updates of shared cache lines in transaction processing, modern scalable in-memory DBMSs employ a deferred memory reclamation named *epoch-based reclamation* (EBR). EBR avoids updating values on shared memory areas when getting and releasing referred objects so that the in-memory DBMSs can achieve high scalability under read-heavy workloads.

Despite the importance of deferred memory reclamation, its trade-off between performance improvements and drawbacks is not the primary focus of prior studies. Such deferred memory reclamation causes high peak memory usage, making it quite difficult to achieve accurate capacity planning and sometimes resulting in disruptive performance degradation. The memory usage of the EBR-based DBMSs fluctuates by incoming request sequences and an interval for which stale objects can reside in memory, known as the grace period. At worst, the malicious sequence of requests can lead to memory exhaustion. Additionally, the EBR's memory usage poses a tail latency problem due to object reclamation. The literature [14] reports that a pause time for reclaiming stale objects in Masstree-based KVS [12] causes large latency spikes.

We argue that gracefully controlling the peak memory usage is a key to achieving the stable throughput and latency of scalable EBR-based DBMSs. In this paper, we conduct a quantitative analysis and evaluation of a representative EBR-based DBMS, Silo [22], from the viewpoint of memory management. Experimental results demonstrate the advantages and disadvantages of the state-of-the-art designs and techniques of in-memory DBMSs. The most remarkable result is that a system combining both reference counting and EBR can improve throughput, latency and peak memory usage simultaneously. Moreover, in the case of the YCSB benchmark, Glasstree-based Silo can achieve a 20% higher throughput, latencies characterized by an 81% lower standard deviation, and 34% less peak memory usage than Masstree-based Silo even under read-majority workloads.

Our contributions are the quantitative analysis and evaluation of the drawbacks of the advanced lifetime management scheme that realizes multicore scalable in-memory DBMSs. To the best of our knowledge, our study is unique mainly because of these points:

(1) We analyze the effect of peak memory usage in scalable in-memory DBMSs and show how the effect influences their throughput and latency. As an example, we show that a carefully designed index structure that reduces peak memory usage offers better throughput and stable latency. The results show that reducing peak memory usage is essential for developing new advanced techniques to increase their scalability.

---

[1] https://www.voltdb.com/.

(2) Our study proposes the idea that physical memory management is also important for in-memory DBMSs [23]. Our evaluation includes workloads that involve the dynamic allocation and reclamation of large amounts of memory area that are suitable for evaluating physical memory management strategies. Our study shows that analyzing methodologies related to physical data management are also valuable for in-memory DBMSs.

The remainder of this paper is organized as follows: Sect. 2 describes the background and related work of our study. We explain the details of the problems and solutions for solving them in Sect. 3. We show the effectiveness of the proposed carefully designed memory lifetime management technique in Sect. 4. We conclude this paper in Sect. 5.

2 Background and Related Work

2.1 Memory Reclamation Mechanism for Achieving Scalability

To avoid the degradation of the scalability achieved by concurrency control techniques, most multicore scalable in-memory DBMSs employ deferred memory reclamation mechanisms [1, 2, 5, 7]. The deferred memory reclamation mechanisms are alternatives to naive resource lifetime management techniques such as reference counting. Traditionally, deferred reclamation is used in an OS kernel to protect mostly read data structures [13]. Typical OS kernels have many such data structures, e.g., a list of loadable modules. Therefore, protecting them with a naive reader-writer lock or managing their lifetime with a naïve reference counting degrades the multicore scalability of the kernel and its user space programs. Read-Copy Update (RCU) [13] is successfully used as an alternative to these naive techniques: the most widely known use case of RCU is the Linux kernel. For implementing RCU in the Linux kernel, quiescent state-based reclamation (QSBR) is used for detecting reader-side critical sections [17]. The mechanism of QSBR is based on the privilege of the kernel space, including enabling and disabling preemption and sending an interprocessor interrupt (IPI) to remote cores. Then, a writer thread can ensure that there are no reader threads that have a reference to an object and should be reclaimed.

Unlike an OS kernel, the DBMSs are usually user space programs, and they do not have the privileges of thread scheduling. For such programs, EBR is more suitable than QSBR as a foundation for a memory reclamation mechanism[2].

Implementing EBR requires a global epoch number e, and each thread that can read objects whose lifetime is managed by EBR has its own epoch number e_w. e is periodically incremented, and e_w of each thread can be synchronized by a global epoch manager with e when the thread is not working[3]. Deleted objects must be registered to the limbo list [5], a temporal place for objects that are awaiting safe reclamation, of the deleting thread with

---

[2] Using QSBR in a user space program requires a specialized system call of the underlying OS, e.g. membarrier(2) in the Linux kernel.

[3] In such a situation, the thread is considered to be in a *quiescent state*.

its e_w. With this rule, a thread can determine that an object in the limbo list with an epoch number that is less than the minimum e_w can be reclaimed safely.

2.2 Using EBR in Multicore Scalable in-Memory DBMSs

Similar to other programs that use scalable data structures, the DBMSs follow the design to exclude cache line contentions as much as possible. In the case of the DBMSs, the data structures managed with EBR are accessed in transactions; each transaction is a critical section of the EBR, and an interval between transactions is a quiescent state. To present such an example, Fig. 1 briefly describes the transaction commit protocol of Silo [22]. If a transaction does not involve write operations, it does not update shared cache lines because the validation phase (from line 5 to line 9) and transaction ID (TID) generation (line 10) do not write to globally shared memory areas; only locking (line 3 and 13) and writing (line 12) operations update the shared area. Therefore, the concurrency control of Silo successfully minimizes the scalability bottleneck.

```
input : txn, e
1 begin
2   for tuple, new-value ∈ sort(txn.writeset) do
3     lock(tuple);
4   end
    // Validation phase
5   for tuple, read-tid ∈ txn.readset ∪ txn.nodeset do
      // Access to the tuple for validation with TID. It
         does not involve shared cache line updates.
6     if read-tid ≠ tuple.tid then
7       abort();
8     end
9   end
    // The validation succeeded, txn can be committed.
10  commit-tid ← generate-tid(txn, e);
11  for tuple, new-value ∈ txn.writeset do
12    write(tuple, new-value, commit-tid);
13    unlock(tuple);
14  end
15  destruct-txn(txn);
    // Now the grace period finishes and quiescent state
       begins. e_w can be updated.
16 end
```

Fig. 1. The transaction commit process of Silo. txn is a transaction to commit. e is a global epoch number. If txn.writeset is empty, lock(), unlock() and write(), which updates shared cache lines, are not called.

This design for avoiding the shared cache line updates is also shared with the memory lifetime management scheme (line 15). If the tuples pointed to by readset, nodeset and writeset are managed with reference counting, then the destruction process must update the cache lines that contain the reference counts of the tuples for the decrement operations. The cache line updates for the decrement operations are avoided because Silo is based on EBR[4]. After the destruction of a transaction, the e_w of the worker thread can be updated by the global epoch manager because the thread is in the quiescent state. This means that the lifetimes of the tuples are updated implicitly; if they already belong to a limbo list of a thread, they can be reclaimed after the destruction of the transaction.

2.3   Drawbacks to Using EBR and Its Alternatives

Although EBR contributes to the scalable performance of in-memory DBMSs, it introduces significant drawbacks caused by its high memory usage. The impact of these drawbacks on in-memory DBMSs is widely acknowledged. In the context of the DBMSs, for example, Wu et al. provided an empirical study on MVCC DBMSs [23]. The scope of the study included physical data management techniques, and they noted that memory reclamation is an important factor for modern in-memory DBMSs. Hekaton used a lock-free memory allocator and asynchronous memory reclamation mechanism to mitigate these problems. Some research systems provide techniques for efficient memory management [10, 11]. However, the drawbacks of EBR and the solutions have yet to be thoroughly analyzed despite their importance.

Previous studies have shown that QSBR contributes to the multicore scalability of kernels but introduces serious drawbacks, such as performance degradation and memory exhaustion, when a large number of deleted objects are generated [18]. Only a few solutions have been proposed, and the solutions can be used only in OS kernels. Our prior work showed that EBR can cause problems similar to QSBR [14]. Therefore, the drawbacks and solutions must be analyzed carefully in the context of in-memory DBMSs.

Another candidate alternative is a GC mechanism of the language runtime (e.g., GC of Java VM). Such GC mechanisms still introduce high overhead as a drawback in their general design. Improving the performance of the language GC is still a popular research topic, especially in the context of multicore scalability [6].

As discussed above, EBR seems to be the only high-performance memory reclamation mechanism that can be used for in-memory DBMSs. Although reference counting introduces scalability limitations, its property of immediate reclamation is attractive for reducing the peak memory usage. Conversely, EBR increases the lifetimes of deleted objects to increase peak memory usage. The trade-off between the two techniques is depicted in Fig. 2.

---

[4] Getting the references of the tuples also avoids the cache line updates due to the lock-free lookup of Masstree used in Silo.

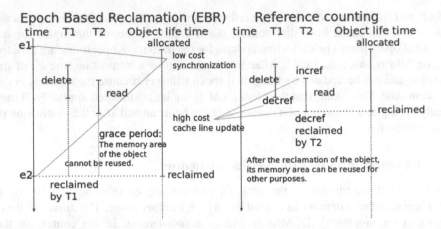

Fig. 2. Comparison of EBR and reference counting from the perspective of memory allocation. e_1 and e_2 denote the epoch of EBR. T_1 and T_2 are threads.

3   Potential Sources of the Drawbacks Caused by EBR and Possible Solutions

Memory objects managed under EBR cannot be immediately reclaimed even when no threads are referencing them.[5] This means that the peak memory usage of systems based on EBR tends to be high. In addition, the hit ratio of the memory allocator's thread-local cache tends to be lower because long-lived objects prevent the recycling of the space. In the context of OS kernels, this low hit ratio is known to cause performance degradation of the entire system [18].

Therefore, the observation requires analyzing the impact of different memory allocators on the memory usage of in-memory DBMSs. Additionally, the impact of memory reclamation on memory usage should be investigated. If not, it is not clear whether a high peak memory load degrades throughput and makes latencies unstable. The drawbacks were not carefully analyzed in past studies to identify the actual sources of the drawbacks. This section first discusses whether a widely used traditional solution may overcome the drawbacks. Then, we consider some possible solutions to overcome high peak memory usage. The paper will present detailed experimental evidences for showing the effects of the solutions in Sect. 4.

3.1   Impact of Memory Reclamation

EBR defers the reclamation of unused objects. The unused objects are periodically reclaimed in an amortized manner. In the case of Silo, the default interval of the memory reclamation is 1 s, and the configuration can cause high latency spikes [14].

---

[5] The time interval required for ensuring the end of all reader-side critical sections is called the grace period.

The widely used and straightforward approach to decrease unstable latencies is to use asynchronous memory reclamation; asynchronous memory reclamation executes a thread that cleans deleted objects that can be reclaimed safely. This approach is employed in both research systems [10] and production systems [4]. Because of this mechanism, these systems do not need to stall active transactions. However, there is no analysis of how asynchronous memory reclamation impacts memory usage in past studies.

We added asynchronous memory reclamation to Silo and investigated the technique with a workload that involves dynamic memory loads. The purpose of the investigation was to understand how effective the technique is and what type of new drawbacks would be introduced. We hypothesized that asynchronous memory reclamation contributes to reducing latency spikes but amplifies peak memory usage. This is because the asynchronous nature of this technique can defer the reclamation so the system cannot guarantee when the reclamation can be completed, and the peak memory usage makes the behavior unstable. The motivation to raise the question is that using asynchronous memory reclamation overcomes some issues of EBR, but it does not solve essential issues that are raised by the drawbacks of EBR.

3.2   Reducing High Peak Memory Usage Caused by EBR

EBR makes high peak memory usage of in-memory DBMSs because of its deferred reclamation. This property makes capacity planning difficult. Unlike EBR, reference counting introduces frequent incrementing and decrementing of count values when it is naively used. This limits multicore scalability; on the other hand, this accurate counting of referring threads enables immediate reclamation after the last reference is finished. The trade-off between the two techniques to reduce high peak memory usage was not carefully considered in past studies.

We could not find a straightforward solution to this problem. As we describe in the next section, using asynchronous memory reclamation cannot be an essential solution. Glasstree (*Garbage-Less Masstree*) is a new carefully designed index structure, which is an enhancement of Masstree, to reduce peak memory usage [15]. The purpose of this new index structure is to understand whether it is possible to reduce the peak memory usage without sacrificing multicore scalability. Its design and implementation were not trivial; thus, we describe these aspects in the following.

Table 1. Breakdown of the types and numbers of reclaimed objects during Silo's TPC-C

Type	# of reclamation	Percentage
Internode	767505	2.3%
Leaf	7734210	24.0%
Value (tuple)	23813904	73.7%

The design of Glasstree exploits a simple but fundamental property of balanced trees, namely, data structures used as ordered indexes; if the access to values is equally distributed, shallower nodes near a root node are accessed frequently, whereas deeper nodes near the values are not accessed frequently. From the perspective of multicore scalability, this means that updating shared cache lines when accessing shallower nodes will likely produce scalability bottlenecks. Conversely, update operations when accessing deeper nodes will produce fewer bottlenecks. The design also respects the principle of deferred reclamation; deferred reclamation is suitable for protecting mostly read and long-lived objects [17]. In general, the internodes and the leaves live longer than the values. The values can be reclaimed if they are updated or deleted. Conversely, internodes and leaves are only reclaimed when reshaping occurs when deleting values. This assumption can also be validated by actual experiments. Table 1 summarizes memory reclamation under Silo's TPC-C benchmark. As shown in this table, a large portion of memory reclamation is dominated by a value object that functions as a tuple.

Based on the above observations, Glasstree employs both EBR and reference counting as its memory lifetime management scheme. As depicted in Fig. 3, Glasstree manages the lifetimes of internodes and leaves using EBR, where the design strategy is shared with Masstree [12]. Conversely, Glasstree manages the lifetimes of values with reference counting as previously described. Therefore, the read operations (get and scan) need to increment before passing the results to their callers and decrement after their usage. This means that Glasstree sacrifices the complete invisibility of the readers. In contrast to the read operations, the write operations (insert, put and remove) are not changed from Masstree.

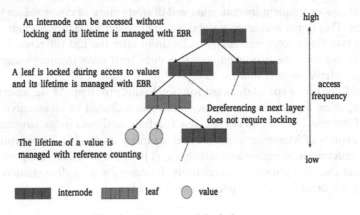

Fig. 3. Abstract model of glasstree

Because of the design, the reclaimed values of Glasstree are directly returned to its allocator. For Masstree, they are registered to a limbo list of each thread once and then returned to its allocator after the end of the grace periods.

4 Evaluation and Analysis

In this section, we analyze the properties of EBR for in-memory DBMSs by comparing the original version of Silo [20][6], which uses Masstree as its index structure (denoted as Silo), and our modified versions of Silo, which uses Glasstree.

We executed the YCSB benchmark of Silo on a machine equipped with dual Intel Xeon E5-2690 v4 CPUs (2.60 GHz, total of 28 physical cores) and 128 GB of DRAM. Hyperthreading was disabled, and each worker thread was pinned on a dedicated physical core. Every workload was executed 3 times for 60 s. The average scores were plotted with lines. The maximum and minimum scores are depicted with error bars.

To simplify the analysis, we also disabled the logging functionality of Silo to focus on memory management components. Therefore, the baseline corresponds to *MemSilo* of the original paper on Silo [22].

We used YCSB [3], a benchmark suite for evaluating various DBMSs, including RDBMSs and KVSs. Silo includes an implementation of YCSB in its benchmark. The benchmark mimics a workload that can be generated when managing a database of web services that stores user information. It is suitable for measuring the performance of key access because the workloads are relatively simple. The simplicity allows us to focus on the physical data management functionalities.

The implementation of YCSB in Silo allows worker threads to issue requests based on a configured probability. The probability is given in the following form: (R, W, RAW, S), where $R + W + RAW + S = 100$. There are four types of requests: read (representing $R\%$ of all requests), write (representing $W\%$ of all requests), read after write (representing $RAW\%$ of all requests), and scan (representing $S\%$ of all requests). Every request is issued to a single table (corresponding to a single instance of Masstree or Glasstree) in its own transaction. However, these requests are not suitable for evaluating memory management components because they only cause read or in-place update operations. As in the original YCSB, we added two more requests: create (representing $C\%$ of all requests) and delete (representing $D\%$ of all requests). The create request creates a new tuple, and the delete request deletes an existing tuple. The combination of these two requests generates deleted tuples, and thus, it is suitable for evaluating physical data management functionalities. We configured the size of the tuples to be 2 KB.[7] Additionally, we chose the uniform workload of YCSB because the original paper on Silo used the same workload, so it is effective to compare Silo and our approach.

4.1 Workload with Creation and Deletion Operations

Original Silo (Silo) and the modified version of Silo that performs memory reclamation in an asynchronous manner (Silo-AsyncGC) were investigated. The purpose of investigating this comparison is to examine how asynchronous memory reclamation influences reducing latencies but causes another drawback.

---

[6] The base commit ID of git used in our evaluation is *cc11ca1* [20].

[7] The reason for choosing the size is that the size clearly reveals the differences according to the measured results of the original Silo paper.

Next, we investigated our modified version of Silo that uses Glasstree as its index structure (denoted as Silo-Glass). Silo-Glass must produce fewer deleted tuples and can keep the peak memory usage low due to the design principle.

In the investigation, we used three memory allocators: the standard libc malloc, jemalloc and SSMalloc. Note that the original version of SSMalloc demands pages from the OS in an extremely aggressive manner (the number of the required pages grows quadratically). We changed the policy because it results in needlessly high memory usage, and our modified SSMalloc requires pages with linear speed.

We present the results of the evaluation of workloads that include create ($C = 15$) and deletion ($D = 15$) operations (note that even with these operations, the workloads are read majority because 70% of the operations are read operations). The initial database size is 10 GB (500,000 tuples)[8].

Figure 4 shows the benchmark scores with three different memory allocators: libc malloc, jemalloc and SSMalloc.[9] As shown in [4, 10], adopting asynchronous memory reclamation significantly reduces the latency spikes, reducing the drawbacks of EBR. As shown in Fig. 4(h), using asynchronous memory reclamation improves latency stabilities. Conversely, the latency stabilities of Silo that use jemalloc are increased in proportion to the number of processor cores. Thus, it seems that asynchronous memory reclamation overcomes the drawbacks if we do not carefully analyze the evaluation with other memory allocators and if we do not investigate the peak memory usage during the execution of the workloads.

Our hypothesis is that a serious source of the drawbacks of EBR is from the high peak memory usage. Thus, if asynchronous memory reclamation cannot reduce the high peak memory usage, the technique can only partially overcome the drawbacks.

As depicted in Fig. 4(h), asynchronous memory reclamation influences latency stabilities when using jemalloc, but Fig. 4(i) shows that the latency stabilities do not change when using SSMalloc. This means that the improvement of latency stabilities when using jemalloc comes from the scalability of memory allocators not using asynchronous memory reclamation.

Moreover, as shown in Fig. 4(k) and (l), using asynchronous reclamation increases peak memory usage, which makes the drawbacks of EBR worse. Thus, as shown in Fig. 4(b), (c), (e), and (f) using asynchronous memory reclamation does not heavily influence the average latency and the throughput. Thus, the results show that the effect of asynchronous memory reclamation is limited in terms of the drawbacks of EBR.

As depicted in Fig. 4(j), (k) and (l), Silo-Glass presents the lowest peak memory usage in every configuration. This is quite natural because of the qualitative property of reference counting, as shown in Fig. 2. Reference counting contributes to the immediate reclamation of unused tuples.

---

[8] Readers may notice that the peak memory usage depicted in this figure is much higher than in Fig. 4. The difference comes from the memory area that is used for storing the commit latencies of the transactions. All the latencies are required for calculating standard deviation, which requires a large memory area because the benchmarks produce tens of millions of transactions.

[9] In the figures, the results are suddenly bigger when the number of threads is changed from 24 to 28. This is due to the conflict between a thread to process the epoch and a thread to perform a transaction. Thus, as shown in the results, the influence of the effect is smaller in Silo-Glass.

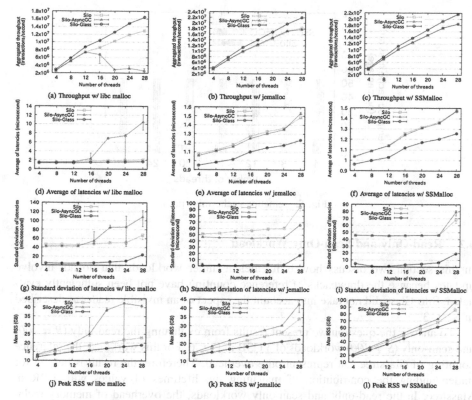

Fig. 4. The benchmark scores of Silo, Silo-AsyncGC and Silo-Glass under an $R = 70$, $C = 15$, $D = 15$ workload with three different memory allocators.

Interestingly, as shown in Fig. 4(a), (b) and (c), Silo-Glass outperforms Silo and Silo-AsyncGC with respect to throughput. This can be explained by the low miss ratio of the thread local cache of malloc. We measured the cache hit ratio for the case of jemalloc, as depicted in Fig. 5. The miss ratio was calculated as *nmalloc/nrequest* for all bins and arenas [8, 19]. As the graph shows, Silo and Silo-AsyncGC suffer from high miss ratios, larger than 50% in every case. Conversely, Silo-Glass achieves a low miss ratio, lower than 10% in every case.

In addition, Silo-Glass achieves better latency scores than Silo and Silo-AsyncGC. This is because a large number of deleted tuples can be reclaimed with the reference counting of Glasstree without waiting for the reclamation process of EBR.

The above discussion shows that using Glasstree significantly reduces peak memory usage and directly contributes to overcoming the drawbacks of EBR.

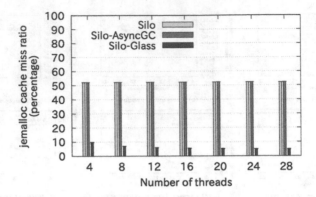

Fig. 5. Jemalloc cache miss ratio

4.2 Read-Only and Scan-Only Workload

In this section, we discuss how is it efficient under workloads that do not involve updates because the drawbacks of reference counting have not been sufficiently evaluated. Silo-Glass did not take into account the aspect of an invisible reader as described in Sect. 3.

To answer the question we present results from evaluating the read-only ($R = 100$) and scan-only ($S = 100$) workloads of YCSB in Fig. 6. For this workload, the reference counting of Glasstree can represent pure overhead. Therefore, the result is helpful for understanding the contribution of managing the lifetimes of values with EBR in Masstree. In the read-only and scan-only workloads, the overhead of memory reclamation can be ignored; thus, we omit the scores of the latencies and the results of Silo-AsyncGC. We provide the throughput of the workloads with various database sizes: 10 GB (500,000 tuples), 20 GB (1,000,000 tuples) and 40 GB (2,000,000 tuples). In the case of read-only benchmarks, worker threads read a single tuple in their transactions. In the case of scan-only benchmarks, worker threads read 50 tuples in their transactions.

In this benchmark, we also evaluated Silo-GlobalTID, which emulates the bottleneck of centralized TID generation. The benchmark results of Silo-GlobalTID are helpful in understanding how the decentralized overhead of Glasstree is different from the centralized overhead that can be found in systems such as Hekaton.

As shown in Fig. 6(a), Silo outperforms Silo-Glass for the case of small (10 GB) databases (Silo achieves a 4% higher throughput than Silo-Glass at 28 threads). However, the differences in the scores decrease as the database size increases to 20 GB, as shown in Fig. 6(b) (Silo achieves a 3% higher throughput at 28 threads). In addition, the result of Fig. 6(c) implies that managing tuples with EBR does not contribute to the read performance under practically large (40 GB) databases (Silo achieves only 1% higher throughput at 28 threads). This difference expresses the contribution of managing tuples with EBR rather than reference counting.

Conversely, Silo-GlobalTID introduces significant overhead because of its centralized bottleneck. The overhead is not affected by the diverged size of the databases. This is because the centralized TID generation mechanism is accessed by every

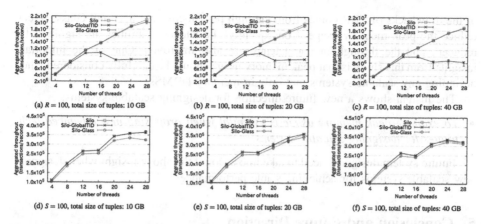

Fig. 6. Throughput of Silo, Silo-GlobalTID and Silo-Glass under read-only and scan-only workloads.

transaction unconditionally, unlike accessing tuples. Therefore, it can be a critical bottleneck for short transactions.

Figure 6(d), (e), and (f) show the throughput of scan-only workloads. The size of each transaction is larger than in the case of read-only benchmarks. Therefore, global TID generation is not a serious bottleneck for scan workloads. As in the case of read-only workloads, we find that the performance difference between Silo and Silo-Glass decreases with the database size. Silo achieves a 13% higher throughput with a 10 GB database, a 5% higher throughput with a 20 GB database, and a 1% higher throughput with a 40 GB database than Silo-Glass at 28 threads.

From the results, we can conclude that the contribution of EBR in Masstree is particularly effective for handling small databases. In such workloads, the frequent updating of shared cache lines can degrade scalability. However, the significance of the contribution decreases for larger databases. It can be considered that the portion of other costs in read operations, e.g., tracking pointers, increases, and the overhead of the shared cache line updates becomes relatively smaller. Moreover, the cost of Masstree's read operations is dominated by the latency of DRAM access [12]. Therefore, managing the lifetimes with reference counting will not introduce centralized bottlenecks, such as global TID generation, if the number of frequently accessed tuples is not small.

4.3 Design for Minimizing Peak Memory Usage

In previous advanced techniques for scalable in-memory DBMSs, avoiding resource contentions among concurrent activities is essential to increasing the system's scalability. Various advanced techniques have been proposed for increasing the scalability, and recent in-memory DBMSs express outstanding scalability. These techniques offer excellent high throughput and low latency.

The analyses shown in the paper reveal that peak memory usage is also an important factor in increasing throughput and decreasing latency. In particular, decreasing peak memory usage significantly reduces latency stabilities. This paper shows that the source of the drawbacks of EBR comes from the high peak memory usage, and Glasstree can reduce the peak memory usage. Consequently, the latency stabilities are significantly reduced, and the throughput and latency are also improved. One of the important insights from this study is that peak memory usage should be carefully considered when designing any scalable systems, not only in-memory DBMSs.

The insight shows a new future direction for designing scalable systems:

- *To examine peak memory usage is essential when proposing new advanced techniques for designing scalable systems.*

Future system designers need to take into account the above insight when designing new scalable systems for achieving stable performance.

5   Conclusion and Future Direction

In this paper, we analyzed the performance properties of EBR, a deferred reclamation mechanism widely used by many state-of-the-art in-memory DBMSs. Our analysis revealed that widely known and straightforward techniques have limited performance benefits. Surprisingly, throughput, latency and memory efficiency can be improved simultaneously by carefully designing lifetime management schemes. We proved this by designing a new index structure, Glasstree that utilizes both EBR and reference counting. As a result, our study revealed the importance of and opportunity for additional research on physical data management techniques, including lifetime management schemes and memory allocators, in the context of in-memory DBMSs. To establish truly robust multicore scalable in-memory DBMSs, more detailed studies on the topic are required. We hope that our analysis will be informative for future studies.

5.1   Optimization for Performance and Energy Consumption

We revealed that EBR introduces high peak memory usage. High peak memory usage can cause various problems. The most straightforward example is capacity planning. The safety margin that cannot be exhausted by the objects deleted and awaiting reclamation must be preserved. This is a critical problem for scale-up systems because unlike scale-out systems which can increase their storage capacity by adding nodes to a cluster during runtime, scale-up systems cannot increase their capacity dynamically. Therefore, in managing scale-up systems, it is important to accurately predict the safety margin. Clearly, EBR makes this task difficult. We also note that this safety margin can cause another problem: high energy costs. The energy consumption of DRAM grows exponentially with the DRAM capacity. Therefore, keeping the peak memory usage lower is valuable, not only for ease of management but also for reducing energy consumption. We believe that research on lifetime management schemes and physical data management for these metrics has substantial opportunities for improvement.

5.2 Optimization for Read Performance

Glasstree improves throughput, latency and memory efficiency simultaneously for workloads that involve dynamic memory loads. However, it introduces a small overhead for read-only workloads because of the newly introduced lock and atomic counting in the reference counting mechanism. We believe that this overhead can be reduced by utilizing existing techniques, e.g., scalable locking, scalable reference counting that enables immediate reclamation and tree-based techniques.

Of course, we do not believe that it is possible to construct a perfect index structure or lifetime management scheme that provides better performance than any other alternatives under every workload because every technique must introduce trade-offs. Combining the various techniques will result in tuned index structures that can handle specific workloads with high performance and predictable drawbacks.

The source code used in this paper is publicly available at https://gitlab.com/mitake1/silo-glass.

References

1. Alistarh, D., Leiserson, W., Matveev, A., Shavit, N.: Forkscan: conservative memory reclamation for modern operating systems. In: Proceedings of the Twelfth European Conference on Computer Systems, EuroSys 2017, pp. 483–498 (2017)
2. Brown, T.A.: Reclaiming memory for lock-free data structures: there has to be a better way. In: Proceedings of the 2015 ACM Symposium on Principles of Distributed Computing, PODC 2015, pp. 261–270 (2015)
3. Cooper, B.F., Silberstein, A., Tam, E., Ramakrishnan, R., Sears, R.: Benchmarking cloud serving systems with YCSB. In: Proceedings of the 1st ACM Symposium on Cloud Computing, SoCC 2010, pp. 143–154 (2010)
4. Diaconu, C., et al.: Hekaton: SQL server's memory- optimized OLTP engine. In: Proceedings of the 2013 ACM SIGMOD International Conference on Management of Data, SIGMOD 2013, pp. 1243–1254 (2013)
5. Fraser, K.: Practical lock-freedom. Ph.D. thesis, Cambridge University. Technical Report UCAM-CL-TR-579 (2004)
6. Gidra, L., Thomas, G., Sopena, J., Shapiro, M.: A study of the scalability of stop-the-world garbage collectors on multicores. In: Proceedings of the Eighteenth International Conference on Architectural Support for Programming Languages and Operating Systems, ASPLOS 2013, pp. 229–240 (2013)
7. Hart, T.E., McKenney, P.E., Brown, A.D., Walpole, J.: Performance of memory reclamation for lockless synchronization. J. Parallel Distrib. Comput. 67(12), 1270–1285 (2007)
8. Jemalloc wiki: Use Case: Basic Allocator Statistics. https://goo.gl/GdX1FB. Accessed 2 June 2019
9. Kemper, A., Neumann, T.: HyPer: A hybrid OLTP & OLAP main memory database system based on virtual memory snapshots. In: Proceedings of the 2011 IEEE 27th International Conference on Data Engineering, ICDE 2011, pp. 195–206 (2011)
10. Kim, K., Wang, T., Johnson, R., Pandis, I.: ERMIA: Fast memory- optimized database system for heterogeneous workloads. In: Proceedings of the 2016 International Conference on Management of Data, SIGMOD 2016, pp. 1675–1687 (2016)

11. Lim, H., Kaminsky, M., Andersen, D.G.: Cicada: dependably fast multi-core in-memory transactions. In: Proceedings of the 2017 ACM International Conference on Management of Data, SIGMOD 2017, pp. 21–35 (2017)
12. Mao, Y., Kohler, E., Morris, R.T.: Cache craftiness for fast multicore key-value storage. In: Proceedings of the 7th ACM European Conference on Computer Systems, EuroSys 2012, pp. 183–196 (2012)
13. McKenney, P.E., Sarma, D., Arcangeli, A., Kleen, A., Krieger, O., Russell, R.: Read-copy update. In: Proceedings of Ottawa Linux Symposium (2002)
14. Mitake, H., Yamada, H., Nakajima, T.: Analyzing the tradeoff between throughput and latency in multicore scalable in-memory database systems. In: Proceedings of the 7th ACM SIGOPS Asia-Pacific Workshop on Systems, APSys 2016, pp. 17:1–17:9 (2016)
15. Mitake, H., Yamada, H., Nakajima, T.: A highly scalable index structure for multicore in-memory database systems. In: Proceedings of the International Conference on the 13th International Symposium on Intelligent Distributed Computing (2019)
16. Ousterhout, J., et al.: The RAMCloud storage system. ACM Trans. Comput. Syst. **33**(3), 7:1–7:55 (2015)
17. McKenney, P.E.: Is parallel programming hard, and, if so, what can you do about it? https://goo.gl/THnFNb. Accessed 24 Aug 2018
18. Prasad, A., Gopinath, K.: Prudent memory reclamation in procrastination-based synchronization. In: Proceedings of the 21st International Conference on Architectural Support for Programming Languages and Operating Systems, pp. 99–112 (2016)
19. Scalable memory allocation using jemalloc. https://goo.gl/skcwvS. Accessed 24 Aug 2018
20. Silo. https://github.com/stephentu/silo. Accessed 2 June 2019
21. Stonebraker, M., Madden, S., Abadi, D.J., Harizopoulos, S., Hachem, N., Helland, P.: The end of an architectural era: (it's time for a complete rewrite). In: Proceedings of the 33rd International Conference on Very Large Data Bases, pp. 1150–1160 (2007)
22. Tu, S., Zheng, W., Kohler, E., Liskov, B., Madden, S.: Speedy transactions in multicore in-memory databases. In: Proceedings of the Twenty-Fourth ACM Symposium on Operating Systems Principles, SOSP 2013, pp. 18–32 (2013)
23. Wu, Y., Arulraj, J., Lin, J., Xian, R., Pavlo, A.: An empirical evaluation of in-memory multi-version concurrency control. Proc. VLDB Endow. **10**(7), 781–792 (2017)
24. https://www.samsung.com/semiconductor/global.semi.static/Data_Processing_with_Samsung_Advanced_DRAM_and_SSD_Solutions_Whitepaper-0.pdf. Accessed 2 2019
25. https://aws.amazon.com/ec2/instance-types/high-memory/?nc1=h_ls. Accessed 2 June 2019

Energy Efficient Data Placement and Buffer Management for Multiple Replication

Satoshi Hikida[✉], Hieu Hanh Le, and Haruo Yokota

Department of Computer Science, Tokyo Institute of Technology,
2-12-1 Ookayama, Meguro-ku, Tokyo 152-8552, Japan
{hikida,hanhlh}@de.cs.titech.ac.jp, yokota@cs.titech.ac.jp

Abstract. Increasing data replication improves the reliability and availability of the large-scale storage systems. However, multiple replication required much more storage capacity and disk I/O frequency that cause of increasing the power consumption of the storage systems. To address this issue, we propose two data placement policies, *Disk Group Aggregation* and *Cache Striping*. These data placement policies employ different data mapping between buffers (memory) and disk drives to control buffer overflow timing of each replica to reduce the disk access frequency. In addition, we also propose two buffer flush algorithms, *WithAllSpins* and *SpinupEE*. *WithAllSpins* flushes buffered data to currently rotating disks, whereas *SpinupEE* forces disks to spin up based on the estimated energy efficiency, and writes buffered data to the disk to make the buffer space fresh. We evaluated the effectiveness of our proposals using a simulation program and demonstrated that they can reduce power consumption, even if the data are replicated multiply.

Keywords: Large storage · Power reduction · Data redundancy · Data replication

1 Introduction

Data replication is very important to ensure the reliability and availability of large-scale storage systems. For example, the Google File System (GFS) [3] and the Hadoop Distributed File System (HDFS) [11] have three replicas on separated storage nodes by default. Although increasing the number of replicas of data enhances the reliability and availability of storage systems, it also increases the power consumption of the storage systems. Specifically, increasing the number of replicas generally increases the number of disk accesses, thus reduces the opportunity to keep the disks in standby mode. As a result, reducing the power consumption of the storage systems with more than three replicas has become an important issue.

Many studies have been proposed to reduce the power consumption of storage systems. MAID [2] is a well-known power saving method for the large-scale near-line storage systems. MAID keeps a small number of disk drives rotating at

© Springer Nature Switzerland AG 2019
S. Hartmann et al. (Eds.): DEXA 2019, LNCS 11707, pp. 19–29, 2019.
https://doi.org/10.1007/978-3-030-27618-8_2

all times; these are used as a cache (cache disk), allowing the majority of the disk drives (data disk) to spin down. However, MAID does not consider the handling of the multiple replication. Thus, the disk access frequency increases as increasing replicas and it decreases the opportunity to keep the disks in standby state. Whereas, RAPoSDA [10] utilizes a primary backup configuration on both buffers (memories) and disk drives to ensure system reliability. It dynamically controls the timing and targeting of disk access based on individual disk rotation status to reduce the number of unnecessary disk accesses and spin-ups to reduce the power consumption of the storage systems. However, RAPoSDA only works with two replicas configuration.

To solve this issue, we investigate energy efficiency data placement policy and buffer flush algorithms for multiple replication, especially, more than three replicas. The contributions of this paper are; (a) We propose two data placement policies between the buffers and the disks for the storage systems with multiple replicas. Each policy provides different buffer overflow timing for each replica to reduce the disk access frequency, (b) we propose two buffer flush algorithms for the storage systems with multiple replicas. One algorithm considers the disk rotation status and aggressively flushes the buffer data to the spinning disks, while the other forces the disk that has the largest buffer to spin up in order to maintain the largest free space in the buffer. (c) We evaluate the efficiency of the data placement policies and buffer flush algorithms with simulation based evaluation.

The remainder of this paper is as follows. In Sect. 2, we describe an approach to reduce the power consumption of storage systems with multiple replicas. We describe the data placement policy in Sect. 3, and the buffer flush algorithms in Sect. 4. Then, the proposed methods are evaluated in Sect. 5. In Sect. 6 we discuss related works. Lastly, Sect. 7 concludes this paper.

2   A Power-Saving Approach for Two-Way Replicated Storage Systems

This section describes an approach to reduce the power consumption of two-way replicated storage systems. RAPoSDA [10] is chosen as an example to explain how to reduce the power consumption of a storage system.

The storage system assumed in this paper consists of several buffers (memories) and disk drives. It is important to maintain reliability when the data are in the volatile memory, so the buffer is provided with a primary backup configuration, and each buffer is connected to corresponding Uninterruptible Power Supply (UPS) to avoid data loss via outage. In addition, a small number of disks can be used as cache optionally, whereas it is mandatory in MAID [2]. Buffers are dedicated corresponding to the data disks. One buffer shared by multiple disks as a logical group, defined as a *Disk Group* (DG). This configuration appears in the typical distributed storage systems such as Cassandra [8].

In RAPoSDA, data are assigned to a corresponding data disk as primary, and then written to the primary area of a buffer that binds to the same DG.

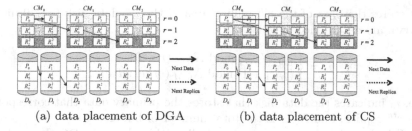

(a) data placement of DGA   (b) data placement of CS

Fig. 1. An example of the data layout of DGA and CS ($N_{CM} = 3$, $N_{DG} = 2$, $N_{DD} = 6$, $N_R = 3$)

The backup data is then written to the backup area of another buffer which assigned randomly. Thus, the disk access can be avoided until the buffer threshold is exceeded. When the data are flushed, it tries to select a disk that is currently rotating or one that has been in the standby state longer than the break-even time if both the primary and backup disks are on standby.

The data on the data disks placed with chained declustering [6] manner.

3   Energy-Efficient Data Placement

This section describes the data placement policies to reduce the disk access frequency of storage systems with more than three replicas. For simplicity regarding the storage system, we only explains the case of three replicas ($N_r = 3$). However, this simplified case can be generalized to cases with more than three replicas.

3.1   Data Placement Policy on Data Disks

The data placement policy on data disks with three replicas extends the chained declustering [6]. The rth replica of primary data ($r = 0$ is primary) on i^{th} data disk is represented as R_i^r, is assigned to the $((i + r) \bmod N_{DD})$-th data disk.

3.2   Data Placement Policy on Buffers

To reduce the disk access frequency with more than three replicas, two data placement policies with different assignments of buffers corresponding to primary data, named Disk Group Aggregation (DGA) and Cache Striping (CS), were utilized. Figure 1 depicts an example of the data placement of DGA and CS. In this figure, D_i represents the ith data disk and CM_j represents the jth buffer. In addition, P_i represents the area of primary data of the D_i-th data disk, and R_i^r represents the area of the rth replica of the primary data, P_i.

In DGA, primary data belonging to the same DG are written in the primary area of the identical buffer. Then, the replicas are assigned to a replica area of another buffer by the chained declustering strategy in the unit of the primary area of the buffer. In other words, the buffer number j and its rth replica area, which stores the data of the rth replica area of the ith data disk ($r = 0$ means primary), are determined as follows.

$$j_r = \begin{cases} \lfloor i_0/N_{DG} \rfloor \bmod N_{CM} \ (r = 0) \\ (j_{r-1} + 1) \bmod N_{CM} \ (otherwise) \end{cases} \tag{1}$$

where, i_0 indicates the data disk that stores the primary data that corresponds to the rth replica, and N_{CM} is the total number of buffers.

In CS, the primary data are not in bulk within the same DG. Instead, CS assigns a buffer in a striping manner. That is, the buffer number j and its rth replica area, which stores the data of the rth replica area of the ith data disk ($r = 0$ means primary), is determined as follows.

$$j_r = \begin{cases} i_0 \bmod N_{CM} \ (r = 0) \\ (j_{r-1} + 1) \bmod N_{CM} \ (otherwise) \end{cases} \tag{2}$$

4 Energy-Efficient Buffer Flush Algorithm

This section describes energy-efficient buffer flush algorithms that we propose.

In respect to the buffer flush, one of the naive method to accomplish this goal is to flush all the data on buffer to disk when a buffer overflow occurs. However, flushing all of the buffered data may access many data disks, causing many disks to spin up, thereby requiring a large amount of power to accommodate the temporarily large peek power. To address this issue, we propose two buffer flush algorithms WithAllSpins and SpinupEE.

The basic buffer flush procedure is presented as following.

1. The disk which has largest buffer data in the flush target buffer region is marked as the target disk. Then the disk is spun-up if it is in standby state.
2. The buffer data on the target disk are flushed. In addition, the buffer data in the same buffer region are flushed to the corresponding disks that are in active state.
3. The some other buffer data are flushed to the corresponding disks according to the each buffer flush algorithm, described in next two sub sections.

4.1 WithAllSpins

After the second step of the basic buffer flush procedure, WithAllSpins flushes all of the buffered data of currently rotating disks. This approach does not only require no disk spin-up, but also can make a lot of free space in the buffer. Thus, the time for the next buffer overflow becomes longer. Therefore, the disks can keep the standby state as long as possible.

Algorithm 1. Flush the Buffer with the Maximum Energy Efficiency

---
1: B_{base} := Get all buffered data chunks of spinning data disks
2: T_{nextOF} := $|B_{base}|/\lambda$ // Calculate the time to the next buffer overflow by using B_{base} and λ
3: B_{max} := B_{base}
4: E_{min} := $1/T_{nextOF}$
5: **for all** $SubD_{spindown}$ which subset of the data disks is in the standby state **do**
6:     B_{SubD} := Get all buffered data chunks of $SubD_{spindown}$
7:     T_{nextOF}^{SubD} := $(|B_{SubD}| + |B_{base}|)/\lambda$
8:     E_{SubD} := $|SubD_{spindown}| \times P_{standby}/T_{nextOF}^{SubD}$
9:     **if** $E_{SubD} < E_{min}$ **then**
10:         B_{max} := $B_{base} \cup B_{SubD}$
11:         E_{min} := $min\{E_{min}, E_{SubD}\}$
12:     **end if**
13: **end for**
14: Flush all buffered data chunks of B_{max}

4.2 SpinupEE

SpinupEE flushes the disks with the buffered data considered as the most energy efficient. The energy efficiency is defined as the ratio of the total energy when the buffered data of a certain combination of disks are flushed and the predicted time to the next buffer overflow. The detail of the SpinupEE is shown in Algorithm 1.

At first, the algorithm separates the data disks into two groups: spinning disks and standby disks. Then, the algorithm places all the buffered data chunks on buffers that correspond to spinning disks and aggregates these chunks into one buffer. In line 1, B_{base} denotes the aggregated buffered data chunk. Next, the algorithm calculates the expected time to the next buffer overflow T_{nextOF} if only the buffered data of spinning disks were flushed by using B_{base} and the average arrival rate of write requests λ. As shown in line 2, the value of T_{nextOF} is calculated as the size of the buffered data chunk (the sum of blocks in the chunk) $|B_{base}|$ divided by λ, and then the algorithm calculates the energy efficiency of the buffer being flushed from the only spinning disks and calculates the value set into the E_{min} temporarily (line 4). The energy efficiency is calculated as the value of the extra energy consumption when the buffer is being flushed divided by the predicted time to the next buffer overflow. The energy efficiency is the required energy per second to get the time to the next buffer overflow. Thus, if flushing buffer data to only the spinning disks, since there is no need for extra energy for the disk spin-up, the energy efficiency can be presented as E_{min} and calculated as $1/T_{nextOF}$. In line 5 to line 13, for the arbitrary combination of the remaining standby disks, the algorithm calculates the energy efficiency by using the time to the next buffer overflow and its energy overhead due to the spinning up of these disks in a similar manner, as discussed in the presented steps. In other words, considering the set of disks in the standby state, $D_{spindown}$ = $\{D_i|\ D_i$ is a data disk $\land\ D_i$ is in a standby state$\}$ and an arbitrary subset $SubD_{spindown}$ of $D_{spindown}$. Thus, the algorithm calculates the energy efficiency

Table 1. Parameters of the hard disk drive used in the simulation

Parameters	Value
Capacity (TB)	2
Number of platters	5
Rotations per minute (RPM)	7200
Disk cache size (MB)	32
Data transfer rate (MB/s)	134
Active power (W)	11.1
Idle power (W)	7.5
Standby power (W)	0.8
Spin-down energy (J)	35.0
Spin-up energy (J)	450.0
Spin-down time (s)	0.7
Spin-up time (s)	15.0

Table 2. Configuration of each storage system

	Normal	RAPoSDA
Buffer	-	24 GB (8 GB × 3)
# of replicas	3	3
# of cache disks	-	6, 9
# of data disks	60, 90	60, 90

Table 3. Configuration of the storage for comparing buffer flush algorithms

Component	Value
# of cache disks	6
# of data disks	30
# of buffers	3 (4GB/memory)
Block size	64 KB
Data mapping policy	CS

E_{SubD} of all spinning disks and the standby disks that belong to one of the subsets $SubD_{spindown}$ with energy overhead to spin up the standby disks.

To calculate the energy efficiency E_{SubD}, first, the algorithm calculates the expected time to the next buffer overflow to merge the buffered data of the spinning disks and the standby disks belonging to the selected subset $SubD_{spindown}$ when flushing their buffered data (line 7). Then, the extra energy required for this case is the sum of the spin-up energy of the standby disks belonging to a selected subset which is derived as $|SubD_{spindown}| \times P_{standby}$. Therefore, the energy efficiency of this case E_{SubD} is calculated as $|SubD_{spindown}| \times P_{standby}/T_{nextOF}^{SubD}$ (line 8). After calculating the E_{SubD} for all subsets of standby disks, except the empty set ϕ, the most energy efficient combination of disks to flush the buffered data is determined. If the energy efficiency of any combination of standby and spinning disks is larger than the case of only using spinning disks, the algorithm flushes the buffered data of only spinning disks.

5   Evaluation

In this section, we evaluate the effectiveness of the data placement policies, DGA and CS (Evaluation 1), and the buffer flush algorithms, WithAllSpins and SpinupEE (Evaluation 2). In the evaluation, a simulation program [12] which we developed is used. We employ RAPoSDA as the base storage configuration through the experiment. And Normal configuration which does not employ any power reduction techniques is also used for the comparison. The simulation uses Hitachi Deskstar 7K2000 [5] as the hard disk drive model. The parameters of

Table 4. Parameters of the synthetic workload

Workload parameter	Evaluation 1	Evaluation 2
Simulation time	5 h	Same
Read:write ratio	3:7	0:10, 3:7
Number of files	10,000,000 (32 KB/file)	10 M (1 MB/Object)
Total file size	960 GB (3*replicas*)	30 TB (3*replicas*)
Number of requests	$\lambda \times 3600 \times 5$ (h)	Same
Data access distribution	Zipf	Same
Request arrival distribution	Poisson arrival	Same
Zipf factor s	1.2	Same
Mean arrival rate (λ)	30 (request/s)	Same

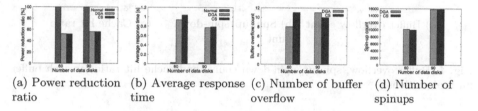

(a) Power reduction ratio   (b) Average response time   (c) Number of buffer overflow   (d) Number of spinups

Fig. 2. Simulation results

this model are described in Table 1. In addition, Tables 2 and 4 depict each of storage system configuration and the parameters of the synthetic workload in this evaluation (Table 3).

5.1   Comparison of Cache Striping and Disk Group Aggregation

Simulation Results. The simulation results are depicted in Fig. 2. In Fig. 2(a), DGA achieved a slightly better power reduction ratio than CS in the case of 60 data disks. Whereas, DGA and CS showed almost the same in the case of 90 data disks. Thus, regarding the power reduction ratio, there are no salient differences between DGA and CS.

Figure 2(b) depicts the average response time. Normal is the fastest, but consumes the most power. In RAPoSDA, the average response time of CS is 1.03 s, while the average response time of DGA is 0.94 s in the case of 60 data disks. For 90 data disks, the average response times of CS and DGA are 0.78 s and 0.77 s, respectively. Thus, DGA has a slightly faster average response time than CS. Also, Fig. 2(c) and (d) depict the total number of buffer overflows of cache memory and the total number of disk spin-ups of RAPoSDA. In Fig. 2(c), DGA has three times fewer overflows than CS with 60 data disks. By contrast, CS has one time fewer overflows than DGA with 90 data disks. From Fig. 2(d), although CS can suppress the spin-up counts slightly more than DGA with 60 data disks, it shows the inverse tendency with 90 data disks.

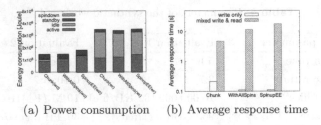

(a) Power consumption   (b) Average response time

Fig. 3. Power consumption and average response time for each buffer flush algorithms

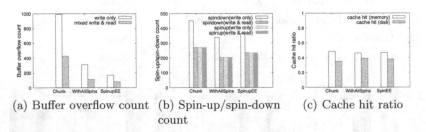

(a) Buffer overflow count   (b) Spin-up/spin-down   (c) Cache hit ratio
count

Fig. 4. Buffer overflow, spin-up/spin-down count, and cache hit ratio for each buffer flush algorithms

5.2   Comparison of Each Buffer Flush Algorithms

In this comparison, two buffer flush algorithms are compared. In addition, we introduce *Chunk*, which represents the original RAPoSDA's buffer flush procedure described in Sect. 4 to compare with the proposed algorithms.

Simulation Results. Figure 3(a) depicts the power consumption of the three buffer flushing algorithms with each workload. With the write only workload (represented as "(wo)" in the figure), the Chunk algorithm shows the lowest power consumption, and the WithAllSpins algorithm consuming the second-most amount of power. By contrast, the SpinupEE algorithm consumes the most power, In particular, it consumes about 23 % more power than the Chunk algorithm. With the mixed write and read workload (represented as "(wr)" in the figure) shows that the WithAllSpins algorithm consumes the least amount of power, 2.8% less than the Chunk algorithm.

Figure 3(b) shows the average response time of the three buffer flush algorithms with each workload. Chunk algorithm has the largest response time with the write only workload; the reason for this is that Chunk algorithm suffered from waiting to the standby disks because it has substantially more buffer overflows and spin-ups/spin-downs than the others. By contrast, with the mixed write and read workload, the result shows that the Chunk algorithm has the fastest average response time. Accordingly, regarding the cache hit rate (in Fig. 4(c)), the Chunk algorithm leaves the largest amount of buffer data in buffer memory. Therefore, the Chunk algorithm has the lowest average response time, and the

SpinupEE algorithm shows the lowest response performance; the reason for this is because access to the disks that are spun up is delayed owing to the buffer flush algorithm described in Sect. 4. In addition, the WithAllSpins and SpinupEE algorithms suffer from relatively large latency.

Figure 4(a) shows that the SpinupEE algorithm achieves the lowest overflow count, as expected. However, according to Fig. 4(b), the lowest spin-up(spin-down) count is achieved by the WithAllSpins algorithm; the reason for this is that the WithAllSpins algorithm does not have any extra disks spin-up during buffer overflow, and it flushes buffered data to all the spinning disks. Thus, the WithAllSpins algorithm can reduce the number of spin-up/spin-down count more than the others.

5.3   Discussion

Regarding data placement policies, the differences between DGA and CS are small. Also, this result could have been affected by the workload characteristics. It is anticipated that CS is effective when the write accesses show a random distribution. However, the workload of this simulation has higher access locality. Therefore, DGA is advantageous compared with CS.

Regarding buffer flushing algorithms, WithAllSpins and SpinupEE significantly reduce the buffer overflow count and spin-up/spin-down count compared with the RAPoSDA approach (Chunk). With the mixed read and write workload, all three algorithms achieve almost the same amount of power. However, the average response times of the WithAllSpins and SpinupEE are more than double that of the Chunk, because of the buffer cache hit for the read requests. By contrast, with the write-only workload, the average response times of the WithAllSpins and SpinupEE algorithms were twice as fast as that of the Chunk. The reason is that Chunk was suffering a longer delay penalty of many spin-up/spin-down compared with the other algorithms.

6   Related Work

Lang et al. [9] studied on both load balancing and power reduction by using a chained declustering data placement strategy. However, since it only has two replicas, they do not consider more than three replicas.

Rabbit [1], Sierra [13], FREP [7], and Accordion [4] determine the *power proportionality*. These studies purpose that the ratio of their power consumption and performance can be kept constant by increasing and decreasing the number of active nodes along with the current load. Since they assume large-scale distributed storage systems, most of the nodes activate to assure the primary data are available, even if they are relatively large.

REED [14] and RESS [15] are designed for large-scale distributed storage systems to reduce the power consumption and ensure the reliability of the storage by integrating SSD and HDD. These studies use HDD as the primary storage and SSD as cache. If the workload is heavy, all HDDs are active and serve the maximum IO performance, whereas, during the light workload, SSD is activated and HDDs are deactivated if their idle time exceeded a certain threshold time. However, they do not consider the data availability, such as MTTDL.

7   Conclusion

In this paper, we proposed two data placement policies, Cache Striping (CS) and Disk Group Aggregation (DGA) for storage systems with more than three replicas. CS saves a little more energy than DGA, while DGA slightly reduces the overhead of the response time more than CS.

In addition, we also proposed and evaluated two buffer flushing algorithms for storage systems with more than three replicas. In the read/write mixed workload, WithAllSpins and SpinupEE reduce the number of spin-ups/spin-downs and the number of buffer overflows more than Chunk. Furthermore, WithAll-Spins reduced the power consumption the most among the three algorithms. In the evaluation, we demonstrated considering disk rotation status when buffer flushing can reduce the power consumption more. However, the proposed algorithms suffer a greater penalty of latency compared with Chunk since they tend to increase I/Os when buffer flushing. Thus, there are design positives and negatives for power reduction and performance, especially response time.

References

1. Amur, H., et al.: Robust and flexible power-proportional storage. In: Proceedings of the 1st ACM Symposium on Cloud Computing, SoCC 2010 (2010)
2. Colarelli, D., Grunwald, D.: Massive arrays of idle disks for storage archives. In: Proceedings of the ACM/IEEE Conference on Supercomputing, pp. 1–11 (2002)
3. Ghemawat, S., Gobioff, H., Leung, S.T.: The Google file system. SIGOPS Oper. Syst. Rev. **37**(5), 29–43 (2003)
4. Le, H.H., Hikida, S., Yokota, H.: Accordion: an efficient gear-shifting for a power-proportional distributed data-placement method. IEICE Trans. Inform. Syst. 1013–1026 (2015)
5. Hitachi Global Storage Technologies: Hard disk drive specification, hitachi deskstar 7k2000. http://www.hgst.com/tech/techlib.nsf/products/Ultrastar_7K4000
6. Hsiao, H.I., DeWitt, D.J.: Chained declustering: a new availability strategy for multiprocessor database machines. In: Proceedings of the 6th ICDE, pp. 456–465 (1990)
7. Kim, J., Rotem, D.: Energy proportionality for disk storage using replication. In: Proceedings of the 14th EDBT (2011)
8. Lakshman, A., Malik, P.: Cassandra: a decentralized structured storage system. SIGOPS Oper. Syst. Rev. **44**(2), 35–40 (2010)
9. Lang, W., Patel, J.M., Naughton, J.F.: On energy management, load balancing and replication. SIGMOD Rec. **38**(4), 35–42 (2010)
10. Hikida, S., Le, H.H., Yokota, H.: A power saving storage method that considers individual disk rotation. In: Lee, S., Peng, Z., Zhou, X., Moon, Y.-S., Unland, R., Yoo, J. (eds.) DASFAA 2012. LNCS, vol. 7239, pp. 138–149. Springer, Heidelberg (2012). https://doi.org/10.1007/978-3-642-29035-0_10

11. Shvachko, K., Kuang, H., Radia, S., Chansler, R.: The Hadoop distributed file system. In: 2010 IEEE 26th MSST, pp. 1–10, May 2010
12. Storage system simulator. https://github.com/reddikih/spsim2
13. Thereska, E., Donnelly, A., Narayanan, D.: Sierra: practical power-proportionality for data center storage. In: Proceedings of the 6th Conference on Computer Systems. ACM (2011)
14. Yin, S., Li, X., et al.: REED: a reliable energy-efficient raid. In: 2015 44th International Conference on Parallel Processing, pp. 649–658, September 2015
15. Yin, S., Xiao, Z., et al.: RESS: a reliable energy-efficient storage system. In: IEEE 22nd International Conference on Parallel and Distributed Systems, pp. 1193–1198, December 2016

Querying Knowledge Graphs
with Natural Languages

Xin Wang[1(✉)], Lan Yang[2], Yan Zhu[1], Huayi Zhan[3], and Yuan Jin[4]

[1] Southwest Jiaotong University, Chengdu, China
{xinwang,yzhu}@swjtu.edu.cn
[2] Sichuan ChangHong Electric Co. Ltd, Mianyang, China
lan.yang@changhong.com
[3] Xi'an Jiaotong University, Xi'an, China
huayizhan2012@stu.xjtu.edu.cn
[4] Nuclear Power Institute of China, Chengdu, China
jinyuan1377@sina.com

Abstract. With the unprecedented proliferation of knowledge graphs, how to process query evaluation over them becomes increasingly important. On knowledge graphs, queries are typically evaluated with graph pattern matching, *i.e.*, given a pattern query Q and a knowledge graph G, it is to find the set $M(Q,G)$ of matches of Q in G, where matching is defined with subgraph isomorphism. However querying big knowledge graphs brings us challenges: (1) queries are often issued with natural languages, hence can not be evaluated directly; (2) query evaluation is very costly and match results are often difficult to inspect. In light of these, this paper studies the problem of querying knowledge graphs with natural languages. (1) We extend pattern queries by designating a node u_o as "query focus", and revise the matching semantic based on the extension. (2) We develop techniques to understand natural language queries, and generate pattern queries with "query focus". (3) We develop efficient techniques to identify top-k matches of "query focus". (4) We experimentally verify that our techniques for query understanding perform well, and our query algorithm is able to find diversified top-k matches efficiently.

1 Introduction

The ever-increasing knowledge graphs impose an urgent demand of effective query techniques for end users. Typically, queries over knowledge graphs are evaluated with pattern matching, *i.e.*, given a pattern query Q and a knowledge graph G, it is to find $M(Q,G)$, the set of matches of Q in G. There are two key issues for querying knowledge graphs. (1) Users' queries are often expressed with natural languages, which are unstructured and can not be evaluated over graphs directly, instead, one needs to bridge the gap by constructing structured pattern queries from corresponding questions. (2) Knowledge graphs are often very big, *e.g.*, DBpedia [2] has more than 4.58 million entities, and 3 billion RDF triples,

© Springer Nature Switzerland AG 2019
S. Hartmann et al. (Eds.): DEXA 2019, LNCS 11707, pp. 30–46, 2019.
https://doi.org/10.1007/978-3-030-27618-8_3

and query semantic is typically defined in terms of subgraph isomorphism, which is an NP-complete problem [7], these together bring following challenges: (a) query evaluation is cost prohibitive, (b) it is a daunting task to understand query results, as there may exist excessive matches of Q in G, and (c) users are often interested in top-k matches of a specific node u_o as "query focus" of Q, that are not only relevant to u_o, but are also as diverse as possible, simultaneously [9].

These highlight the need for (1) *query structuring*, which transforms natural language queries Q_{NL} into pattern queries Q with node u_o as "query focus", and (2) *diversified top-k graph pattern matching*, that identifies diversified top-k matches of the "query focus" u_o of Q.

Example 1. A fraction of a movie knowledge base IMDB consists of a set of triples (s, p, o), indicating subject, predicate and object, respectively. For example, $(m_1, \text{hasRating}, 5.7)$ says that a movie with id m_1 has rating 5.7. This triple set can be modeled as a graph G (shown in Fig. 1(a)), in which s and o are nodes, that are connected by an edge from s to o with label p. Note that the prefix of the id of each entity indicates the type of the entity, *e.g.*, m (resp. p) represents that the entity is of the type "movie" (resp. "performer").

Due to the schema-less characteristic of graph data, one prefers issuing a natural language query Q_{NL} to find movies that he is interested in, *e.g.*, "Find comedy movies played by You Ge and another actress who played comedy movies before". Obviously, Q_{NL} can not be used directly to query on knowledge graphs, it should be transformed into a pattern query taking a designated node u_o as "query focus". To this end, we generate such a pattern query Q, which is shown in Fig. 1(b), by first choosing a pattern structure that is extracted from query log, and can best capture the relationship among entities in Q_{NL}, and then mapping entities and relationship in Q_{NL} to corresponding nodes (including query focus) and edges in the pattern structure.

As there may exist excessive matches of u_o, while users may only be interested in best k ones, that are as diverse as possible. It is hence unnecessary and too costly to find all the matches of u_o, an algorithm with the *early termination property* is desired. To rank matches of u_o, one may consider the following criteria. (1) Influences. Observe that m_9 can form multiple matches of Q with different nodes surrounded, hence is considered more influential. (2) Diversity. Consider sets $\{m_4, m_9\}$ and $\{m_4, m_6\}$, m_4 is more "dissimilar" to m_9 than m_6 is, since m_4 and m_9 have less common neighbors as matches of pattern nodes of Q than m_6 and m_9 have. Putting these together, when $k = 2$, $\{m_4, m_9\}$ makes a good candidate for top-k matches in terms of both influence and diversity. $\square$

This example shows that natural language can simplify query expression on knowledge graphs, and diversified top k graph pattern matching can not only improve user's satisfaction, but also may improve query efficiency. However, to query knowledge graphs with natural languages, two fundamental problems have to be settled. (1) How to generate a pattern query Q with output node u_o from a natural language query? (2) How to develop efficient algorithms, better with *early termination* property, for computing diversified top-k matches of u_o?

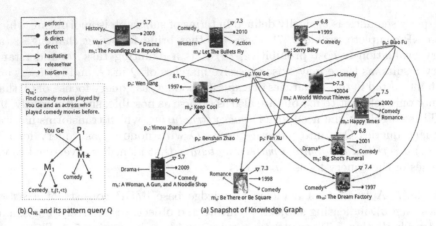

(b) Q_{NL} and its pattern query Q   (a) Snapshot of Knowledge Graph

Fig. 1. Knowledge graph, natural language query and its corresponding pattern query

Contributions. This paper investigates following questions for querying knowledge graphs. We focus on graph pattern matching defined in terms of subgraph isomorphism, since it is commonly used in knowledge graph search.

(1) Knowledge graph search often has specific focuses [4], we hence revise pattern query by designating a node u_o as "query focus". Then given Q and G, it is to compute $M_u(Q, G, u_o)$, the set of matches of u_o in G via Q. To rank matches of u_o, we develop two functions, namely, *relevance function* $w()$ and *distance function* $d()$. Based on both, we define a bi-criteria *diversification function* $F()$ that aims to maximize relevance and diversity simultaneously (Sect. 2).

(2) We propose a novel method to transform natural language queries Q_{NL} into pattern queries (Sect. 3). The approach extracts typical pattern structures from query log as class labels, trains a classifier with labeled query log, and employs the classifier to find a pattern structure that best captures relationship among keywords occurred in the query sentence.

(3) We investigate the diversified top-k graph pattern matching problem. It is to find top-k matches of u_o based on the diversification function $F()$. We show that the decision version of the problem is NP-hard. Nevertheless, we develop an algorithm TopKET in $O(|Q|^2 + |Q||G|^2 + |G|^2|G|!)$ time, with *the early termination property* (Sect. 4).

(4) Using both real-life and synthetic data, we experimentally verify the efficiency and effectiveness of our methods (Sect. 5). We find that (a) our query structuring technique is very effective, with accuracy ratio R over 95%, on average; and (b) TopKET is not only effective, but also efficient. Taking knowledge graph Movie as example, when top-5 matches are required, its F-measurecan reach 80% over patterns with size $|Q| = (3, 2)$, and moreover it only needs less than 1 second to find top-10 matches, and accounts for, on average, 20% and 18.5% of the time used by TopKApx and TopKNaive, respectively.

These results yield a promising approach to querying big knowledge graphs. Our technique can understand natural language queries well, and our query algorithm can efficiently identify diversified top-k matches of "query focus".

Related Work. We categorize the related work as follows.

Query Structuring. Keyword search may not be expressive enough to query structured data, *e.g.,* knowledge graphs, since only a few words are not sufficient to specify the query intention, and structured queries are in demand for querying structured data. In light of this, various techniques have been developed to understand natural language queries, among these are [13,16,17,19]. [17] trains a classifier by mining frequent structured queries from query log, and combining them with linguistic structure in keyword queries. With the classifier, keyword queries can be translated to structured queries. [19] translates users' questions into an extended form of structured SPARQL queries, with text predicates attached to triple patterns. [13] presents MING, a principled method for extracting an informative subgraph for given query nodes, and hence form a structured query representation. When users do not know how to describe the specifications of the items of interest, but does know some examples, [16] proposed methods to understand the structure of exemplar queries. [21] proposes an approach to understanding questions and generating structured queries through talking between the data (*i.e.,* the knowledge graph) and the user.

Top-k Graph Pattern Matching. Top-k graph pattern matching is to retrieve k best matches from the match set. There has been a host of work on this topic. For example, [8,22] propose to rank matches, *e.g.,* by the total node similarity scores [22], and identify k matches with the highest scores. [11] investigates top-k query evaluation for twig queries, which essentially computes isomorphism matching between rooted graphs. To provide more flexibility of top-k pattern matching, [6] extends matching semantics by allowing edge to path mapping, and proposes to rank matches based on their compactness. Instead of matching with subgraph isomorphism, graph simulation [12] is applied as matching semantic, and pattern graph is designated an output node in [9], then match result is a set of nodes that are matches of the output node.

Diversified Graph Pattern Matching. Result diversification is a bi-criteria optimization problem for balancing result relevance and diversity [5,10], with applications in *e.g.,* social searching [3]. Following the idea, diversified graph pattern matching has been studied in, *e.g.,* [9,20]. [9] takes both diversity and relevance into consideration, and proposes functions to capture both relevance and diversity. In contrast, [20] considers diversity only, and measures diversity by the number of vertices covered by all the matches in the result.

Our work differs from prior work in the following: (1) our matching semantic is quite different from that in [9,20], where [9] extends pattern queries with output node and applies graph simulation as matching semantic, and [20] finds matches of pattern queries rather than its "query focus", via subgraph isomorphism. (2) [17] translates natural language queries into structured queries without output node, while ours can not only structure natural language queries but also specify query focus. (3) [20] only considers match diversity, while ours considers both relevance and diversity.

2 Preliminary

In this section, we review notions about knowledge graph, pattern queries (Sect. 2.1), graph pattern matching (Sect. 2.2), and query structuring (Sect. 2.3), respectively.

2.1 Knowledge Graphs and Pattern Queries

We start with notions of knowledge graphs and pattern queries.

Knowledge Graphs. Assume a set $\mathcal{E}$ of entities, a set $\mathcal{D}$ of values, a set $\mathcal{P}$ of predicates (labels), and a set Θ of types. Each entity e in $\mathcal{E}$ has a unique ID and a type in Θ. A *knowledge graph* (or simply a graph) consists of a set of triples (s, p, o), where *subject* s is an entity in $\mathcal{E}$, p is a *predicate* in $\mathcal{P}$, and o is either an entity in $\mathcal{E}$ or a value d in $\mathcal{D}$. It can be represented as a directed edge-labeled graph $G = (V, E, L)$, such that V is the set of nodes consisting of s and o for each triple (s, p, o); and E includes edges $e = (u, v)$ with label $L(e) = p$, for each triple (s, p, o), where $L()$ is the edge labeling function.

Two types of equality are considered: (a) node identity on $\mathcal{E}$: if entities e_1 and e_2 have the same ID; and (b) value equality on $\mathcal{D}$: $d_1 = d_2$ if they are the same value.

Pattern Queries. A *pattern query* (or simply a pattern) Q is a set of triples (s_Q, p_Q, o_Q), where s_Q is a variable z, o_Q is one of a value d or z, and p_Q is a predicate in $\mathcal{P}$. Here variable z has one of three forms: (a) *entity variable* y, to map to an entity, (b) *value variable* $y*$, to map to a value, and (c) *wildcard* $\_y$, to map to an entity. Here s_Q can be either y or $\_y$, while o_Q can be y, $y*$ or $\_y$. Entity variables and wildcard carry a *type*, denoting the type of entities they represent. A *pattern query* can also be represented as a graph $Q = (V_p, E_p)$, such that two variables are represented as the same node if they have the same name of y, $y*$ or $\_y$; similarly for values d.

We denote $|V_p| + |E_p|$ as $|Q|$ (the size of Q), and $|V| + |E|$ as $|G|$ (the size of G).

2.2 Graph Pattern Matching Revised

We first propose the notion of *valuation*, followed by graph pattern matching problem (GPM). We next introduce result diversification. Consider a knowledge graph G and a pattern $Q(u_o)$.

Valuation. A valuation of Q in a set T of triples is a mapping ν from Q to T that preserves values in $\mathcal{D}$ and predicates in $\mathcal{P}$, and maps variables y and $\_y$ to entities of *the same type*. More specifically, for each triple (s_Q, p_Q, o_Q) in Q, there exists (s, p, o) in T, written as $(s_Q, p_Q, o_Q) \mapsto_\nu (s, p, o)$ or simply

$(s_Q, p_Q, o_Q) \mapsto (s, p, o)$, where (a) $\nu(s_Q) = s, p = p_Q, \nu(o_Q) = o$; (b) o is an entity if o_Q is a variable y or $\_y$; it is a value if o_Q is $y*$, and $o = d$ if o_Q is a value d; and (c) entities s and s_Q have the same type; similarly for entities o and o_Q if o_Q is y or $\_y$. We say that ν is a *bijection* if ν is one-to-one and onto mapping.

Graph Pattern Matching [7]. Consider a knowledge graph G and a pattern query Q. We say that G matches Q if there exist a set T of triples in G and a valuation ν of Q in T such that ν is a bijection between Q and T. We refer to T as a match of Q in G at e under ν. Intuitively, ν is an isomorphism from Q to T when Q and T are represented as graphs. The *answer to Q in G*, denoted by $M(Q, G)$ is the set of matches T of Q in G. To search *query focus* of a pattern query, we extend a pattern query as $Q = (V_p, E_p, u_o)$, where u_o is a node in Q labeled with $*$, referred to as the *output node* of Q, Then the matches of u_o in G is defined to be $M_u(Q, G, u_o) = \{v_o | \nu(u_o) = v_o, v_o \in T\}$, i.e., all the matches of the output node u_o.

Example 2. Recall G, Q in Example 1. It can be easily verified that match set $M(Q, G)$ includes 6 matches: (p_4, p_5, m_{10}, m_4), (p_4, p_5, m_4, m_6), (p_4, p_5, m_{10}, m_6), (p_4, p_5, m_4, m_9), (p_4, p_5, m_6, m_9), (p_4, p_5, m_{10}, m_9). One may further infer $M_u(Q, G, u_o)$ (u_o is M, marked with $*$) from $M(Q, G)$, and obtain three movies, i.e., m_4, m_6 and m_9.   □

For a pattern query Q, we use $\text{dist}_u(u, u')$ to denote the length of the shortest path between u and u' in Q, then its *radius* r is the largest $\text{dist}_u(u_o, u')$ between u_o and any other node u' in V_p, when Q is treated as an undirected graph.

Result Diversification. It is recognized that search results should be relevant, and at the same time, be as diverse as possible [10]. This gives the rise of diversification measurement.

Relevance Function. We define the relevance set $R_s(v_o)$ as $\{v_k | v_k \in T, T \in M(Q, G), \nu(u_o) = v_o, v_o \neq v_k\}$. Then the relevance function is defined as $w() = |R_s(v_o)|$.

Intuitively, for a match v_o of u_o, if the more nodes that can make up distinct matches of $Q(u_o)$ with v_o, the more important v_o is. That's the relevance function favors those matches that connect more other matches. For example, $R_s(m_4) = \{p_4, p_5, m_{10}\}$, then $|R_s(m_4)| = 3$, while $R_s(m_9) = \{p_4, p_5, m_4, m_6, m_{10}\}$, indicating $|R_s(m_9)| = 5$, hence m_9 is considered a more relevant match of M than m_4.

Distance Function. To characterize the diversity of a match set, we define a distance function to measure the "dissimilarity" of two matches. Given two matches v_1, v_2 of u_o, we define their distance $d(v_1, v_2)$ as

$$d(v_1, v_2) = 1 - \frac{|R_s(v_1) \cap R_s(v_2)|}{|R_s(v_1) \cup R_s(v_2)|}$$

Diversification Function. To measure the diversification of a match set $\mathcal{U} = \{v_1, v_2, \cdots, v_k\}$ of the output node u_o, a function $F()$ is defined as

$$F(\mathcal{U}) = (1 - \lambda) \sum_{v_i \in \mathcal{U}} w'(v_i) + \frac{2 \cdot \lambda}{k - 1} \sum_{v_i \in \mathcal{U}, v_j \in \mathcal{U}, i < j} d(v_i, v_j),$$

where $w'(v_i)$ is a normalized relevance function defined as $\frac{w(v_i)}{w(v_m)}$, and $w(v_m)$ is the largest relevance value among all matches of u_o, $d()$ is the distance function, and $\lambda \in [0, 1]$ is a parameter set by users. The diversity metric is scaled down with $\frac{2 \cdot \lambda}{k-1}$, since there are $\frac{k(k-1)}{2}$ numbers for the difference sum, while only k numbers for the relevance sum.

Diversified Top-k Matching Problem. With the diversification function $F()$, we next state the diversified top-k matching problem, denoted by TopKM. Given G, Q, a positive integer k, and a parameter $\lambda \in [0, 1]$, it is to find a set of k matches $\mathcal{U} \subseteq M_u(Q, G, u_o)$ such that

$$F(\mathcal{U}) = \underset{\mathcal{U}' \subseteq M_u(Q, G, u_o)}{\arg \max} F(\mathcal{U}'),$$

i.e., for all k-element sets $\mathcal{U}' \subseteq M_u(Q, G, u_o)$, $F(\mathcal{U}) \geq F(\mathcal{U}')$.

Example 3. Recall Example 2. As the match set $M_u(Q, G, u_o) = \{m_4, m_6, m_9\}$, when $k = 2$ and $\lambda = 0.5$, three two-element sets $\{m_4, m_9\}$, $\{m_4, m_6\}$ and $\{m_6, m_9\}$ have diversification values 1.2, 0.95 and 1.1, respectively. Hence $\{m_4, m_9\}$ makes a diversified top-2 match set.   □

2.3   Query Understanding

Due to unstructured character, natural language queries Q_{NL} can not be directly used to query knowledge graphs. It needs to be correctly translated into a structured pattern query. Typically, translation of Q_{NL} consists of two tasks, *i.e.,* phrase identification, and query structuring.

Phrase Identification. A natural language query is consisted of a sequence of tokens, $Q_{NL} = (t_0, t_1, \cdots, t_n)$. A phrase is a contiguous subsequence of tokens $\mathcal{P} = (t_i, t_{i+1}, ..., t_{i+l}) \subseteq Q_{NL}$ $(0 \leq i, 0 \leq l \leq n)$. Phrases can denote entities (*e.g.,* the city of Casablanca or the movie Casablanca), types (*e.g.,* actresses, movies), or relations/attributes (*e.g.,* played in between people and movies). Phrase identification is to identify a set of phrases $\mathcal{P}$ from Q_{NL} such that each of them potentially corresponds to semantic items such as "movie" and "played in". To simplify presentation, we denote an ordered list of phrases of Q_{NL} as L_P.

Query Structuring. Querying a knowledge graph with identified phrases still can not well capture user's query intention [17], instead, a structured pattern query with query focus is required. However, it is a challenging task to structure Q_{NL} due to that relationship among phrases may not be explicitly mentioned in a Q_{NL}, which calls for techniques to infer implicit relations and construct a pattern query.

Example 4. A natural language query Q_{NL} and its corresponding pattern query Q is shown in Fig. 1(b). Note that nodes in Q represent either entities or their contents; edges indicate relationships among nodes, *e.g.,* perform, hasGenre and releaseYear; and the *query focus* M is marked with "*" as "output node".   □

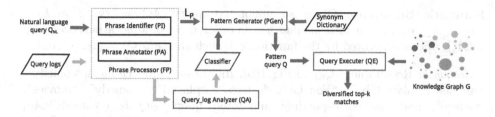

Fig. 2. Overview of the approach

2.4   Approach Overview

Figure 2 presents the overview of our approach, which consists of four modules, *i.e.,* Phrase Processor (FP), Query_log Analyzer (QA), Pattern Generator (PGen) and Query Executer (QE). The main functions of the modules are as follows. (1) Module FP contains two submodules, *i.e.,* phrase identifier (PI) and phrase annotator (PA, Sect. 3.1), which are in charge of phrase identification and annotation, respectively. The output of module PI is an ordered list of phrases of Q_{NL}, denoted as L_P, and will be used by QA and PGen for classifier training and pattern query generation, respectively. (2) Using query log, module QA trains a classifier, which will be used by module PGen to generate pattern queries Q (Sect. 3.2). (4) Module QE computes diversified top-k matches of Q in G (Sect. 4). It deserves mentioning that module QE takes advantage of a synonym dictionary to eliminate ambiguities of entities or relations.

3   Structuring Natural Language Queries

In this section, we first introduce method to annotate queries (Sect. 3.1), then provide techniques to generate pattern queries (Sect. 3.2).

3.1   Query Annotation

Given a natural language query Q_{NL}, one can identify its phrases with techniques, *e.g.,* [14], and generate a phrase sequence $L_P = k_1, \cdots, k_n$, where k_i in L_P is a phrase that refers to an entity, a value or a relation. Based on L_P, a semantically annotated query is defined as following.

Annotated Query. Given a sequence of phrases $L_P = k_1, k_2, \cdots, k_n$, a semantically annotated query $\mathcal{AQ}$ is a sequence of "phrase:semantic annotation" pairs:

$$\mathcal{AQ} = \langle k_1 : a_1, k_2 : a_2, \cdots, k_n : a_n \rangle,$$

where each a_i is a semantic annotation from annotation set Φ.

Following [17], we let Φ take five semantic annotations, *i.e.*, entity, type, value, relation and attribute, where entity, type, value and relation, attribute are used for annotating nodes and edges, respectively. To ease presentation, we denote entity, type, value, relation and attribute by $\mathcal{A}_{\mathcal{E}}$, $\mathcal{A}_{\mathcal{T}}$, $\mathcal{A}_{\mathcal{V}}$, $\mathcal{A}_{\mathcal{R}}$ and $\mathcal{A}_{\mathcal{A}}$, respectively.

Semantic Summary. Given an annotated query $AQ = \langle k_1 : a_1, k_2 : a_2, \cdots, k_n : a_n \rangle$, a *semantic summary* is an ordered list of annotations occurring in AQ, and is generated by the function $g()$, such that $g(AQ) = \langle a_1, \cdots, a_n \rangle$.

Example 5. Recall query Q_{NL} in Fig. 1(b). Its phrase sequence L_P is "comedy", "movies", "played by", "You Ge", "actress", "played", "comedy", "movies", "before", and the corresponding annotated query AQ is $\langle$ "comedy":$\mathcal{A}_{\mathcal{V}}$, "movies":$\mathcal{A}_{\mathcal{T}}$, "played by":$\mathcal{A}_{\mathcal{R}}$, "You Ge":$\mathcal{A}_{\mathcal{E}}$, "actress":$\mathcal{A}_{\mathcal{E}}$, "played":$\mathcal{A}_{\mathcal{R}}$, "comedy":$\mathcal{A}_{\mathcal{V}}$, "movies":$\mathcal{A}_{\mathcal{T}}$, "before":$\mathcal{A}_{\mathcal{V}}\rangle$, hence the semantic summary $f(AQ) = \langle \mathcal{A}_{\mathcal{V}}, \mathcal{A}_{\mathcal{T}}, \mathcal{A}_{\mathcal{R}}, \mathcal{A}_{\mathcal{E}}, \mathcal{A}_{\mathcal{E}}, \mathcal{A}_{\mathcal{R}}, \mathcal{A}_{\mathcal{V}}, \mathcal{A}_{\mathcal{T}}, \mathcal{A}_{\mathcal{V}} \rangle$.

Similar to the semantic summary defined on keyword queries, we define *structured semantic summary* for pattern queries as following.

Structured Semantic Summary. A *structured semantic summary* Q_s is a structured representation of the semantic summary of query Q_{NL}, in which each node carries an annotation from $\{\mathcal{A}_{\mathcal{E}}, \mathcal{A}_{\mathcal{T}}, \mathcal{A}_{\mathcal{V}}\}$, each edge takes annotation from $\{\mathcal{A}_{\mathcal{R}}, \mathcal{A}_{\mathcal{A}}\}$, and one node is particularly specified as *output node*, and marked with $*$.

3.2 Pattern Query Generation

An annotated query reveals parts of the latent structure of a phrase-based query by indicating the semantic role represented by various parts of the query. However, query annotation alone can not describe how these semantic annotations interact to model the underlying query intention. To bridge the gap between annotated queries and structured semantic summaries, we next introduce the *query structuring problem*.

Query Structuring Problem. Given a natural language query Q_{NL}, the query structuring problem is to find the most probable *structured semantic summary* Q_s, such that

$$Q_s = \mathsf{argmax}_{Q'_s} \Pr(Q'_s | f(AQ)),$$

where $f(AQ)$ is the semantic summary of Q_{NL}, and Q'_s is a structured semantic summary, extracted from labeled query log $\mathcal{L}og$.

Indeed, the calculation of $\Pr(Q'_s | f(AQ))$ can be achieved by estimating from labeled training data in query log following the definition of conditional probabilities and using the semantic summary as a high level representation of the annotated query. Specifically, given a semantic summary $f(AQ)$, and a set of labeled query log $\mathcal{L}og$, one can evaluate the probability of a structured semantic summary Q_s that's mapped from $f(\mathcal{A}\mathcal{Q})$, with below formula:

$$\Pr(Q_s|AQ) = \frac{|S_2|}{|S_1|},$$

where $S_1 = \{Q_{s_i} \mid f(AQ) = f(AQ_i), \langle Q_{s_i}, AQ_i \rangle \in \mathcal{L}og\}$, $S_2 = \{Q_{s_j} \mid f(AQ) = f(AQ_j), Q_s = Q_{s_j}, \langle Q_{s_j}, AQ_j \rangle \in \mathcal{L}og\}$.

In a nutshell, the probability of a structured semantic summary Q_s given a semantic summary $f(\mathcal{A}\mathcal{Q})$ can be evaluated by the proportion of structured semantic summary with the same semantic summary as $f(\mathcal{A}\mathcal{Q})$ and the same structure along with *output node* as Q_s, versus the total number of structured semantic summary with the same semantic summary as $f(\mathcal{A}\mathcal{Q})$ but any structure. Following the idea, module QA employs algorithm Bayes (not shown) to train a classifier, *i.e.*, the conditional probability, by using labeled training data.

Example 6. Consider two $\mathsf{Q_{NL}}$: "Find operas of Korea of Huiqiao Song." and "Find operas of Huiqiao Song of Korea." One may verify that they have the same semantic summaries, *i.e.*, $f(AQ) = \langle \mathcal{A}_T, \mathcal{A}_A, \mathcal{A}_A \rangle$, while they correspond to different pattern queries. Thus, the most probable pattern query can be used to structure the query. □

Pattern Query Construction. With the classifier, we show how module PGen construct a pattern query for a given $\mathsf{Q_{NL}}$ as follows: (1) extract a semantically annotated query $\mathcal{A}\mathcal{Q}$ from $\mathsf{Q_{NL}}$; (2) find a Q_s such that $\Pr(Q_s|f(\mathcal{A}\mathcal{Q}))$ is maximum among all structured semantic summaries with the same $f(\mathcal{A}\mathcal{Q})$; and (3) for each "phrase:semantic annotation" pair $(k_i : a_i)$ in $\mathcal{A}\mathcal{Q}$, replace v_i in Q_s with k_i if v_i is annotated with a_i.

4 Querying Knowledge Graphs

In this section, we investigate the diversified top-k graph pattern matching problem. The main result of this section is as follows.

Theorem 1. *The* TopKM *problem (1) is* NP-*hard (decision problem); (2) has a heuristic that is in* $O(|Q|^2 + |Q||G|^2 + |G|^2|G|!)$ *time, and preserves early termination property.*

Proof: As TopKM problem embeds subgraph isomorphism problem, which is known an NP-complete problem, Theorem 1 (1) then follows. To see Theorem 1 (2), we present an algorithm, denoted as TopKEI, as a constructive proof.

Before illustrating the algorithm, we first introduce notions *ball* and *dual simulation*, followed by a lemma used by the algorithm for fast pruning.

Balls. Given a node v in a graph G, a pattern query Q and a non-negative integer r, the ball with center v and radius r is a subgraph of G, denoted by $G[v, r]$, such that (1) for all nodes $v' \in G[v, r]$, the shortest distance $\mathrm{dist}_u(v, v') \leq r$, (2)

Input: Pattern query $Q = (V_p, E_p, u_o)$, graph $G = (V, E, L)$, k, λ.
Output: A top-k match set of u_o.

1.   set $\mathsf{T} := \emptyset$, $\mathsf{U} := \emptyset$;
2.   compute $r := \max(\{\mathrm{dist}_u(u_o, u') | u' \in V_p\})$;
3.   **for each** $v \in V$ **do**
4.   **if** v is a candidate match of u_o **then**
5.   compute *ball* $G(v, r)$; $S := \mathsf{DualSim}(Q, G(v, r))$;
6.   **if** $S \neq \emptyset$ **then**
7. $R_a(v) := \{v' | (u', v') \in S, u' \in V_p, u' \neq u_o\}$; $\mathsf{T} := \mathsf{T} \cup \{\langle v, R_a(v) \rangle\}$;
8.   sort v in T in descending order of $|R_a(v)|$;
9.   **for each** $\langle v, R_a(v) \rangle \in \mathsf{T}$ **do**
10.   **if** v is a true match of u_o **then**
11. $\mathsf{update}(v, \mathsf{U})$;
12.   **if** termination condition is satisfied **then break** ;
13. **return** U;

Fig. 3. Algorithm TopKET

for each node v in $G[v, r]$, there must exist a node u in V_p such that $u = v$, and (3) it has exactly the edges that appear in G over the same node set.

Dual Simulation [15]. Given graph G and pattern query Q, we say a *dual simulation* relation between Q and G exists, if there exists a binary relation $S \subseteq V_p \times V$ such that (1) for each $(u, v) \in S$, $u = v$; and (2) for each node u in V_p, there exists v in V such that (a) $(u, v) \in S$, (b) for each edge $(u, u') \in E_p$, there exists an edge (v, v') in E such that $(u', v') \in S$, and (c) for each edge $(u'', u) \in E_p$, there exists an edge (v'', v) in E such that $(u'', v'') \in S$.

The benefits of computing a *dual simulation* relation are threefold: (1) a set $R_a(v)$, inferred from S, can be treated as approximate relevant set of v since each v in $R_a(v)$ is also possibly in $R_s(v)$; and (2) the relation can be evaluated by an algorithm in $O((|V_p| + |E_p|)(|V| + |E|))$ time, which is much more efficient than the exponential algorithms for subgraph isomorphism. with S, one can safely prune candidate matches by applying the lemma given below.

Lemma 1. *If the dual simulation relation S between Q and G is an empty set, then G can not match Q via subgraph isomorphism.*

Proof sketch: To see the correctness of Lemma 1, observe that when $S = \emptyset$, there must exist at least one node u in V_p such that no element $(u, v) \in S$ for any v in V. This indicates that there does not exist a node v as a match of u, such that for each child u' (resp. parent u'') of u, there is a child v' (resp. parent v'') of v, and moreover (u', v') (resp. u'', v'') in S. This further indicates that there does not exist a *bijective function* h from V_p to a subgraph G_s of G.

Based on above notions and Lemma 1, we are ready to illustrate algorithm TopKET.

Algorithm. The algorithm TopKET takes Q, G and k as input, and works as following. It first initializes empty sets T and U to keep track of a set of $\langle v, R_a(v) \rangle$, where v is a match candidate of u_o, and $R_a(v)$ is a superset of $R_s(v)$ and will be used as approximate relevance set (line 1). It then treats Q as an undirected graph, and computes its radius r, i.e., the largest distance between u_o and any other node u in V_p (line 2). TopKET next computes a set $R_a(v)$ for each candidate match v of u_o as following (lines 3–7). For each node v in V, if v has the same node label as u_o, TopKET first computes the *ball* $G(v, r)$, and then invokes Procedure DualSim to compute the dual simulation relation S between Q and $G(v, r)$ (lines 4–5). If S is not empty, TopKET derives $R_a(v)$ from S, and expands T with $\langle v, R_a(v) \rangle$ (lines 6–7). Observe that when S is empty, there exists no subgraph of $G(v, r)$ that is isomorphic to Q [15], hence v can not be a match of u_o, and can be safely excluded from T. After T is initialized, TopKET sorts nodes v in T in descending order of $R_a(v)$ (line 8), and iteratively verifies match candidate, expands U with valid match until termination condition reaches (lines 9–11). Specifically, TopKET first verifies whether a candidate match v is a true match of u_o by applying an algorithm that revises VF2 [7] by terminating enumeration as soon as v is verified a valid match of u_o (line 10). For a valid match v, it invokes procedure update to update U with v (line 11). The procedure finds a node v_r in U and replaces v_r with v if $F(U \cup \{v\} \setminus \{v_r\}) > F(U)$. TopKET then checks whether the termination condition, i.e., there do not exist a match v_r in U and a candidate v_c with largest $|R_a(v_c)|$ in T such that $F(U \cup \{v_c\} \setminus \{v_r\}) > F(U)$, is encountered, and breaks the loop if so (line 12). After loop completes, TopKET returns U as final result (line 13).

Example 7: Consider G and Q in Example 1. Let $\lambda = 0.5$, TopKET finds top-2 matches of u_o as follows. It first identifies the radius r as 4. For each match candidate, i.e., all comedy movies m_2-m_{10}, it computes the *ball*, and dual simulation for each of them. Take m_9 as example, its *ball* includes nodes: p_4, p_5, m_2, m_4, m_5, m_6, m_7, m_8, m_9 and m_{10} (only entity nodes are considered here) and exactly the same edges among these nodes as in graph G, the dual simulation $S = \{(\text{You Ge}, p_4), (\text{Fan Xu}, p_5), (M_1, m_4), (M_1, m_6), (M_1, m_{10}), (M, m_9)\}$ and $R_a(v) = \{p_4, p_5, m_4, m_6, m_{10}, m_9\}$. After all the candidates are processed and sorted, set T includes following nodes, i.e., m_9, m_6, m_4, along with their approximate relevance sets. TopKET then iteratively confirms validity for each candidate, and updates U as following. It first initializes set M with m_9 followed by m_6, as the termination condition can not be met, TopKET next updates U by replacing m_6 with m_4. After the loop, U is returned as a top-2 match set. □

Correctness and Complexity. We show the correctness of TopKET by proving that (1) TopKET always terminates, and (2) when TopKET terminates, it returns a set of at most k matches of u_o.
(1) TopKET first repeats the **for** loop (lines 3–7, Fig. 3) $|V|$ times, and in each iteration, it computes a *dual simulation* relation between Q and $G(v, r)$, for a match candidate v. TopKET next iteratively verifies whether a match candidate v in T is valid or not, and selects v if it is a valid match (lines 9–12). As $|T|$ is bounded by $|V|$, hence the iteration repeats no more than $|V|$ times. As two rounds iteration repeats limited times, thus TopKET always terminates.

(2) Termination condition ensures that when algorithm terminates, at most k matches will be returned.

For the complexity, TopKET first computes radius r of Q in $O(|V_p|(|V_p| + |E_p|))$ time (line 1). TopKET then iteratively computes a ball $G(v, r)$ and the dual simulation between Q and $G(v, r)$ (lines 3–7). As a single iteration takes TopKET $O(|V| + |E|)$ time to compute a *ball* $G(v, r)$ with node set V_b and edge set E_b, and $O((|V_p| + |E_p|)(|V_b| + |E_b|))$ time to compute the dual simulation relation via Procedure DualSim (lines 3–7), and the **for** loop repeats at most $O(|V|)$ time, hence it is in total $O(|V|((|V| + |E|) + (|V_p| + |E_p|)(|V_b| + |E_b|)))$ time. The sorting process takes TopKET $O(|V|log(|V|))$ time as $|\mathsf{T}|$ is bounded by $|V|$ (line 8). The second **for** loop repeats at most $|V|$ times, and in each iteration, it takes TopKET $O(|V||V|!)$ time to verify whether v is a match or not (line 10), thus, it is in $O(|V|^2|V|!)$ time to verify and select matches. Summing these up, the total time is in $O(|Q|^2 + |Q||G|^2 + |G|^2|G|!)$.

Early Termination. Algorithm TopKET has the *early termination property*, as it leverages one strategy for early termination, thus no need to verify validity for all candidate matches. As will be verified in Sect. 5, TopKET can find a set of k matches with F-measure reaching 80%, and better still, it is more efficient than its counterpart.

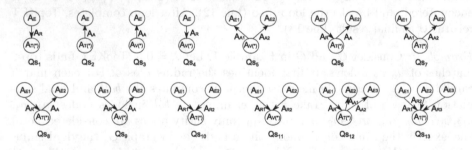

Fig. 4. Typical structured semantic summaries

5   Experimental Study

We next present an experimental study of our query structuring method (Sect. 5.1) and diversified top-k graph pattern matching method (Sect. 5.2), using real-life data. The experiments were conducted on a machine with an Intel Core (TM) 2 Dual Core 3.00 GHz CPU and 8 GB of RAM. Each experiment was run at least 5 times, and the average is reported.

5.1   Experiments for Query Structuring

We first evaluate the effectiveness of our query structuring method.

Experimental Setting. We used real-life query log from our industrial collaborator in our experiments. The query log consists of 97801 query sentences (all in

Chinese), with search intention focusing on movies or TV programs. To evaluate our approaches, we choose 20% of the sentences in query log as training data, and leave remaining sentences as testing data.

Implementation. We implemented the training algorithm Bayes, in Java.

Experimental Results. We next report our findings.

Exp-1: Effectiveness. To measure the effectiveness of our query structuring method, we define the ratio $R = \frac{|CS|}{|TS|}$, where TS is the set of testing sentences, and CS includes those sentences whose pattern queries are correctly generated. Before computing R, we used CRF++ [1] to recognize entities and explicit relations for each query sentence, inspected training data and manually extracted a set of structured semantic summaries, as shown in Fig. 4. Taking Q_{s_1} as example, it represents a class of pattern queries, that searches objects (marked with "*") with type $\mathcal{A}_T$, *e.g.*, movie and attribute (with edge labeled as $\mathcal{A}_A$) that represents an entity $\mathcal{A}_\mathcal{E}$, *e.g.*, country. These typical structured semantic summaries are treated as class labels of training data. We then ran Bayes on training data to obtain a classifier, and applied the classifier to determine whether a test sentence can be correctly structured. We find that our method is very effective, *i.e.*, the R value is over 95%, on average.

5.2   Experiments for Diversified Top-k Graph Pattern Matching

Experimental Setting. We used real-life data in our experiments.
(1) *Real-life data.* We used two real-life graphs: (a) Movie, a crawled knowledge graph with $87K$ nodes and $167K$ edges. Each node has attributes such as name, genre and rating, and each edge from a person to a movie indicates that the person played in (resp. directed) the movie. (b) Youtube, a network with $1.4M$ nodes and $3M$ edges. Each node is a video with attributes such as category, rate, and edge edge from v to v' indicates that v' is in the related list of v.

Implementation. We implemented the following algorithms, all in Java: (1) TopKApx, an algorithm that first computes a match set $M_u(Q, G, u_o)$ with algorithm [7], and then selects k matches from $M_u(Q, G, u_o)$ by following the strategy introduced by [5]. (2) TopKNaive, which computes all the matches of Q, identifies matches of u_o along with their relevance sets, and finds diversified top-k matches via exhaustive search. (3) our algorithm TopKET.

Experimental Results. We denote $(|V_p|, |E_p|)$ as the size $|Q|$ of Q. To measure how pattern size influences performances of matching algorithms, we used three kinds of pattern queries with size $(2, 1)$, $(3, 2)$ and $(4, 3)$. We next present our findings.

Exp-1: Effectiveness. We first evaluated the effectiveness of our diversified top-k matching algorithms, *i.e.*, TopKET vs. TopKApx and TopKNaive. We measured effectiveness by computing the F-measure [18], which is defined as $\frac{2\cdot(\text{recall}\cdot\text{precision})}{(\text{recall}+\text{precision})}$), where recall $= \frac{|\text{true\_match\_found}|}{|\text{true\_matches}|}$, and precision $= \frac{|\text{true\_matches\_found}|}{|\text{matches\_found}|}$.

Varying $|Q|$. Fixing $k = 10$, $\lambda = 0.5$, we varied $|Q|$ from $(2,1)$ to $(4,3)$, and evaluated F-measure over Movie. The results shown in Fig. 5(a) tell us the following. (1) TopKET is very effective, especially on *tree* patterns, *e.g.*, it achieves higher F-measure than TopKApx when $|Q| = (2,1)$. (2) TopKNaive identifies all the matches of Q, hence has F-measure= 100%. (3) TopKApx has higher F-measure than TopKET for larger patterns, as it pays more time to find approximate matches.

In the same setting as in Fig. 5(a), we evaluated F-measure on Youtube with patterns ranging from $(4,6)$ to $(8,16)$. As shown in Fig. 5(b), the F-measure of TopKET is slightly worse than TopKApx on larger patterns, and TopKNaive has F-measure=1 in all the cases.

Varying k. Fixing $|Q| = (3,2)$ (resp. $|Q| = (4,6)$) and $\lambda = 0.5$, we varied k from 5 to 30 in 5 increments, and reported F-measure on Movie (resp. Youtube). As shown in Figs. 5(c) and (d), when k gets larger, more true and false matches are identified by TopKET, hence the F-measure varies between 69% and 80% on Movie and 55% and 68% on Youtube, respectively.

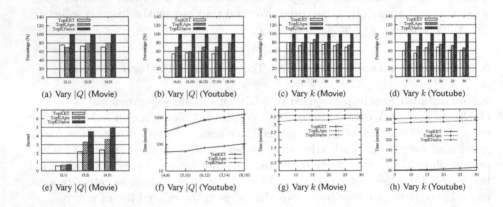

(a) Vary $|Q|$ (Movie)   (b) Vary $|Q|$ (Youtube)   (c) Vary k (Movie)   (d) Vary k (Youtube)

(e) Vary $|Q|$ (Movie)   (f) Vary $|Q|$ (Youtube)   (g) Vary k (Movie)   (h) Vary k (Youtube)

Fig. 5. Performance evaluation

Exp-2: Efficiency. We evaluated efficiency of the algorithms, in the same setting as in Exp-1.

Varying $|Q|$. As shown in Fig. 5(e), (1) TopKET takes only 20.2% (resp. 18.5%) time of TopKApx (resp. TopKNaive), on average, on Movie. This verifies the effectiveness of the *early termination* property that TopKET preserves. In the same setting as in Fig. 5(e) but with larger patterns, we evaluated efficiency of the algorithms on Youtube. As shown in Fig. 5(f), (1) all the algorithms spend more time on larger patterns, (2) TopKET is the most efficient one among three algorithms. These results also verify the observations on Movie.

Varying k. Fixing $|Q| = (3,2)$ (resp. $|Q| = (4,6)$) and $\lambda = 0.5$, we varied k from 5 to 30 in 5 increments, and tested the efficiency of the algorithms on Movie (resp.

Youtube). The results shown in Fig. 5(g) (resp. Fig. 5(h)) tell us following. (1) All the algorithms spend more time for larger k. (2) TopKET is more sensitive to the increase of k than TopKApx and TopKNaive, as TopKApx and TopKNaive spend a large part of time to compute entire match set.

6   Conclusion

We have introduced and studied query structuring and diversified top-k graph pattern matching problems. We have proposed techniques for structuring natural language queries based on their semantic summaries. We have also provided techniques for computing diversified top-k matches based on the diversification function, with *early termination* property. As verified analytically and experimentally, our methods indeed remedy the limitations of prior algorithms, by eliminating excessive matches and improving efficiency on big real-life knowledge graphs.

Acknowledgement. Xin Wang is supported in part by the NSFC 71490722, and Fundamental Research Funds for the Central Universities, China.

References

1. Crf++. https://taku910.github.io/crfpp/
2. Dbpedia. https://en.wikipedia.org/wiki/DBpedia
3. Alonso, O., Gamon, M., Haas, K., Pantel, P.: Diversity and relevance in social search. In: DDR (2012)
4. Bendersky, M., Metzler, D., Croft, W.B.: Learning concept importance using a weighted dependence model. In: WSDM 2010, pp. 31–40 (2010)
5. Borodin, A., Lee, H.C., Ye, Y.: Max-sum diversification, monotone submodular functions and dynamic updates. In: PODS, pp. 155–166. ACM (2012)
6. Cheng, J., Zeng, X., Yu, J.X.: Top-k graph pattern matching over large graphs. In: ICDE (2013)
7. Cordella, L.P., Foggia, P., Sansone, C., Vento, M.: A (sub) graph isomorphism algorithm for matching large graphs. TPAMI **26**(10), 1367–1372 (2004)
8. Ding, X., Jia, J., Li, J., Liu, J., Jin, H.: Top-k similarity matching in large graphs with attributes. In: DASFAA, pp. 156–170 (2014)
9. Fan, W., Wang, X., Wu, Y.: Diversified top-k graph pattern matching. PVLDB **6**, 1510–1521 (2013)
10. Gollapudi, S., Sharma, A.: An axiomatic approach for result diversification. In: WWW (2009)
11. Gou, G., Chirkova, R.: Efficient algorithms for exact ranked twig-pattern matching over graphs. In: SIGMOD (2008)
12. Henzinger, M.R., Henzinger, T., Kopke, P.: Computing simulations on finite and infinite graphs. In: FOCS (1995)
13. Kasneci, G., Elbassuoni, S., Weikum, G.: MING: mining informative entity relationship subgraphs. In: CIKM, pp. 1653–1656 (2009)
14. Lafferty, J.D., McCallum, A., Pereira, F.C.N.: Conditional random fields: probabilistic models for segmenting and labeling sequence data. In: ICML, USA, pp. 282–289 (2001)

15. Ma, S., Cao, Y., Fan, W., Huai, J., Wo, T.: Capturing topology in graph pattern matching. PVLDB **5**(4), 310–321 (2011)
16. Mottin, D., Lissandrini, M., Velegrakis, Y., Palpanas, T.: Exemplar queries: a new way of searching. VLDB J. **25**(6), 741–765 (2016)
17. Pound, J., Hudek, A.K., Ilyas, I.F., Weddell, G.E.: Interpreting keyword queries over web knowledge bases. In: CIKM, pp. 305–314 (2012)
18. Wikipedia: F-measure. http://en.wikipedia.org/wiki/F1_score
19. Yahya, M., Berberich, K., Elbassuoni, S., Weikum, G.: Robust question answering over the web of linked data. In: CIKM, pp. 1107–1116 (2013)
20. Yang, Z., Fu, A.W., Liu, R.: Diversified top-k subgraph querying in a large graph. In: SIGMOD, pp. 1167–1182 (2016)
21. Zheng, W., Cheng, H., Zou, L., Yu, J.X., Zhao, K.: Natural language question/answering: let users talk with the knowledge graph. In: CIKM, pp. 217–226 (2017)
22. Zou, L., Chen, L., Lu, Y.: Top-k subgraph matching query in a large graph. In: Ph.D. workshop in CIKM (2007)

Explaining Query Answer Completeness and Correctness with Partition Patterns

Fatma-Zohra Hannou, Bernd Amann[⊠], and Mohamed-Amine Baazizi

Sorbonne Université, CNRS, LIP6, Paris, France
{Fatma.Hannou,Bernd.Amann,Mohamed-Amine.Baazizi}@lip6.fr

Abstract. Information incompleteness is a major data quality issue which is amplified by the increasing amount of data collected from unreliable sources. Assessing the completeness of data is crucial for determining the quality of the data itself, but also for verifying the validity of query answers over incomplete data. In this article, we tackle the issue of efficiently describing and inferring knowledge about data completeness w.r.t. to a complete *reference data set* and study the use of a *partition pattern algebra* for summarizing the completeness and validity of query answers. We describe an implementation and experiments with a real-world dataset to validate the effectiveness and the efficiency of our approach.

1 Introduction

Information incompleteness is a major data quality issue that is exacerbated by the growing number of applications, collecting data from distributed, open, and unreliable environments. Sensor networks and data integration are significant examples in which data incompleteness naturally arises due to hardware or software failures, data incompatibility, missing data access authorizations etc. In all these situations, querying and analyzing data can lead to deriving partial or incorrect answers.

Extensive effort has been devoted to representing and querying incomplete databases [2,6,7,10,11,14]. The common characteristics of these approaches is the use of some intensional or extensional information about completeness for deciding whether a query returns complete answers and, in some cases, for annotating the query answers with some completeness meta-data. Despite these efforts, reasoning about data completeness remains tricky due to the complexity of exhaustively representing and deriving information about available and missing data in large datasets.

In many situations, datasets and query results are explicitly or implicitly depend on some reference (or master) datasets to describe their expected full extent. Reference datasets can be obtained in different ways. A first way is to analyze the completeness of some attribute M in some table $T(\underline{A}, M)$ with key A by separating T into a table $D(\underline{A}, M)$ which contains all tuples where M is not null and a reference table $R(\underline{A})$ which contains all keys of T. Another solution is to exploit external reference tables for identifying the tuples missing in measure

© Springer Nature Switzerland AG 2019
S. Hartmann et al. (Eds.): DEXA 2019, LNCS 11707, pp. 47–62, 2019.
https://doi.org/10.1007/978-3-030-27618-8_4

table D. For instance, sensor databases are usually construed within a spatio-temporal reference delimiting the coverage of the captured data. According to [8] 80% of enterprises maintain master data with their analytic databases. In other data-centric applications, a reference is defined by domain experts during database design and updated when necessary. Finally, it may also sometimes be useful to use an existing table or query result as a reference for deriving a comprehensive representation about available and missing information in some specific context.

To understand the use of reference datasets for assessing data completeness, consider the database in Table 1c which depicts an example of a sensor database. The table D_E reports on daily energy measurements for some locations specified by floor (fl) and room (ro). For various reasons, the current database misses some values with respect to some reference dataset R_E. These values are pinpointed in grey in the D_E table. The reference dataset is built by taking the Cartesian product of a spatial reference **MAP** describing all locations in some building and a calendar **CAL** indicating the expected temporal coverage (Table 1a). Observe that we also could have built a reference dataset by extracting the keys of a measure dataset with null values for measure kWh.

Table 1. Reference tables and data table

(a) Reference tables

MAP

fl	ro
f1	r1
f1	r2
f2	r1

CAL

we		da
w1	×	Mon
w2		Tue

(b) Reference table

R_E	fl	ro	we	da
r_0	f1	r1	w1	Mon
r_1	f1	r1	w1	Tue
r_2	f1	r1	w2	Mon
r_3	f1	r1	w2	Tue
r_4	f1	r2	w1	Mon
r_5	f1	r2	w1	Tue
r_6	f1	r2	w2	Mon
r_7	f1	r2	w2	Tue
r_8	f2	r1	w1	Mon
r_9	f2	r1	w1	Tue
r_{10}	f2	r1	w2	Mon
r_{11}	f2	r1	w2	Tue

(c) Data table

D_E	fl	ro	we	da	kWh
t_0	f1	r1	w1	Mon	10
t_1	f1	r1	w1	Tue	12
t_2	f1	r1	w2	Mon	10
m_0	f1	r1	w2	Tue	$\perp$
t_3	f1	r2	w1	Mon	8
t_4	f1	r2	w1	Tue	10
m_1	f1	r2	w2	Mon	$\perp$
m_2	f1	r2	w2	Tue	$\perp$
t_5	f2	r1	w1	Mon	12
t_6	f2	r1	w1	Tue	7
t_7	f2	r1	w2	Mon	8
t_8	f2	r1	w2	Tue	8

Assume that an analyst wants to gain a full knowledge about the segments of the data that are available or missing. To facilitate her understanding of the data, the analyst would like a *summarized* version of the completeness information and may opt for a *pattern-based* representation like the one presented in Table 2. This figure shows two partition pattern tables $\mathbf{P}_E$ and $\overline{\mathbf{P}_E}$ capturing the available and the missing information of table D_E respectively. More exactly, complete pattern table $\mathbf{P}_E$ contains pattern tuples that capture all partitions which are complete w.r.t. the reference and empty

Table 2. Minimal covers of D_E

$\mathbf{P}_E$	fl	ro	we	da
p_0	*	*	w1	*
p_1	f2	*	*	*
p_2	f1	r1	*	Mon

$\overline{\mathbf{P}_E}$	fl	ro	we	da
p_3	*	r2	w2	*
p_4	f1	*	w2	Tue

pattern table $\overline{\mathbf{P}_E}$ contains pattern tuples that capture all partitions which are empty in D_E but not in the reference dataset R_E. For instance, pattern tuple p_0 indicates that all measurements pertaining to week $w1$ are available, whatever the values of fl, ro or da are. Pattern p_3 in table $\overline{\mathbf{P}_E}$ reports that no measure can be found for room $r2$ and week $w2$. This representation is *compact* as it only reports on the largest possible partitions that are complete (resp. missing) in the data. It is also *covering* as it reports on every possible maximal complete (resp. missing) partition of the data.

In this article we introduce a *pattern algebra* for querying pattern summaries and reasoning about query results. This pattern algebra can be considered as a generalization of the relational algebra where tuples are replaced by partition patterns. This algebra can be applied to complete and empty pattern summaries for reasoning about the completeness and the correctness of query answers. The idea of the pattern algebra is to define for each relational operator op a pattern counterpart $\boldsymbol{op}$ which reflects the transformation of the underlying pattern tables. To illustrate this idea, consider Q_1 which retrieves all measures referring to week $w2$ on table D_E:

Q_1: **select** $*$ **from** D_E **where** we $=$ 'w2';

This query corresponds to a simple selection $Q_1 = \sigma_{we='w2'}(D_E)$ and the completeness of the result of this query is defined with respect to the result of the same selection applied to the reference table R_E. The pattern queries which reflect the corresponding transformations on the complete and missing pattern tables are obtained by using the pattern counterpart of Q_1: $\boldsymbol{Q_1} = \boldsymbol{\sigma}_{we='w2'}(\mathbf{P}_E)$ summarizes all complete patterns and $\overline{\boldsymbol{Q_1}} = \boldsymbol{\sigma}_{we='w2'}(\overline{\mathbf{P}_E})$ summarizes all empty partitions as shown in Table 3. Observe that both pattern queries do not only choose those patterns where $we =' w2'$ but must generate all patterns that summarize partitions in $\sigma_{we='w2'}(D_E)$ which are complete with respect to partitions in $\sigma_{we='w2}(R_E)$.

Table 3. Partition patterns for Q_1 and Q_2

Q_1	fl	ro	we	da		$\overline{Q_1}$	fl	ro	we	da		Q_2	fl	ro		$\overline{Q_2}$	fl	ro
	f2	*	*	*			f2	*	*	*			*	*			*	r2
	f1	r1	*	Mon			f1	r1	*	Mon							f1	r1

As another example take a projection $Q_2 = \pi_{fl,ro}(D_E)$ on attributes fl and ro. The corresponding pattern query $\boldsymbol{Q_2} = \boldsymbol{\pi}_{fl,ro}(\mathbf{P}_E)$ returns the pattern table summarizing all floors and rooms where some measures are avaaible and $\overline{\boldsymbol{Q_2}} = \boldsymbol{\pi}_{fl,ro}(\overline{\mathbf{P}_E})$ returns the pattern table summarizing all floors and rooms where some measures are missing (see Table 3). The pattern completeness model also plays a useful role for validating the correctness of aggregation queries answers. When such queries are applied on incomplete data, the values resulting from aggregating incomplete partitions are simply incorrect and there is no means to notify this fact to the user. To illustrate the role of the pattern model in detecting potential problems with aggregation queries, consider Q_3 which sums the energy consumption over all da values.

Q_3: **select** fl , ro , we, **sum**(kWh) **as** kWh **from** D_E
         where fl=' f1 ' **group by** fl , ro , we

This query returns both valid and non-valid answers produced by complete and incomplete partitions respectively. As before pattern table $Q_3 = \pi_{fl,ro,we}(\sigma_{fl='f1'}(\mathbf{P}_E))$ summarizes all complete and partially complete partitions whereas set $\overline{Q_3} = \pi_{fl,ro,we}(\sigma_{fl='f1'}(\overline{\mathbf{P}_E}))$ summarizes all empty or partially empty partitions. To obtain the summaries of all complete, empty and partially complete (empty) partitions, we can apply pattern table difference and intersection as shown in Table 4 where $Q_3 - \overline{Q_3}$ summarizes all correct results, $\overline{Q_3} - Q_3$ summarizes all missing results and $Q_3 \cap \overline{Q_3}$ summarizes all incorrect results.

Table 4. Partition patterns for Q_3.

Q_3		
fl	ro	we
*	*	w1
f1	r1	*

$\overline{Q_3}$		
fl	ro	we
*	r2	w2
f1	r1	*

$Q_3 - \overline{Q_3}$		
fl	ro	we
*	r2	w2
f1	r1	*

$\overline{Q_3} - Q_3$		
fl	ro	we
*	r2	w2

$Q_3 \cap \overline{Q_3}$		
fl	ro	we
*	r2	w2
f1	r1	*

Since data tables are pattern tables without wildcards, it is then possible to use the pattern algebra for annotating query results. For example, Table 5 shows the annotated result of query Q_3 where completeness information is directly extracted from tables $Q_3 - \overline{Q_3}$ (correct results generated by complete partitions) and $Q_3 \cap \overline{Q_3}$ (incorrect results generated by incomplete partitions). This result can be obtained by rewriting Q_3 into a union of two queries Q_{ok} and Q_{nok} separating the correct and incorrect answers. For example, the following pattern query generates the subset of all annotated results by applying a pattern join between the data table and the pattern table identifying all complete partitions:

Table 5. Result of Q_3

fl	ro	we	kWh	Annot
f1	r1	w1	22	ok
f1	r1	w2	10	nok
f1	r2	w1	18	ok

$$\sigma_{fl='f1'}(D_E) \bowtie (\pi_{fl,ro,we}(\sigma_{fl='f1'}(\mathbf{P}_E)) - \pi_{fl,ro,we}(\sigma_{fl='f1'}(\overline{\mathbf{P}_E}))) \quad (1)$$

Contributions and Paper Outline. The main contributions of this article are (1) a new data completeness model based on the notion of *partition patterns* for summarizing relative completeness information, (2) a new sound and complete pattern algebra for generating and transforming partition pattern covers, (3) a practical implementation of the algebra based on SQL and (4) an experimental evaluation on a real-world sensor dataset, on top of a standard relational DBMS. The rest of the article is structured as follows. Section 2 discusses related work. Section 3 introduces the pattern model as well as the notions of completeness and correctness of SQL queries. The pattern algebra and some applications of pattern queries are presented in Sect. 4. Section 5 describes our solution for processing and optimize pattern queries using standard relational database technology and presents two algorithms for generating pattern tables. The experimental evaluation presented in Sect. 6 validates our approach on real-world and synthetic sensor datasets.

2   Related Work

Our work is reminiscent to the work on *relative information completeness* [2] using *materialized* reference (master) datasets to model information completeness. Given a database DB and a master database DB_C, deciding whether DB is *complete* for a query Q resorts to finding a set of containment constraints V of the form $q(DB) \subseteq p(DB_C)$ where q is a query on DB and p is a projection on DB_C. The complexity bounds obtained for different languages used for expressing queries and containment constraints demonstrate the difficulty of reasoning about information completeness [2]. Within this formal setting, our pattern tables can be considered as exhaustive sets of conjunctive containment constraints and our goal is not only to decide if the answer is complete, but also to compute all containment constraints (patterns) satisfied by the query answer.

The seminal work of [10] suggests the use of meta tuples to describe data integrity (completeness and correctness) constraints. Meta relations are similar to our pattern tables, where meta tuples are used to define available, valid and invalid data and to encode logical views over virtual complete and correct data tables. Query completeness checks if there exists a rewriting of the query using only complete views. Another idea we adopt from this early work is the definition of an algebra that manipulates meta tuples for producing sound (but not complete) sets of meta tuples satisfied by an input query. More recently, [11] presents an approach which consists in associating completeness patterns to data tables and an algebra for querying patterns to produce query answer completeness information. From this work we adopt the approach of using patterns and of defining an algebra to manipulate these patterns. In [11] completeness patterns describe the extent of data completeness as views over a *virtual* complete database, whereas we suppose that the completeness of a data table or query answer is automatically assessed w.r.t. a *materialized* reference table. This introduces an additional practical and semantic dimension for analyzing quality issues of data and query results related to information incompleteness. Our completeness patterns and pattern algebra can also be assimilated to the m-tables model [14] (inspired from *c-tables* [5]) which introduces extended tuples for representing completeness information and an algebra over m-tables for annotating query answers with certainty information. The work in [6] analyzes different types of partial result anomalies engendered by data incompleteness.

The data completeness model distinguishes between cardinality (incomplete, phantom, indeterminate) and correctness (credible and non-credible) anomalies at different granularity levels (input, operator, column, partition). The authors also study how these anomalies are propagated within a query plan. We follow the same approach regarding completeness propagation using operators at the granularity of partition (down to individual tuples). We derive raw completeness information from reference data whereas [6] derives completeness information from observed physical access anomalies.

In our setting, correctness does not deal with the validity of data tuples w.r.t. logical constraints as in [7,10], but is more related to the concept of *summarizability* [9]. The notion of summarizability was first introduced by [12] in the context of statistical databases, where it refers to the correct computation of aggregate values with a coarser level of detail from aggregate values with a finer level of detail. One of the summarizability conditions defined in [12] is *completeness* which checks if all elements in a cluster (coarser level) exist and are attached to some cluster. In our setting, this mainly corresponds to the constraint that the partitions (clusters) generated by the `group by` clause of an analytic query are complete. As we will show in Sect. 4, our pattern model and algebra also allows us to identify and filter incomplete partitions. Finally, other existing work deals with deriving explanations for missing answers [4] or with answering why-not questions [1,15]. These works assume the data to be complete and focus on understanding the behaviour of queries rather than on evaluating the impact of incomplete data on the completeness and the validity of queries with aggregation.

3 Pattern Model: Definitions

Our data model starts from the standard relational data model extended by the possibility to define *reference tables* for representing completeness constraints over *data tables*.

Definition 1. *Let D and R be two relational tables and A the set of attributes of R, called* reference attributes. *If A is a key in table D, table R is called a* reference table *for data table D and the pair $T = (D, R)$ is called a* constrained table.

For example, table R_E in Table 1a is a reference table of D_E with reference attributes $A = \{fl, ro, we, da\}$. Observe that any table $S(\underline{A}, M)$ with key A and *with null values for attribute M* can be decomposed into a constrained table $\Delta(S) = (D, R)$ where measure table $D \subseteq S$ contains all tuples in S *without null values* and $R = \pi_A(S)$ contains all key values in S. Similarly, we can build from any constrained table $T = (D, R)$ a relational table $\Gamma(T) = R \ltimes D$ with *null* values such that $\Delta(\Gamma(T)) = T$.

Definition 2. *A constrained table $T = (D, R)$ with reference attributes A is* complete *iff $R \subseteq \pi_A(D)$.*

For example, the constrained table $T = (D_E, R_E)$ in Table 1c is not complete whereas $T' = (D_E, \sigma_{we='w1' \wedge ro='r1'}(R_E))$ is complete. We introduce the notion of pattern as a comprehensive description of all complete and empty data partitions in a constrained table.

Definition 3. *Let $A = \{a_1, a_2, ..., a_n\}$ be a set of reference attributes where the domain of each attribute is extended by a distinguished value $*$ called wildcard. A partition pattern $p = [a_1 : v_1, a_2 : v_2, ..., a_n : v_n]$ over A is a tuple which assigns to each reference attribute $a_i \in A$ a value $v_i \in dom(a_i) \cup \{*\}$ in the extended domain of a_i. A set of partition patterns $P(A) = \{p_1, p_2, ..., p_k\}$ over a set of reference attributes A is called a* pattern table.

In the following we will denote by $[*]$ the *wildcard pattern* where all attributes are assigned to wildcards. Observe that a pattern table might contain only data tuples, i.e. pattern tuples without any wildcards. Partition patterns are part of a generalization/specialization hierarchy defined as follows.

Definition 4. *A pattern p_1 generalizes a pattern p_2 if p_1 can be obtained from p_2 by replacing zero or more constants by wildcards. Inversely, p_1 specializes p_2 if p_1 can be obtained from p_2 by replacing zero or more wildcards by constants.*

The generalization/specialization hierarchy forms a semi-lattice with the wildcard pattern as top-element and data tuples as bottom elements.

Definition 5. *The* instance $\vartriangleleft(p, S)$ *of a pattern p in some table S is the subset of tuples $t \in S$ which are specializations of p.*

The instance $\vartriangleleft(p, S)$ of a pattern $p = [a_1 : v_1, a_2 : v_2, ..., a_n : v_n]$ in some table S can be computed by a relational selection $\vartriangleleft(p, S) = \sigma_{cond}(S)$ with filtering condition $cond = \bigwedge(a_i = p.a_i \vee p.a_i = *)$. It is then easy to show that (1) $\vartriangleleft([*], S) = S$, (2) $\vartriangleleft(p, \vartriangleleft(p, S)) = \vartriangleleft(p, S)$, and (3) $S \subseteq S' \Rightarrow \vartriangleleft(p, S) \subseteq \vartriangleleft(p, S')$. The notion of instance can naturally be extended from patterns to pattern tables P and constrained tables $T = (D, R)$: $\vartriangleleft(P, S) = \bigcup_{p \in P} \vartriangleleft(p, S)$ and $\vartriangleleft(p, T) = (\vartriangleleft(p, D), \vartriangleleft(p, R))$. In the following, we define several properties and relationships connecting pattern tables to constrained tables which are necessary to define the final notion of *minimal pattern cover*.

Definition 6. *A constrained table $T = (D, R)$* satisfies *a partition pattern p, denoted by $T \models p$, if $\vartriangleleft(p, R) = \vartriangleleft(p, D)$. By extension, T satisfies P if T satisfies all patterns in P.*

It is easy to show that a constrained table T is *complete* if it satisfies the wildcard pattern $[*]$.

Definition 7. *A partition pattern p_1* subsumes *a partition pattern p_2, denoted $p_2 \sqsubseteq p_1$ if for all constrained tables T: $T \models p_1 \Rightarrow T \models p_2$.*

Proposition 1. *If p_2 is a specialization of p_1, then p_1 subsumes p_2.*

Definition 8. *A pattern table P covers a constrained table T if $T \models P$ and for all patterns p satisfied by T there exists a pattern $p' \in P$ such that $p \sqsubseteq p'$.*

Pattern table $\mathbf{P}_E$ in Table 2 covers the constrained table $T_E = (D_E, R_E)$. When replacing $p_0 = [*, *, w_1, *]$ by two patterns $p_a = [f_1, *, w_1, *]$ and $p_b = [f_2, *, w_1, *]$ this is not true anymore, since pattern $p_0 = [*, *, w_1, *]$ is satisfied by T but not subsumed by any pattern in $P - \{p_0\} \cup \{p_a, p_b\}$.

Definition 9. *A pattern table P is* reduced *if there exists no pair of distinct patterns $p \in P$ and $p' \in P$ such that p is a specialization of p'.*

Proposition 2. *For each constrained table T, there exists a unique reduced cover $\triangleright(T)$ called the* minimal pattern cover *of T.*

For example, pattern table $\mathbf{P}_E$ in Table 1c is the minimal pattern cover of constrained table $T = (D_E, R_E)$.

4   Pattern Algebra: Folding and Unfolding

Let $T = (D, R)$ be a constrained table and Q be a relational query which can be applied to D and R. Our goal is to define a set of operators which allow us to compute the minimal cover $\triangleright(T')$ of the result of $T' = Q(T)$. One solution is to implement an operator $\triangleright$ for computing the minimal cover $\triangleright(T')$ of constrained table T' (see red dashed lines in Fig. 1). An alternative way is to rewrite $Q(D)$ into a new query $\mathbf{Q}(\triangleright(T))$ over a minimal cover $\triangleright(T)$ to produce $\triangleright(T')$ (see blue solid line in Fig. 1).

In the following we extend the relational algebra RA with two operators *unfold* ($\triangleleft$) and *fold* ($\triangleright$) for transforming pattern tables and use this extended algebra $RA_{ext} = RA \cup \{\triangleright, \triangleleft\}$ to define a new *pattern algebra* RA_{patt} over pattern tables.

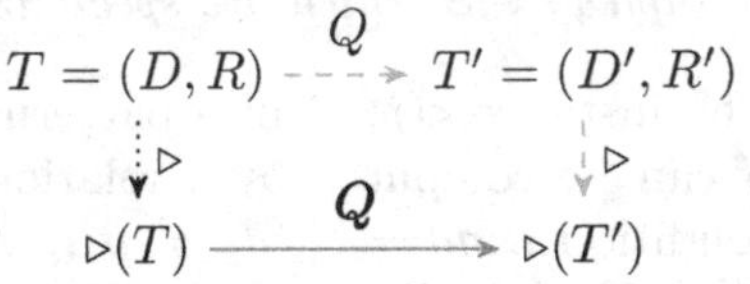

$$T = (D, R) \dashrightarrow^{Q} T' = (D', R')$$
$$\downarrow \triangleright \qquad\qquad \downarrow \triangleright$$
$$\triangleright(T) \xrightarrow{\quad Q \quad} \triangleright(T')$$

Fig. 1. Pattern queries (Color fig online)

Definition 10. *The* unfold *operator $\triangleleft_A(P, R)$ generates for a given pattern table P and reference table R an equivalent[1] pattern table P' where all values of attributes $a_i \in A$ are constant values.*

The unfolding $\triangleleft_A(P, R)$ of a pattern table P on some attribute set A w.r.t. its reference table R can be defined by the following relational algebra expression:

$$\triangleleft_A (P, R) = \pi_{R.A, P.\neg A}(P \bowtie_{\theta_\triangleleft} R) \tag{2}$$

where $\theta_\triangleleft = \bigwedge_{a_j \in A}(P.a_j = * \vee P.a_j = R.a_j)$ for all attributes in A and $\pi_{R.A, P.\neg A}$ denotes the projection on attributes A of R and on all attributes of P except A.

Consider the pattern table P in Table 6. Its unfolding $\triangleleft_{\{fl\}}(P, R)$ over attribute fl obviously is not minimal. For example, the third pattern subsumes the second one in $\triangleleft_{\{fl\}}(P, R)$.

---

[1] Two pattern tables are equivalent if their instances in R are equal.

Table 6. Example

P

fl	ro	we	da
*	*	w1	*
f2	*	*	*
fl	r1	*	Mon

$\lhd_{\{fl\}}(P,R)$

fl	ro	we	da
fl	*	w1	*
f2	*	w1	*
f2	*	*	*
fl	r1	*	Mon

Operator fold $\rhd_{a_i}$ is the inverse operator of $\lhd_{a_i}$ and *generalizes*, when possible, all subsets S of patterns $p \in S$ which are equal for all attributes except for attribute a_i into a single pattern $p_{a_i:*}$ with a wildcard value for attribute $a_i = *$:

Definition 11. *Given a pattern table P and a reference table R, the* fold *operator $\rhd_{a_i}$ generates a pattern table $\rhd_{a_i}(P,R) = P'$ which is equivalent to P and where there exists no subset $S \subseteq P'$ and pattern $p \notin S$ where $p.a_i =' *'$ and p is equivalent to S: $P' \equiv_R P \wedge \not\exists S \subseteq P', p \notin S, : p.a_u = * \wedge \{p\} \equiv_R S$.*

Operator $\rhd_{a_i}$ can be expressed in the relational algebra (see the extended version of this article [3]). As for unfold, the fold operation is associative and can be generalized on a set of attributes $A = \{a_1, a_2, ..., a_n\}$:

$$\rhd_A (P,R) = \begin{cases} P & \text{for } A = \emptyset \\ \bigcup_{a_i \in A_h} (\rhd_{a_i}(\rhd_{A-a_i}(P,R), R)) & \text{otherwise} \end{cases} \tag{3}$$

We show in Sect. 5.2 two folding algorithms based on this definition. In the following, $\lhd(P,R)$ (unfold all) and $\rhd(P,R)$ (fold all) will denote the unfold and fold operations over *all* reference attributes in P (and R). Using this extended relational algebra RA_{ext}, we can now define a *pattern algebra* RA_{patt} which contains for each data table operator $\diamond \in= \{\sigma, \pi, \bowtie, \cup, \cap, -\}$ its pattern-table counterpart $\diamond \in \{\sigma, \pi, \bowtie, \cup, \cap\}$.

Definition 12. *Let P and P' be two minimal covers of constrained tables $T = (D, R)$ and $T' = (D', R')$. Then we define the following pattern algebra $RA_{patt} = \{\sigma, \pi, \bowtie, \cup, \cap, \}$ where each operator $\diamond$ is defined by using its relational counterpart $\diamond$ and operators $\rhd$ and $\lhd$:*

$$\diamond(P) = \rhd (\diamond(\lhd(P,R)), \diamond(R)) \tag{4}$$
$$P \diamond P' = \rhd (\lhd(P,R) \diamond \lhd(P',R'), R \diamond R') \tag{5}$$

Observe that the previous definition does not include set difference. Instead, we introduce a *complement* operator $\overline{P}$ which generates the "complement" of a partition table P. Using this operator and intersection we define pattern difference as follows:

$$\overline{P} = \rhd (R - \lhd(P,R), R) \tag{6}$$
$$P - P' = P \cap \overline{P'} \tag{7}$$

Theorem 3. RA_{patt} *is sound and complete: for all relational operators* $\diamond \in \{\sigma, \pi, \bowtie, \cup, \cap\}$, *constrained tables* $T = (D, R)$ *and* $T' = (D', R')$ *with covers* P *and* P' *respectively, the following equations are true:*

$$\diamond(P) = \triangleright (\diamond(D), \diamond(R)) \tag{8}$$

$$P \diamond P' = \triangleright (D \diamond D', R \diamond R') \tag{9}$$

Proof. The proof is given in the extended version of this article [3].

5   Pattern Query Processing and Folding Algorithms

5.1   Pattern Query Rewriting and Optimization

As shown in Sect. 4 unfolding $\triangleleft$ can be expressed in the relational algebra (RA), whereas folding $\triangleright$ over a set of attributes is not expressible in RA (see Sect. 5.2 for implementations of $\triangleright$). It then is possible to rewrite any pattern query *without folding* into a relational query over pattern tables and reference tables. We will illustrate this by two examples with selection and projection.

Let pattern table P be a cover of constrained table $T = (D, R)$. Let $\sigma_\theta(D)$ be a filtering query over data table D with a predicate θ using only reference attributes. Then, the following pattern selection query generates a cover for the result of Q:

$$Q_a = \sigma_\theta(P, R) = \triangleright (\sigma_\theta(\triangleleft(P, R)), \sigma_\theta(R)) \tag{10}$$

Unfolding is necessary to check the existence of tuples in pattern instances. For example, in order to check if a pattern $p = (a_1 : v_1, \ldots, a_i : *, \ldots, a_n : v_n)$ satisfies a filtering condition $a_i = c_i$, p must be unfolded on attribute a_i. The subexpression $E = \sigma_\theta(\triangleleft(P, R))$ can be translated into the relational algebra and optimized using standard relational query rewriting. Starting from the expression E we can apply several rewriting steps to obtain a more optimal expression in relational algebra which can be translated into SQL:

$$Q_a = (\{[A : *]\} \times (\pi_{P, \neg A}(P \bowtie_{\theta_\triangleleft} (\sigma_\theta(R))), \sigma_\theta(R))) \tag{11}$$

Fold ($\triangleright$) and unfold ($\triangleleft$) comprise costly joins with reference tables. In many real world settings, reference tables $R = R_1 \times R_2 \times \ldots \times R_n$ are defined by the Cartesian product of independent reference tables R_i corresponding to spatial, temporal and other dimensions. These reference tables R_i are obviously much smaller than the generated reference table R and introduce optimization opportunities for reducing unfolding/folding costs. Consider the following more complex query expression over the same constrained table $T = (D, R)$:

$$Q_B = \pi_{fl, ro, we, da}(\sigma_{fl = 'fl'}(D)) \tag{12}$$

Let P be a minimal cover of T and $R = \textbf{MAP} \times \textbf{CAL}$. By applying several rewriting steps, we can obtain the following pattern query which generates a (but not minimal) cover of the constrained query result $(Q(D), Q(R))$:

$$Q_B = \{fl : *\} \times \pi_{P.ro, P.we, P.da}(P \bowtie_{cond} (\sigma_{fl = 'fl'}(\textbf{MAP}))) \tag{13}$$

Observe that Q_B only refers to table **MAP** for unfolding attribute fl and reference table **CAL** can be ignored.

5.2 Folding Algorithms

This section will present two optimized folding algorithms. The first algorithm *FoldData* computes minimal covers for *data* tables and the second algorithm *FoldPatterns* directly folds *pattern* tables into minimal covers without a preliminary unfolding step.

Algorithm *FoldData* computes for a given constrained table $T = (D, R)$ a strict cover $\triangleright(T)$ following a set of attributes A. If A is the set of all attributes in T, *FoldData* produces the minimal cover of T. The algorithm explores the data table by searching for complete partitions. It starts from the most general pattern *i.e.* wildcard pattern $[*]$ (level 0) and explores *top-down* and *breadth-first* the pattern subsumption lattice L_D generated by the active attribute domains in the data table D. Each level l corresponds to all patterns p with l constants. For checking if some pattern p is satisfied by D, the algorithm compares the cardinality of p in D and R using SQL. After each level, all specializations of the found complete patterns p are by definition also complete and the tuples covered by p can be pruned from D before executing the next level. The exact algorithm is defined in the extended version of this article [3].

Algorithm *FoldData* operates exclusively on data tables and cannot be applied to fold *pattern* tables without a preliminary complete unfold. This unfolding obviously reduces the compression ratio of pattern tables, in particular for pattern tables with a high compactness ratios. A pattern table P is not minimal for two main reasons. First, it might not be reduced, *i.e.* it contains two patterns p_1 and p_2 such that $p_1 \sqsubseteq p_2$. Second, it might not be a cover, *i.e.* there might exist a subset of patterns $S \subseteq P$ which could be merged into a single equivalent pattern $p \notin P$. For example, [f1, r1, w1, *] and [f1, r2, w1, *] from P can be merged into [f1, *, w1, *] $\notin P$. Algorithm *FoldPatterns* explores the pattern lattice *bottom-up* starting from the most specialized pattern (at the lowest level) and by recursively merging sets S of patterns which differ only on the constant of one attribute. The *merge* step first solves the second issue and checks if the instance $\triangleleft(S, R)$ of a subset $S \subseteq P$ is equivalent to the instance $\triangleleft(p, R)$ of a pattern $p \notin P$. As soon as S can be merged into one pattern p, the latter is added to P (it might be merged with a higher level pattern at the next iteration). Notice that one pattern can take part in several pattern merges and merged patterns are removed only after all level merges are performed. For example, [f1,r1,w1,*] can merge first with [f1, r2, w1, *] to generate [f1, *, w1, *], and merge a second time with [f2, r1, w1, *] to produce [*, r1, w1, *]). The algorithm is described in more detail in the extended version of this article [3].

6 Experimentation

We ran our experiments on a standard Linux machine equipped with a 2.4 GHz dual core CPU, 8 GB of RAM and 350 GB of standard storage.

The algorithms are implemented in Python 2.6 whereas data and patterns were managed in PostgresSQL [13] and accessed using the psycopg2+ library of Python. We did not define any indexes to accelerate filters and joins. We use temperature measures collected in 12 out of 96 buildings and refer to this *measure table* with $Temp$. The reference table R_{Temp} includes all campus spatial localities equipped with a *temperature* sensor. Measure table $Temp$ has key $(building, floor, room, year, month, day, hour)$ and an additional attribute $value$. The reference tables only contains the key attributes of $Temp$. The size of table $Temp$ is $1.3M$ tuples and table R_{Temp} contains $24.6M$ tuples.

We perform a preliminary experiment to measure the completeness of different datasets D and the compactness ratio of the corresponding complete and missing pattern tables P and $\overline{P}$. We define the *compactness ratio* $\Gamma(P, D)$ of a pattern table P by the ratio $|D|/|P| \in [1, D]$ where $|P|$ is the size (cardinality) of the pattern table and $|D|$ is the size of the data table. The *completeness* $\Omega(D)$ of a measure table D with respect to its reference table R is defined by the ratio $|D|/|R| \in [0, 1]$. In addition to $Temp$, we consider a subset $OneBlg$ of all measures in building 2232 and a subset $OneMonth$ of all measures collected during the month of January. The corresponding reference tables are built by extracting the reference subsets corresponding to the same building and month respectively.

The completeness ratio Ω is significantly higher for the dataset $OneBlg$ (restricted to building 2232) than for the overall campus average which can be explained by a better sensor coverage in this building. Completeness is not uniformly distributed over months of the year, many sensors experience periods of no recording activity (failure) or are installed after January, leading to a lower monthly completeness rate than for other months. Observe from the Table 7 that the completeness ratio and the data size are not sufficient to explain the compactness ratio since the compactness ratio is governed by the distribution of missing data over the spatial and temporal localities.

Table 7. Patterns tables sizes and compactness ratios

D	$\Omega(D)$	$|D|$	$|P|$	$\Gamma(P, D)$	$|\overline{P}|$	$\Gamma(\overline{P}, R - D)$
Temp	5.36%	1,321,686	11,269	117	10,777	2,161
OneBlg	21.43%	341,640	39	8760	55	22,776
OneMonth	4.23%	88,536	119	744	370	5,390

We define two "real" measure datasets $Temp\_0$ (empty temperature table), $Temp\_50\%$ (containing the first 50% of $Temp$ sorted by time and space), $Synthetic\_0\%$ (empty table) and two "synthetic" datasets $Synthetic\_30\%$ (containing a random 30% sample of the reference table). Starting from these four datasets with a fixed completeness ratio, we build two series of datasets obtained by successively inserting and deleting tuples from the dataset. The insertion and deletions follow two strategies: (i) a sequential strategy which selects the (inserted or deleted) tuples using their spatial and temporal domain order preserving the original data distribution in $Temp\_0\%$ and $Temp\_50\%$, and (ii) a random strategy which picks these tuples in a random fashion for $Synthetic\_0\%$ and $Synthetic\_30\%$.

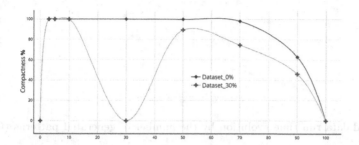

Fig. 2. Compactness versus completeness: synthetic dataset (random)

Figures 2 and 3 depict the evolution of compactness w.r.t. completeness for each dataset. In the synthetic datasets (Fig. 2), the compactness of a random dataset with 30% completeness evolves symmetrically in both directions (insertion and deletion): successive insertions/deletions generate/remove tuples which give raise to new patterns. At some point, these insertions/deletions will cause the merging of fine-grained patterns to coarser-grained ones increasing the compactness ratio to achieve maximum compactness at both extremities. In the real datasets we observe the same trend with a lower amplitude for a dataset with 50% initial completeness: insertions lead to a faster completion of the partial partitions (thanks to order sensitive updates) and thus to faster derivation of coarser patterns without deriving all their subsumed patterns. In the following experiment we evaluate the performance of algorithm *FoldData*. From the original dataset Temp, we derived 30 datasets grouped into three categories, each with approximately the same completeness rate, but different dataset sizes. Figure 4 shows the run time of *FoldData* for all datasets according to the number of generated patterns. Categories are represented by points of different colors

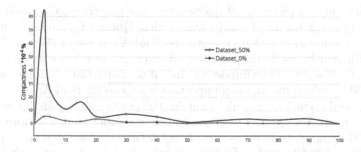

Fig. 3. Compactness versus completeness: real dataset (sensor failures)

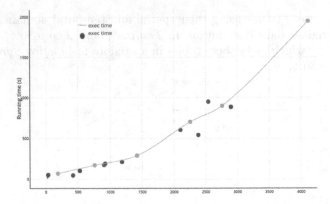

Fig. 4. Fold data run time evolution by the number of generated patterns (Color fig online)

($orange = 15\%$, $violet = 10\%$ and $green = 3\%$ completeness rate). Notice that execution time is not impacted by the data completeness but grows exponentially with the number of generated patterns.

The following experiment measures the efficiency of processing pattern queries for producing minimal covers for queries over constrained tables. We compare the pattern-based query plans (blue solid path in Fig. 1) using the techniques described in Sect. 5 by comparing it with the "naive" strategy of computing the minimal cover from the results of the query applied to the data and reference tables (red dashed path in Fig. 1). We tested both approaches on the queries below and report the result in Table 8. The reported execution times correspond to the queyr answer completeness pattern table generation cost (Fold Answer), and to the sum of pattern query evaluation cost and the Fold Pattern cost necessary to produce a minimal pattern set (Pattern Algebra).

$Q_1: \sigma_{b=2223}(Temp)$	$Q_2: \sigma_{b=2223 \wedge f=1}(Temp)$
$Q_3: \sigma_{b \in (1213,3334) \wedge (m \in (11,12)}(Temp)$	$Q_4: \pi_{b,f,r,m}\sigma_{bin(1213,2324)}(Temp)$
$Q_5: \pi_{f,r,m,d}(Temp)$	$Q_6: \Pi_{b,f,r,area,temp}Temp \bowtie_b LocArea$
$Q_7: OneBlg - OneMonth$	

Assessing the completeness of queries with the pattern algebra outperforms the naive approach for all of the tested queries. Queries Q_1 and Q_2 only refer to the spatial dimension and both methods (Fold Answer and Pattern Algebra) can be optimized by exploiting attribute domain independence as described in Sect. 5. For Q_3 the gain is less important since it needs partial unfolding over both reference tables which incurs in an important overhead for Fold Answer. Queries Q_4 and Q_5 need no unfolding which explains the performance gain of the pattern algebra approach. For Q_5, the pattern algebra evaluation is much more efficient because of the compactness of the pattern covers and the fact that corresponding pattern query doesn't need unfolding (no selection) in contrast with Q_4. The performance gain for the last two queries Q_6 and Q_7 is less significant, since both imply accessing two tables, leading to performing joins between corresponding pattern tables. Pattern queries are independent of the data size: for Q_7 the data size is much larger than for Q_6, but the pattern queries have similar run time since both queries have pattern tables of similar size.

Table 8. Complete and missing query answer patterns

Query	Complete		Missing		Execution time (sec)	
	$|Answer|$	$|P(Q_i)|$	$|Missing|$	$|\overline{P}(Q_i)|$	Fold answer	Pattern algebra
Q_1	96,360	11	1,103,760	66	7.410	0.091
Q_2	8,760	1	191,808	15	0.250	0.002
Q_3	16,025	217	584,250	91	156.060	13.700
Q_4	144	12	3,228	46	1.700	0.140
Q_5	25,342	114	101,678	763	143.920	9.890
Q_6	327	11	10,415	578	10.090	8.630
Q_7	312,624	39	1,146,288	143	23.520	9.870

7 Conclusion

In this paper, we presented a pattern-based approach for representing and summarizing relative completeness information. We proposed a formal model and characterized a powerful reasoning mechanism for inferring and analyzing exhaustive sets of completeness statements about data and query answers. We validated our approach experimentally and confirm the efficiency of the pattern algebra and its usefulness in evaluating query completeness and correctness. Extending the model with statistical information about data completeness is a challenging future direction. A natural extension under study is the use of a map-reduce platform like Apache Spark [16] to compute minimal pattern covers and implement the pattern algebra. In parallel to this work we also study the use of pattern covers for reasoning about data completeness and implement rule-based data imputations strategies.

References

1. Bidoit, N., Herschel, M., Tzompanaki, A.: Efficient computation of polynomial explanations of why-not questions. In: Proceedings of the 24th ACM International on Conference on Information and Knowledge Management, pp. 713–722 (2015)
2. Fan, W., Geerts, F.: Relative information completeness. ACM Trans. Database Syst. **35**(4), 27:1–27:44 (2010)
3. Hannou, F.Z., Amann, B., Baazizi, M.A.: Explaining query answer completeness and correctness using partition patterns (long version). Technical report (2019). http://www-bd.lip6.fr/wiki/site/recherche/articles/start
4. Herschel, M., Hernández, M.A.: Explaining missing answers to SPJUA queries. Proc. VLDB Endow. **3**(1–2), 185–196 (2010)

5. Imieliński, T., Lipski, W.: Incomplete information in relational databases. In: Readings in Artificial Intelligence and Databases, pp. 342–360. Elsevier (1988)
6. Lang, W., Nehme, R.V., Robinson, E., Naughton, J.F.: Partial results in database systems. In: International Conference on Management of Data, SIGMOD, pp. 1275–1286. Snowbird, USA, June 2014
7. Levy, A.Y.: Obtaining complete answers from incomplete databases. In: Proceedings of the 22th International Conference on Very Large Data Bases, VLDB 1996, pp. 402–412. Morgan Kaufmann Publishers Inc., San Francisco (1996)
8. Loshin, D.: Master Data Management. Morgan Kaufmann, Burlington (2010)
9. Mazón, J.N., Lechtenbörger, J., Trujillo, J.: A survey on summarizability issues in multidimensional modeling. Data Knowl. Eng. **68**(12), 1452–1469 (2009)
10. Motro, A.: Integrity = validity + completeness. ACM Trans. Database Syst. **14**(4), 480–502 (1989)
11. Razniewski, S., Korn, F., Nutt, W., Srivastava, D.: Identifying the extent of completeness of query answers over partially complete databases. In: Proceedings of the 2015 ACM SIGMOD International Conference on Management of Data, Melbourne, Victoria, Australia, pp. 561–576, 31 May–4 June 2015
12. Shoshani, A.: OLAP and statistical databases: similarities and differences. In: Proceedings of the Sixteenth ACM SIGACT-SIGMOD-SIGART Symposium on Principles of Database Systems, pp. 185–196. ACM (1997)
13. Stonebraker, M., Rowe, L.A.: The design of postgres. SIGMOD Rec. **15**(2), 340–355 (1986)
14. Sundarmurthy, B., Koutris, P., Lang, W., Naughton, J.F., Tannen, V.: m-tables: representing missing data. In: 20th International Conference on Database Theory, ICDT, Venice, Italy, pp. 21:1–21:20, March 2017
15. Tran, Q.T., Chan, C.Y.: How to conquer why-not questions. In: Proceedings of the 2010 ACM SIGMOD International Conference on Management of Data, pp. 15–26. ACM (2010)
16. Zaharia, M., et al.: Apache spark: a unified engine for big data processing. Commun. ACM **59**(11), 56–65 (2016)

Information Retrieval

Research Paper Search Using a Topic-Based Boolean Query Search and a General Query-Based Ranking Model

Satoshi Fukuda[✉], Yoichi Tomiura, and Emi Ishita

Kyushu University, Fukuoka 819-0395, Japan
{s.fukuda, tom}@inf.kyushu-u.ac.jp,
ishita.emi.982@m.kyushu-u.ac.jp

Abstract. When conducting a search for research papers, the search should return comprehensive results related to the user's query. In general, a user inputs a Boolean query that reflects the information need, and the search engine ranks the research papers based on the query. However, it is difficult to anticipate all possible terms that authors of relevant papers might have used. Moreover, general query-based ranking methods emphasize how to rank the relevant documents at the top of the results, but require some means of guaranteeing the comprehensiveness of the results. Therefore, two ranking methods that consider the comprehensiveness of relevant papers are proposed. The first uses a topic-based Boolean query search. This search converts every word in the abstract set and query into a topic via topic analysis by Latent Dirichlet Allocation (LDA) and conducts a search at the topic level. The topic assigned to synonyms of a search term is expected to be the same as that assigned to the search term. Each paper is ranked based on the number of times it is matched with each topic-based Boolean query search executed for various LDA parameter settings. The second is a hybrid method that emphasizes better results from our topic-based ranking result and a general query-based ranking result. This method is based on the observation that the paper sets retrieved by our method and by a general ranking method will be different. Through experiments using the NTCIR-1 and -2 datasets, the effectiveness of our topic-based and hybrid methods are demonstrated.

Keywords: Latent Dirichlet Allocation · Research paper search · Search recall

1 Introduction

When searching for research papers, it is important that the search returns comprehensive results related to a user's information needs. In many cases, a user inputs a Boolean query that reflects his/her information needs to an academic search engine, and acquires the research papers that are most closely related to the query. However, there are two main problems when searching for papers in such a search engine.

The first problem is a construction of query. When a user searches research papers for a method of extracting a hierarchical relationship between words, for example, the user first defines a Boolean query that expresses his/her information needs as follows:

© Springer Nature Switzerland AG 2019
S. Hartmann et al. (Eds.): DEXA 2019, LNCS 11707, pp. 65–75, 2019.
https://doi.org/10.1007/978-3-030-27618-8_5

"hierarchical relationship" AND extract. If this query produces any hits, then the results are likely to satisfy the information needs. However, there are many synonyms and similar expressions for "hierarchical relationship" and "extract." Therefore, the user might extend the query to ("hierarchical relationship" OR "hierarchical structure") AND (extract OR acquire). Unfortunately, this query does not return relevant papers in which, for example, "hierarchical relationship" or "hierarchical structure" appears but neither "extract" nor "acquire" appears. This query also misses relevant papers if the author uses terms such as "obtain" or "get" instead of "extract" and "acquire." Similarly, the above query also misses relevant papers that use terms such as "superordinate-subordinate relation" and "part-whole relationship," which are specific words for "hierarchical relationship" in the field. Thus, it is difficult for a user to fully predict how representations of search terms might have been expressed by other authors.

The second problem concerns the ranking model to find the relevant papers. In daily search, it is important that several documents ranked highly satisfy the information needs, because we need only check some highly ranked documents to find those that satisfy the information needs. However, in an academic search, it is not sufficient to check only highly ranked documents: the comprehensiveness of relevant documents is also important. More specifically, completeness when confirming across a certain amount, that is, the cumulative recall up to rank r, is important, and a high recall is desirable for realistic values of r (e.g., $r = 1,000$ for 90% recall).

Therefore, we propose two ranking methods and model a Boolean query-based search by considering the comprehensiveness of relevant papers.

(1) Search method based on a topic-based Boolean query

We propose a search method in which every word in each abstract in the research paper set (determining of the research paper set is described in Sect. 3) and each search term in the Boolean query is converted to a topic using Latent Dirichlet Allocation (LDA) [4], and then topic-level matching is conducted. By converting words to topics, abstracts that do not include the search terms but include synonyms or similar expressions are matched. In the topic analysis by LDA, the same topic is assigned to words that tend to appear together in many abstracts. For example, if words w, w_1, w_2, and w_3 appear together in many abstracts and words w', w_1, w_2, and w_3 appear together in many other abstracts, then the same topic tends to be assigned to words w, w', w_1, w_2, and w_3 in such abstracts. In this case, w and w' have the potential to be semantically similar; that is, when word w is specified in the query, a research paper that matches a Boolean query in which w is replaced by w' could be a relevant search result. Using this characteristic of LDA, we construct a query search system that conducts a search based on the topics assigned to the search terms and collects papers that exactly match the topic-based Boolean query from the topics assigned to each word in the abstracts.

Using the topic-based Boolean query search described above, we propose a new ranking method. First, we set the range of parameters assigned to LDA, and execute a topic-based Boolean query search for each parameter setting. The system then sorts research papers in descending order of the number of times they matched the query search for each parameter setting. This approach is based on the following observation. In LDA, it is necessary to provide the hyper-parameters α, β, and number of topics K in advance. In many empirical studies using LDA, a symmetric Dirichlet distribution with

$\alpha = 50/K$, $\beta = 0.1$ is used [4]. However, there may not be truly optimal parameters for a given document set. On the other hand, we know empirically that a group of words generated from the same topic with a certain high probability and a group of documents containing common topics at a certain ratio or more are retained even if the number of topics and hyper-parameters assigned to LDA changes slightly, and such word groups and document groups represent the stable relationships between words and between documents that do not depend on the slight differences in parameters. The stable relationships found through integrating multiple topic analysis results achieve a comprehensive search without specifying all possible search terms related to the query.

(2) Hybrid method to integrate the ranking results of our topic-based Boolean query search and general query-based ranking model

Our topic-based search (ranking) method focuses on collecting a comprehensive set of relevant papers. Therefore, if a user searches for research papers using our method and a general ranking method (as represented by a query likelihood model), the set of highly ranked papers may be significantly different, because the latter model emphasizes how several documents satisfying the user's information needs should be ranked at the top of the results. We examined the precision and overlap ratio in two types of research paper set ranked by our topic-based ranking method and Wei and Croft's ranking method, which is a query likelihood model using LDA [20]. The results using a search task of the NTCIR-1 dataset [7] are shown in Fig. 1. The horizontal axis represents the rank of papers sorted by each method. The left vertical axis represents the ratio of relevant papers contained in the paper set within a certain interval on the horizontal axis, i.e., the precision, and the right vertical axis represents the ratio of overlapping papers within two types of paper sets given by each method within certain intervals on the horizontal axis, i.e., the overlap ratio. Figure 1 indicates that many relevant papers are included in the papers ranked highly by each method; however, the overlap ratio between each high-ranked paper sets is relatively low at 0.349, and the overlap rates in the subsequent intervals are also low. From these results, it is highly likely that the research papers ranked highly by each method will be relevant for the user's information needs. Therefore, we expect that more effective paper search results can be obtained by integrating the paper sets output by two different ranking approaches.

2 Related Work

2.1 Query-Based Academic Search

The challenge for an academic search is to comprehensively collect research papers related to a user's information needs (i.e., recall-oriented) [9, 19]. Many academic searches engines require multiple queries when a user comprehensively collect relevant papers, however, constructing queries manually is a heavy burden for users. Therefore, not only systems using word-based queries but also specific systems using elements other than words have been developed. For example, some systems use the body of a paper [2, 12, 23], the URL of a web page [16], and user profiles [1, 5, 6, 15], and however, general academic search engines such as Google Scholar, Web of Science, and Scopus require word-based queries, and we also consider a word-based search

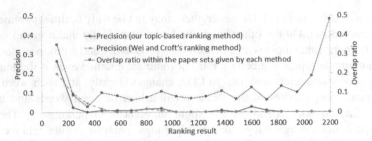

Fig. 1. Precision and overlap ratio given by our topic-based ranking method and by Wei and Croft's method using a search task ("0006") of the NTCIR-1 dataset. This search task datum has 2,196 relevance judgment data. We ranked using all of these data, and set the intervals at 5% (i.e., top 1–121, 122–243, …). In Wei and Croft's method we set $\alpha = 0.1$, $\beta = 0.1, K = 10$, $\lambda = 0.7$, and $\mu = 50$. The overlap ratio was calculated by Dice coefficient.

engine. For systems using a word-based search query, there have been many studies on query term suggestion for estimating alternative queries using the initial query [9, 19] and query expansion for automatically expanding the initial query [17, 21]. To estimate or extend the query, collections ranked highly by the initial query [9, 19] or thesaurus [17] are used, but relevant papers with words and phrases that are not included in these resources may be missed. Our topic-based search method does not require such resources because of the expansion to words that are potentially similar to the search word using an LDA topic model.

Recently, along with the development of an academic database, an academic search system using research paper-specific meta-information such as the title, author, proceedings, and venue has been actively studied [12, 21]. Our task uses only the abstract, so we can retrieve research papers that do not include meta-information.

2.2 Query-Based Ranking Model

A general search model based on a query forms a ranking by calculating the degree of association with the query based on the document model in each document. For the document modeling, there have been many attempts to develop a mathematically descriptive framework by introducing a stochastic language model. A stochastic language model using a query is called a query likelihood model [14], and is often adopted in the ranking module for academic search tasks [9, 19].

In query likelihood models, a multinomial distribution model is widely used when expressing a document by a language model. This can be expressed as:

$$P(Q|D) = \prod_{q \in Q} P(q|D) \tag{1}$$

where D is a document, q is a query term in query set Q, and $P(Q|D)$ is the likelihood of the document model generating the query terms. For the estimation of $P(q|D)$, Zhai and Lafferty [22] proposed a query likelihood model using Dirichlet smoothing to assign a probability value to words that do not appear in the document, as shown below:

$$P(w|D) = \frac{N_D}{N_D + \mu} P'(w|D) + \left(1 - \frac{N_D}{N_D + \mu}\right) P'(w|coll) \tag{2}$$

where N_D is the number of word tokens in D, $P'(w|D)$ is the maximum likelihood estimate of w in D, and $P'(w|coll)$ is the maximum likelihood estimate of w in the entire collection. μ is a smoothing parameter. Hereafter, we call this method LM (Language Model) Search. The LM search uses only superficial language information appearing in the document. Several query likelihood models using the potential relationships among words have been studied. These use document clusters obtained by analyzing a document collection [11, 20] and word embedding [3, 13]. One study on a query likelihood model using LDA, similar to our topic-based search method, is that of Wei and Croft [20], who proposed a ranking model incorporating topic analysis results into a language model, as shown below:

$$P(w|D) = \lambda \left(\frac{N_D}{N_D + \mu} P'(w|D) + \left(1 - \frac{N_D}{N_D + \mu}\right) P'(w|coll) \right)$$
$$+ (1 - \lambda) \left(\sum_{t=1}^{K} \frac{n_{-i,j}^{(w_i)} + \beta_{w_i}}{\sum_{v=1}^{V} n_{-i,j}^{(v)} + \beta_v} \times \frac{n_{-i,j}^{(D_i)} + \alpha_{z_i}}{\sum_{t=1}^{T} n_{-i,t}^{(D_i)} + \alpha_t} \right). \tag{3}$$

where λ is a smoothing parameter. $n_{-i,j}^{(w_i)}$ is the number of instances of word w_i assigned to topic j, not including the current token, and $n_{-i,j}^{(D_i)}$ is the number of words in D_i assigned to topic j, not including the current token. $\sum_{v=1}^{V} n_{-i,j}^{(v)}$ is the total number of words assigned to topic j and $\sum_{t=1}^{T} n_{-i,t}^{(D_i)}$ is the total number of words in D_i, not including the current word. Hereafter, we call this search method LDA + LM Search. We integrate our topic-based search method with the LM (LDA + LM) Search, and attempt to improve the ranking performance by emphasizing better results from individually ranked results.

3 Search Method Based on a Topic-Based Boolean Query Using Multiple Topic Analysis Results

Our topic-based search method consists of five steps: (1) a user defines the Boolean query and collects research papers; (2) the system conducts the preprocessing of the abstract set; (3) the system performs topic analysis of the abstract set using LDA with various parameter settings; (4) the system performs the topic-based Boolean query search using a topic analysis result for each parameter setting; (5) the system ranks papers based on the number of times it matched each topic-based Boolean query search. We call this method "Topic Search." In the following, we describe each step in detail.

Query Definition and Collection of Research Paper Set. In step (1), our system requires the user to define a Boolean query in the following form:

$$(w_1 \text{ OR } w_2 \text{ OR} \ldots w_m) \text{ AND } (w'_1 \text{ OR } w'_2 \text{ OR} \ldots w'_n) \text{ AND} \ldots$$

where w is a search term, the words in parentheses connected by OR comprise a concept unit, and each word inside the parentheses expresses a synonym of the same concept.

Simultaneously with the definition of the query, the system requires the user to comprehensively collect research papers that avoid missing relevant papers. At this time, various approaches are conceivable for collecting research papers more exhaustively, such as the system asks the user to specify a research field or an academic journal name and the system requires the user to construct another more comprehensive query. The research paper set collection used in our experiment is described in Sect. 5.1.

Preprocessing. Step (2) comprises two modules, one for the conversion of search terms and one for the removal of unnecessary words. In the conversion of search terms module, all search terms in the concept unit are converted to the same special symbol not appearing in the abstract set, and each search term appearing in each abstract in the database is converted to the corresponding symbol, as shown in Fig. 2. This process ensures that the same topics are assigned to all terms in each concept unit in the topic analysis. In the removal of unnecessary words module, we use TreeTagger [18] to tokenize and convert the original form, and retain only nouns, verbs, and adjectives that occur two or more times in the abstract set to be analyzed as features.

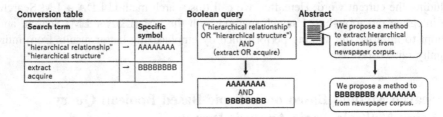

Fig. 2. Example of the conversion of a Boolean query and abstract.

Topic Analysis Using LDA. The LDA used in step (3) supposes that an abstract d is a sequence of words and each word is generated from a topic. Let $\mathbf{w}^{(d)} = \left(w_1^{(d)}, \cdots, w_{l_d}^{(d)} \right)$ be the sequence of words for the d-th abstract, $\mathbf{z}^{(d)} = \left(z_1^{(d)}, \cdots, z_{l_d}^{(d)} \right)$ be the sequence of topics, $\mathbf{w} = \left(\mathbf{w}^{(1)}, \cdots, \mathbf{w}^{(D)} \right)$ be the abstract set, and $\mathbf{z} = \left(\mathbf{z}^{(1)}, \cdots, \mathbf{z}^{(D)} \right)$ be the sequence of topics for the whole collection. Following [4], $\mathbf{z}$ is generated using Gibbs Sampling according to:

$$P(\mathbf{z}|\mathbf{w}, \alpha, \beta) = \frac{P(w, z|\alpha, \beta)}{\sum_z P(w, z|\alpha, \beta)}. \qquad (4)$$

What is needed for our method is to assign a topic to every word in the abstract set. Therefore, using Gibbs Sampling, $\mathbf{z}$ with a relatively high probability of Eq. (4) are intensively generated, and $\mathbf{z}$ that maximizes Eq. (4) among them is determined. Finally, $\mathbf{z}$ that maximizes Eq. (4) that can be reached from this $\mathbf{z}$ is found using gradient method.

Topic-Based Boolean Query Search. In step (4), the system first constructs a topic-based Boolean query from a symbol-based query using the topic analysis result. The format of the topic-based Boolean query is as follows:

$$\text{AND}_{i=1}^{I}\left(t_{i,1} \text{ OR } t_{i,2} \text{ OR } \cdots \text{ OR } t_{i,J_i}\right)$$

where I is the number of concept units constituting the query, and $\{t_{i,1}, t_{i,2}, \cdots, t_{i,J_i}\}$ is the topic set assigned to the special symbol with which the i-th concept unit is replaced. The procedure for constructing the topic-based Boolean query is shown in Fig. 3. The system seeks the topics assigned to each special symbol in the abstract set that exactly match the symbol-based Boolean query, and then converts those symbols into their corresponding topics. Note that if different topics are assigned in different abstracts, they are joined by OR according to the above format. Research papers that exactly match the Boolean query constructed in step (1) are likely to satisfy the information needs, and so a topic-based Boolean query constructed using these research papers will properly represent the information needs.

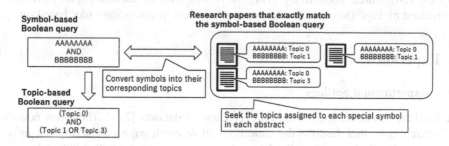

Fig. 3. Example of query conversion.

After constructing a topic-based Boolean query, the system examines the topic types for all words in each abstract given by LDA. Finally, abstracts that exactly match a topic-based Boolean query are identified and these research papers are returned. For example, a topic-based Boolean query "Topic 0 AND (Topic 1 OR Topic 3)" for which research paper A contains topics 0 and 1, research paper B contains topics 0, 1 and 3, research paper C contains topics 0 and 3, research paper D contains topic 2, and research paper E contains topics 1, 2 and 3 will return research papers A, B, and C.

Topic-Based Ranking Method. In step (5), after executing the topic-based Boolean query search for each of the parameter settings assigned to LDA, the system counts the number of times each paper is included in each search result. Based on the number of matches for all papers, the system ranks each paper in descending order. For example, we suppose that research papers A, B, and C are included in the result of a topic-based Boolean query search with parameters $(\alpha, \beta, K) = (0.1, 0.1, 10)$, research papers A and C are included in the result of a topic-based search with parameters $(\alpha, \beta, K) = (0.1, 0.01, 10)$, and research papers A, B, C, and D are included in the result of a topic-based search with parameters $(\alpha, \beta, K) = (0.01, 0.1, 10)$. The system counts 3, 2, 3, 1, and 0 as the number of matches for research papers A, B, C, D, and E, respectively, and ranks the papers in the order A, C, B, D, and E. Even if relevant papers have not been retrieved by a topic-based Boolean query search because of the inappropriate parameter setting, this influence can be reduced by integrating multiple topic analysis results.

4 Hybrid Search

In this section, we describe a ranking method that integrates the ranking results of the Topic Search with topics from the LM (or LDA + LM) Search. Hereafter, we call this method "Hybrid Search." In the design of this method, we use two simple approaches to re-rank the papers based on the ranking results given by two type of search methods.

$$r_3 = \min(r_1, r_2) \tag{5}$$

$$r_3 = r_1 + r_2 \tag{6}$$

where r_1 and r_2 are the ranks of a research paper determined by Topic Search and LM (LDA + LM) Search, respectively. After the rankings for all research papers have been determined by Eqs. (5) or (6), the system sorts papers in ascending order based on r_3.

5 Experiment

5.1 Experimental Settings

We used the test collections of the NTCIR-1 and -2 datasets [7, 8]. These sets contain 132 search tasks that describe the conditions of research papers satisfying the information needs. Each search task datum has approximately 1,000–4,000 relevance judgment data that determine whether it satisfies the information needs for its task from the research papers collected using the pooling method [10], and each paper is manually judged according to the following criteria: highly-relevant, relevant, partially relevant, and non-relevant. In this experiment, we regarded "highly relevant," "relevant," and "partially relevant" to be equivalent to "relevant," and used 40 search tasks that have approximately 10–100 relevant papers. The Boolean queries in each search task were manually constructed by one subject by reading the contents of the search tasks.

We compared the following two baseline methods and five versions of our method:

- **LM Search (baseline)**: rank papers calculated by Eqs. (1) and (2).
- **LDA + LM Search (baseline)**: rank papers calculated by Eqs. (1) and (3).
- **Topic Search**: rank papers by multiple topic analysis results.
- **Hybrid (Topic & LM) Search (min)**: integrate the ranking results of Topic Search and LM Search using Eq. (5).
- **Hybrid (Topic & LM) Search (sum)**: integrate the ranking results of Topic Search and LM Search using Eq. (6).
- **Hybrid (Topic & LDA + LM) Search (min)**: integrate the ranking results of Topic Search and LDA + LM Search using Eq. (5).
- **Hybrid (Topic & LDA + LM) Search (sum)**: integrate the ranking results of Topic Search and LDA + LM Search using Eq. (6).

For the evaluation, we used the cumulative recall, where the recall is calculated as (number of relevant papers in the top $n\%$ research paper sets of the ranking result)/ (number of relevant papers in the paper set). We set n to 1–100% with increments of 1%. The macro averages of the cumulative recall for the top $n\%$ of research papers in the ranking results of 40 search tasks are calculated. The method that has the highest rank that achieves a specific cumulative recall is best at its recall point. We evaluated Topic Search with parameter $\alpha \in \{0.01, 0.02, 0.05, 0.1, 0.2, 0.5\}$, $\beta \in \{0.01, 0.02, 0.05, 0.1, 0.2, 0.5\}$, and $K \in \{6, 7, 8, 9, 10, 11, 12, 13, 14, 15\}$. We also evaluated LM Search, LDA + LM Search and each Hybrid Search using 40-fold cross validation. At this time, we performed a grid search using $\alpha \in \{0.01, 0.02, 0.05, 0.1, 0.2, 0.5\}$, $\beta \in \{0.01, 0.02, 0.05, 0.1, 0.2, 0.5\}$, $K \in \{6, 7, 8, 9, 10, 11, 12, 13, 14, 15\}$, $\lambda \in \{0.1, 0.3, 0.5, 0.7, 0.9\}$, and $\mu \in \{10, 20, 30, 40, 50\}$ to determine the optimal setting of parameters in each cross-validation fold. For the Gibbs Sampling, we set the number of iterations to 10,000.

5.2 Experimental Results and Discussion

The ranks when achieving a specific cumulative recall are shown in Table 1. First, we compare Topic Search and LM Search. From Table 1, when the target recall is 0.75 or more, Topic Search significantly improves the rank that achieves the desired recall in the ranking results. In particular, Topic Search shows an improvement in the ranking result of 15% or more compared with LM Search when the target recall is 0.90 and 0.95. This means that the number of search results including the same number of relevant papers can be reduced by 15% or more by using Topic Search instead of LM Search. We next compare the results of Topic Search and LDA + LM Search. When the target recall is set to 0.85 or more, Topic Search returns a higher rank that achieves the desired recall than LDA + LM Search. From these results, when ranking research papers based on queries in a research paper search that requires comprehensive results for relevant papers, we can confirm that our topic-based search method outperforms the query likelihood methods with the same rank in the ranking results.

Next, we examine the effectiveness of our hybrid search. From Table 1, when the target recall ranges from 0.65–0.95, Hybrid (Topic & LDA + LM) Search (sum) achieves the recall with the highest rank. In particular, when the target recall is from 0.65–0.95, Hybrid (Topic & LDA + LM) Search (sum) shows an improvement in

Table 1. Comparison of the top $n\%$ rank in the ranking result when achieving a specific cumulative recall.

	Target recall							
	0.650	0.700	0.750	0.800	0.850	0.900	0.950	1.000
Topic search	20%	25%	29%	33%	39%	46%	60%	95%
LM search	20%	24%	30%	36%	46%	61%	77%	99%
LDA + LM search	18%	22%	27%	33%	41%	51%	62%	95%
Hybrid (Topic & LM) search (min)	17%	21%	28%	31%	36%	47%	57%	97%
Hybrid (Topic & LM) search (sum)	14%	18%	22%	27%	32%	42%	57%	95%
Hybrid (Topic & LDA + LM) search (min)	17%	21%	26%	30%	37%	46%	57%	97%
Hybrid (Topic & LDA + LM) search (sum)	13%	17%	21%	27%	33%	40%	55%	97%

the ranking result of 5% or more compared with LDA + LM Search. Also, Hybrid (Topic & LDA + LM) Search (sum) improves the ranking result by 5–8% compared with Topic Search when the target recall is from 0.65–0.95. From these results, we can confirm the effectiveness of integrating two types of query-based ranking methods that have a different purpose and analytical approach for the collection of relevant papers.

6   Conclusion

We have proposed two kinds of research paper search methods using a word-based Boolean query constructed by a user. The first method uses a topic-based Boolean query search. This search converts every word in the abstract set and query into a topic via topic analysis and conducts a search at the topic level. Using this search method, we ranked research papers by combining the results of multiple topic-based searches for each parameter setting. The second method integrates the ranking results using our topic-based ranking method and a query likelihood ranking method. In future work, we will integrate our topic-based method with other query-based ranking methods such as word embedding [3, 13] in the Hybrid Search.

Acknowledgements. This work was supported by JSPS KAKENHI Grant Number JP15H01721. We thank Stuart Jenkinson, PhD, from Edanz Group (www.edanzediting.com/ac) for editing a draft of this manuscript.

References

1. Amami, M., Pasi, G., Stella, F., Faiz, R.: An LDA-based approach to scientific paper recommendation. In: Métais, E., Meziane, F., Saraee, M., Sugumaran, V., Vadera, S. (eds.) NLDB 2016. LNCS, vol. 9612, pp. 200–210. Springer, Cham (2016). https://doi.org/10.1007/978-3-319-41754-7_17

2. Dhanda, M., Verma, V.: Recommender system for academic literature with incremental dataset. Procedia Comput. Sci. **89**, 483–491 (2016)
3. Ganguly, D., Roy, D., Mitra, M., Jones, G.J.F.: A Word embedding based generalized language model for information retrieval. In: SIGIR, pp. 795–798 (2015)
4. Griffiths, T.L., Steyvers, M.: Finding scientific topics. In: National Academy of Sciences, pp. 5228–5253 (2004)
5. Hassan, H.A.M.: Personalized research paper recommendation using deep learning. In: UMAP, pp. 327–330 (2017)
6. Hong, K., Jeon, H., Jeon, C.: Personalized research paper recommendation system using keyword extraction based on userprofile. Convergence Inf. Technol. **8**(16), 106–116 (2013)
7. Kando, N., et al.: The NTCIR workshop: the first evaluation workshop on Japanese text retrieval and cross-lingual information retrieval. In: Information Retrieval with Asian Languages Workshop (1999)
8. Kando, N.: Overview of the second NTCIR workshop. In: NTCIR Workshop, pp. 35–43 (2001)
9. Kim, Y., Seo, J., Croft, W.B.: Automatic Boolean query suggestion for professional search. In: SIGIR, pp. 825–834 (2011)
10. Kuriyama, K., Kando, N., Nozue, T., Eguchi, K.: Pooling for a large-scale test collection: an analysis of the search results from the first NTCIR workshop. Inf. Retrieval **5**(1), 41–59 (2002)
11. Liu, X., Croft, W.B.: Cluster-based retrieval using language models. In: SIGIR, pp. 186–193 (2004)
12. Mai, G., Janowicz, K., Yan, B.: Combining text embedding and knowledge graph embedding techniques for academic search engines. In: SemDeep–4 at ISWC (2018)
13. Masumura, R., Asami, T., Masataki, H., Sadamitsu, K., Nishida, K., Higashinaka, R.: Hyperspherical query likelihood models with word embeddings. In: IJCNLP, pp. 210–216 (2017)
14. Ponte, J.M., Croft, W.B.: A language modeling approach to information retrieval. In: SIGIR, pp. 275–281 (1998)
15. Sugiyama, K., Kan, M.-Y.: Scholarly paper recommendation via user's recent research interests. In: JCDL, pp. 29–38 (2010)
16. Takaku, M., Egusa, Y.: Simple document-by-document search tool "fuwatto search" using web API. In: Tuamsuk, K., Jatowt, A., Rasmussen, E. (eds.) ICADL 2014. LNCS, vol. 8839, pp. 312–319. Springer, Cham (2014). https://doi.org/10.1007/978-3-319-12823-8_32
17. Tannebaum, W., Rauber, A.: Using query logs of USPTO patent examiners for automatic query expansion in patent searching. Inf. Retrieval **17**(5–6), 452–470 (2014)
18. TreeTagger. http://www.cis.uni-muenchen.de/ ~ schmid/tools/TreeTagger/
19. Verberne, S., Sappelli, M., Kraaij, W.: Query term suggestion in academic search. In: de Rijke, M., et al. (eds.) ECIR 2014. LNCS, vol. 8416, pp. 560–566. Springer, Cham (2014). https://doi.org/10.1007/978-3-319-06028-6_57
20. Wei, X., Croft, W.B.: LDA-based document models for ad-hoc retrieval. In: SIGIR, pp. 178–185 (2006)
21. Xion, C., Power, R., Callan, J.: Explicit semantic ranking for academic search via knowledge graph embedding. In: WWW, pp. 1271–1279 (2017)
22. Zhai, C., Lafferty, J.: A study of smoothing methods for language models applied to information retrieval. ACM Trans. Inf. Syst. **22**(2), 179–214 (2004)
23. Zhao, W., Wu, R., Liu, H.: Paper recommendation based on the knowledge gap between a researcher's background knowledge and research target. Inf. Process. Manage. **52**(5), 976–988 (2016)

Extractive Document Summarization
using Non-negative Matrix Factorization

Alka Khurana$^{(\boxtimes)}$ (iD) and Vasudha Bhatnagar (iD)

Department of Computer Science, University of Delhi, Delhi, India
{akhurana,vbhatnagar}@cs.du.ac.in

Abstract. Effectiveness of Non-negative Matrix Factorization (NMF) in mining latent semantic structure of text has motivated its use for single document summarization. Initial promise shown by the method provokes further research in this field to advance state-of-the-art.

In this paper, we propose two methods to improve the performance of NMF based document summarization method for mining important sentences from the text to construct summary. We use Non-negative Double Singular Value Decomposition (NNDSVD) method to initialize NMF factor matrices, which begets summary stability and improves quality. Next, we propose two novel sentence scoring methods that use parts-based representation of text obtained after NMF decomposition. Both variations exploit information contained in feature and co-efficient matrices to achieve improvement in summary quality. Quality of summaries mined by the proposed methods is evaluated for four public data-sets using standard ROUGE tool.

The proposed method is unsupervised, agnostic to the language of the document and does not use external knowledge. It is also generic, independent of domain and collection. These features of NMF based summarization along with additional advantage of speed make our method a potent candidate for online extractive summarization tool.

Keywords: Non-negative Matrix Factorization · Extractive summarization · NNDSVD

1 Introduction

Summarization is an important task in view of overload of text data, born digital on the Web. Manual summarization of text is an arduous task that demands time and intellect, beside being prone to human subjectivity. Mining text documents for automatic summarization is the process of condensing text while ensuring that the condensed text faithfully conveys the content of original document. Extensive research has been carried out in the area of automatic document summarization recently [1,3,7,11,24,34].

© Springer Nature Switzerland AG 2019
S. Hartmann et al. (Eds.): DEXA 2019, LNCS 11707, pp. 76–90, 2019.
https://doi.org/10.1007/978-3-030-27618-8_6

Extractive summarization techniques mine documents to select relevant sentences from a document to produce corresponding summary, while ensuring coherence and minimum redundancy. Contrastingly, *abstractive summarization* techniques create summary by paraphrasing sections of the text. Extractive summarization methods are more popular [2,3,8,11,14,18,22,29,30,34] than abstractive summarization methods [13,20,23] since the latter are more complex in terms of language understanding and demand natural language generation.

Typically, unsupervised techniques for extractive summarization entail sentence scoring based on pre-specified criteria and include top scoring sentences in the summary. These techniques span over statistical, probabilistic, optimization, graph-based, machine learning and matrix decomposition [2,3,18,21,22,24,29].

Statistical techniques use attributes such as frequency of significant words, cue words, position of the sentence, length of the sentence, presence of proper noun etc. as features for scoring and selecting sentences to be included in the summary [10,12,21]. These techniques are simple and computationally efficient but completely ignore the semantics and context. Probabilistic methods for summarization include Hidden Markov Models (HMMs) and Conditional Random Fields (CRFs), which consider sentence dependencies and interactions to extract important sentences [8,29]. These techniques take cognizance of document context but are computationally expensive.

Optimization based techniques solve document summarization problem by modeling it in terms of an objective function based on summary attributes such as length, relevance and redundancy [2,3,15]. Besides optimizing various summary features, these approaches endure high computational complexity due to expense incurred for simultaneous optimization of parameters. Another popular technique for extractive summarization is based on graph representation of text [11,22,34]. These methods transform text into a graph, where each sentence is a node and edges connect semantically related sentences. Subsequently, node based ranking algorithms are employed for selecting summary sentences. Language and domain independent features of these techniques make them attractive.

Recently, neural network and reinforcement learning based techniques have gained research focus [1,7,24,36]. Although neural network based approaches produce high quality summaries, these methods demand large size data-sets for training, extensive computation and suffer from low interpretability. These methods are often domain-, collection- dependent and hence are less amenable for generic document summarization.

Matrix decomposition approaches like PCA (Principal Component Analysis), LSA (Latent Semantic Analysis) and NMF (Non-negative Matrix Factorization) exhibit promising results for document summarization [14,18,31]. These methods are attractive because of frugal resource requirement (ordinary commodity machine with basic setup). Further, matrix decomposition techniques for document summarization are language and genre independent, which makes them attractive for generic document summarization.

In this work, we focus attention on non-negative matrix factorization for document summarization. Non-negative Matrix Factorization [18] is preferred over Singular Value Decomposition (SVD) based Latent Semantic Analysis (LSA) [14] because of better interpretability of factor matrices in latent space [17]. NMF factors are parts-based representation of the complete document, which can be meaningfully combined. Non-negativity constraint on NMF factor matrices results in useful interpretation of the latent semantic space. Major caveat of NMF based summarization is random initialization of NMF factors, which results in ambivalent summaries of the document.

We overcome the stochastic variations inherent in Non-negative Matrix Factorization method by using Non-negative Double Singular Value Decomposition (NNDSVD) for initializing NMF factors [6]. We also propose two novel sentence scoring methods that take cognizance of the interplay between terms and topics in the document. Specifically, our research contributions are as follows:

i. We propose an efficient method for mining important sentences from text using Non-negative Matrix Factorization. The method is generic in the sense that it is language-, domain- and collection- independent (Sect. 4).
ii. We use NNDSVD initialization for NMF factor matrices and empirically demonstrate improvement in summary quality compared to fixed value initialization as proposed earlier [18] (Sect. 3).
iii. We use both NMF feature and co-efficient matrices, thereby effectively utilizing term and topic contributions in the document (Sect. 4).
iv. We compare the proposed sentence scoring methods on four public datasets. Though the performance of the proposed methods is not the best for all four corpora, the domain-, language- and collection- agnostic features of the methods in addition to the fast execution time makes it a candidate for online generic summarization tool (Sect. 6).

2   Background and Motivation

Matrix Factorization is recognized as an eminent approach for dimension reduction and to uncover latent features in data objects. In the context of text mining, matrix decomposition techniques have potential to reveal latent semantic structure of a text document in the reduced rank space. Matrix factorization techniques such as Latent Semantic Analysis (LSA), Principal Component Analysis (PCA) and Non-negative Matrix Factorization (NMF) have shown high quality results when used for summarization [14,18,31], topic modeling [5,28] and document clustering [33].

Gong and Liu [14] successfully applied LSA for single document summarization by factorizing term-sentence matrix and ranking sentences for inclusion in summary. LSA decomposes input matrix into factors that contain both positive and negative entries, making understanding and interpretation of semantic space non-intuitive. Lee et al. [18] successfully used non-negative matrix factorization for automatic single document summarization. Since the factors obtained by NMF decomposition are non-negative, interplay of the topics and terms in the document potentially increases understanding of the latent space in a perceptive way [17].

2.1 Non-negative Matrix Factorization

Non-negative Matrix Factorization (NMF) is a matrix decomposition method for reduced rank approximation of non-negative matrix $A \in \mathbb{R}^{m \times n}$ as $A \approx WH$, such that $W \in \mathbb{R}^{m \times r}$ and $H \in \mathbb{R}^{r \times n}$ are non-negative factor matrices ($r \ll \min\{m, n\}$). Initializing W and H factors with non-negative seed values, NMF algorithm iteratively improves both factor matrices to converge locally and approximate A by product of W, H such that Frobenius norm $\| A - WH \|_F^2$ is minimized.

2.2 Non-negative Matrix Factorization for Document Summarization

Non-negativity constraint in NMF factors ($W \& H$) provides better interpretability and enhances understanding of the text in latent semantic space than SVD [17]. Lee et al. [18] propose NMF based method for automatic document summarization by computing Generic Relevance Score (GRS) for each sentence as described below.

Consider document D consisting of n sentences ($S_1, S_2, \ldots, S_n$) for summarization. Let $T = \{t_1, t_2, \ldots, t_m\}$ be the set of m terms in D after removal of stop-words. Further, let A denote $m \times n$ term-sentence matrix for D, where columns of A correspond to document sentences and rows represent terms. Element a_{ij} in A denotes the weight of term t_i in sentence S_j.

Suppose k is the number of sentences required to create summary of desired length. Decomposing matrix A using NMF results into two non-negative factor matrices W and H, where W is $m \times r$ term-topic (feature) matrix and H is $r \times n$ topic-sentence (co-efficient) matrix. Columns in W correspond to r document topics ($\tau_1, \tau_2 \ldots \tau_r$) in the latent semantic space and columns in H correspond to sentences in D. Element w_{ij} in W signifies the contribution of term t_i in topic τ_j, and element h_{ij} in H represents the strength of topic τ_i in sentence S_j.

Post factorization, NMF co-efficient matrix H is used for calculating Generic Relevance Score (GRS) of a sentence [18], as given below:

$$GRS(S_j) = \sum_{i=1}^{r} (h_{ij} * \Theta(i*)), \text{where} \tag{1}$$

$$\Theta(i*) = \frac{\sum\limits_{q=1}^{n} h_{iq}}{\sum\limits_{p=1}^{r} \sum\limits_{q=1}^{n} h_{pq}} \tag{2}$$

Here, $\Theta(i*)$ is the normalized strength of topic τ_i in n sentences. It specifies relative contribution of each topic in n sentences. Top-k scoring sentences are selected for creating summary.

3   NNDSVD Initialization

Presence of stochastic elements in the initialization phase of NMF results in variable summaries of a document. To overcome this problem, Lee et al. used fixed initial values for NMF factors as a straight forward solution [18]. This is not an appealing proposition because same fixed initialization value may not produce quality summary for all documents.

We propose to use NNDSVD initialization of NMF factor matrices as used by Belford et al. [5] for topic modeling. NNDSVD chooses initial factors in a deterministic manner based on a sparse SVD approximation of the original data matrix (here, term-sentence matrix) [6]. Eliminating initial stochasticity not only results in stable summaries, but also yields significant improvement in performance.

We empirically demonstrate the effectiveness of NNDSVD initialization using documents in four data-sets[1]. These data-sets have been extensively used in earlier works for evaluating the performance of data summarization methods [3, 11,22,24,32]. System summaries are generated using fixed value of 0.5 (following [18]) and NNDSVD initialization for NMF[2] followed by GRS scoring [18]. For each system summary of a document in DUC2001, DUC2002 data-sets, recall scores for ROUGE-1, ROUGE-2 and ROUGE-L are calculated against reference summaries. F-measure of the three ROUGE variations is calculated for CNN and DailyMail documents following earlier works [1,7,24]. Macro averaged ROUGE scores for each data-set are reported in Table 1.

Table 1. Comparison of summary quality for fixed and NNDSVD initialization for NMF decomposition using four data-sets.

	ROUGE-1		ROUGE-2		ROUGE-L	
	Fixed	NNDSVD	Fixed	NNDSVD	Fixed	NNDSVD
DUC2002	44.7	**46.3**	16.9	**18.2**	40.4	**41.5**
DUC2001	40.9	**42.1**	13.0	**14.2**	36.5	**37.4**
CNN+DailyMail	28.8	**30.8**	8.5	**10.4**	26.0	**27.7**

Higher average ROUGE scores for all corpora indicate that use of NNDSVD initialization significantly boosts summary quality compared to fixed initialization of NMF factor matrices.

Convinced by the leverage, we perform all subsequent experiments reported in Sect. 6 using NNDSVD initialization.

---

[1] We apologize for the forward reference to the data-set overview and the metric description in Sect. 5.

[2] We create binary incidence term-sentence matrix (A) for NMF decomposition.

4 Proposed Summarization Method

Summary sentences are expected to well represent the content covered in the document. Sentences that are strong contributors to important latent topics, or those having high contribution of important terms in each topic are considered good candidates for document summary. Based on this idea, we propose two novel sentence scoring methods which take complete cognizance of information contained in both NMF factors W & H.

We decompose the input text into individual sentences and create binary incidence term-sentence matrix A. Each element a_{ij} in A is 1 if term t_i appears at-least once in sentence S_j, otherwise a_{ij} is set to 0. Using NNDSVD initialization, we decompose term-sentence input matrix into feature matrix (W) and co-efficient matrix (H). We explain the two sentence scoring methods below.

4.1 Term-oriented Sentence Scoring

Intuitively, a term with higher contribution in the latent topics of the text is a better descriptor of the document than that with lower contribution. Assuming that the importance of a sentence is an additive function of terms' contribution, the sentence with terms having higher contribution in latent topics is a better candidate for inclusion in summary.

Based on this conjecture, we propose an approach for scoring a sentence to explicitly employ the contribution of terms in latent topics. Element w_{ij} in W quantifies the contribution of term t_i in latent topic τ_j. Row sum $(\sum w_{i*})$ in matrix W represents aggregate contribution of term t_i in r latent topics and Eq. 3 computes normalized contribution ϕ_i of the term t_i in all latent topics. Additive contribution of unique terms in a sentence (Eq. 4) quantifies its importance in the document.

$$\phi_i = \frac{\sum\limits_{q=1}^{r} w_{iq}}{\sum\limits_{p=1}^{m} \sum\limits_{q=1}^{r} w_{pq}} \tag{3}$$

$$Score(S_q) = \sum\limits_{i=1}^{m} a_{iq}\phi_i \tag{4}$$

Top-k scoring sentences are selected to create summary of desired length. The computational complexity of the scoring method is $O((n + r)m)$ where n is number of sentences, m is number of terms (after removing stop-words) and r is number of latent topics in the document considered for summarization.

4.2 Topic-oriented Sentence Scoring

This approach for sentence scoring is based on the idea that topic importance in the document should be reflected proportionately in the summary. In this method, sentence scoring takes into account how well the sentence represents the important topics in the document. Sentences that contribute heavily to important topics get higher score by this method. Topic strength in this method implicitly uses terms' contribution to latent topics unlike the previous method, which uses terms' contribution explicitly.

Element h_{ij} in H quantifies the contribution of topic τ_i in document sentence S_j. Column sum in the feature matrix W ($\sum w_{*i}$) denotes aggregate contribution of all terms in the document for the topic τ_i. Normalizing this aggregate yields relative dominance ω_i of topic τ_i as follows:

$$\omega_i = \frac{\sum\limits_{q=1}^{m} w_{qi}}{\sum\limits_{p=1}^{m} \sum\limits_{q=1}^{r} w_{pq}} \tag{5}$$

Thus, score of sentence S_q is calculated as:

$$Score(S_q) = \sum_{i=1}^{r} \omega_i h_{iq}, \tag{6}$$

where $\omega_i h_{iq}$ is weighted contribution of topic τ_i in sentence S_q. Higher the score of a sentence, more capable it is of representing latent topics in the document. Top-k scoring sentences are selected to create document summary of desired length. Computational complexity of the scoring method is $O((n+m)r)$.

5   Experimental Setup

In this section, we describe the experimental setting for performance evaluation of the proposed sentence scoring methods. Algorithms were coded in Python 3.6.1 using Natural Language Toolkit (NLTK), *Scikit-learn* toolkit and textmining package of python. All experiments were performed on Windows (64-bit) machine with Intel Core $i3$ processor and 4 GB memory.

Data-Sets. We use four public corpora viz. DUC2001[3], DUC2002 (See Footnote 3), CNN[4] and DailyMail (See Footnote 4) for performance evaluation of proposed algorithms, which have been used in similar studies for performance comparison [3,11,24,27,32,35].

DUC2001 data-set consists of 308 documents while DUC2002 data-set consists of 567 documents[5] for the task of generic single document summarization. Reference summaries for both DUC2001 and DUC2002 documents are abstractive and are approximately 100 words long. Each document in DUC2001, DUC2002 data-set has 1–3 reference summaries.

---

[3] http://duc.nist.gov.
[4] CNN and DailyMail corpora contain news articles and were originally constructed by [16] for the task of passage-based question answering, and later re-purposed for the task of document summarization.
[5] However, we experimented on 533 unique documents from the data-set.

CNN and DailyMail data-sets documents are partitioned into training, validation and test sets. CNN data-set consists of 92,579 (90,266 + 1,220 + 1,093) documents whereas DailyMail data-set comprises of 219,506 (196,961 + 12,148 + 10,397) documents. Following earlier studies, we evaluate the proposed sentence scoring methods using test set for both corpora taken together. Each document in CNN and DailyMail corpora has associated story highlights, which are used as gold standard reference summary. For comparison with state-of-the-art methods [26,36], which report results on CNN and DailyMail corpora, we extract three sentences for each CNN document summary and extract four sentences for each DailyMail document summary.

Number of Latent Topics(r). Optimal number of latent topics for a document cannot be predicted. We follow the method proposed by Aliguliyev et al. [4] to determine the number of latent semantic topics in a document. This approach is based on the distribution of words in the document sentences. Following [4], parameter r for non-negative factorization of term-sentence matrix A is computed as:

$$r = n\frac{|\bigcup_{i=1}^{n} S_i|}{\sum_{i=1}^{n}|S_i|} \qquad (7)$$

i.e. number of latent topics is n (number of sentences) times the ratio of the total number of unique terms in the document to total number of terms in the document (after removing stop-words).

Performance Metrics. Performance of the proposed methods is evaluated using ROUGE toolkit [19]. ROUGE (Recall-Oriented Understudy for Gisting Evaluation) is an evaluation toolkit for evaluating the performance of system (algorithmic) summary against a set of reference summaries. ROUGE evaluates system summary against a set of reference summaries by generating recall, precision and F-measure. While *recall* measures the extent to which the system summary captures reference summary, *precision* measures how much of the system summary is relevant. However, limit on choice of summary length makes precision less meaningful.

We use three ROUGE variations viz. ROUGE-1, ROUGE-2 and ROUGE-L for our experiments. ROUGE-1 identifies overlapping uni-grams between system and reference summaries, ROUGE-2 calculates bi-grams, while ROUGE-L depicts longest common sub-sequence (LCS) and identifies longest N-gram that co-occurs both in system and reference summaries. Recall scores for these three ROUGE variations are computed for DUC2001, DUC2002 documents while F-measure is reported for CNN and DailyMail documents. We report macro averaged ROUGE score along with standard deviation for proposed methods.

Competing Methods. We use NNDSVD initialization of NMF factors with *general relevance scoring* method proposed by Lee et al. [18] as the baseline method (NMF-GRS). We compare the two proposed scoring methods - NMF-TR (term-oriented sentence scoring), NMF-TP (topic-oriented sentence scoring) against the baseline to assess the quantum of improvement attained by augmenting the computation of sentence score with the information contained in both

NMF factor matrices. We also pit the proposed methods against each other to gauge the significance of term-topic interplay for document summarization.

Furthermore, we compare the proposed sentence scoring methods with several state-of-the-art supervised and unsupervised methods. Interestingly, no published work on extractive summarization evaluates performance on all four datasets. We refrain from implementing competing algorithms and testing on all corpora to preserve transparency. Instead, we choose to compare the performance with the published results for each corpora. Accordingly, we dedicate one section for presenting quantitative assessment of summaries for each data-set.

6   Experimental Results

In this section, we report experimental results for performance evaluation of the proposed sentence scoring methods with baseline NMF-GRS method and state-of-the-art supervised as well as unsupervised methods.

6.1   DUC2002 Corpus

Table 2 shows the result of comparative evaluation for DUC2002 corpus. The proposed scoring methods outperform the baseline demonstrating the validity of our conjecture that both NMF factor matrices (feature and co-efficient) contain information that can be efficiently and effectively exploited for sentence scoring. It is observed that NMF-TR performs better than NMF-TP suggesting that the use of explicit contribution of terms in latent topics is more effective than the latent topics themselves. Relatively weaker performance of NMF-TP can be explained by the nature of the documents, which are news articles where topics are not clearly distinguishable. We envisage that scientific documents which are better structured will be better summarized by topic-oriented NMF-TP method.

Unsupervised Methods. We observe that NMF-TR method achieves competitive ROUGE-1 and ROUGE-2 performance compared to URANK method [32]. However, performance of NMF-TP is lower by 1%. Since ROUGE-L metric for URANK method is not reported, its superiority is not conclusive. Furthermore, URANK follows unified approach for summarizing single and multiple documents simultaneously, which makes it unsuitable for standalone task of single document summarization.

ROUGE-1 score for NMF-TR summaries is also comparable with that of TGRAPH summaries [27]. TGRAPH performs topic modeling using external knowledge (corpus) and represents the document as weighted graph. Subsequently, it employs Integer Linear Programming based optimization to simultaneously optimize importance, coherence and redundancy for single document summarization. There is a dip of approximately 3% in ROUGE-2 score for NMF-TR, while ROUGE-L score for TGRAPH is not available. Thus it is indiscreet to judge superiority of either method. It is noteworthy that TGRAPH method is collection- and domain- dependent as it uses external knowledge to generate topics for graph creation. This makes the method unsuitable for generic document summarization.

Table 2. Performance comparison of the proposed methods for DUC2002 data-set on basis of ROUGE recall scores. CI: Collection Independent, LI: Language Independent, DI: Domain Independent

		ROUGE-1	ROUGE-2	ROUGE-L	CI	LI	DI
Baseline	NMF-GRS	46.3 ± 0.1	18.2 ± 0.1	41.5 ± 0.1	✓	✓	✓
Proposed methods	NMF-TR	49.0 ± 0.1	21.5 ± 0.1	44.1 ± 0.1	✓	✓	✓
	NMF-TP	47.6 ± 0.1	19.7 ± 0.1	42.4 ± 0.1	✓	✓	✓
Unsupervised methods	URANK [32]	48.5	21.5	-	✗	✓	✓
	TGRAPH [27]	48.1	24.3	-	✗	✓	✗
	CoRank [11]	50.7	24.0	43.4	✓	✓	✓
	CoRank+ [11]	**52.6**	**25.8**	45.1	✓	✓	✓
	COSUM [3]	49.08	23.09	-	✓	✗	✓
	iGraph [34]	48.5 ± 0.4	22.0 ± 0.2	43.7 ± 0.3	✗	✗	✗
	iGraph-R [34]	49.2 ± 0.3	23.1 ± 0.2	44.1 ± 0.2	✗	✗	✗
Supervised methods	NN-SE [7]	47.4	23.0	43.5	✗	✓	✗
	Deep-Classifier [25]	46.8 ± 0.9	22.6 ± 0.9	43.1 ± 0.9	✗	✓	✗
	SummaRuNNer [24]	46.6 ± 0.8	23.1 ± 0.9	43.03 ± 0.8	✗	✓	✗
	HSSAS [1]	52.1	24.5	**48.8**	✗	✓	✗
	DQN [35]	46.4	22.7	42.9	✗	✓	✗

CoRank and CoRank+ [11], augment sentence-sentence relationship with word-sentence relationship using graph-based approach for scoring sentences. Both methods are unsupervised, language-, collection- and domain- independent like the proposed methods. CoRank+ outperforms almost all methods listed in Table 2. However, when we compared the average execution time per document (reported as 30 s per document for DUC2002 data-set in Sect. 4.3 of [11]), we found proposed methods to be much faster with average execution time per document as 0.218 s for NMF-TR and 0.175 s for NMF-TP for DUC2002 data-set.

COSUM [3] formulates summarization as clustering based optimization problem. ROUGE-1 and ROUGE-2 recall scores of COSUM are better than ROUGE scores of both proposed variations. However, in absence of ROUGE-L recall score and evaluation on other data-sets, overall performance of COSUM is not decisively better. Further, the method uses stemming and hence is not language independent.

iGraph and its variation iGraph-R [34] are graph based methods which use an enhanced embedding model to detect the inherent semantic properties at the word level, bigram level and trigram level. Words with part-of-speech (POS) tags, bigrams and trigrams are extracted to train the embedding models. Embedding model is used to calculate similarity between sentences which act as graph edges and then TextRank [22] is used to rank document sentences. Performance of NMF-TR method is almost at par with the performance of iGraph method. Both iGraph and iGraph-R use external knowledge to train the enhanced embedding model used for generating sentence similarities which discourages their use for generic document summarization.

Supervised Methods. These methods adopt deep neural network based approach, which are known to be effective for NLP tasks. Results show that NMF-
TR performs better than NN-SE [7], Deep-Classifier [25], SummaRuNNer [24]
and DQN [35] for ROUGE-1 and ROUGE-L, but slightly lower for ROUGE-
2. However, the method is outperformed by HSSAS [1] on all three metrics.
It is noteworthy that the competing methods in this category are supervised,
collection-, domain- dependent [24] unlike NMF-TR. Consequently, their suitability for generic document summarization is arguable.

6.2   DUC2001 Corpus

In this section, we report comparative evaluation on DUC2001 corpus (Table 3).
Proposed sentence scoring methods outperform baseline for DUC2001 corpus
also. Since DUC2001 corpus also contains news articles, as expected, NMF-TR
performs slightly better.

Table 3. Performance comparison of the proposed methods for DUC2001 data-set on
basis of ROUGE recall scores. CI: Collection Independent, LI: Language Independent,
DI: Domain Independent

		ROUGE-1	ROUGE-2	ROUGE-L	CI	LI	DI
Baseline	NMF-GRS	42.1 ± 0.1	14.2 ± 0.1	37.4 ± 0.1	✓	✓	✓
Proposed methods	NMF-TR	44.7 ± 0.1	15.9 ± 0.1	39.3 ± 0.1	✓	✓	✓
	NMF-TP	43.7 ± 0.1	15.6 ± 0.1	38.5 ± 0.1	✓	✓	✓
Unsupervised methods	URANK [32]	45.4	17.6	-	✓	✓	✓
	COSUM [3]	**47.3**	**20.1**	-	✓	✗	✓

Unsupervised Methods. Both unsupervised methods outperform NMF based
methods, with COSUM [3] leading the pack. There is no clear winner because of
unavailability of ROUGE-L scores for both URANK [32] and COSUM methods.
TGRAPH [27], CoRank [11] and iGraph [34] have not been evaluated on this
data-set. Interestingly, none of the supervised methods listed in Table 2 have
been evaluated on this corpus.

6.3   CNN/DailyMail Corpora

Table 4 presents comparative evaluation of the proposed methods, competing
methods for combined CNN and DailyMail corpora. Quality of NMF-TR summaries is clearly better than the baseline and NMF-TP. None of the unsupervised methods considered in Table 2 have reported performance evaluation on
this data-set.

Supervised Methods. We report comparison of the proposed methods with
NN-SE [7], NEUSUM [36], SummaRuNNer [24], REFRESH [26], HSSAS [1],
DQN [35] and BANDITSUM [9] methods, for combined CNN and DailyMail
corpora. All these methods follow some or the other variation of deep learning

Table 4. Performance comparison of the proposed methods for combined CNN and DailyMail data-sets on basis of ROUGE F-measure scores. CI: Collection Independent, LI: Language Independent, DI: Domain Independent.

		ROUGE-1	ROUGE-2	ROUGE-L	CI	LI	DI
Baseline	NMF-GRS	30.8 ± 0.1	10.4 ± 0.1	27.7 ± 0.1	✓	✓	✓
Proposed methods	NMF-TR	34.2 ± 0.1	13.2 ± 0.1	31.0 ± 0.1	✓	✓	✓
	NMF-TP	30.4 ± 0.1	10.9 ± 0.1	27.4 ± 0.1	✓	✓	✓
Supervised methods	NN-SE[a] [7]	35.5	14.7	32.2	✗	✓	✗
	NEUSUM [36]	41.59	**19.01**	**37.98**	✗	✓	✗
	SummaRuNNer [24]	39.6	16.2	35.3	✗	✓	✗
	REFRESH [26]	40.0	18.2	36.6	✗	✓	✗
	HSSAS [1]	**42.3**	17.8	37.6	✗	✓	✗
	DQN [35]	39.4	16.1	35.6	✗	✓	✗
	BANDITSUM [9]	41.5	18.7	37.6	✗	✓	✗

[a] Combined results for this model has been taken from [26].

approach. These are supervised methods trained on CNN/DailyMail corpora, with hyper-parameters tuned using the validation set.

Table 4 shows that all competing methods create markedly better quality summaries than NMF-TR summaries. Though, effectiveness of deep learning approaches for extractive document summarization is clearly indicated, performance of the trained models on other out-of-domain documents need to be further investigated. E.g. SummaRuNNer [24], which is trained on CNN/DailyMail corpora exhibits better performance than NMF-TR in Table 4, but the same model reveals relatively lower performance on out-of-domain DUC2002 data-set.

7   Discussion

Our empirical study reveals some interesting observations. It emerges from Tables 2, 3 and 4 that no method performs best on all four data-sets. CoRank+ [11] is a clear winner for DUC2002 data-set among unsupervised methods, but it is not tested for other data-sets. Hence, calling it state-of-the-art is arguable. HSSAS [1] has good performance for DUC2002 and combined CNN & DailyMail data-sets but is not tested for DUC2001. Hence, its status as state-of-the-art is also ambiguous.

The proposed NMF-TR has slightly degraded performance among unsupervised methods for DUC data-sets but its performance is completely overshadowed by neural network and reinforcement based methods for combined CNN & DailyMail data sets. Please note that no supervised method is evaluated on DUC2001 data-set, even though it was released earlier. Since there are no quality issues with this data-set, the possible reason could be its small size, which is unfavorable for effective model training. CNN+DailyMail data-set has been favored for evaluation of all neural based deep methods possibly because of large data available for training.

Superior performance of deep methods for extractive summarization establishes their high effectiveness within domain. However, since the models were trained, tuned and tested on the same data-set, and it is difficult to predict their performance for (out-of-domain) DUC data-sets. Nallapati et al. have clearly observed degraded performance of recurrent neural network based sequence method for out-of-domain DUC2002 data-set (Sect. 4.7 in [24]). Since preparing large training data for specific domains, and training the models used by deep neural network methods is an arduous and expensive task, we argue for efficient and effective, unsupervised generic text summarization methods that are language-, domain-, collection- independent. There is a vast space of use-cases where such methods are much needed.

Another observation is related to the use of NMF-TR vs. NMF-TP, even though both are unsupervised, fast and language-, domain- and collection- independent methods. NMF-TR is superior for online generic document summarization because there is no underlying design assumption, whereas NMF-TP is designed for longer documents, which usually have clear topic oriented discourse structure. NMF-TR is expected to deliver reasonable performance for all types of general documents, even when topics are not intensely and clearly demarcated within the discourse. Since all evaluation corpora are news articles where the discourse is not topic oriented, NMF-TP has under-performed on all data-sets. Furthermore, both NMF-TR and NMF-TP methods are computationally inexpensive. LA073089-0118 in DUC2001, the longest document among all four data-sets (29 KB, 1509 terms, 227 sentences) was summarized in 12.48 s by NMF-TR and 11.98 s by NMF-TP. Timing measurements were averaged over 20 runs. Summarization time per document, averaged over all four data-sets was recorded as 0.39 s for NMF-TR and 0.31 s for NMF-TP. As expected, NMF-TP is faster in comparison to NMF-TR.

8 Conclusion

In this paper, we leverage NMF based extractive summarization by (i) initializing NMF factors using NNDSVD, and (ii) proposing two novel sentence scoring methods that utilize both feature matrix and co-efficient matrix. The document is transformed to term-sentence binary incidence matrix and decomposed using Non-negative Matrix Factorization. Use of NNDSVD initialization for NMF factors eliminates stochastic variations in the summaries, leading to stable and quality summaries.

The sentence scoring methods exploit information contained in both factors of the decomposed matrix, thereby attending to the importance of terms and topics in the latent semantic space. The first method (NMF-TR) considers the explicit contribution of terms appearing in a sentence, whereas the second method (NMF-TP) takes into account contribution of terms implicitly by acknowledging topic importance. The computations are simple, elegant and fast.

Our experiments reveal that NNDSVD initialization pays-off well. Extensive comparative performance evaluation is reported on four public data-sets.

The results are presented data-set wise, because no existing work reports result on all four data-sets. The major insight obtained by analysis of comparative evaluation is that neural network based extractive summarization methods create high quality summaries for the documents within the domain. However, their performance on out-of-domain documents is yet to be explored. Some unsupervised methods yield superior scores for some data-sets than the proposed methods, yet there is no clear winner in absence of exhaustive evaluation on all four data-sets. We aim to improve the quality of summaries by reducing redundancy.

References

1. Al-Sabahi, K., Zuping, Z., Nadher, M.: A hierarchical structured self-attentive model for extractive document summarization (HSSAS). IEEE Access **6**, 24205–24212 (2018)
2. Alguliev, R.M., Aliguliyev, R.M., Hajirahimova, M.S., Mehdiyev, C.A.: MCMR: maximum coverage and minimum redundant text summarization model. Expert Syst. Appl. **38**(12), 14514–14522 (2011)
3. Alguliyev, R.M., Aliguliyev, R.M., Isazade, N.R., Abdi, A., Idris, N.: COSUM: text summarization based on clustering and optimization. Expert Syst. **36**(1), e12340 (2019)
4. Aliguliyev, R.M.: A new sentence similarity measure and sentence based extractive technique for automatic text summarization. Expert Syst. Appl. **36**(4), 7764–7772 (2009)
5. Belford, M., Mac Namee, B., Greene, D.: Stability of topic modeling via matrix factorization. Expert Syst. Appl. **91**, 159–169 (2018)
6. Boutsidis, C., Gallopoulos, E.: SVD based initialization: a head start for nonnegative matrix factorization. Pattern Recogn. **41**(4), 1350–1362 (2008)
7. Cheng, J., Lapata, M.: Neural summarization by extracting sentences and words. arXiv preprint arXiv:1603.07252 (2016)
8. Conroy, J.M., O'leary, D.P.: Text summarization via hidden Markov models. In: 24th ACM SIGIR, pp. 406–407. ACM (2001)
9. Dong, Y., Shen, Y., Crawford, E., van Hoof, H., Cheung, J.C.K.: BanditSum: extractive summarization as a contextual bandit. arXiv:1809.09672 (2018)
10. Edmundson, H.P.: New methods in automatic extracting. J. ACM (JACM) **16**(2), 264–285 (1969)
11. Fang, C., Mu, D., Deng, Z., Wu, Z.: Word-sentence co-ranking for automatic extractive text summarization. Expert Syst. Appl. **72**, 189–195 (2017)
12. Fattah, M.A., Ren, F.: GA, MR, FFNN, PNN and GMM based models for automatic text summarization. Comput. Speech Lang. **23**(1), 126–144 (2009)
13. Genest, P.E., Lapalme, G.: Framework for abstractive summarization using text-to-text generation. In: Proceedings of the Workshop on Monolingual Text-To-Text Generation, pp. 64–73. Association for Computational Linguistics (2011)
14. Gong, Y., Liu, X.: Generic text summarization using relevance measure and latent semantic analysis. In: 24th ACM SIGIR, pp. 19–25. ACM (2001)
15. He, Z., et al.: Document summarization based on data reconstruction. In: AAAI (2012)
16. Hermann, K.M., et al.: Teaching machines to read and comprehend. In: Advances in Neural Information Processing Systems, pp. 1693–1701 (2015)

17. Lee, D.D., Seung, H.S.: Learning the parts of objects by non-negative matrix factorization. Nature **401**(6755), 788 (1999)
18. Lee, J.H., Park, S., Ahn, C.M., Kim, D.: Automatic generic document summarization based on non-negative matrix factorization. Inform. Process. Manage. **45**(1), 20–34 (2009)
19. Lin, C.Y.: ROUGE: a package for automatic evaluation of summaries. Text Summarization Branches Out (2004)
20. Lloret, E., Romá-Ferri, M.T., Palomar, M.: COMPENDIUM: a text summarization system for generating abstracts of research papers. Data Knowl. Eng. **88**, 164–175 (2013)
21. Luhn, H.P.: The automatic creation of literature abstracts. IBM J. Res. Dev. **2**(2), 159–165 (1958)
22. Mihalcea, R., Tarau, P.: Textrank: bringing order into text. In: Proceedings of the 2004 Conference on Empirical Methods in Natural Language Processing (2004)
23. Moawad, I.F., Aref, M.: Semantic graph reduction approach for abstractive text summarization. In: ICCES 2012, pp. 132–138. IEEE (2012)
24. Nallapati, R., Zhai, F., Zhou, B.: SummaRuNNer: a recurrent neural network based sequence model for extractive summarization of documents. In: Thirty-First AAAI Conference on Artificial Intelligence (2017)
25. Nallapati, R., Zhou, B., Ma, M.: Classify or select: neural architectures for extractive document summarization. arXiv:1611.04244 (2016)
26. Narayan, S., Cohen, S.B., Lapata, M.: Ranking sentences for extractive summarization with reinforcement learning. arXiv preprint arXiv:1802.08636 (2018)
27. Parveen, D., Ramsl, H.M., Strube, M.: Topical coherence for graph-based extractive summarization. In: Proceedings of the 2015 EMNLP, pp. 1949–1954 (2015)
28. Qiang, J., Li, Y., Yuan, Y., Liu, W.: Snapshot ensembles of non-negative matrix factorization for stability of topic modeling. Appl. Intell. **48**, 1–13 (2018)
29. Shen, D., Sun, J.T., Li, H., Yang, Q., Chen, Z.: Document summarization using conditional random fields. In: IJCAI, vol. 7, pp. 2862–2867 (2007)
30. Steinberger, J., Ježek, K.: Text summarization and singular value decomposition. In: Yakhno, T. (ed.) ADVIS 2004. LNCS, vol. 3261, pp. 245–254. Springer, Heidelberg (2004). https://doi.org/10.1007/978-3-540-30198-1_25
31. Vikas, O., Meshram, A.K., Meena, G., Gupta, A.: Multiple document summarization using principal component analysis incorporating semantic vector space model. IJCLCLP **13**(2), 141–156 (2008)
32. Wan, X.: Towards a unified approach to simultaneous single-document and multi-document summarizations. In: Proceedings of the 23rd International Conference on Computational Linguistics, pp. 1137–1145. ACL (2010)
33. Xu, W., Liu, X., Gong, Y.: Document clustering based on non-negative matrix factorization. In: 26th ACM SIGIR, pp. 267–273. ACM (2003)
34. Yang, K., Al-Sabahi, K., Xiang, Y., Zhang, Z.: An integrated graph model for document summarization. Information **9**(9), 232 (2018)
35. Yao, K., Zhang, L., Luo, T., Wu, Y.: Deep reinforcement learning for extractive document summarization. Neurocomputing **284**, 52–62 (2018)
36. Zhou, Q., Yang, N., Wei, F., Huang, S., Zhou, M., Zhao, T.: Neural document summarization by jointly learning to score and select sentences. arXiv:1807.02305 (2018)

Succinct BWT-Based Sequence Prediction

Rafael Ktistakis[1]($\boxtimes$), Philippe Fournier-Viger[2], Simon J. Puglisi[3],
and Rajeev Raman[1]

[1] Department of Informatics, University of Leicester, Leicester, UK
{crk15,r.raman}@leicester.ac.uk
[2] Harbin Institute of Technology (Shenzhen), Shenzhen, China
philfv@hit.edu.cn
[3] Department of Computer Science, University of Helsinki, Helsinki, Finland
puglisi@cs.helsinki.fi

Abstract. Sequences of symbols can be used to represent data in many
domains such as text documents, activity logs, customer transactions and
website click-streams. Sequence prediction is a popular task, which consists of predicting the next symbol of a sequence, given a set of training
sequences. Although numerous prediction models have been proposed,
many have a low accuracy because they are lossy models (they discard
information from training sequences to build the model), while lossless
models are often more accurate but typically consume a large amount of
memory. This paper addresses these issues by proposing a novel sequence
prediction model named SUBSEQ that is lossless and utilizes the succinct Wavelet Tree data structure and the Burrows-Wheeler Transform
to compactly store and efficiently access training sequences for prediction. An experimental evaluation shows that SUBSEQ has a very low
memory consumption and excellent accuracy when compared to eight
state-of-the-art predictors on seven real datasets.

1 Introduction

Sequences of symbols (strings) are a type of data found in many domains. For
instance, they can be used to represent sequences of words in a text, events in a
business process log, purchases made by customers, or point-of-interests visited
by tourists. An important task in data mining is sequence prediction. Given
a multi-set of training *strings* (or sequences) $\hat{D} = \{x_1, \ldots, x_d\}$ defined over a
finite ordered *alphabet* of symbols, sequence prediction consists of predicting the
next symbol of the prefix of an unknown query sequence Q. The underlying
assumption is that all the strings are created by a same underlying process.
To perform sequence prediction, a predictor can be trained using the training
strings. Then the predictor can perform predictions.

Various sequence prediction models have been proposed, having various characteristics. They have been used in many domains to perform tasks such as predicting heart failure [18], human activities [19] and webpage prefetching [6].

© Springer Nature Switzerland AG 2019
S. Hartmann et al. (Eds.): DEXA 2019, LNCS 11707, pp. 91–101, 2019.
https://doi.org/10.1007/978-3-030-27618-8_7

Although numerous prediction models have been proposed, many are lossy models [3,9,15,16,21]. In other words, they discard information from training sequences to build small models. But the drawback of this approach is that they may lack information when its time to make a prediction, which can result in low prediction accuracy [7]. Some models such as DG [15] also adopt simplifying assumptions such that each symbol of a string only depends on the previous one. But this assumption often does not hold in real life applications.

The aforementioned limitations of lossy predictors have recently been addressed by proposing lossless models, which keep all information about training sequences in memory to perform more accurate predictions. The assumption is that a lossless model should be more accurate because they can use all the available information to make each prediction. Some of the best models of this type is CPT [7], which was then extended as CPT+ [6]. These models store training sequences in a trie-based structure, and were shown to be more accurate than multiple state-of-the-art lossy models. However, the CPT/CPT+ have several important drawbacks:

- To perform a prediction, the CPT/CPT+ models utilize the bag-of-words model, which does not consider the order between symbols. But for some domains, the order is important.
- The CPT/CPT+ models require choosing several dataset-specific parameters. The prediction accuracy can vary greatly depending on how these parameters are set. Setting these parameters is not trivial and requires to have background knowledge or use a trial-and-error approach to find optimal parameter settings.
- All lossless predictors end up storing the entire training sequence in main memory. Thus, it is essential that a lossless predictor should store the training sequence *space-efficiently*. We use the following variables to denote the size of the sequence database D: d is the number of sequences, M is the total length of all the sequences and σ is the alphabet size. We note that the information-theoretic lower bound for storing D is $M \log \sigma$ bits[1] in the worst case. On the other hand:
    - CPT+ uses σ bit-strings of length d to represent the sets of symbols contained in each sequence. This alone takes $d\sigma$ bits, which can be much larger than $M \log \sigma$ bits if σ is large.
    - CPT+ stores the training dataset in a trie. In the worst case, there could be $\Omega(M)$ trie nodes, and each trie node contains three (64-bit) pointers, a significant overhead.
    - CPT+ uses ideas such as Patricia compression and replacing frequently occurring sub-sequences by a single symbol to try to minimize the number of trie nodes [6]. However, success is unpredictable, and the frequent pattern mining slows down the training phase.
- During the prediction phase, given a query Q of k symbols, CPT+ performs several bitwise-and of up to k bit-strings of length d each to find sequences

---

[1] Logs are to base 2 unless stated otherwise.

containing a subset of symbols in Q. This takes $O(f(k) \cdot d)$ time where $f(k)$ can be as large as 2^k. In practice, many fewer than 2^k combinations are tried, and the constants in the $O()$ are small. However, as we show, the query time of CPT+ grows linearly with d.

This paper addresses drawbacks of the CPT/CPT+ models by proposing a novel sequence predictor named SUBSEQ. This model adopts the succinct Wavelet Tree data structure and the Burrows-Wheeler Transform to store training sequences in a very compact way, while still allowing fast access to training sequences for prediction. An experimental evaluation shows that SUBSEQ has a very low and predictable memory consumption (the space usage varies between 1.6 and 2.2 times the binary size of D) and excellent accuracy when compared to state-of-the-art predictors on real datasets. Last but not least, SUBSEQ is largely parameter-free.

The rest of this paper is organized as follows. Section 2 introduces preliminaries about sequence prediction. Section 3 presents the proposed SUBSEQ predictor. Section 4 presents the performance evaluation. Finally, a conclusion is drawn and future work is discussed.

2   Preliminaries

Strings. A string $x = x[0..n-1] = x[0]x[1]\ldots x[n-1]$ is a sequence of $|x| = n$ symbols drawn from a constant ordered alphabet of size σ. For $i = 0, \ldots, n-1$ we write $X[i..n-1]$ to denote the *suffix* of X of length $n-i+1$, that is $X[i..n-1] = X[i]X[i+1]\ldots X[n-1]$. We will often refer to suffix $X[i..n-1]$ simply as "suffix i". Similarly, we write $X[0..i]$ to denote the *prefix* of X of length $i+1$. We write $X[i..j]$ to represent the *substring* $X[i]X[i+1]\ldots X[j]$ of X that starts at position i and ends at position j.

In this paper we consider a multiset of d strings $\hat{D} = \{x_1, x_2, \ldots x_d\}$. We represent $\hat{D}$ as a single string by concatenating the strings in D into a single string $D = x_1 \$ x_2 \$ \ldots \$ x_d$, using a special symbol \$ to delineate individual strings, which does not occur in any string x_i. We let $M = |D|$ denote the length of D.

Suffix Arrays. We make use of several standard data structures built from D. The first of these is the suffix array [10], denoted SA, which is an array $SA[0..M-1]$ containing a permutation of the integers $0..M-1$ such that $D[SA[0]..M-1] < D[SA[1]..M-1] < \cdots < D[SA[M-1]..M-1]$. In other words, $SA[j] = i$ iff $D[i..M-1]$ is the jth suffix of D in ascending lexicographical order.

The Burrows-Wheeler Transform [2,11], denoted BWT is a string $BWT[0..M-1]$ is a permutation of D defined by SA, such that $BWT[i] = D[SA[i]-1]$, except when $SA[i] = 0$, in which case $BWT[i] = D[M]$. See Fig. 1 for an example.

Backward Search. The FM-index is a compressed text index (see [13]) that consists of two main components: a wavelet tree build from the BWT string, and an array C of σ integers such that $C[c]$ gives the total number of symbols in the BWT string that are less than symbol c. Searching with an FM-index is based on a procedure called *backward search*, which finds the range of SA containing all suffixes that begin with a given query pattern Q. This range then contains the positions of occurrence of Q in D. Figure 2 shows how backward search is used for counting the number of occurrences (the count query). In the algorithm, $C[c]$ is the position of the first occurrence of the symbol c in F, and the function $\mathrm{rank_L}$ is defined as $\mathrm{rank_L}(c, j) \equiv |\{i \mid i < j \text{ and } L[i] = c\}|$. The main difference between the members of the FM-family is how they implement the $\mathrm{rank_L}$-function. The best ones use wavelet trees.

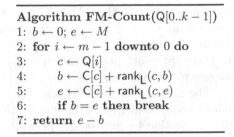

L	SA	
A	6	$
N	5	A $
N	3	A N A $
B	1	A N A N A $
$	0	B A N A N A $
A	4	N A $
A	2	N A N A $

Algorithm FM-Count$(Q[0..k - 1])$
1: $b \leftarrow 0;\ e \leftarrow M$
2: **for** $i \leftarrow m - 1$ **downto** 0 **do**
3:    $c \leftarrow Q[i]$
4:    $b \leftarrow C[c] + \mathrm{rank_L}(c, b)$
5:    $e \leftarrow C[c] + \mathrm{rank_L}(c, e)$
6:    **if** $b = e$ **then break**
7: **return** $e - b$

Fig. 1. SA and BWT string L for string D = BANANA$.

Fig. 2. Counting pattern occurrences using backward search.

Wavelet Tree. The wavelet tree [12] of string D over an alphabet Σ is a binary tree with leaves labelled by the symbols of Σ. Each node v is associated with the subsequence of D consisting of those symbols that appear in the subtree rooted at v. The associated strings are not stored; instead each internal node v stores a bitvector $B(v)$ that tells for each character in the associated string whether it is in the left or right subtree of v.

In a wavelet tree the total length of the bitvectors is $|D|\lceil \log |\Sigma| \rceil$, which is exactly the length of D in bits using the standard representation.

A rank query $\mathrm{rank_D}(c, r)$ over a wavelet tree is evaluated by a traversal from the root to the leaf labelled by c. Wavelet trees answer rank queries in $O(\log \sigma)$ time. A similar procedure enables one to access a given symbol $D[i]$ in $O(\log \sigma)$ time, or to enumerate all the distinct symbols in a range of the string, as well as compute the frequency of each of those symbols. Wavelet trees answer these distinct(i, j) queries in $O(k \log \sigma)$ time, where k is the number of distinct symbols in $D[i..j]$. Wavelet trees also support the query select(c, i) in $O(\log \sigma)$ time, which returns the position of the ith occurrence of symbol c in D. The queries rank, select, access, and distinct involve rank (or select) queries over the bitvectors stored on the root-to-leaf path. There are many data structures for representing bitvectors so that rank and select queries can be answered in constant time [14, 17]. These data structures are a standard component in succinct data structure design. Recent experimental studies of these bitvectors can be found in [5,8].

3   Succinct BWT-Based Sequence Prediction Model

The *Succinct BWT-based Sequence prediction* model (SUBSEQ) is a new lossless predictor. Its main distinctive characteristics are that (1) efficiently stores the entire input training data without any loss (2) fetches training sequences similar to a given sequence (query prefix) (3) it does not depend in any parameter-set fine-tuning in order to be accurate (4) SUBSEQ keeps into account the item order of a given query prefix. The latter is the main key difference to the CPT+ prediction model. CPT+ searches for sequences using the bag-of-words model. This model does not take into account the items order of a prefix for matching it in the training data (which might be important aspect for some domain applications, as discussed).

3.1   Algorithm Description

The SUBSEQ prediction algorithm is consisted of two main phases; the train phase and the ready-for-prediction phase. A multiset $\hat{D}$ of training sequences is given as an input. During the train phase, SUBSEQ will use the D to produce the FM-index and store BWT in memory using a wavelet tree. During the ready-for-prediction phase, SUBSEQ is ready to answer query prefixes. The answers that SUBSEQ returns can further be evaluated with the *query suffix* (see Sect. 4.2).

For every query prefix SUBSEQ will try to give an answer by finding similar sequences in its training data sequences. This is done through the given query prefix and a generated collection of *sub-queries*. Due to the fact that SUBSEQ is only able to locate exact matches of a given pattern in its training data, it is essential to have a mechanism that expands our prediction model coverage to more training data. The collection of sub-queries plays the role of this mechanism. Every sub-query comes from the initial query prefix. These are produced by allowing operations of deletion and substitution. The deletions are always at the start of the query or sub-query and the substitutions are limited to two.

Example. For a given $Q = [a, b, c, d]$, SUBSEQ will try to find exact matches for $Q_1 = [a, b, c, d]$, $Q_2 = [¿, b, c, d]$, $Q_3 = [a, ¿, c, d]$, $Q_4 = [a, b, ¿, d]$, $Q_5 = [b, c, d]$, $Q_6 = [¿, c, d]$, $Q_7 = [b, ¿, d]$, $Q_8 = [c, d]$, $Q_9 = [¿, d]$.

On the example above we denote with $¿$ the place where we can replace with any symbol from our alphabet. Assuming our alphabet as $\Sigma = \{a, b, c, d\}$ then SUBSEQ can match Q_6 with some example training sequences like: $[a, c, d, a, d]$, $[b, c, d, c, a]$, $[c, c, d, b, b]$, $[d, c, d, a, b]$.

After SUBSEQ has found the similar sequences, it uses them to produce possible answers and eventually order them according to a *weight*. Producing possible answers is done through the *consequents* of the similar sequences. The consequent of a similar sequence s is considered the subsequence from the item common to both s and the current (sub-)query used, and up to the last item of s. For SUBSEQ we will be using consequents of length up to two items long. Every time a (sub-)query is used to find similar training sequence, we come up with consequents. The items of the consequents are put into a *Frequency Array* and they are ordered by a weight. A final prediction answer is the item in the array with the highest weight value. The final answer is given either (a) when SUBSEQ has collected all possible consequents for both the initial query prefix and its all produced sub-queries or (b) when a threshold of confidence is met.

Finally, when an item of a consequent is inserted to the frequency array, it is assigned a weight value. If the item exists in the array then the new value is added-up on the old value. The weight formula is defined as $w = y/Y + (2 - sub)/2 + 1 + r$. We consider y to be the suq-query length, Y the initial query length, sub the number of substitutions and $r = \frac{1}{index+1}$. The later indicates the index of the item in the consequent.

3.2   Implementation Using FM-Index

We mainly need four core functions; (1) *backwardSearch* (2) *forwardSearch* (3) *neighbourExpansion* (4) *getConsequents*.

The **backwardSearch** can be implemented by tweaking the **FM-Count** (see Fig. 2) to return the (b, e) for a query item at a time.

The **forwardSearch** does the opposite of the *backwardSearch* for a given i. It gives the index $i' = \mathsf{C}[c] + rank_\mathsf{L}(c, i)$ where $c = \mathsf{L}[i]$, and $c' = \mathsf{L}[i']$ occurs after c in D.

The **neighbourExpansion** constitutes the key function of our prediction model. Using the FM-index, one can only find exact matches for a given pattern. This creates a twofold issue; (1) there is no way to locate similar training sequences (2) usually in sequence prediction, searching only for exact matches does not give an enough coverage (if any) for confident predictions. The main idea of neighbour expansion is that for a given query prefix, it will perform a normal *backwardSearch* if the prefix does not have any substitutions in place or for any substitution that it meets it will recursively expand to all possible symbols that might follow. Taking into account our previous example of sub-queries, Q_3, we will make the following assumption; before a $[c, d]$ all of the $\{a, b, c, d\}$ appear in the training data. This can be figured out with a distinct call for a range in L. Then Q_3 will be expanded to $[a, a, c, d]$, $[a, b, c, d]$, $[a, c, c, d]$ and $[a, d, c, d]$ for a normal *backwardSearch* each.

The **getConsequents** utilises the *forwardSearch* definition to obtain the consequents for ranges that have been acquired through the *neighbour Expansion*. Expanded sub-queries which result in patterns that have already been used, are excluded. We do this by utilising a bit-vector of length M. Every index of successful *neighbourExpansion* ranges, is a set bit in the bit-vector. Thus, consequents from sub-queries that have been prior utilised, will not be re-used and only new consequent information will added in Frequency Array.

A C++ implementation of our prediction model can be found on github.com/rafkt/SUBSEQ.

4   Evaluation

We split this section as: the set-up environment, our experimental aims, the competition to our prediction model and finally the discussion of accuracy and performance evaluation. For this section, full details about our experimental data and about our results can be found on github.com/rafkt/SUBSEQ.

4.1   Experimental Setup

Environment. Experiments were performed under macOS 10.14.1 with an Intel Core i7 (4 Cores, 256 KB L2 per Core, 8 MB L3), 32 GB DDR3 1867 MHz RAM and a 8.0 GT/s Link speed SSD. The lossless predictors, CPT+, CPT, were ran using *IPredict* framework [6] under java version 1.8.0_112 with *JIT* enabled which allows the bytecode to be compiled into native machine code, allowing a fair comparison with native implementations. The SUBSEQ Predictor was compiled under clang-1000.11.45.5, while SPiCe baseline [1] was compiled and run under Python 2.7.10. We used the *sdsl-lite* library [4] for implementing SUBSEQ.

Aims. To measure and compare different prediction models in terms of their accuracy and their performance. Performance is measured in terms of the execution time a prediction model needs to train itself; the execution time it needs to complete answering a testing set; the memory usage it utilises after the training phase is complete.

Competition. We compare SUBSEQ with a variety of state-of-the-art lossy and lossless predictors. These are: All-K-order Markov (AKOM) [16], LZ78 [21], Transition Directed Acyclic Graph (TDAG) [9], Prediction by Partial Matching (PPM) [3] and Dependency Graphs (DG) [15]. We also included a spectral learning prediction model from SPiCe competition [1]. We also compare SUBSEQ with CPT+ [6] as it is the current state-of-the-art lossless prediction model.

Data. For our experiments we used datasets with various characteristics from *SPMF* library[2] library. In addition, we used synthetic data[3] which was generated by IBM *QUEST* data generator [20].

4.2   Accuracy of Prediction

Each dataset is read in memory, and then is split into a training set and a testing set using the k-fold cross validation. Once a predictor has been trained, each sequence of the testing set is split into two parts, the *query prefix* and the *query suffix*. The size of each can be defined through a parameter in advance. Then a trained prediction model is called to give answers for every prefix in the testing set. A prediction answer for a query prefix is accurate if it appears within the query suffix[4]. The accuracy rate is the ratio of accurate predictions

---

[2] Available at http://www.philippe-fournier-viger.com/spmf.

[3] Details about QUEST exported data, are available at github.com/rafkt/SUBSEQ.

[4] Same evaluation approach was followed for CPT+[6].

to the total number of test sequences. Each prediction model has been trained and tested using k-fold cross validation with k = 14 to obtain a low variance for each run.

Accuracy results are shown in Table 1. Our prediction model provides better accuracy than any other lossy predictor for SIGN, KOSARAK and FIFA datasets. At the same time, we can observe that SUBSEQ has an overall better accuracy than any predictor for MSNBC and BIBLE_CHAR. However, if we take into consideration the accuracy variation of CPT+ (as show in the Table 1 at CPT+ column in a [min-max] range) based on its different possible parameter tunes, then SUBSEQ provides an overall better accuracy performance for KOSARAK and FIFA as well. Thus, CPT+ gets less competitive if it is not finely tuned making SUBSEQ more attractive.

Table 1. Prediction models and their accuracy in %. First and second best performers are in bold

Datasets	DG	TDAG	CPT+	subSeq	Mark1 (PPM)	AKOM	LZ78	SPiCe baseline
BMS	**36**	7	[30–**38**]	33	30	31	33	0.19
SIGN	2	0	[**26–34**]	**23**	4	7	5	4
MSNBC	55	31	[49–59]	**64**	38	48	43	30
BIBLE_WORD	6	23	[0–22]	**29**	11	**32**	18	2
BIBLE_CHAR	3	79	[1–80]	**88**	16	**81**	65	6
KOSARAK	30	1	[**31–37**]	**34**	23	20	20	0.6
FIFA	25	7	[18–**34**]	**29**	23	26	25	0.38

4.3   Performance

The Memory of SUBSEQ was measured by using the relevant api in sdsl library. The memory for the rest of the predictors was measured through IPredict. We compared the different prediction models through the ratio of their memory usage over the training set binary size. In the Table 2, SUBSEQ is the most consistent and most memory efficient prediction model. It uses an average memory of up to 2.2 times the memory of the input training set binary size. Prediction models like TDAG and CPT+ appear to be highly inconsistent. TDAG can utilise space between 70 to 2500 times the input binary size while CPT+ between 0.5 to 80 times; indicating an unpredictable performance.

The running time of SUBSEQ was directly compared to CPT+ for various datasets (Fig. 3c) in respect of the testing-phase (and training-phase). Evaluations also included input data of an increasing σ, n, d using the QUEST generator. The results showed competitive and consistent performance for SUBSEQ in comparison to CPT+.

Table 2. Ratio of prediction model memory to training binary size ($M * \lceil \log(\sigma) \rceil$)

Datasets	DG	TDAG	CPT+	CPT	subSeq	Mark1 (PPM)	AKOM	LZ78
BMS	4.87	136.34	9.01	15.58	2.14	1.63	26.05	5.60
SIGN	2.96	124.51	0.54	10.86	1.73	1.69	38.07	5.08
MSNBC	0.06	176.29	3.19	5.42	2.14	0.06	13.71	4.14
BIBLE_WORD	6.07	77.72	11.10	12.74	1.90	1.70	20.83	3.40
BIBLE_CHAR	0.68	2689.15	3.38	6.46	2.18	0.25	51.69	42.77
KOSARAK	6.76	126.92	81.49	86.43	1.67	21.17	30.62	4.86
FIFA	2.98	90.74	4.88	6.64	1.60	1.15	23.40	3.59

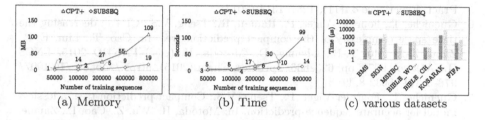

(a) Memory   (b) Time   (c) various datasets

Fig. 3. Testing time performance of CPT+ and SUBSEQ

4.4   Optimisation Discussion

Our current implementation of SUBSEQ is not fully optimised yet. Experimental evaluation showed that 90% of the time needed from SUBSEQ to answer a query, it is spent for neighbour expansion. Further experiments revealed that in average only a 45% of the executed rank operations are unique per query. Thus, preventing neighbour expansion from performing excessive rank calls in the wavelet tree, would optimise the speed performance of SUBSEQ for datasets with large σ. Figure 3c shows that for a dataset like KOSARAK ($\sigma = 654,987$), SUBSEQ performance is less competitive. One way to minimise excessive rank calls is to store (retrieve) each rank result in (from) a trie-based data structure.

5   Conclusion

Lossless sequence predictors are often very accurate but can consume a large amount of memory. To address this issue, this paper presented a novel predictor named SUBSEQ that is lossless and utilizes the succinct Wavelet Tree data structure and the Burrows Wheeler Transform to compactly store and efficiently access training sequences for prediction. Experimental results have shown that SUBSEQ has a very low and predictable memory consumption (varying 1.6 to 2.2 times the binary size of D) and excellent accuracy in comparison to state-of-the-art predictors on real datasets. Moreover, SUBSEQ is mostly parameter-free. Future work includes optimising SUBSEQ neighbour expansion along with its overall speed performance.

References

1. Balle, B., Eyraud, R., Luque, F.M., Quattoni, A., Verwer, S.: Results of the sequence PredIction ChallengE (SPiCe): a competition on learning the next symbol in a sequence. In: Proceedings 13th International Conference in Grammatical Inference, vol. 57. JMLR W&CP, Delft (2016)
2. Burrows, M., Wheeler, D.: A block-sorting lossless data compression algorithm. Technical report, 124, Digital Equiptment Corporation (1994)
3. Cleary, J., Witten, I.: Data compression using adaptive coding and partial string matching. IEEE Trans. Commun. **32**(4), 396–402 (1984)
4. Gog, S.: simongog/sdsl-lite (2015). https://github.com/simongog/sdsl-lite
5. Gog, S., Petri, M.: Optimized succinct data structures for massive data. Softw. Pract. Experience **44**(11), 1287–1314 (2014)
6. Gueniche, T., Fournier-Viger, P., Raman, R., Tseng, V.S.: CPT+: decreasing the time/space complexity of the compact prediction tree. In: Cao, T., Lim, E.-P., Zhou, Z.-H., Ho, T.-B., Cheung, D., Motoda, H. (eds.) PAKDD 2015. LNCS (LNAI), vol. 9078, pp. 625–636. Springer, Cham (2015). https://doi.org/10.1007/978-3-319-18032-8_49
7. Gueniche, T., Fournier-Viger, P., Tseng, V.S.: Compact prediction tree: a lossless model for accurate sequence prediction. In: Motoda, H., Wu, Z., Cao, L., Zaiane, O., Yao, M., Wang, W. (eds.) ADMA 2013. LNCS (LNAI), vol. 8347, pp. 177–188. Springer, Heidelberg (2013). https://doi.org/10.1007/978-3-642-53917-6_16
8. Kärkkäinen, J., Kempa, D., Puglisi, S.J.: Hybrid compression of bitvectors for the FM-index. In: Proceedings DCC, pp. 302–311. IEEE (2014)
9. Laird, P., Saul, R.: Discrete sequence prediction and its applications. Mach. Learn. **15**(1), 43–68 (1994)
10. Manber, U., Myers, E.W.: Suffix arrays: a new method for on-line string searches. SIAM J. Comput. **22**(5), 935–948 (1993)
11. Manzini, G.: An analysis of the Burrows-Wheeler transform. J. ACM **48**(3), 407–430 (2001)
12. Navarro, G.: Wavelet trees for all. J. Discrete Algorithms **25**, 2–20 (2014)
13. Navarro, G., Mäkinen, V.: Compressed full-text indexes. ACM Comput. Surv. **39**(1) (2007). Article 2
14. Okanohara, D., Sadakane, K.: Practical entropy-compressed rank/select dictionary. In: Proceedings ALENEX, pp. 60–70. SIAM (2007)
15. Padmanabhan, V.N., Mogul, J.C.: Using predictive prefetching to improve world wide web latency. SIGCOMM Comput. Commun. Rev. **26**(3), 22–36 (1996)
16. Pitkow, J., Pirolli, P.: Mining longest repeating subsequences to predict world wide web surfing. In: Proceedings of the 2nd Conference on USENIX Symposium on Internet Technologies and Systems - Volume 2, USITS 1999, pp. 139–150 (1999)
17. Raman, R., Raman, V., Rao, S.S.: Succinct indexable dictionaries with applications to encoding k-ary trees, prefix sums and multisets. ACM Trans. Algorithms **3**(4), 43 (2007)
18. Rjeily, C.B., Badr, G., Al Hassani, A.H., Andres, E.: Predicting heart failure class using a sequence prediction algorithm. In: 2017 Fourth International Conference on Advances in Biomedical Engineering (ICABME), pp. 1–4. IEEE (2017)
19. Tax, N.: Human activity prediction in smart home environments with LSTM neural networks. In: 2018 14th International Conference on Intelligent Environments (IE), pp. 40–47. IEEE (2018)

20. Zheng, Z., Kohavi, R., Mason, L.: Real world performance of association rule algorithms. In: Proceedings ACM SIGKDD, pp. 401–406. ACM (2001)
21. Ziv, J., Lempel, A.: Compression of individual sequences via variable-rate coding. IEEE Trans. Inf. Theor. **24**(5), 530–536 (1978)

TRR: Reducing Crowdsourcing Task Redundancy

Sh. Galal[(⊠)] and Mohamed E. El-Sharkawi

Faculty of Computers and Information, Department of Information Systems,
Cairo University, Giza, Egypt
{sh.galal,m.elsharkawi}@fci-cu.edu.eg

Abstract. In this paper, we address the problem of task redundancy in crowdsourcing systems while providing a methodology to decrease the overall effort required to accomplish a crowdsourcing task. Typical task assignment systems assign tasks to a fixed number of crowd workers, while tasks are varied in difficulty as being easy or hard tasks. Easy tasks need fewer task assignments than hard tasks. We present TRR, a task redundancy reducer that assigns tasks to crowd workers on several work iterations, that adaptively estimates how many workers are needed for each iteration for Boolean and classification task types. TRR stops assigning tasks to crowd workers upon detecting convergence between workers' opinions that in turn reduces invested cost and time to answer a task. TRR supports Boolean, classification, and rating task types taking into consideration both crowdsourcing task assignment schemes of anonymous workers task assignments and non-anonymous workers task assignments. The paper includes experimental results by performing simulating experiments on crowdsourced datasets.

Keywords: Crowdsourcing task redundancy ·
Crowdsourcing HITs redundancy · Crowdsourcing tasks

1 Introduction

Crowdsourcing was coined as a new problem solving paradigm allowing integrating people into the computational process to enhance problem solving techniques as well as providing solutions to unsolvable problems. Crowdsourcing allows humans to perform tasks in the form of questions that is called human intelligent tasks (HITs) in compensation to monetary incentives in most cases. Various crowdsourcing platforms [1, 2] provide a wide range of tasks such as data entry, and image classification where a task requester submits batches of tasks for a small incentive to be selected by workers. Many computer applications have made use of crowdsourcing to enhance computer algorithms by providing humans with a set of questions to be answered. The produced task assignments are getting overwhelmingly large jeoparding the tasks cost and elapsed time as humans are expensive and time-consuming. As a consequence, this introduced the data redundancy problem [3] that is defined as the problem of estimating

© Springer Nature Switzerland AG 2019
S. Hartmann et al. (Eds.): DEXA 2019, LNCS 11707, pp. 102–117, 2019.
https://doi.org/10.1007/978-3-030-27618-8_8

how many workers should answer a single question (task or HIT) with stable quality, taking into consideration that some questions may be easy while others may be more difficult in terms of recognizing the answer by the assigned crowd workers (e.g. a question asking 'Is "iPhone XS 5.8-inch" and "iPhone XS Max" are the same entities?' –is an easy question it needs only 2 or 3 reliable crowd workers to recognize the correct answer, however, entity resolution questions are assigned to 10 workers as a practice). The essential dilemma of this problem is that we cannot automatically detect whether we are encountering an easy or hard task prior to task assignment process. Determining a task hardness is challenging due to the following: (1) Annotating the task hardness before task assignment is impractical. (2) Whenever a task is announced by the crowdsourcing platform, a random set of workers will volunteer to answer this task making it complex to predict whether it is an easy or hard task for the assigned workers. This paper introduces a model that automatically reduces task assignments in crowdsourcing platforms, that is crucial to reduce the invested cost, time, and effort devoted to answer a set of crowdsourcing questions.

2 Related Work

Previous researches [4–7] have provided solutions to determine the number of task assignments needed for a *classification task*. Work presented in [4, 5] has utilized a belief-propagation technique that represents workers and tasks in a bipartite graph. The model learns the worker correctness and the task probable correct answer upon receiving task answers in an iterative manner. The model continually accepts workers' answers once the task has been declared targeting minimization of the total task price guided by the final answer correctness. Work presented in [6] starts the task assignment process by estimating the number of workers needed for each task (the prediction phase). The prediction phase enumerates all the possible available workers' correctness in providing an answer (combinations) deploying the sum-product probabilistic algorithm targeting the group of workers maximizing the answer correctness. The model stops assigning a task to workers upon reaching predefined answer correctness. The most recent work presented in [7] has proposed an iterative exploration technique to collect task answers by assigning each task to one worker at a time and wait for the worker to reply back, then assign the task to another worker till a certain stopping condition. The stopping condition is satisfied by a certain bias in the task answer's voting distribution. The previously depicted researches illustrate effective solutions to the task redundancy problem however, it presented the following few shortcomings:

(1) All models provides handling for Boolean and classification tasks *only*, despite of the variety of crowdsourcing task types that need different handling methods such as rating tasks. (2) Prediction models presented in [4–6] are worker quality-sensitive answering models, however not all crowdsourcing platforms provide such information. For example the AMT crowdsourcing platform records the approval rate (i.e. the percentage of questions approved by task requesters), however the approval rate does

not measure the worker quality as the worker quality differs from a task to another, and some requesters automatically approves all answers. Work in [6] suggested a sampling based technique to provide initial estimate to the user quality through the use of gold-questions, and [4] suggested to learn the user quality during currently assigned tasks. However, there is still a need to provide handling of anonymous workers where gold questions are costly, and users' quality varies over different topics.

Work in [7] presented few shortcomings that are: (1) The proposition of one by one task assignment can cause a latency problem as the platform must wait till workers volunteer to answer tasks. In some cases, this is a matter of seconds causing no latency at all and in other cases, this might take days initiating a considerable latency problem. Moreover, few crowd opinions may not be enough to evaluate a task's answer. (2) Utilization of complex priors such as probabilities bias that should be determined by a data expert in order to achieve the targeted goal of reducing task assignments.

This paper introduces the task redundancy reducer (TRR); a model to reduce task assignments while overcome the aforementioned problems.

Contributions of this paper are: (1) TRR reduces the aforementioned latency problem via assigning the task to few crowd members on several iterations, each iteration's collected answers guide TRR whether to assign more workers to the task (as no consensus agreement on the answer among the assigned workers) utilizing the diversity measure. (2) Providing dichotomy handling for both crowdsourcing platforms based on hiring anonymous and non-anonymous workers. (3) Providing a method to handle rating tasks. (4) Applicability to crowdsourcing platforms proving a way to understand the representation of the model priors allowing non-experts to interact with the platform. (5) We conducted experiments to demonstrate the efficiency and effectiveness of applying the TPR model via performing experiments on real crowdsourced datasets.

3   An Overview of Task Redundancy Reducer (TRR)

Before the elaboration of TRR in details we present two important design considerations:

1. Supported tasks: TRR targets constrained mini-tasks [8] only for the meanwhile leaving the macro-tasks [9] as they are different in task nature. Constrained tasks are Boolean, classification, and rating tasks.

2. Diversity measure: Diversity measure (index) [10] is a statistical index intended to measure the variety of a set of various classifications. TRR utilizes the diversity measure to observe the diversity level of workers' answers recognizing whether there is a consensus on the task answer with some percentage. The observed variables in our context are the crowd members' answers, and thus we can mathematically compute the task answers' diversity level. We have chosen to utilize the diversity measure rather than the entropy measure as the later provides a number that we cannot interpret whether there is high or low entropy. However, entropies are used to compare entropies of different states, it is useful to recognize whether there is a reduction or increase in a presented entropy level such as work in [11].

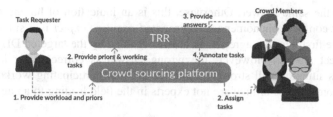

Fig. 1. Task redundancy reducer workflow

Figure 1 depicts the TRR workflow as a set of steps. At the first step, a task requester provides a workload (i.e. a set of questions) to the crowdsourcing platform. The requester also provides the required priors guiding the platform through the task assignment process. The priors are *the minimum and the maximum number of crowd members to be assigned to a task (NC) and (MC)*, respectively. *The required answer diversity level (DL)* identifying a percentage of how a task answers are different where Zero% diversity means that all crowd members have provided the same answer, and 100% diversity means we are completely unsure of what is the answer, this happens when each worker provides a different answer. An optional prior *answer-confirming crowd (ACC)* represents crowd members who are needed to confirm the answer reached so far. The confirming crowd is advised to be experts (of high reliabilities) for crowdsourcing platforms considering workers' reliabilities (i.e. qualities), to avoid inaccurate answer deviation. However, for crowdsourcing platforms that do not consider workers' reliability, we do not recommend assigning confirming-crowd as we cannot distinguish experts.

TRR utilizes NC task assignments as fewer assignments less than the minimum workers will not provide an informative answer. Moreover, it reduces the introduced latency problem discussed in Sect. 1. On the other hand, determining the task maximum assignments will prevent infinite task assignment iterations in case the task answer did not encounter convergence of workers' opinions. At the second step, the platform starts to assign the workload questions (tasks) to *NC* number of crowd members as an initial iteration of a task answers collection. At the third step, crowd members provide answers back to the TRR; TRR then checks whether the collected answers' diversity level is less than or equal DL (the targeted diversity level). At this step, TRR annotates tasks as '*completed*', '*not completed*' or '*closed*'. Reaching the predetermined DL signifies that all or most of the crowd workers agree on choosing a certain answer (i.e. no further task assignments); the task is annotated as a '*completed*' task reaching the fourth step. On the other hand, a task is annotated as '*not completed*' if its diversity level is greater than the targeted DL and did not reach the maximum number of assignments as well. TRR will then provide an estimate of the needed number of crowd members that might help to reach the targeted DL as a next iteration. Steps from two to four are repeated until there are no remained '*not completed*' tasks. '*Not completed*' tasks reaching the maximum number of task assignments are annotated as '*closed*' tasks. Tasks that are annotated as '*completed*' can be optionally assigned to answer-confirming crowd members (ACC) as an attempt to assure that the task has reached the correct answer when the confirming answer crowd members have *all* agreed to the same answer. Otherwise, this is an indication of having the "aggregated answer conflict" phenomenon. *Aggregated answer conflict* is a situation where workers of the first iteration agree to an answer (guided by the targeted DL) while it is not the correct answer. However, the wrong answer is corrected by hiring more workers. This situation will strongly appear if the first participating workers are performing random choices or they are not experts in the field leading to an agreement by coincidence.

The following methods can mitigate or even eliminate the effect of this conflict:

- Using reputation-based systems such as iCrowd [12], QASCA [13], and Docs system that estimates each worker reliabilities for several domains [14].
- Qualification test support: qualification tests [3, 15] are essential to determine the needed crowd member's qualifications when tasks are related to specific domains.

An intermingled approach using qualification test and reputation-based systems are promised to guarantee the best results of selecting the adequate workers. However, assigning the answer-confirming crowd members is a powerful utility of checking that we get the correct answer whenever there is an emphasis on the answer correctness.

4 Handling Boolean and Classification Tasks

In this section, we give the definitions related to the problem of this paper and the computations behind Boolean and classification task types.

Definition 1 (Boolean and Classification Question and the Question-Answer Set)
Let q_i denote a question, provided a possible set of labels (answers), denoted by $L = \{l_1, l_2, \ldots, l_m\}$. For Boolean questions, the set of labels contains only two labels, while labels extend to more than two labels in classification tasks. $A_{q_i} = \{A_{q_i,j}, \ldots, A_{q_i,k}\}$ denotes the answer set of all assigned workers to question q_i, where $A_{q_i,k}$ is the answer of worker k to q_i.

Definition 2 (Observed Question Answers Distribution Vector)
Vector V_{q_i} is a "1 × m" row matrix representing the observed probability distribution for the answers of question q_i, where m is the number of labels. V_{q_i} is computed based on workers' answers set A_{q_i}. Each cell $V_{q_i,j}$ denotes the observed probability distribution that label j is the correct answer for the question q_i, the summation of probabilities for any question q_i will equal one and the vector is initialized to zero.

Crowdsourcing platforms are either assigning tasks to anonymous workers such as [1] in which the worker quality is not evaluated nor considered or assigning tasks to non-anonymous workers such as [2] in which the worker quality is acquainted and computed automatically by the platform. Therefore, the entries of the observed data vector $V_{q_i,j}$ are computed differently according to the assigned worker model. The probabilities computation is provided in the following sub-sections.

4.1 Observed Probabilities Distribution in Anonymous Workers Model

In this working scheme workers request working on tasks, then task requester approves or rejects the worker; in case of rejection the platform re-announce the task for other workers, and in case of approval the worker is assigned the task. The answer set of question q_i for a worker is an ordered tuple with two components that are the worker Id and the worker answer. Upon answers submission, TRR starts to update the V_{q_i} vector according to support votes for each label. Consider L_{js} is the number of crowd members that have chosen label j as a correct answer for the question q_i. Thus $V_{q_i,j} = L_{js}/T_w$ (where T_w – total workers- is the total number of crowd members worked on question q_i so far).

4.2 Observed Probabilities Distribution in Non-anonymous Workers Model

In this working scheme workers' qualities are considered during the task assignment process that is performed automatically in the online task assignment systems such as [13, 14]. The value of V_{q_i} entries consider the worker's quality. q_i answer set is a set of ordered tuples with three components that are the worker's Id, worker's quality, and worker's answer. Work in [13, 14, 16] considered calculating $V_{q_i,j}$ as follows:

$$WA_{n,j} = \begin{cases} q^n, & for\ j = the\ worker\ n\ answer \\ \frac{1}{m-1}(1 - q^n), & for\ j \neq the\ worker\ n\ answer \end{cases}$$

$$V_{q_i,j} = \prod_{n=1}^{r} WA_{n,j}, \ r\ is\ the\ assigned\ workers\ number$$

Where q^n is worker n quality, m is the number of labels, and V_{q_i} vector is normalized.

4.3 Measuring Diversity for Boolean and Classification Tasks

TRR utilizes Gini-Simpson diversity index [10] to measure a question's answers diversity. A major reason for utilizing Gini-Simpson is its non-parametric nature (it doesn't require priors), in addition it takes into consideration evenness (workers' answers distribution) and richness (number of labels). A task diversity level denotes the level of divergence between the assigned workers' answers, and is defined as follows:

Definition 3 (Boolean and Classification Question Answers' Diversity)
For Boolean and classification questions Gini-Simpson diversity index is defined by·

$$DL_{q_i,r} = 1 - \sum_{j=1}^{m} \left(V_{q_i,j}\right)^2$$

Where m is the labels number of q_i, and r is the number of assigned workers to q_i.

4.4   TRR Redundancy Estimation Algorithm for Boolean and Classification Tasks

Tasks annotated as *'not completed'* (i.e. the task did not reach the required diversity level, and did not reach the maximum task assignments as well) are expected to be assigned to more crowd workers for another work iteration (i.e. new redundancy). TRR performs a simulation to compute how many assignments are minimally needed to reach the targeted diversity level (seeking convergence), assuming that the prospect next assigned crowd members will give the same answer with the highest probability to be the correct answer, as this is the minimum number of needed workers (i.e. the optimistic case). Algorithm 1 depicts how TRR estimates the next iteration redundancy for both working schemes. The algorithm starts with computing dl (i.e. the current question diversity) in lines 1–3. k (line 4) represents the allowable number of task assignments for q_i. k is identified as the difference between current redundancy (i.e. number of workers that have worked on the question q_i through previous work iterations) and the allowable maximum number of workers. In the non-anonymous working scheme, the algorithm maintains an active worker queue W_{q_i} providing qualities of workers that are willing to work on this question. W_{q_i} is considered the set of workers that have submitted a working request for q_i, otherwise, W_{q_i} is an empty input set when dealing with anonymous workers (lines 5–6). iCrowd [12] discussed a set of methods to keep track of the active workers' set such as considering workers currently holding tasks as active workers, otherwise, they are considered inactive workers. Another method is to consider a time window (e.g. 1 h), if the working request is less than the time window, the worker is considered an active worker. The algorithm assumes that the label with the highest probability is the correct label and there is a high probability that next workers will choose the same label. Algorithm1 updates the new probability distribution for the question answers (lines 7–22), then checks the answers divergence till reaching the targeted DL or reaching the allowable number of workers k. The algorithm outputs r (i.e. the new iteration estimated redundancy), in addition to the selected active workers who should work on the task in case of non-anonymous crowdsourcing platforms. A special case is that where the probability vector is equally distributed (i.e. each task answer have been voted for the same number of times). In such case TRR, assigns an extra single worker to explore the probable correct answer (line 9). Another solution is to make task assignments an odd number of workers to guarantees answer distinction. Our proposed model employs majority voting (MV) for truth inference (i.e. the task final answer) for generality, however, the Expectation maximization (EM) technique [3] provided a tuning for the management of human errors and recognizing the possible correct answer (i.e. truth). We leave the adoption of EM technique to future work. However, the impact of both techniques provides improved results for reducing the task assignments.

Algorithm 1 Redundancy Estimator $(W_{q_i}, V_{q_i}, MC, CR, DL$

Input: W_{q_i}: Active worker queue for question q_i; V_{q_i} observed probability vector; MC: maximum assignments; CR: current number of assigned workers to q_i; DL: required diversity level;

Output: r: worker redundancy; W_s: selected non-anonymous workers (for the non-anonymous working scheme);

```
1    Begin
2        r ← 0 , V_s ← V_{q_i}, W_s = ∅
3        dl ← Compute initial diversity of V_s vector
4        k ← MC – CR
5        If W_{q_i} ≠ ∅ Then // for non-anonymous working scheme
6            W_{q_i}← select top-k active workers sorted descendingly
7        While r <= k and dl >DL
8            If dl = 1 Then r = 1
9            Else
10               L_p ←select the highest probability in V_s
11               r ← r +1
12               If W_{q_i} ≠ ∅ Then        // Non-anonymous workers
13                   w ← Pop first worker in W //worker quality
14                   add w to W_s
15                   simulate w answers the same label of L_p
16                   update V_s according to w quality
17               Else                         // Anonymous workers
18                   simulate w votes for the label with L_p
19                   update V_s
20               compute dl
21           return r and W_s if not empty;
22   End
```

5   Handling Rating Tasks

Rating questions [17] are asking workers to rate answers for several classifications. Unlike Boolean and classification questions the TRR model provided previously cannot be utilized for rating questions as the worker rates several choices for the same task disallowing representing the answer selection as a discrete probability distribution. However, a rating question answers represent fuzzy sets [18] and the diversity of fuzzy sets can be measured [19].

Definition 4 (Rating Question and the Question-Answer Set)
Given a rating question q_i provided with a set of possible labels $L = \{l_1, l_2, \ldots, l_m\}$, where each label is rated given a min and max scale. $A_{q_i} = \{A_{q_i,j}, \ldots, A_{q_i,k}\}$ denotes the answer set of all assigned workers to question q_i. The worker j answer set for q_i is represented as $A_{q_i,j} = \{\mu_{j,1}, \ldots, \mu_{j,m}\}$, where $\mu_{j,i}$ is the rating of label i (membership degree) divided by the max rating scale to guarantee that $\mu_{j,i}$ lies between [0, 1].

5.1 Measuring Diversity for Anonymous Workers Model

The diversity of several rating answer sets represents "How diverse crowd answers are for a certain question". We adapted the normalized hamming distance in rating questions diversity computation due to its computational efficiency and distances interval lies between [0, 1] that can be later averaged to the same interval producing a percentage of answers' diversity. The diversity of a worker answers is measured by the distance between this worker answers and the average answers of all the assigned workers to the task [20]. The task diversity level is considered the average diversity of the assigned workers' answers diversities.

Definition 5 (Anonymous Worker Rating Answers Diversity)
Given the rating question q_i provided in Definition 4, the normalized hamming distance measuring worker j answers' diversity is defined by:

$$d\left(A_{q_i,j}, \bar{A}_{q_i}\right) = \frac{1}{m} \sum_{x=1}^{m} \left| \mu_{j,x} - \mu_{\bar{A}_{q_i},x} \right|$$

Where $\bar{A}_{q_i}$ is the average workers' answers set and $\mu_{\bar{A}_{q_i},x}$ is the average rating of label x in $\bar{A}_{q_i}$.

Definition 6 (Anonymous Rating Question Answers Diversity)
The diversity of q_i provided in Definition 5 of k assigned workers is defined by:

$$DL_{q_i,k} = \frac{1}{k} \sum_{j=1}^{k} d\left(A_{q_i,j}, \bar{A}_{q_i}\right)$$

Sharp rating labels that have the same value of all answer sets are discarded while computing the worker answers' diversity as it causes deviation of the actual value.

5.2 Measuring Diversity for Non-anonymous Workers Model

In the non-anonymous working scheme, the worker quality and the worker subjective rates influence the final question's diversity in two ways. The first way, they influence the average answer introduced in Definition 5 to be a weighted average answer [21] as the high-quality workers are expected to provide the more reliable answers and vice versa. The second way, the worker answers' diversity of a question is influenced to be a weighted diversity reflecting the worker's reliability.

Definition 7 (Rating Question-Answer Weighted Average) [21]
Given a rating question q_i of m labels provided the workers' quality vector of k workers $W_q = \{q^1, \ldots, q^k\}$, *where* $\sum_{j=1}^{k} q^j = 1$. A label l_i weighted average answer is defined by $\bar{l}_{wi} = \sum_{j=1}^{k} q^j \mu_{j,i}$, and the question q_i answer weighted average is provided by $\bar{A}_{wq_i} = \{\bar{l}_{w1}, \bar{l}_{w2}, \ldots, \bar{l}_{wm}\}$.

TRR measures the non-anonymous workers' answers diversity utilizing Definition 5, however, it evaluates the worker answers' against the answer weighted average provided in Definition 7 instead of the regular answer average. The question's diversity is measured by the total weighted workers' answers diversities.

Definition 8 (Non-anonymous Rating Question-Answers Diversity)
The diversity of the rating question q_i provided in Definition 7 is defined by:

$$DL_{q_i,k} = \sum_{j=1}^{k} q^j d\left(A_{q_i,j}, \bar{A}_{wq_i}\right)$$

Utilizing the workers' qualities has a threefold impact on TRR. The first, by collecting more accurate final question answer via being biased to high-quality workers who tend to provide more accurate ratings. The second, by providing a more accurate diversity measure. The third, downsizing the outlier worker's ratings influence to the final question answer, and the measured question diversity as well.

For Boolean and classification tasks TRR predicts the needed number of workers for next work iteration as it can predict the correct answer based on the assigned workers' choices. However, TRR cannot perform this prediction for rating tasks. The correct rating question answer is considered an average of the assigned workers' ratings. However, performing a simulation of assigned workers with an average answer does not make the diversity converge. As a consequence, TRR performs monotonic worker assignment after the first iteration by performing one by one worker assignment.

6 Experiments

The goal of our experiments is to understand the effectiveness and efficiency of applying TRR to crowdsourcing platforms. This is achieved by analyzing the workload cost and the elapsed time to resolve questions for both anonymous and non-anonymous crowdsourcing platforms against crowdsourcing platforms utilizing fixed task redundancy. We seek to provide an answer to the following inquiries: (1) Whether the algorithm will produce the same task "answer" when it halts the assignment process early in the case of "completed tasks" against the model assigning fixed redundancy to tasks (Answer in Sect. 6.4 for rating tasks. The detailed experiment for Boolean and classification tasks is not included due to space limitation, however, results are highlighted in Sect. 6.2). (2) We need to study the cost and latency in both systems (Answer in Sects. 6.3, and 6.4). (3) Conclude how to provide priors in real crowdsourcing platforms, particularly the average diversity level (Answer in Sect. 6.2).

6.1 Datasets

We experimented three real datasets, each of which represents one of the aforementioned task types. In all the experimented datasets a task is only one question.

Dataset 1 (Boolean Tasks): Crowdsourced Web Relevance Judgments Dataset
In this dataset [22], AMT workers judged the relevance of a set of Web pages. Relevance is judged on a binary scale: relevant, and non-relevant representing Boolean tasks. The primary version of the dataset produced in 2010 contains full information about the assignment process such as (task submission time, accept and rejection time) that are critical in our experiment to verify the model. The experiment elicited workers answers in the same order they occurred on the crowdsourcing platforms. The dataset contains 1000 task of 100 distinct questions posed to 149 workers where each question has been assigned to 10 workers (i.e. the fixed task assignment redundancy is 10). The dataset does not include worker quality, thus, we implemented the majority voting technique to compute and consider worker quality in our model.

Dataset 2 (Classification Tasks): Weather Tweets' Sentiment Analysis Dataset
The weather tweets' sentiment analysis dataset [23] is collected using AMT. The dataset contains 6000 classifications of the sentiment of 300 tweets, with gold-standard sentiment labels, answered by 110 workers. The sentiment judgments were provided in the following categories: negative (0), neutral (1), positive (2), tweet not related to weather (3) and cannot tell (4). Each task was assigned to 20 workers as a fixed data redundancy.

Dataset 3 (Representing Rating Tasks): News Headlines Emotion Analysis Dataset
The news headlines emotion analysis dataset [22] is used to study the effect of the TRR model to rating tasks. For each task crowd workers were presented with a news headline, the worker task is to score each headline for how much it holds regarding six specific emotions: anger, disgust, fear, joy, sadness, and surprise. Each of these is to be judged on a scale of 0–100, with 0 meaning "not at all", and 100 meaning "maximum emotion". The dataset presents 100 distinct questions that were posed to different 38 workers with total 1000 task assignments, with gold-standard ratings. The dataset represents rating tasks with fixed data redundancy of 10 workers per each task. The dataset does not include worker quality, thus, we implemented a "quality adjust" technique to induce workers' qualities. A worker answers a question correctly if he/she has provided ratings for the same classifications of the gold-standard answer with a limit of 30% rating deviation from the gold rating.

6.2 Estimating Priors

An accurate estimation for the TRR priors is a key to the model success; we conducted a study of the three datasets to estimate the adequate priors' values as follow:

1. The Minimum number of crowd members to be assigned to a task (NC): by inspecting datasets 1 and 2, setting NC to three assigned workers in the first dataset and four workers in the second dataset was sufficient to have an early correct consensus answer. Such an answer is equivalent to the same final answer collected from the fixed number of workers in each dataset.

2. The Maximum number of crowd members to be assigned to a task (MC): it is the fixed redundancy that was considered in the first place, that is 10 workers for datasets 1&3, and 20 workers for dataset 2.
3. The required answer diversity level (DL): We analyzed the diversity of questions by applying the fixed redundancy working model to reveal the average diversity that questions expose. Boolean, classification and rating tasks datasets have shown average diversity interval of (20%–42%) (i.e. on average around one-third the crowd members will deviate from the correct answer whatever the assigned number of crowd workers). Figure 2 illustrates the median diversity index computed while assigning a fixed redundancy of 10 workers for Boolean and rating task datasets, and 20 workers for the classification task dataset.

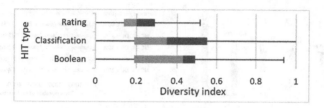

Fig. 2. Task's diversity index

4. Answer-confirming crowd (ACC): one or two high-quality crowd members are advised to be assigned in order to review the 'completed' tasks to avoid answer deviation.

6.3   Boolean and Classification Tasks Analysis

Task Cost Analysis
In this section, we explore the experimental evaluation of the model behavior for Boolean and classification task types' datasets while utilizing different NC values. We have compared our model to the baseline algorithm that utilizes fixed data redundancy for all tasks.
Boolean tasks' cost analysis
Dataset 1 provides 1000 Boolean task of varied prices (0.01$ for 800 tasks and 0.02 $ for 200 tasks) with total cost 1200 cent. Applying TRR to this dataset have coined a significant change in cost reduction.

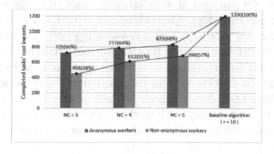

Fig. 3. Boolean tasks' cost analysis

Figure 3 depicts the total cost of the two working modes (anonymous, and non-anonymous workers) while varying different values of NC (3, 4, and 5), fixing the values of MC to 10 workers and DL to 35% (representing the median diversity). We did not include NC = 2 as two workers are not enough to provide informative judgment and caused the aggregated answer conflict problem. The cost reduction varied from 62% to 32% of the task's total cost utilizing different parameters' values.

Classification tasks' cost analysis

Dataset 2 utilizes fixed redundancy of 20 workers per task. Tasks' cost analysis of applying TRR to dataset 2 is shown in Fig. 4. The experiment presents varied values of NC, DL = 35%, and MC = 20 workers. Numbers bars represents the workload cost.

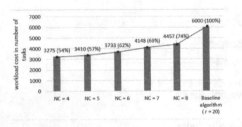

Fig. 4. Classification tasks' cost analysis

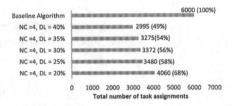

Fig. 5. Classification tasks' cost analysis of varied DL values

The relationship between the total needed task assignments and the diversity level is an inverse relationship. Lowering the required diversity level requires more task assignments to reach such opinion convergence. Figure 5 depicts the total cost (in terms of task assignments) of dataset 2 under different DL levels.

Task Working Time Analysis

TRR reduces task assignments, as a consequence, the working time to produce the workload answers will be reduced as well. Dataset 1 of web relevance judgments provided critical timing information of the task assignment process including task creation time, task acceptance time, task assignment approval time, worker elapsed time to finish the task, and the task expiration time. Timing information allowed us to study how much saved time has been accomplished by applying TRR during the task assignment process. Dataset 1 tasks consumed eight working days starting from Thursday '18th February 2010' till 'Thursday 25th February 2010' of total 18 working hours, tasks lifecycle elapsed time varied from few hours to few days. The dataset tasks were announced on the platform at simultaneous times thus, there was no latency saving in terms of working days. Meanwhile, we would run into situations where workers would take a long time to pick up a task causing latency that can be avoided by applying our TRR model. The intrinsic time saving was the human working time that was not assigned to this task alleviating work pressure on the available crowd members as well as shortening the working time for task requesters. Figure 6 depicts the elapsed working time of the 1000 task for dataset 1 while setting DL = 35%. The baseline algorithm of fixed data redundancy = 10 workers per task has a cost of 18.12 working hours. Numbers above each bar represents the total working time per hours.

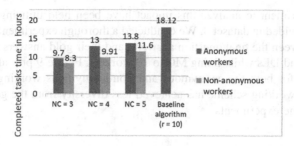

Fig. 6. Tasks' working time analysis

6.4 Rating Tasks Analysis

Task Cost Analysis

An experiment of assigning varied values of NC = 5, 6, 7, 8, and 9 where MC = 10 and DL = 10%, 15%, 20%, 25% and 30% (20% is the median diversity for rating tasks) has been performed to experiment the anonymous and non-anonymous working scheme. A comparison has been held between the final consensus answer when assigning 10 workers as fixed redundancy (baseline algorithm), and when applying TRR. We computed the correct task answer classification utilizing majority voting for the supported classifications by the assigned workers and considered its average rate. Rating tasks showed an aggregated final answer conflict upon assigning NC to two and three workers. The conflict down-streamed for NC = 4 and completely disappeared for setting NC = 5. Thus, we recommend setting NC as five on the first work iteration to avoid that problem.

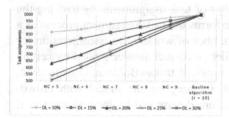
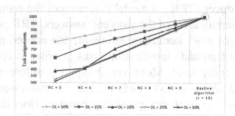

Fig. 7. Rating tasks' assignments for anony-   **Fig. 8.** Rating tasks' assignments for non-
mous workers   anonymous

Figure 7 depicts the cost saving between the baseline algorithm, and TRR anonymous working scheme for a variety of NC and DL values while, Fig. 8 depicts the non-anonymous working scheme.

Final Answer Correctness Analysis

Utilizing the workers' qualities has an impact on the aggregated final answer correctness via being biased to high-quality workers who tend to provide more accurate ratings. An experiment to analyze this impact have been held utilizing the questions gold answers provided in dataset 3. We conducted a thorough experiment that measures the diversity between the final question answers' and their gold answers using different values of NCs, and DLs while fixing MC to 10 workers. Figure 9 provides the average diversity results for both the anonymous and non-anonymous working scheme. The non-anonymous working scheme has achieved less diversity from the gold answers in almost most of the experiments.

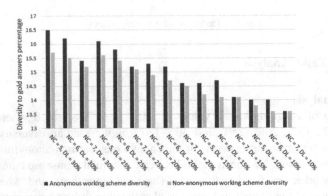

Fig. 9. Final answer diversity percentage against gold answers varying NC, and DL values

7   Conclusion and Future Directions

Reducing the number of crowdsourcing task assignments significantly reduces the overall cost, effort, and time to accomplish a task. We presented Task Redundancy Reducer (TRR), a model that reduces the number of task assignments by tracking the diversity level of the workers' answers. TRR performs the task assignment process as a set of work iterations, estimating the needed number of workers for each iteration for Boolean and classification tasks in order to collect a correct answer guided by certain diversity level. Moreover, it supports different types of tasks (Boolean, classification, and rating tasks). TRR provides a framework to integrate the model with crowdsourcing platforms providing concise priors that are determined by the task requester.

However, the model would benefit from several extensions: (1) Providing a technique to predict the needed number of crowd members of subsequent iterations of rating tasks instead of monotonic increase (one-by-one assignments). (2) Extending the model to other task types such as macro-tasks where it is difficult to define and recognize a consensus answer. (3) Utilizing other truth inference algorithms (i.e. optimization, and the probabilistic graphical model).

References

1. Amazon Mechanical Turk (2005). https://www.mturk.com/
2. CrowdFlower (2009). http://www.crowdflower.com/

3. Zheng, Y., Li, G., Li, Y., Shan, C., Cheng, R.: Truth Inference in Crowdsourcing : Is the Problem Solved ? PVLDB 3(3.0), 541–552 (2016)
4. Karger, D.R., Oh, S., Shah, D.: Iterative learning for reliable crowdsourcing systems. In: Advances in Neural Information Processing Systems 24 (NIPS) (2011)
5. Karger, D.R., Oh, S., Shah, D.: Budget-optimal task allocation for reliable crowdsourcing systems. Oper. Res. 62(1), 1–24 (2014)
6. Liu, X., Lu, M., Ooi, C., Shen, Y., Wu, S., Zhang, M.: CDAS: a crowdsourcing data analytics p system. PVLDB 5(10), 1040–1051 (2012)
7. Abraham, I., Alonso, O., Kandylas, V., Patel, R., Shelford, S., Slivkins, A.: How many workers to ask? adaptive exploration for collecting high quality labels. In: ACM SIGIR, p. 473 (2016)
8. Franklin, M.J., Kossmann, D., Kraska, T., Ramesh, S., Xin, R.: CrowdDB: answering queries with crowdsourcing. In: SIGMOD, pp. 61–72 (2011)
9. Haas, D., Ansel, J., Gu, L., Marcus, A.: Argonaut: macrotask crowdsourcing for complex data processing. PVLDB 8(12), 1642–1653 (2015)
10. Jost, L.: Entropy and diversity. Oikos 113(2), 363–375 (2006)
11. Zhang, C.J., Zhao, Z., Chen, L., Jagadish, H.V.: Cao CC. CrowdMatcher: crowd-assisted schema matching. In: SIGMOD, pp. 721–724 (2014)
12. Fan, J., Tan, K.: iCrowd : an adaptive crowdsourcing framework. In: SIGMOD, pp. 1015–1030 (2015)
13. Zheng, Y., Wang, J., Li, G., Cheng, R., Feng, J.: Berkeley UC, pp. 1031–1046. A quality-aware task assignment system for crowdsourcing applications. SIGMOD, QASCA (2015)
14. Zheng, Y., Li, G., Cheng, R.: DOCS: domain-aware crowdsourcing system. PVLDB 10(4), 361–372 (2016)
15. Jain, A., Das Sarma, A., Parameswaran, A., Widom, J.: Understanding workers, developing effective tasks, and enhancing marketplace dynamics: a study of a large crowdsourcing marketplace. PVLDB 10(7), 829–840 (2017)
16. Guo, S., Parameswaran, A., Garcia-molina, H.: So who won ? dynamic max discovery with the crowd. In: SIGMOD (2012)
17. Khanfouci, M., Nicolas, G.: Consensus-based techniques for range-task resolution in crowdsourcing systems. In: EDBT/ICDT Workshops (2017)
18. Dubois, D., Prade, H.: Fundamentals of Fuzzy Sets. Springer (2000)
19. Szmidt, E., Kacprzyk, J.: Distances between intuitionistic fuzzy sets. Fuzzy Sets Syst. 114, 505–518 (2000)
20. Solanas, A., Selvam, R.M., Leiva, D.: Common indexes of group diversity: upper boundaries. Psychol. Rep. 111(3), 777–796 (2012)
21. Zeng, S.: Some intuitionistic fuzzy weighted distance measures and their application to group decision making. Group Decis. Negot. 22(2), 281–298 (2013)
22. Crowd database group. http://dbgroup.cs.tsinghua.edu.cn/ligl/crowddata/
23. Weather sentiment analysis dataset. https://eprints.soton.ac.uk/376543/

Software Resource Recommendation for Process Execution Based on the Organization's Profile

Miller Biazus[1] , Carlos Habekost dos Santos[1] , Larissa Narumi Takeda[1] ,
José Palazzo Moreira de Oliveira[1] , Marcelo Fantinato[2] , Jan Mendling[3] ,
and Lucinéia Heloisa Thom[1(✉)]

[1] Institute of Informatics, Postgraduate Program in Computer Science,
Federal University of Rio Grande do Sul, Porto Alegre, Brazil
{mbiazus,cfhsantos,lntakeda,palazzo,lucineia}@inf.ufrgs.br
[2] School of Arts, Sciences and Humanities, University of São Paulo, São Paulo, Brazil
m.fantinato@usp.br
[3] Vienna University of Economics and Business, Vienna, Austria
jan.mendling@wu.ac.at

Abstract. Lack of information on the infrastructure resources needed
to execute business processes may interfere with the execution flow of the
BPM lifecycle phases. If an organization recognizes that it does not have
the resources needed to execute a process as planned, it might have to
redesign the process. This paper presents an approach to recommending
the infrastructure resources needed to execute a process. The recommendation relies on the task labels of the process model and comprises two
phases: *resource type classification* and *resource recommendation*.

The approach contributes to the redesign phase as it provides the
process analyst with information on the resources needed to execute the
process. It also supports decision-making process before the implementation phase regarding, for example, remodeling, project cancellation,
resource procurement etc. The developed approach was validated based
on a set of real processes of a public university through a cross-fold validation that reached 83% of accuracy.

Keywords: Business processes · BPMN · Recommender systems ·
Process mining · Machine learning · Systematic literature review

1   Introduction

Business Process Management (BPM) is a discipline that aims to support organizations in carrying out their work to ensure the expected results. According
to Dumas et al. [7], the BPM lifecycle comprises the following phases: *process
identification, process discovery, process analysis, process redesign, process implementation,* and *process monitoring.* Organizations do not always have the proper
infrastructure to support the execution of *to-be* processes as modeled [6]. In this

© Springer Nature Switzerland AG 2019
S. Hartmann et al. (Eds.): DEXA 2019, LNCS 11707, pp. 118–128, 2019.
https://doi.org/10.1007/978-3-030-27618-8_9

case, process analysts and systems engineers should be aware of the *infrastructure resources* needed to execute a process before reaching the implementation phase. In this paper, we use only the word *resources* to refer to the expression *infrastructure resources*. Resources needed for process execution may be software, hardware, and people. In this paper, we consider only the recommendation of software. Recommending the needed resources at the end of the *redesign phase* would overcome the lack of information at the *implementation phase*. Recommender systems comprise tools and techniques that suggest items most likely to be interesting to specific users [17]. Recommender systems often use machine learning and data mining techniques to predict a user's preference for a particular item or help users deal with information overload by filtering information.

In this paper, we present a *semi-automatic approach to recommending the resources needed to execute a process* aiming to improve the process implementation phase. The approach helps prevent system engineers from attempting to implement a modeling process without having the resources needed for the subsequent process execution. If the process analyst identifies a recommended resource is neither available in the organization's profile nor is targeted for acquisition, the process model must be redesigned to meet the available resources.

The paper is structured as follows: Sect. 2 presents related work. Section 3 details the proposed recommendation approach with its two phases of resource type classification and resource recommendation. Section 4 presents the approach validation. Section 5 concludes the paper.

2   Related Work

This section presents an analysis of the work related to the approach proposed herein. A quasi-systematic literature review was conducted to find studies on resource recommendation in BPM.

2.1   Search Protocol and Found Studies

We looked for studies that could help us answer the following research questions: *(i)* What *types of resources* have been addressed in approaches to recommending resources in BPM? and *(ii)* What *techniques or algorithms* have been applied in approaches to recommending resources in BPM?

The applied *search string* was: ("business process" OR "process model*" OR BPM OR workflow) AND ("resource recommend*" OR "recommending resource*" OR "recommendation of resource*"). The search was carried out using Scopus engine, in December of 2018, which indexes papers from the most important publishers such as IEEE Xplore Digital Library, Springer Link and ACM Digital Library. As the inclusion criterion, we selected the papers returned in the search whose main goal is resource recommendation in BPM published in any date. For the exclusion criteria, we removed papers that: *(i)* are not published in journals, conference proceedings or as book chapters; *(ii)* are secondary or tertiary studies; *(iii)* are not in English; *(iv)* are not available in digital libraries; *(v)* are duplicate or similar to another more complete paper; and *(vi)* are not full papers, i.e., short papers, poster abstracts etc. The references of the papers found in the search were added in the evaluation to complement the ones found by the applied research string.

We obtained 13 primary studies. In the oldest work identified, Liu et al. [14] presented an approach to reduce the amount of manual staff assignment performed at run-time workflow instantiation and execution stages. The approach uses three supervised machine learning algorithms to obtain a classifier: C4.5 decision tree, Naïve Bayes and Support Vector Machine (SVM). Several works have been proposed afterwards [2–5,8,9,11,13,18–20].

The most recent work, Abdulhameed et al. [1] used co-working history to predict human resources based on event log history. The approach determines the criteria and the metrics from event logs for resource recommendation, considering as criteria frequency and duration. The approach then gets information from the resources used to perform the same tasks in previous runs for the instance of the currently running process to recommend resources.

2.2   Requirements for Supporting Redesign

First, in terms of *types of resources* that have been addressed in the research on recommending resources in BPM: the results of the systematic review carried out clearly show that the vast majority of studies published in the last ten years have been mainly concerned with recommending human resources. Of the 13 related works found, nine approaches refer exclusively to human resources (people) while the other four approaches work with other types of resources, but mainly considering human resources. This strong feature of the existing recommendation approaches for BPM seems to be quite natural as there is a high cost in the human resources associated with the tasks of the business processes.

As for the techniques or algorithms that have been applied in the works oriented to recommend resources in BPM: the results of the systematic review show that, except for one paper, all 12 other papers refer to approaches based on data mining or process mining. Only one research does not depend on mining event logs to allow for the recommendation; instead, it applies a prioritization approach based on user preferences [5]. Using process mining means recommending resources through the support of techniques or algorithms that analyze data and extract relevant information from event logs derived from past executions of business processes. No pattern of techniques or algorithms was identified in these 12 papers proposing approaches to recommend resources based on process mining. In terms of technique for this event mining, although we tested decision trees, Naïve Bayes and SVM, the results of our tests indicated the Naïve Bayes as the best choice.

Our approach is oriented to the recommendation of the most appropriate types of resources to perform the tasks of the process according to what is available in the organization. We aim to reduce costs, not by optimizing the use of resources but avoiding the need to redesign the process to choose new resources. In this context, one natural choice is to address software resources, which is the kind of resource covered in this paper. In fact, according to our review's results, this is the first work to address the recommendation of the needed infrastructure resources to execute the tasks of business processes.

3   The Resource Recommendation Approach

This section introduces the semi-automatic methodology developed to recommend the resources needed to execute a process. The recommendation occurs at the end of the process redesign phase and is based on other process models of the organization and on the resources available in the organization.

Our research suggest software applications to be invoked by a BPM System (BPMS) to support the execution of a *to-be* process previously modeled. The recommendation occurs in the redesign phase when the process analyst can verify whether the organization has the resources needed to execute the process and then implement it. If the needed resources are not available, the organization may consider acquiring them or alternative resources need to be considered. Ultimately, the process model should not follow for implementation.

To recommend the resources, the following items must be defined before: (i) *Resource types*: identification of the type of each resource that can be used to execute tasks, such as browsers, printers, spreadsheet editors etc; (ii) *Organization's resources*: the list of all the resources available in the organization, accompanied by their types, acquisition cost and attributes. The attributes are defined at the discretion of each organization; (iii) *Organization's process models*: a set of process models, obtained from a historical base of the organization, with the definition of the resources used to execute their tasks.

The set of organization's process models must follow the 7PMG [15] guideline on task naming, which defines that tasks should be labeled in the verb-object format (e.g. "create the trip itinerary" instead of "creation of trip itinerary").

The resource recommendation occurs in two phases: *(i) resource type classification*, when process model tasks are classified by the types of resources needed to execute them; and *(ii) resource recommendation*, when, based on the types of resources needed for each task, the resources available in the organization are analyzed and recommended according to the organization interests.

3.1   Resource Type Classification

The resource type classification is carried out in four steps: *(i)* reading the process model and, for each task, extracting its types (e.g. manual task) and labels; *(ii)* normalizing task labels; *(iii)* applying the text classification algorithm; and *(iv)* calculating the classification accuracy.

In the first step, the *to-be* process model, designed in BPMN 2.0, should be parsed. For this, the model should be exported from the modeling tool to a format suitable for processing, what is usually done via XML (Extensible Markup Language). Several libraries can be used to manipulate XML files. File on *.bpmn* format is represented using XML, which is structured following specifications defined in OMG [16].

As for the second step, the normalization of task labels begins by *removing stop words*, which are less useful to index documents. As the process models considered in this work are in Portuguese, then the stop words are also in Portuguese in the next list. Two types of stop words can be considered [12]: *(i) Frequently used **function** stop words*: words with little relevance to represent meaning, e.g., *que* (that), *é* (is) and *em* (in). *(ii) Frequently used **content** stop words*: words with more semantic information than functional words, but commonly with a very high incidence, e.g., *querer* (want) and *através* (through).

Considering the scenario of this work, we removed only the function stop words. Since the average number of words in task labels is low (around three to five words), also removing content stop words could lead to an exaggerated decrease in the size of the labels. To support the removal of stop words, a natural language tool, the NLTK library was applied. Therefore, the process is structured as follows: for each word of task labels, we verified if it belongs to stop words. Normally, natural language tools already define the most common stop words of a specific idiom the library NLTK specifies words such as '*a*' (to), '*aquela*' (that), '*aquelas*' (those), as stop words.

Still about normalization, *stemming* should follow. Stemming refers to the removal of affixes of words (e.g. gerunds), keeping their stem (i.e. radical), to reduce a word to its essence It is assumed that a typical word has a stem that refers to a central idea or meaning and that certain affixes are added to change the meaning of the word or adjust it to its syntactic role. The morphological rules of natural languages lead to the use of different stemming algorithms according to the corresponding language. If applied in document pre-processing, stemming helps index and search documents, allowing to increase the level of document retrieval.

Stemming is used here to standardize words with the same stem, indexing them as equals. For example, the words *agenda* (schedule), *agendar* (to schedule) and *agendamento* (scheduling) are all indexed as *agend* (schedul), preventing text classification algorithms from interpreting them as different words and thus improving the classification accuracy. We applied the stemming procedure to all words that make up each task label.

As the third step, the tasks should be classified according to the resource types. For this, first the type of all tasks should be identified. Only the following BPMN task types should be considered for classification: *user*, *send*, *receive*, *business rule* and *abstract* tasks. BPMN *manual* tasks should not be classified as they are not executed with the support of software applications. BPMN *service* tasks should also not be classified as they are executed through web services or automated applications, which are not within the scope of this work.

Several algorithms for text classification are proposed such as Naïve Bayes, decision trees and SVM. Naïve Bayes stands out for its high accuracy even though it is conceptually simple. Depending on the data being processed, the algorithms' accuracy changes, and hence the choice of the algorithm is strongly related to the data to be classified [10]. To choose the best algorithm for this approach, we conducted a comparison among Naïve Bayes, decision trees and SVM in terms of classification accuracy. We used 521 task labels related to real-world processes of an university.

We used Weka to compare the accuracy of the classifiers. We measured the accuracy using all possible classes (i.e. all possible types of resources) for the instances. Thus, all the correctly classified samples were checked and split by the total number of instances that were to be classified.

The three algorithms had accuracy close to 80%. Naïve Bayes presented the highest accuracy, recall and precision, having, then, been chosen for this work.

As highlighted before, the algorithm execution assumes the existence of both: *(i)* pre-defined *types of resources* (such as spreadsheet, text editor, electronic agenda etc.) and *(ii)* a set of organization's *process models* with the definition of which *resources* (and hence *types of resources*) were used to execute their tasks, which is needed for the Naïve Bayes' supervised learning. As an example, consider that for a given process, the resource type *agenda* was used to perform the task *agendar entrevista* (schedule interview). The more accurate the classification of the process models that precede the execution of the algorithm, the better the training and hence the more accurate the classification of the algorithm.

Therefore, considering the Naïvee Bayes algorithm, first, the tasks of the set of organization's process models were used for training. After the training, the classification was performed on the tasks of another specific process. The algorithm calculates the probability that a task label word belongs to a given class (in this case, the class represents a resource type).

In the fourth step, to calculate the accuracy of the resource type classification, we used *k-fold cross-validation*, dividing the task labels into ten groups. In each round of validation, nine groups were used for training and the other for testing. The mean predictive accuracy was 81.7%.

3.2 Resource Recommendation

Based on the types of resources needed to execute the process tasks according to the classification carried out in the previous phase, the most appropriate resources for each type are recommended considering only those available in the organization. To perform the recommendation, a content-based filtering is performed, analyzing the resource attributes and the organization's profile.

The recommendation relies on the organization's resource profile. To define the profile of a particular organization, the relevant attributes (e.g. supplier, platform, license type etc.) of its resources must be identified. Based on the organization's profile, it is possible to identify which of its resources are best suited to support the execution of the process tasks and thus recommend them.

For each resource type identified for the tasks in the process, our approach recommends the most appropriate resource in the organization's profile, considering the similarities between the resource types and the resources available. To identify these similarities, an N-dimensional space is used to represent the organization's profile and the process for which the resources should be recommended. The organization's profile is represented by one point in the N-dimensional space defined as follows: considering all the resources available in the organization, we

calculate the *arithmetic mean* for each one of its attributes, generating a point in
the N-dimensional space. Each attribute represents one of the space dimensions.
A similar procedure is followed for the process to receive recommendations, i.e.,
for each task for which a resource type was identified, a point is calculated in
the N-dimensional space for the resources related to the corresponding type.

As an example, consider for training, three resources and their attributes, as
shown in Table 1. In this case, the organization's profile would be a set of five
points derived from the arithmetic mean of the three values of each attribute,
as presented in the last line of the table. The organization's profile would be
therefore a point with the coordinates [0.333, 0.333, 0.666, 0.666, 0.000].

Table 1. Example of attribute values for the resources

Resource	Attributes				
	Collabor.	Open-sour.	Web ver.	Cross-plat.	Intern
Microsoft Word	0	0	0	0	0
Google Agenda	1	0	1	1	0
Firefox	0	1	1	1	0
Arithmetic mean	0.333	0.333	0.666	0.666	0.000

The similarity for recommendation is then calculated using the *Euclidean
distance* between the organization's profile and some resource type of the process
for which the resources are to be recommended. The similarity is, therefore,
calculated through the distance between the point in the dimensional space that
represents the organization's profile and the points referring to the candidate
resources to be recommended, i.e., related to the resource types of the tasks
in the process. For example, assuming that a candidate resource is identified
by the point [0, 0, 0, 0, 1] in the dimensional space, then, its distance to the
organization's profile would be:

$$d(p,q) = \sqrt{(0 - 0.333)^2 + (0 - 0.333)^2 + (0 - 0.666)^2 + (0 - 0.666)^2 + (1 - 0)^2}$$

We need to compare this result with the results of the other resources candidates for recommendation considering the resource type for the same task in the
process. The resource with the lowest value is recommended (i.e. the one that
represents the shortest distance and hence more similar). The recommendation
should be carried out iteratively, for all the tasks of the process model, until all
the resources needed to execute the process are recommended.

4   Evaluation of the Recommendation Approach

This section presents the results of an experiment conducted to test the proposed
resource recommendation approach. The experiment was based on a set of real

process models related to one department of a public university. These process models were developed by students of a BPM course and semantically validated by actual process users. In addition, the course lecturer verified the robustness properties of the process models. The prototype developed to evaluate the resource recommendation approach was developed with Python language and Django framework for development. Python provides a wide variety of libraries to support natural language processing. Django provides features that facilitate rapid development, which is useful for proof-of-concept purpose. The resources available in the university department, as well as a set of other possible resources for acquisition, were all registered. Thereafter, we defined their types, as follows: E-mail client (Microsoft Outlook), Browser (Firefox); File manager (Google Drive), Bank app (Banco do Brasil App), Spreadsheet (Microsoft Excel), Allocation system (University Room Allocation System), Schedule (Google Agenda), Text editor (Microsoft Word), E-printer (PDF Printer).

The following attributes were considered to characterize the resources (cf. Table 1), based on the software features: *(i)* collaborative (or non-collaborative); *(ii)* open-source (or proprietary); *(iii)* having a web version (or not); *(iv)* cross-platform (or specific platform); and *(v)* internally developed at university (or not). Further, a cross-validation on the department's processes was carried out with 10 folds $(k = 10)$. In each cross-validation round, we created the organization's profile for the university department based on the process tasks used for *training* and using the means of the attribute values. As the attributes used are all binary (i.e., they assume 1 for positive and 0 for negative), then the profile for each round is composed of five values of 0 and 1.

Considering the *test* tasks as the recommendation target, the prototype recommended resources for the tasks in each of the cross-validation rounds. The resources were related to the resource types that the classifier Naïve Bayes suggested for the tasks. The recommendation was made by similarity between the organization's profile of the university department generated in each round and the resources registered in the prototype. The accuracy of the recommendation among all 521 tasks of the training processes after cross-validation was 83%.

Although it has been shown the possibility of recommending resources, to execute process tasks, based on the organization's profile, there are some remarks: a higher number of processes and hence tasks for training would tend to improve the accuracy of classification of resource types by Naïve Bayes; Naïve Bayes does not consider the relative position of the words on the task label, and we did not evaluate whether there would be any impact when using a classifier that considers such a relative position; as different students have developed the process models, there may be a higher linguistic bias than if there was only one modeler; we did not define the degree of relevance of the attributes; as a result, the weight was the same for all of them.

5   Conclusion

We have developed an approach to recommend resources aiming to minimize the problem of the compatibility of process models modeled in the process redesign phase and the organization's infrastructure. The recommendation could support the process analyst about the resources that are needed to execute a process, and with to the process owner, can take appropriate action, such as continuing Process Redesign, by adapting the process to the organization's infrastructure. With the implementation of the proposed approach, we can evaluate the accuracy of the recommendation, using cross-validation, with accuracy resulting as 83%.

The main contributions of this paper are as follow. First, we provide a comparative analysis among three text classifier algorithms in the context of process task labels. We used processes from a federal public university, and the accuracy of Naïve Bayes stood out compared to the other two analyzed algorithms (Supporting Vector Machines and Decision Trees). Second, we developed an approach for recommending resources needed to implement a specific process, based on process models, wherever it is proposed to train with the process task labels. Third, we conducted a systematic literature review, aiming to obtain the state of art about approaches to estimate or recommend BPM resources. The literature review also identifies the characteristics of the resources and the tasks, for a recommendation. As result, there is a discussion about the use of the term *resource*, in the selected publications.

As limitations, we identified that different students modeled the process models that we used to classifier and to test the recommendation, which can influence the classification, considering that there is no synonyms normalization, for example. We performed tests about the approach, realized in a single university department. This observation implies that meaning that context of the other organizations and process were not considered in this work.

As future work, we suggest customizing the approach, focusing on using collaborative filtering in the recommendation, i.e., recommends based on the organization with similar characteristics. Also, we suggest investigating the accuracy of the recommendation for process modeled in other idioms, besides Portuguese. We indicate to explore a recommendation focused on the most efficient resource for the organization, based on attributes and the organization's profile.

Acknowledgments. Lucinéia Heloisa Thom is a CAPES scholarship holder, Program *Professor Visitante no Exterior*, grant 88881.172071/2018-01; José Palazzo Moreira de Oliveira receive support from CNPq by grants 301425/2018-3 and 400954/2016-8; Carlos Habekost dos Santos and Larissa Narumi Takeda are scholarship holders from CNPq; Marcelo Fantinato is funded by FAPESP, grant 2017/26491-1; this study was financed in part by the CAPES - Brazil - Finance Code 001.

References

1. Abdulhameed, N., Helal, I., Awad, A., Ezat, E.: A resource recommendation approach based on co-working history. Int. J. Adv. Comput. Sci. Appl. **9**(7), 236–245 (2018)
2. Arias, M., Munoz-Gama, J., Sepúlveda, M., Miranda, J.: Human resource allocation or recommendation based on multi-factor criteria in on-demand and batch scenarios. Eur. J. Ind. Eng. **12**(3), 364–404 (2018)

3. Arias, M., Rojas, E., Munoz-Gama, J., Sepúlveda, M.: A framework for recommending resource allocation based on process mining. In: Reichert, M., Reijers, H.A. (eds.) BPM 2015. LNBIP, vol. 256, pp. 458–470. Springer, Cham (2016). https://doi.org/10.1007/978-3-319-42887-1_37
4. Brander, S., et al.: Refining process models through the analysis of informal work practice. In: Rinderle-Ma, S., Toumani, F., Wolf, K. (eds.) BPM 2011. LNCS, vol. 6896, pp. 116–131. Springer, Heidelberg (2011). https://doi.org/10.1007/978-3-642-23059-2_12
5. Cabanillas, C., García, J.M., Resinas, M., Ruiz, D., Mendling, J., Ruiz-Cortés, A.: Priority-based human resource allocation in business processes. In: Basu, S., Pautasso, C., Zhang, L., Fu, X. (eds.) ICSOC 2013. LNCS, vol. 8274, pp. 374–388. Springer, Heidelberg (2013). https://doi.org/10.1007/978-3-642-45005-1_26
6. Confort, V.T.F.: The BPM Issues in Brazilian Perspective. Master's thesis, Federal University of the State of Rio de janeiro, Brazil (2016)
7. Dumas, M., Rosa, M.L., Mendling, J., Reijers, H.A.: Fundamentals of Business Process Management. Springer, Heidelberg (2013). https://doi.org/10.1007/978-3-662-56509-4
8. Huang, Z., van der Aalst, W., Lu, X., Duan, H.: Reinforcement learning based resource allocation in business process management. Data Knowl. Eng. 70(1), 127–145 (2011)
9. Huang, Z., Lu, X., Duan, H.: Mining association rules to support resource allocation in business process management. Exp. Sys. App. 38(8), 9483–9490 (2011)
10. Jaiswal, R., Lokhande, S.: Analysis of early traffic processing and comparison of machine learning algorithms for real time internet traffic identification using statistical approach. In: Kumar Kundu, M., Mohapatra, D.P., Konar, A., Chakraborty, A. (eds.) Advanced Computing, Networking and Informatics- Volume 2. SIST, vol. 28, pp. 577–587. Springer, Cham (2014). https://doi.org/10.1007/978-3-319-07350-7_64
11. Koschmider, A., Yingbo, L., Schuster, T.: Role assignment in business process models. In: Daniel, F., Barkaoui, K., Dustdar, S. (eds.) BPM 2011. LNBIP, vol. 99, pp. 37–49. Springer, Heidelberg (2012). https://doi.org/10.1007/978-3-642-28108-2_4
12. Li, H., Chen, Q., Wang, X.: An improved method for semantic similarity calculation based on stop-words. In: Wang, X., Pedrycz, W., Chan, P., He, Q. (eds.) ICMLC 2014. CCIS, vol. 481, pp. 339–347. Springer, Heidelberg (2014). https://doi.org/10.1007/978-3-662-45652-1_34
13. Liu, T., Cheng, Y., Ni, Z.: Mining event logs to support workflow resource allocation. Knowl. Based Syst. 35, 320–331 (2012)
14. Liu, Y., Wang, J., Yang, Y., Sun, J.: A semi-automatic approach for workflow staff assignment. Comput. Ind. 59(5), 463–476 (2008)
15. Mendling, J., Reijers, H.A., van der Aalst, W.M.P.: Seven process modeling guidelines (7PMG). Inf. Softw. Technol. 52(2), 127–136 (2010)
16. OMG: Business process model and notation (BPMN), version 2.0 (2011)
17. Ricci, F., Rokach, L., Shapira, B.: Recommender systems: introduction and challenges. In: Ricci, F., Rokach, L., Shapira, B. (eds.) Recommender Systems Handbook, pp. 1–34. Springer, Boston, MA (2015). https://doi.org/10.1007/978-1-4899-7637-6_1
18. Sindhgatta, R., Ghose, A., Dam, H.K.: Context-aware analysis of past process executions to aid resource allocation decisions. In: Nurcan, S., Soffer, P., Bajec, M., Eder, J. (eds.) CAiSE 2016. LNCS, vol. 9694, pp. 575–589. Springer, Cham (2016). https://doi.org/10.1007/978-3-319-39696-5_35

19. Yang, H., Wen, L., Liu, Y., Wang, J.: An approach to recommend resources for business processes. In: Herrero, P., Panetto, H., Meersman, R., Dillon, T. (eds.) OTM 2012. LNCS, vol. 7567, pp. 662–665. Springer, Heidelberg (2012). https://doi.org/10.1007/978-3-642-33618-8_88
20. Zhao, W., Liu, H., Dai, W., Ma, J.: An entropy-based clustering ensemble method to support resource allocation in business process management. Knowl. Inf. Syst. **48**(2), 305–330 (2016)

An Experiment to Analyze the Use of Process Modeling Guidelines to Create High-Quality Process Models

Diego Torales Avila[1] , Raphael Piegas Cigana[1] , Marcelo Fantinato[2] ,
Hajo A. Reijers[3] , Jan Mendling[4] , and Lucineia Heloisa Thom[1]([⊠])

[1] Institute of Informatics, Postgraduate Program in Computing,
Federal University of Rio Grande do Sul, Porto Alegre, Brazil
{dtavila,rpcigana,lucineia}@inf.ufrgs.br
[2] School of Arts, Sciences and Humanities, University of São Paulo, São Paulo, Brazil
m.fantinato@usp.br
[3] Department of Information and Computing Sciences,
Universiteit Utrecht, Utrecht, The Netherlands
h.a.reijers@uu.nl
[4] Vienna University of Economics and Business, Vienna, Austria
jan.mendling@wu.ac.at

Abstract. Process modeling guidelines are an essential tool to help process modelers to create models that are correct and easy to understand. Many guidelines have been proposed in the literature, but there is little empirical evidence to which extent guidelines are effectively used. This paper addresses this research gap by presenting the results of a semi-controlled experiment conducted on two occasions with 21 students from a Business Process Management course. Two successive process modeling tasks were compared, one before and one after the subjects were presented to a set of 20 guidelines, which were collected through a systematic literature review. From the results obtained with the experiment, it was observed that the subjects would be more receptive to the guidelines if they were easier to understand and use.

Keywords: Process modeling · Process modeling guidelines · BPM · BPMN · Experiment

1   Introduction

Business process modeling is a difficult [7] but important task, in which a process analyst studies the business processes of an organization to create a representation – graphical, usually – of its activities, events and control flow logic [4]. The result is a process model, which may be used as a tool for learning, improvement and communication of the business process. While it is important that process models have high quality [13], they often have modeling issues, such as control

© Springer Nature Switzerland AG 2019
S. Hartmann et al. (Eds.): DEXA 2019, LNCS 11707, pp. 129–139, 2019.
https://doi.org/10.1007/978-3-030-27618-8_10

flow errors, badly designed structures and layouts, or incorrect labeling [6], which may significantly impair their understandability.

A frequent cause of these issues is the inexperience of process modelers [9], which can be lightened by the use of process modeling guidelines [6]. Guidelines are simple rules that help in creating more understandable process models and with fewer errors [7]. For example, a common modeling guideline is to use fewer modeling elements. Many guidelines are a result of experimental research that sought to understand what characteristics of process models influence their quality. Despite this, it is still uncertain whether process modelers, especially beginners learning to model, can successfully use guidelines to create better process models.

In this context, this paper reports an experiment in which the use of process modeling guidelines is analyzed for a process modeling task. We asked students of a process modeling course to create two process models, with only the second modeling task being supported by a set of modeling guidelines we collected from the literature. The data collected through this experiment was evaluated via statistical analysis. This experiment was executed twice with two sets of students and both datasets were merged for analysis and reporting. We present in this paper the protocol and the instruments designed for this experiment. We also exhibit the statistical analysis and the discussion of the results.

This paper is organized as follows: Sect. 2 provides background on process modeling guidelines and discusses other work related to this paper. Section 3 defines the protocol of the experiment, its hypotheses, design and instruments. Section 4 presents the results of the experiment, the test of the hypotheses and a discussion of the results. Section 5 concludes this paper with a summary and an outlook for future work.

2   Background

This section presents the fundamental background of our work. First, we present the set of process modeling guidelines that was used during our experiment. Second, we describe the related work on modeling guidelines.

2.1   Process Modeling Guidelines

Prior to our experiment, we have conducted a systematic literature review [1] in search of insights on important characteristics of high-quality process models that were interpreted as or transformed into a set of 45 modeling guidelines. These studies analyzed by the review did not share a common modeling notation among themselves, so all the extracted guidelines were adapted to the Business Process Model and Notation (BPMN) [11], which has been rising in popularity in recent years, as perceived throughout the review.

One characteristic discovered during this review was that not all guidelines were equally valuable or useful. Some of the guidelines we found were not studied in an empirical research to determine if they can improve the understandability of process models without changing their underlying behavior. Thus, these guidelines may be detrimental to the process modeling task, possibly even reducing the quality of the resulting process model. In our experiment, for example, they may have made it considerably longer and more difficult for its subjects. Therefore, we found it necessary to remove these guidelines.

Table 1 shows the set of guidelines we chose to use in our experiment. These guidelines were selected through a manual analysis, removing those that are possibly detrimental to the modeling task. We also removed those that were too similar to the guideline "Use as few elements as possible", since they could be considered redundant. The guidelines are arranged in four categories: *size*, which related to the size of the process model, *topology*, which contains guidelines on how model elements combine with each other, *layout*, which consists of conventions on how the process model should be visually presented, and *labeling*, which has instructions on how to label model elements.

2.2   Related Work

Defining what is process model quality has been a long-standing issue to which theoretical frameworks such as SEQUAL, SIQ and the Guidelines of Modeling (GoM) [2, 5, 13] were created. While insights provided by these frameworks are invaluable, they often define quality categories overly abstractly to be applied by novice modelers. In addition, the frameworks do not provide a straightforward method for their implementation in a process modeling project [7].

Creating more concrete and straightforward guidelines to be used in process modeling may solve this problem. One well-known work on modeling guidelines is the "Seven Process Modeling Guidelines (7PMG)" proposed by Mendling et al. [7]. It is notable for synthesizing a set of guidelines built upon empirical insights and contributing a ranking of them based on the opinions of expert analysts. This ranking solves the issue of when modelers have the opportunity to apply multiple guidelines that guide them to conflicting solutions.

Another important work is from Moreno-Montes de Oca et al. [10], in which a set of 30 modeling guidelines was presented to students that were asked to evaluate each one individually through its perceived ease of use, perceived usefulness and behavioral intention. The results were then compared against each other to find the highest scoring guidelines for these variables and their correlations.

Despite these studies, we found none that analyzes one of the main goals of modeling guidelines, which is to guide inexperienced process modelers to create more understandable process models. Thus, in our work, we use the set of modeling guidelines from Table 1 to evaluate whether this goal can be completed and what are the main challenges faced by inexperienced process modelers when using modeling guidelines.

Table 1. Process modeling guidelines used in this experiment (from literature review)

ID	Guideline	Category
S-1	Use fewer than 37 modeling elements	Size
S-2	Avoid using inclusive (OR) gateways	Size
S-3	Do not use implicit gateways	Size
S-4	Minimize the degree of all gateways	Size
T-1	Model as structured as possible	Topology
T-2	Do not create cycles with multiple exit points	Topology
T-3	Decompose models that are too large	Topology
T-4	Decompose model fragments that occur multiple times or that benefit from being grouped together or hidden	Topology
T-5	Do not overly decompose the process model	Topology
Ly-1	Minimize the drawing area of the model (preferably within a page)	Layout
Ly-2	Make the process flow from left to right	Layout
Ly-3	Minimize the number of bends in sequence flows	Layout
Ly-4	Minimize the crossing of sequence flows	Layout
Ly-5	Make use of symmetry between elements	Layout
Ly-6	Avoid overlapping elements	Layout
Ly-7	Keep model elements related to one another close to each other	Layout
Lb-1	Label everything necessary	Labeling
Lb-2	Use a consistent labeling style, such as: verb-object style for activity labels; object-particle style for event labels; and object-particle question style for gateway labels	Labeling
Lb-3	Avoid labels that are vague or ambiguous	Labeling
Lb-4	Use short labels	Labeling

3   Experiment Protocol

This section presents the research method applied to conduct this study, which is through an experiment. It displays the protocol used to conduct our experiment, which includes the definition of hypotheses and variables, the design of the experiment, the selection of subjects and instruments, and how the data collected was validated.

3.1   Problem Definition and Hypotheses

The influence that modeling guidelines have on process modeling is still an open issue. Since they are an additional concern to the task of modeling, they presumably affect cognitive load [14]. As such, they may increase extraneous cognitive load and block cognitive resources, making process modeling more difficult by requiring modelers to monitor not only the process being modeled but also whether the guidelines are being met. Consequently, if modelers believe they have more difficult modeling while using the guidelines, they may feel discouraged from using them again. Another possible effect would be modelers feeling the need to rely on some method or tool to support the use of guidelines. On the other hand, if the guidelines are formulated as clear instructions on how to model correctly, the increased cognitive load might be a germane cognitive load that helps the modeler in their task.

It is also unclear how effective a modeler can be when using modeling guidelines after being introduced to them. Pragmatically, modeling guidelines should be straightforward and well-founded rules that show how to create a better quality process model [7]. However, some guidelines found in literature have no explicit instructions as to when they can be applied; for example, when to use subprocesses. This imprecision can cause difficulties for modelers. Finally, modelers can perceive their process models with a higher level of understandability after using modeling guidelines, even though they have not been used correctly or other modeling issues still remain.

Considering these issues, we formulated three hypotheses in this paper: [H_1] guidelines increase cognitive load, which leads process modelers to a perception of higher degree of difficulty when modeling with the support of process modeling guidelines than without them; [H_2] process models created with the support of process modeling guidelines have *fewer modeling issues* than those without them; and [H_3] process modelers believe their process models have *higher level of understandability* when using process modeling guidelines than when not using them.

Besides these hypotheses, we searched for how receptive the modelers are to process modeling guidelines. They were specifically asked about how easy to use and how useful the guidelines are and if they intend to continue using them.

3.2 Experiment Variables

Based on the hypotheses, we defined three dependent variables: for H_1, we measured the *perceived level of difficulty* the modelers had during process modeling through a 5-point Likert scale, ranging from "very easy to model" to "very difficult to model"; for H_1, we measured the *perceived level of difficulty* the modelers had during process modeling through a 5-point Likert scale, ranging from "very easy to model" to "very difficult to model"; for H_3, we measured the *perceived level of understandability* of the process models, from the point of view of their modelers. This variable was also measured through a 5-point Likert scale, ranging from "very easy to understand" and "very difficult to understand".

Three additional dependent variables were defined for the modelers' receptivity to the modeling guidelines: perceived ease of use, perceived usefulness and future intended use. For each one, the subjects' opinions were measured using a 5-point Likert scale, ranging from "strongly disagree" to "strongly agree".

Personal factors, such as experience in modeling, are also a possible influence on the understanding and performance of subjects interacting with process models [3]. Therefore, we measured the subjects' experience through three independent variables: process modeling, BPMN and other process modeling notations. Each of these variables was measured using a 5-point Likert scale, ranging from "not experienced" to "very experienced" and their values were averaged to define the subjects' overall modeling expertise. Finally, the subjects were also asked whether they knew some set of process modeling guidelines, as such knowledge could also be an influence.

3.3 Experiment Design and Subjects

The goal of the experiment was to compare the performance of subjects in two process modeling tasks based on having or not the support of modeling guidelines. We gave the subjects textual descriptions of two processes, one for each step of the experiment. In the first step, the subjects were asked to model a first process. In the second step, they were presented to the list of modeling guidelines and encouraged to use them when modeling a second process. Since the order of which process would be modeled first could influence the results, the subjects were randomly separated into two groups, with the order of the processes alternated. Figure 1 shows the design of the experiment.

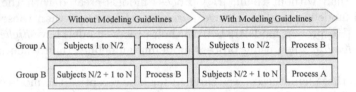

Fig. 1. The experiment design.

The subjects were students enrolled in an introductory course of business process management at a Brazilian public university. We selected students because, due to their inexperience, they might be more motivated to learn how to model processes better, which is a goal of modeling guidelines. The subjects were assumed to be familiar with the basics of process modeling and BPMN. An overview of the experiment was presented to the subjects, with general goals, procedures and time limits for each step. They were encouraged to create process models with quality in mind.

The experiment was executed twice. First, 13 subjects participated (divided into two groups). Then, the experiment was replicated with eight other subjects. Each execution was performed in a single laboratory with all subjects at the same time. The subjects had a limited time to perform each step of the experiment, whose total took an average of 80 min to complete. In addition, any questions the subjects might have about the procedure could be answered by the authors who were controlling the experiment.

3.4   Experiment Instrumentation

Four instruments were used during the experiment. The first one was the list of process modeling guidelines presented in Table 1, along with a small description for each guideline detailing how to apply it. The second instrument was the *Bizagi BPM modeler*[1], a modeling tool that is used during the university's Business Process Management course to learn process modeling. The third instrument was an on-line questionnaire that collected data measuring our independent and dependent variables. It also had open-ended questions where the students could provide reasoning for their answers and their opinions about the modeling guidelines.

The last instrument was the processes that would be modeled during the experiment. They came from a collection of real-world process models from a Brazilian public university. We sought in this collection two process models with complexities similar to each other and that could provide opportunities for the use of the modeling guidelines. The selected process models are medium-sized (i.e., over 20 elements), with at least one loop, a potential sub-process and multiple exclusive (XOR) gateways. We also ensured that the subjects had no in-depth prior knowledge of the selected processes. Finally, the selected process models were manually transcribed into a textual description.

3.5   Data Validation

All 21 subjects completed the experiment, and data collection through the questionnaire was successful. Although 42 process models were collected, eight models were excluded from the analysis of the hypothesis H2 (four from the first part of the experiment and four from the second part) because they contained serious syntax errors. These errors occurred because the subjects were unable to finalize the process modeling in the available time.

4   Data Analysis and Interpretation

This section presents the results of the experiment and its analysis, including some descriptive statistics, the hypothesis testing, and finally the discussion of the results.

4.1   Descriptive Statistics

All subjects reported knowledge of the 7PMG guidelines [7], which was expected by us, as they were introduced in the BPM course from which the subjects were recruited. The overall experience of both groups of subjects, calculated by averaging the three modeling experience variables, was similar. *Group A* had an average experience of 2.88 and standard deviation of 0.5, while *group B* had an

---

[1] www.bizagi.com/en/products/bpm-suite/modeler.

average of 2.97 and standard deviation of 0.67. We have not found any significant outlier, thus we can assume that these groups are homogeneous.

Figure 2 shows the distribution of the responses to the variables related to the hypotheses H_1 and H_3, respectively. After introducing the guidelines, there was a slight worsening in the *perceived level of difficulty*, but there was also an improvement in the *perceived level of understandability*.

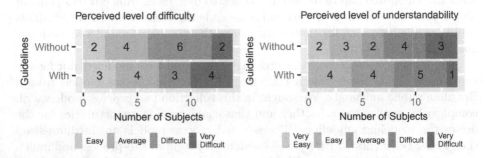

Fig. 2. Data collected for the *perceived level of difficulty* and *understandability*.

Regarding the hypothesis H_2, the subjects had an average of 7.35 *modeling issues* when modeling without guidelines and 7.94 when modeling with them. The standard deviation was 2.74 and 2.73, respectively. The increase in the average when modeling with guidelines goes against our expectations as we had assumed that the guidelines would help process modelers avoid modeling issues.

Figure 3 shows the responses regarding the receptiveness to the modeling guidelines. Although all subjects recognize the usefulness of the guidelines and almost all intend to use them again, some of them do not consider them easy to use. Through the open questions, some subjects addressed their difficulty in understanding how to apply some guidelines. One of the subjects argued that their questions could be clarified with practice and study.

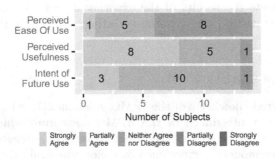

Fig. 3. Receptiveness to the modeling guidelines.

4.2   Hypothesis Testing

To address our three hypotheses, we tested if there was statistical difference between the results of each dependent variable for each step of the experiment, i.e., the two modeling tasks. To select the type of hypothesis test to apply, we first used the Shapiro-Wilk test, a powerful normality test [12], to check if the data collected was normally distributed. Then, the appropriate parametric or non-parametric test was chosen, depending on the type of the dependent variable.

For H_1 and H_2, the Shapiro-Wilk test confirmed that data for the *perceived level of difficulty* and *modeling issues found* are normally distributed. Thus, we applied a *one-sided paired t-test* [8], which is commonly used when the sample data comes from experiments with a paired design, such as this one. For H_3, the variable *perceived level of understandability* was not found to be normally distributed. Therefore, we applied a *one-sided Wilcoxon signed-rank test* [15]. For all three hypothesis, the tests showed that there is no significant difference between the two steps of the experiment. The resulting p-values were 0.6974, 0.2106, and 0.7288 (for H_1, H_2 and H_3 respectively), which are not significant at a significance level of 0.05.

4.3   Discussion

While all 21 subjects fully performed the experiment, it was not possible to find statistical support for the hypotheses pursued. The sample size is possibly a limiting factor for the statistical power of the tests carried out. Nevertheless, the experiment shows that it is possible to analyze the effect that process modeling guidelines have on the process modeling task through the protocol established to investigate our hypotheses.

More detailed information was identified in the responses to the questionnaire's open-ended questions. We realized through them how difficult it was for the subjects to effectively deal with and use the modeling guidelines in a quality-focused process modeling task. Many subjects reported they had to struggle to model the processes because many modeling elements were required. They blamed this over-effort mainly on an over-complexity of the processes. Further, one subject reported that the Bizagi tool impaired their organizational ability when working with a large number of modeling elements.

Subjects did not blame the difficulty to model processes on the use of (or the lack of) modeling guidelines. Instead, they reported that the processes *per se* were difficult to model. This also holds true for the quality of the process models they created. Only when asked directly on the modeling guidelines, some of them reported difficulty also in understanding and using the modeling guidelines.

This analysis is reflected through the data collected to assess the subjects' receptiveness to the guidelines, in which usability and intent of future use received good evaluations while ease of use received moderate ones. These results may mean that modeling guidelines require further refinement to make them easier to understand and use. One option would be to implement modeling guidelines directly in a modeling tool to support the process modelers during their work.

5   Conclusion

This paper reports on an experiment conducted to analyze the effects of using process modeling guidelines. These effects were measured based on the level of difficulty to model and the level of understandability of the resulting process models, both from the perspective of the modeler, as well as the effect on the number of modeling issues in the resulting process models. Two process modeling tasks were compared, one with and one without the support of modeling guidelines. Based on the results, it was not possible to provide significant evidence that the use of process modeling guidelines influences the measured variables. The best likely reason is the small sample size that may have affected the statistical conclusion validity.

In future research, this experiment can be improved to address the identified issues, focusing especially on strengthening the power of the statistical tests. Other approaches to applying the modeling guidelines should also be investigated, such as using a modeling tool to automatically verify whether a process model meets them. Finally, to address the issues modelers had related to ease of use, it seems valuable to analyze which modeling guidelines could be simplified or would demand further training.

Acknowledgments. This study was financed by the CNPq - Brazil and CAPES - Brazil - Finance Code 001. Diego Toralles Avila is a CAPES scholarship holder. Lucineia Heloisa Thom is a CAPES scholarship holder, *Program Professor Visitante no Exterior*, Process Number: 88881.172071/2018-0. We also acknowledge the Graduate Program in Computer Science as well as the Institute of Informatics, UFRGS. Marcelo Fantinato is funded by Fapesp, Brazil (grant number: 2017/26491-1).

References

1. Avila, D.T.: Process Modeling guidelines: systematic literature review and experiment. Master's thesis, Federal University of Rio Grande do Sul, Institute of Informatics, Graduate Program in Informatics, Porto Alegre (2018)
2. Becker, J., Rosemann, M., von Uthmann, C.: Guidelines of business process modeling. In: van der Aalst, W., Desel, J., Oberweis, A. (eds.) Business Process Management. LNCS, vol. 1806, pp. 30–49. Springer, Heidelberg (2000). https://doi.org/10.1007/3-540-45594-9_3
3. Dikici, A., Turetken, O., Demirors, O.: Factors influencing the understandability of process models: a systematic literature review. Inf. Softw. Technol. **93**, 112–129 (2018)
4. Dumas, M., Rosa, M.L., Mendling, J., Reijers, H.A.: Fundamentals of Business Process Management, 2nd edn. Springer, Heidelberg (2018). https://doi.org/10.1007/978-3-662-56509-4
5. Krogstie, J.: Model-Based Development and Evolution of Information Systems, 1st edn. Springer, London (2012). https://doi.org/10.1007/978-1-4471-2936-3
6. Leopold, H., Mendling, J., Gunther, O.: Learning from quality issues of BPMN models from industry. IEEE Softw. **33**(4), 26–33 (2016)
7. Mendling, J., Reijers, H.A., van der Aalst, W.M.P.: Seven process modeling guidelines (7PMG). Inf. Softw. Technol. **52**(2), 127–136 (2010)

8. Montgomery, D.C.: Design and Analysis of Experiments, 9th edn. John Wiley & Sons, Hoboken (2017)
9. Nelson, H.J., Poels, G., Genero, M., Piattini, M.: A conceptual modeling quality framework. Softw. Qual. J. **20**(1), 201–228 (2012)
10. Moreno-Montes de Oca, I., Snoeck, M., Casas-Cardoso, G.: A look into business process modeling guidelines through the lens of the technology acceptance model. In: Frank, U., Loucopoulos, P., Pastor, Ó., Petrounias, I. (eds.) PoEM 2014. LNBIP, vol. 197, pp. 73–86. Springer, Heidelberg (2014). https://doi.org/10.1007/978-3-662-45501-2_6
11. OMG: Business Process Model and Notation (BPMN) version 2.0. Technical report, Object Management Group (2011)
12. Razali, N.M., Wah, Y.B., et al.: Power comparisons of Shapiro-Wilk, Kolmogorov-Smirnov, Lilliefors and Anderson-Darling tests. J. Stat. Modeling Anal. **2**(1), 21–33 (2011)
13. Reijers, H.A., Mendling, J., Recker, J.: Business process quality management. In: vom Brocke, J., Rosemann, M. (eds.) Handbook on Business Process Management 1. IHIS, pp. 167–185. Springer, Heidelberg (2015). https://doi.org/10.1007/978-3-642-45100-3_8
14. Sweller, J.: Cognitive load theory. In: Mestre, J.P., Ross, B.H. (eds.) Psychology of Learning and Motivation, vol. 55, pp. 37–76. Elsevier, San Diego (2011)
15. Wilcoxon, F.: Individual comparisons by ranking methods. Biometrics Bull. **1**(6), 80–83 (1945)

Semantic Web and Ontologies

Novel Node Importance Measures to Improve Keyword Search over RDF Graphs

Elisa S. Menendez[1,3], Marco A. Casanova[1(✉)],
Luiz A. P. Paes Leme[2], and Mohand Boughanem[3]

[1] Department of Informatics, PUC-Rio, Rio de Janeiro, RJ, Brazil
{emenendez, casanova}@inf.puc-rio.br
[2] Computer Science Institute, UFF, Niteroi, RJ, Brazil
lapaesleme@ic.uff.br
[3] Institut de Recherche en Informatique de Toulouse,
IRIT, Toulouse, France
boughanem@irit.fr

Abstract. A key contributor to the success of keyword search systems is a ranking mechanism that considers the importance of the retrieved documents. The notion of importance in graphs is typically computed using centrality measures that highly depend on the degree of the nodes, such as PageRank. However, in RDF graphs, the notion of importance is not necessarily related to the node degree. Therefore, this paper addresses two problems: (1) how to define importance measures in RDF graphs; (2) how to use these measures to help compile and rank results of keyword queries over RDF graphs. To solve these problems, the paper proposes a novel family of measures, called InfoRank, and a keyword search system, called QUIRA, for RDF graphs. Finally, this paper concludes with experiments showing that the proposed solution improves the quality of results in two keyword search benchmarks.

Keywords: Keyword search · RDF · SPARQL · PageRank

1 Introduction

Keyword search is a well-known and convenient way for users to query large amounts of data, whether in Web pages or databases. The user simply types some terms, called *keywords*, and it is up to the system to retrieve the documents that best match the list of keywords. Search engines for Web pages popularized this kind of search. More recently, some of the Information Retrieval techniques used by Web search engines [17] were adapted to query databases to hide from users unfriendly SQL queries.

In the last decade, RDF emerged as a data model that represents data as a set of triples, which in turn induces a graph. This kind of modeling adds flexibility to describe resources and follows W3C standardized formats and ontologies. Considering that RDF graphs are interesting sources of knowledge that are also queried with unfriendly SPARQL queries, keyword search over RDF graphs (or briefly *RDF-KwS*) becomes a relevant research topic.

© Springer Nature Switzerland AG 2019
S. Hartmann et al. (Eds.): DEXA 2019, LNCS 11707, pp. 143–158, 2019.
https://doi.org/10.1007/978-3-030-27618-8_11

In Web Information Retrieval there are two main tasks: (1) matching keywords with indexed documents; (2) ranking the retrieved documents by order of relevance. RDF graphs present a further challenge, when compared to the Web, since the information that a user needs may not be in a single triple, but rather it is distributed over the graph. Hence, an answer for a keyword query over an RDF graph is better formalized as a minimal subgraph of the RDF graph that covers the keywords.

Summarizing, there are three main tasks in *RDF-KwS*: (1) finding pieces of information in the RDF graph; (2) assembling the retrieved pieces of information to compose complete answers; (3) ranking the complete answers. The main motivation of this work is how to construct an *RDF-KwS* system that covers these three tasks.

To achieve a good ranking mechanism, typical information retrieval systems rank the documents based not only on how well they match the keyword query, but also based on how important the documents are. The notion of importance for Web pages is typically computed using centrality measures for graphs created using the hyperlink structure of the Web. PageRank [6] and HITS [23] are some of the most popular centrality measures used in Web Information Retrieval. Their main idea is to assign high scores to pages that are referenced by many other important pages.

Returning to the RDF environment, the majority of the related work test their strategies using some RDF graph that reflects Web pages and their links [12, 15, 18, 21, 26], such as DBpedia[1], or using some dataset about co-authorship of research papers [3, 12, 33], such as DBLP[2]. We argue that PageRank or HITS variations work well for these types of RDF graphs because the incoming or outgoing edges indeed indicate the relevance of a resource. In the Web, it is reasonable that a Web page (or node) with several incoming edges is more important than a Web page with a few incoming edges. Likewise, in an RDF graph about research publications, the importance of an author is proportional to the number of accepted papers.

However, *RDF-KwS* operates over full RDF graphs, where the incoming or outgoing edges of a node do not necessarily indicate the node's importance with respect to any existing node relationship or, at least, it may be hard to detect which relationships would express the notion of importance. Thus, traditional measures may fail to compute the importance of a node. As an example, in an RDF graph representation of the Internet Movie Database – IMDb (www.imdb.com), instances of "common classes" (e.g. Genre, Language, Country, Company) have a high number of incoming edges. Hence, a traditional PageRank algorithm will assign scores to these common instances that are higher than the scores of popular movies and actors. Of course, we could manually assign weights to the object properties in order to capture their semantics, and use a Weighted PageRank or HITS Algorithm, as in [3, 10, 29]. However, one may argue that the manual assignment of weights is bothersome and subjective. Thus, other works focused on strategies to learn weights based on user feedback [1, 24, 27]. In addition to the difficulty of detecting relationships that express the importance of a graph node, it would be interesting to eliminate unwanted relationships that would distort traditional importance measures.

---

[1] http://dbpedia.org/sparql.

[2] http://dblp.uni-trier.de.

Summarizing, the problems addressed in this work are: (1) how to define importance measures in RDF graphs in which the importance of a node is not directly related to its degree; (2) how to use these measures to help compute and rank answers of keyword queries over RDF graphs.

To solve these problems, the first and key contribution of this paper is a novel family of importance measures for RDF graphs, collectively called *InfoRank*, that combine three intuitions: (I) *"important things have lots of information about them"*; (II) *"important things are surrounded by other important things"*; (III) *"few important relations (e.g. friends) are better than many unimportant relations (e.g. acquaintances)"*. They require neither the manual assignment of weights to object properties nor a training dataset to use as input to a learning algorithm.

The second contribution is an *RDF-KwS* system, called *QUIRA* (*QUerying with InfoRAnk*), which uses InfoRank: to narrow the retrieved pieces of information; to choose the best paths to connect the resources (nodes) in the graph; to rank the retrieved answers.

Finally, the third contribution of this paper consists of two enriched datasets, IMDb and MusicBrainz (http://musicbrainz.org), along with keyword search benchmarks adapted to the RDF environment. We use these datasets in our experiments to assess the correctness and the performance of InfoRank in the QUIRA system.

The rest of this paper is organized as follows. Section 2 summarizes related work. Section 3 defines the InfoRank measures. Section 4 describes the QUIRA keyword search system. Section 5 evaluates the performance of InfoRank in the QUIRA system. Finally, Sect. 6 contains the conclusions and suggestions for future work.

2 Related Work

Keyword Search over Structured Databases. Tools that implement keyword-based queries over relational databases [34] and RDF datasets have been investigated for some time. Since both fields have similar challenges, we discuss them together.

We may distinguish between tools that are *schema-based*, that use information about the conceptual schema to compile a keyword-based query into an SQL or SPARQL query, from those that are *graph-based*, which operate directly on the data.

BANKS [5] and BLINKS [16] are examples of relational graph-based tools, and Sindice [28] and *Structured LM* [11] are examples of RDF graph-based tools.

Relational schema-based tools explore the foreign keys declared in the relational schema to compile a keyword-based query into an SQL query with a minimal set of join clauses, based on the notion of *candidate networks* (CNs). This approach was first proposed in DISCOVER [19] and DBXplorer [2] and adopted in quite a few tools, including recent ones [9].

SPARK [37] offers an example of an early RDF schema-based tool. Tran et al. [31] combine the idea of generating summary graphs for the original RDF graph, using the class hierarchy, to generate and rank candidate SPARQL queries. QUICK [36] is a tool designed to translate keyword-based queries to SPARQL queries with the help of the users, who choose a set of intermediate queries, that the tool ranks and executes.

The *QUIOW* tool, our earlier implementation [13, 20], is schema-based and supports both the RDF and the relational environments by translating keyword queries into SPARQL or SQL queries. Although the tool proved efficient for an industrial dataset about petroleum, it had poor performance for an RDF graph representation of IMDb due to the large size and ambiguity of the domain. The importance measures introduced in this paper remediate these problems, as shown in Sect. 5.

Importance Measures for Structured Databases. ObjectRank [3] was one of the first proposals to compute a global importance score for database entities using PageRank. The authors transformed the structure of a relational database (RDB) into a graph, using foreign keys as links between entities, and then applied PageRank with manual weight assignment to different types of links. The authors evaluated their strategy using the DBLP dataset.

In RDF, other works that manually assign weights to use with PageRank are: Swoogle [10], which evaluated their strategy using documents crawled from the Web; Park et al. [29], which performs evaluation using their own small research dataset; and Beagle++ [7], which adapted ObjectRank to an RDF Graph about activity metadata in desktops.

TripleRank [12] represented an RDF graph as a tensor. Then, it used the PARAFAC decomposition of the tensor to induce groups of properties and resources, with authority and hub scores for the particular latent aspect (topic) the group represents. It showed how to use the result of the PARAFAC decomposition to guide a faceted browsing application. Finally, it tested the application in several experiments over RDF datasets with 5 to 55 thousand triples. PARAFAC decomposition proved interesting for faceted browsing exactly because it induces groups of properties and resources, together with authority and hub scores. However, it is not clear how to extend this strategy to the context of keyword search, not to mention the problem of computing the PARAFAC decomposition of tensors with 200+ million non-zero entries, as in the experiments described in Sect. 5.

More recently, FORK [24] adapted ObjectRank to Linked Data. The main contribution of the work is a learning algorithm for property weights based on user relevance feedback, instead of the manual assignment of weights. The authors evaluated their strategy using DBpedia and results showed that FORK achieves the best ranking method when compared to baseline approaches. Similarly, DBtrends [25] uses query logs to improve its ranking function.

As mentioned in the introduction, DBpedia and DBLP are highly influenced by link semantics: DBLP through authorship links, and DBpedia through links derived from Wikipedia, such as *wikiPageRedirects, wikiPageDisambiguates, primaryTopic, etc.* Furthermore, in the LOD cloud (http://lod-cloud.net), DBpedia has many incoming links from other RDF datasets.

For further references that focus on ranking strategies for degree-dependent datasets, such as DBpedia or DBLP, we refer the reader to [4, 30, 35]. We continue our discussion with some alternative strategies that do not highly depend on node degree.

Graves et al. [14] proposes the use of closeness centrality for undirected graphs and evaluates the strategy using three small datasets. The authors compare their strategy with a ranking using the number of incoming edges. The problem with closeness centrality is that it is not efficient for large RDF graphs.

Although the work presented in [22] is not specific to RDF graphs, it proposes the *degree decoupled PageRank* technique that penalizes or boosts the importance of the node degree in recommendation graphs, depending on the domain characteristics. They argue that, in some contexts, the importance of the node can be inversely proportional to its degree. The authors performed an evaluation using graphs extracted from IMDb, Last.fm, DBLP and Epinions. From results for the IMDb dataset, they noticed that, for a movie recommendation graph, traditional PageRank performs better; however, for an actor recommendation graph, the node degree actually needs to be penalized. They argue that, when an actor plays in a large number of movies, he probably is a non-discriminating ("B movie") actor, whereas, when an actor is associated with relatively few movies, he may be a more discriminating ("A movie") actor.

3 The InfoRank Importance Measures

3.1 Background on Importance Measures

Importance measures have as goal to identify the most important or central node in a graph, depending on what importance means. A simple way to compute the importance of a node is just to analyze its degree. However, this returns a local measure of importance, whereas in some contexts a global analysis of the graph is preferable. For instance, the *Betweenness Centrality* counts the number of shortest paths going through a node; hence it is able to identify important connectors in a graph. The *Closeness Centrality* measures the average distance from a node to all other nodes, hence the more central a node is, the closer it is to all other nodes.

Other types of importance measures try to capture the idea that "it is not about what you know, but who you know". That is, the notion of importance is given by how well-connected a node is to other important nodes. PageRank [6] is the most popular importance measure of this type. Using the hyperlink structure of the Web, the basic idea is that, if a Web page has links from other high-quality Web pages, then that is an indication that it is likely to be worth looking at the page.

PageRank can be computed using an iterative method, called *Power Iteration*. Let $G = (V, E)$ be a directed graph and $PR(r, i)$ be the PageRank score of a node $r \in V$ calculated at iteration i. First, the method initializes all scores with the same value:

$$PR(r, 0) = 1/N \tag{1}$$

where N is the total number of nodes in G. Then, for $0 < i < x$, it iterates until the computation of the score converges or exceeds x, the maximum number of iterations:

$$PR(r, i) = \frac{1 - \alpha}{N} + \alpha \sum_{s \in M_I(r)} \frac{PR(s, i - 1)}{d_O(s)} \tag{2}$$

where α is a dumping factor (usually set to 0.85), $M_I(r)$ is the set of nodes that have a link to r and $d_O(s)$ is the number of outgoing links from s.

One variant of PageRank uses link weights to give more importance for certain types of links. The *Weighted PageRank* PR_W is defined as:

$$PR_W(r,0) = 1/N \tag{3}$$

$$PR_W(r,i) = \frac{1-\alpha}{N} + \alpha \sum_{s \in M_I(r)} \frac{PR_W(s,i-1)}{d_O(s)} * w(r,s) \tag{4}$$

where $w(r, s)$ is a weight between 0 and 1 of edge $(r, s) \in E$.

3.2 The Intuitions Behind InfoRank

Following the intuition that *"important things have lots of information about them"* and observing the way that RDF graphs are modeled, we notice that more important nodes are usually associated with more literals (information) through datatype properties than less important nodes. As an example, in IMDb, a movie with international projection, such as *Titanic* (1997), has 205 literals with trivia, 134 literals with quotes said by the characters, 180 triples with tags, and so on. In fact, there are a total of 1,297 literals describing the movie *Titanic*. By contrast, a movie with only national projection, such as the Brazilian movie *O Auto da Compadecida*, has only 70 literals. Furthermore, in a multilingual dataset, such as DBpedia, *Titanic* has the label translated in many languages (e.g. Japanese, Russian, French, Spanish, etc.), while the Brazilian movie has the label only in Portuguese and English.

The second intuition that we follow is inspired by PageRank and says that *"important things are surrounded by other important things"*. For instance, *Titanic* has links through object properties with actors *Kate Winslet* and *Leonardo Dicaprio*, which are also important nodes in the graph. As in [14], we agree that, in RDF graphs, the direction of an object property does not have the same meaning as a Web hyperlink since a property is often found in its inverse form (e.g. directedBy/hasDirector). Given that, we treat an RDF graph as undirected and consider all neighbors of a node (i.e. all other nodes that have an object property linked to it) when propagating the importance with PageRank.

We further improve this intuition by introducing a third one that says "few friends are better than many acquaintances". As discussed in the introduction, the typical centrality measures are highly dependent on the degree of the node. In our work, we do not want to boost (or penalize) the degree importance, but we focus on a strategy that favors the quality of relations, rather than their quantity, that is, we prefer an approach that captures the notion that *"few important relations (e.g. friends) are better than many unimportant relations (e.g. acquaintances)"*.

3.3 Ranking Resources with InfoRank

Let T be a set of RDF triples. Assume that T contains schema information and that it is possible to identify the set C of *classes* defined in T, the set P of *object properties* defined in T, the set L of *literals* defined in T, and the set R of *blank nodes* and *(class) instances* defined in T, i.e., $r \in R$ iff there is a triple $(r, rdf{:}type, c) \in T$ such that $c \in C$.

Instance Informativeness. The level of "informativeness" of a resource measures how informative the resource is. As discussed in the previous section, information is represented as literals in RDF graphs. However, data resources (instances) usually have more literals than metadata resources (classes and properties). Hence, we first focus our strategy on the informativeness of instances.

The *informativeness* of an instance $r \in R$, denoted $IW(r)$, is defined as the number of triples of the form $(r, p, v) \in T$, where $v \in L$.

Ranking Schema Elements. Continuing our strategy based on instance informativeness, we say that "important classes usually have informative instances" and "important properties are usually those connecting informative instances".

The *InfoRank* of a class $c \in C$, denoted $IR(c)$, is defined as the maximum value of $IW(r)$ of all instances of class c. We will rank classes by descending order of $IR(c)$.

Likewise, the *InfoRank* of an object property $p \in P$, denoted $IR(p)$, is defined as the maximum value of $IW(r) + IW(s)$ of all triples of the form $(r, p, s) \in T$. We will rank object properties by descending order of $IR(p)$.

Ranking Data. Note that we used only *Intuition I* in our strategies to rank metadata resources. However, we propose a combination of the three intuitions to rank data, that is, the instances and blank nodes.

Let $r, s \in R$ and $p \in P$. Assume that $(r, p, s) \in T$ or $(s, p, r) \in T$, that is, ignore the direction of the object property p. The *normalized weight* of (r, p), denoted $W(r, p)$, is defined as:

$$W(r,p) = IR(p) / \sum\nolimits_{q \in P \text{ and } ((r,q,t) \in T \text{ or } (t,q,r) \in T)} IR(q) \qquad (5)$$

Note that the normalized weight $W(r, p)$ does not depend on "who" the neighbors of v are, but it depends only on how they are connected to r, that is, it considers the InfoRank scores of properties p and q.

Then, we compute PageRank using $W(r, p)$ as the edge weights:

$$PR_W(r,i) = \frac{1-\alpha}{N} + \alpha \sum\nolimits_{(r,p,s) \in T \text{ or } (s,p,r) \in T} PR_W(s, i-1) * W(r,p) \qquad (6)$$

where, as in Eq. (2), N is the total number of nodes in G and α is a dumping factor.

The InfoRank score of an instance r, denoted $IR(r)$, is the final PageRank score of r after x iterations, $PR_W(r, x)$, weighted by the informativeness of r, $IW(r)$:

$$IR(r) = PR_W(r,x) * IW(r) \qquad (7)$$

4 The QUIRA Keyword Search System

4.1 Overview

Recall that, given a graph G and a set M of nodes of G, a *Steiner tree* S for M is a tree whose nodes contain all nodes in M (and perhaps other nodes of G) and whose edges are edges of G. The Steiner tree S is *minimal* iff no other Steiner tree for M has fewer nodes than S.

As stated in the introduction, an answer of a keyword query over an RDF graph G is one or more minimal subgraphs that cover all keywords. Hence, a naïve approach to address the three main tasks would be: (1) find a set M of nodes of G that match all keywords; (2) find a minimal Steiner tree for M; (3) if there is more than one answer, rank the answers according to some criterion. Note that computing a Steiner tree avoids including unnecessary edges to connect the nodes.

There are two main problems with this approach that make it infeasible for most RDF graphs: (1) the set of nodes that match the keywords can be large; and (2) computing a minimal Steiner tree is an NP-complete problem.

Therefore, in previous work [13, 20], we described a tool, called *QUIOW*, that explores schema information to minimize these problems. The schema information is organized as a *schema graph*, as illustrated in Fig. 1. Without going into the details, in the first stage, *QUIOW* groups the keyword matches around classes, that is, *QUIOW* identities the properties whose values match keywords and creates groups of properties that have the same class as domain. In the second stage, *QUIOW* generates a Steiner tree for the set of classes found in the first stage over the schema graph (which is typically a small graph). In the third stage, *QUIOW* synthesizes a SPARQL query using the Steiner tree. Finally, the triplestore processes the SPARQL query synthesized to actually compute an answer to the keyword query.

In this work, we maintain the idea of grouping the matches in classes/properties to generate SPARQL templates. However, we completely reformulated the strategy to compute the templates to take advantage of InfoRank, as described in what follows.

4.2 Finding Pieces of Information in an RDF Graph

In this section, we present a greedy algorithm that takes keywords as input and returns the best set of class/property groups, as defined in Sect. 4.1.

Table 1 shows examples of groups in an IMDb dataset. The *count* column indicates that there are five movies named Titanic, one actress named *Kate Winslet* and four Episodes also named *Kate Winslet*. The *info_score* column is the aggregation of the InfoRank scores of all resources of a given group. For instance, all resources of group u_1 sum up to 0.0099 of InfoRank scores. Finally, group u_4 indicates that there is an *rdfs:Class* labeled Movie with score 1,468. We define a function *accum_score*(J, v) that simply counts the number of keywords from a set of keywords $J = \{j_1, j_2, ..., j_n\}$ that occurs in a literal value v. As an example, consider the keyword query $K = \{kate, winslet, titanic\}$ and the data in Table 1. The non-zero *accum* scores are:

$$accum\_score(\{kate, winslet, titanic\}, Kate\ Winslet) = 2$$
$$accum\_score(\{kate, winslet, titanic\}, Titanic) = 1$$

Table 1. Example of groups from IMDb.

Group	Class	Property	Value	info_score	Count
u_1	imdb:Movie	rdfs:label	Titanic	0.0099	5
u_2	imdb:Actress	rdfs:label	Kate Winslet	0.0010	1
u_3	imdb:Episode	rdfs:label	Kate Winslet	0.0000068	4
u_4	rdfs:Class	rdfs:label	Movie	1,468	1

Algorithm 1 presents an overview of a greedy strategy to obtain the best groups that satisfy a keyword query K. The strategy first gives priority to class matches. Then, it searches the groups looking for properties and data matches (e.g. *Titanic, Kate Winslet*).

Algorithm 1. Greedy Strategy to return the best set of groups that match a *Keyword Query*.

Input: A keyword query K and the set of groups U
Output: A subset of groups M
J = all keywords in K
M = empty list of groups
While J is not empty
    u = find in U a *class* group with the highest *accum_score* given J, use the highest *info_score* to disambiguate
    If a match is found
        add u to M, remove the keywords matched in u from J
    Else
        u = find in U a property or data group (i.e. *class* is *not rdfs:Class*) with the highest *accum score* given J, use the highest *sum_score* to disambiguate
        If a match is found
            add u to M, remove the keywords matched in u from J
    If J did not change
        break

As an example of the algorithm, consider again $K = \{kate, winslet, titanic\}$. In the first iteration of the while loop, $J = \{kate, winslet, titanic\}$ and the algorithm chooses group u_2. Although groups u_2 and u_3 have the same *accum_score* for J, the *info_score* is higher for u_2. In the second iteration, $J = \{titanic\}$ and the algorithm chooses group u_1, and the loop ends. At the end of this step, we generate SPARQL templates that satisfy the groups retrieved in Algorithm 1. The resulting templates for $K = \{kate, winslet, titanic\}$ are shown in Table 2.

Table 2. Templates generated for $K = \{kate, winslet, titanic\}$.

Template	Interpretation
?r1 rdf:type :Movie. ?r1 rdfs:label ?v1. **filter(contains(**?v1, 'titanic'))	All movies with label *titanic*
?r2 rdf:type :Actress. ?r2 rdfs:label ?v2. **filter(contains(**?v2, 'kate winslet'))	All actresses with label *kate winslet*

4.3   Connecting and Ranking

Connecting. The second task of the *RDF-KwS* process, i.e., connecting pieces of information, consists of finding a minimal Steiner tree between the classes of the groups retrieved in the first task. The Steiner tree is computed over the schema graph, a representation of the schema as in Fig. 1. Since the number of classes in an RDF Dataset is usually not large, it is feasible to compute a minimal Steiner tree.

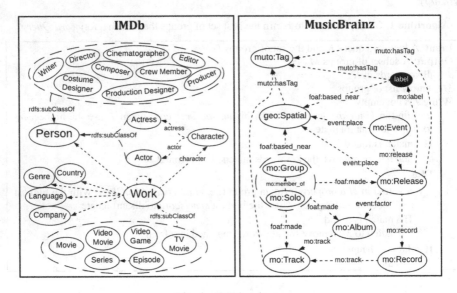

Fig. 1. IMDb schema.

Completing our templates example presented in Table 2, this step generates one more template (*?r1 ?p1 ?r2*), which says that a movie *?r1* and an actress *?r2* are connected through some property *?p1*.

Ranking. In the third task, i.e., ranking the results, we materialize triples together with the InfoRank score (e.g. *:Kate_Winslet :inforank "0.0010"*). Hence, we can generate templates for these triples (e.g. *?r1 :inforank ?s1, ?r2 :inforank ?s2*), and synthesize a SPARQL query with an ORDER BY clause that aggregates the scores of all instances from the templates. Finally, the following SPARQL query is synthesized for $K = \{kate, winslet, titanic\}$.

```
select * where {
    ?r1 rdf:type :Movie . ?r1 rdfs:label ?v1 . filter(contains(?v1, 'titanic'))
    ?r2 rdf:type:Actress . ?r2 rdfs:label ?v2 . filter(contains(?v2, 'kate winslet'))
    ?r1 ?p1 ?r2 . ?r1 :inforank ?s1 . ?r2 :inforank ?s2 . }
order by desc (?s1 + ?s2)
```

5 Evaluation

5.1 Setup

In order to evaluate our strategy, we downloaded the relational IMDb dataset (https://sites.google.com/site/ontopiswc13/home/imdb-mo) in MySQL and used Oracle 12c to transform it to RDF via R2RML. We used an RDF dump of MusicBrainz as our second dataset; however, since the given dump was incomplete, we enriched it with DBpedia information. The IMDb and MusicBrainz datasets have around 200 million triples. Figure 1 shows an overview of the schemas.

All experiments were conducted using a RESTful Web application developed in Java. The app ran on a macOS Sierra, 1,7 GHz Intel Core i5 RAM 4 GB. To store and manage the RDF data, we used Oracle 12c, running on a 2x deca-core Intel(R) Xeon(R) CPU E5-2640 v4 @ 2.40 GHz, 128 GB RAM, 32 KB Cache L1.

The datasets, benchmarks, and a detailed description of the experiments are available at the QUIRA Web page (https://sites.google.com/view/quira/).

5.2 Ranking Experiments

This section presents experiments to assess the potential of InfoRank as an importance measure to be used in a keyword search system over RDF graphs.

Table 3 presents the InfoRank score and the node degree of several classes and properties (i.e., metadata) from IMDb. We argue that, in an IMDb dataset, the most important classes are those that represent works (movies, TV series, etc.) and people (actors, actresses, directors, etc.), which is the result that InfoRank gives. Note that if we ranked the results using the degree, the order of classes would be Character, Person, Work; however, a typical IMDb user is likely to be more interested in movies and other type of works rather than in characters. Furthermore, the top properties are those connecting movies, such as follows/followed_by, which indicates that a movie is a sequence of another.

Table 3. IMDb metadata ranking computed by InfoRank.

#	Class	Info	Degree	Property	Info	Degree
1	imdb:Work	1,619	2,410,207	imdb:follows	2,538	332,551
2	imdb:Person	1,482	3,913,018	imdb:followed_by	2,538	332,548
3	imdb:Character	3	19,419,994	imdb:edited_from	2,538	14,103
4	imdb:Company	3	224,971	imdb:edited_into	2,538	14,103
5	imdb:Language	2	364	imdb:referenced_in	2,509	223,535
6	imdb:Country	2	319	imdb:references	2,509	223,532
7	imdb:Genre	2	46	….		

Table 4 shows the top 10 instances induced by PageRank and InfoRank. With PageRank, the top instances are highly connected nodes, such as countries, language and genres. However, we argue that, when considering a movies dataset, we would expect as top instances popular movies, series, actors, actresses, etc.

To indicate popularity, Table 4 also shows the users' rating of works extracted from the IMDb Web site. In the case of a person, we extracted the most rated work that she stared, directed, produced, etc. InfoRank results show highly rated work/person, such as *Star Wars*, *The Wizard of Oz*, *Titanic* and *Morgan Freeman*. The results show some TV Series with lower rates because they have a considerable level of informativeness (*General Hospital* – 375 literals; *Days of Our Lives* – 232 literals), and also a high degree through property *:episode_of_series*, since they have been on the air for a long time. Likewise, the results show some hosts from TV Shows that also have been on the air for a long time. Although InfoRank results show a few less popular works/people, we argue that InfoRank results correspond better to what users would expect in an IMDb dataset.

A similar scenario happens with MusicBrainz, in which the PageRank top instances also include countries. However, the InfoRank top instances include famous musicians, such as *Elvis Presley*, *Mozart*, *Beethoven*, *Bob Dylan*, etc.

Table 4. IMDb top 10 instances induced by PageRank and InfoRank.

#	PageRank			InfoRank		
	Instance	Class	User rating	Instance	Class	User rating
1	English	Language	–	Star Wars	Movie	8.6
2	USA	Country	–	Dolly Parton	Actress	6.8
3	Short	Genre	–	Jay Leno	Actor	5.3
4	Drama	Genre	–	Morgan Freeman	Actor	8.6
5	Comedy	Genre	–	The Wizard of Oz	Movie	8.0
6	Documentary	Genre	–	General Hospital	Series	6.7
7	UK	Country	–	Days of Our Lives	Series	5.3
8	Spanish	Language	–	Bob Barker	Actor	7.7
9	German	Language	–	Titanic	Movie	7.8
10	France	Country	–	Around the World in 80 Days	Movie	6.8

5.3   Keyword Search Experiments

To evaluate the impact of using InfoRank in a keyword search system over IMDb, we used all 50 queries (adapted to the RDF schema) from Coffman's IMDb Benchmark [8]. We ran versions of *QUIRA* using a variety of ranking measures. Table 5 presents the Mean Average Precision (MAP) [32], the total elapsed time and the number of iterations needed to compute the measures.

The measures in Table 5 include InfoRank, a version of PageRank considering the graph as undirected, the HITS Authorities, which prioritizes nodes with high in-degree, and HITS Hubs, which prioritizes nodes with high out-degree. We also include the Degree-decoupled (DD) PageRank [22] with a penalization parameter of 0.5. Note that we compared InfoRank neither with any approach that uses manually weighted links due to their subjectivity nor with approaches that learn weights from user feedback since we face the cold start problem. Moreover, we eliminated measures that are not computed efficiently in large graphs, such as the closeness centrality.

Table 5. IMDb results.

	Time (min)	Iterations	MAP
InfoRank	28	24	0.82
PageRank	27	30	0.76
HITS Authorities	25	12	0.73
HITS Hubs	25	12	0.30
DD PageRank p = 0, 5	38	37	0.54

Analyzing the results (not shown here for brevity), we noted that PageRank and HITS Authorities fail when choosing class Character, instead of class Work, in queries where a Steiner tree needs to be computed. They also fail in the ranking step for some keyword queries due to the high dependency on the degree. For example, Fig. 2 shows the results for PageRank and InfoRank for the query "actor terminator", whose expected results are the movies stared by *Arnold Schwarzenegger*. PageRank ranks first the voice actor *Jim Cummings* because his node has a high degree, since voice actors are usually cast several times, whereas InfoRank correctly returns the movies *Terminator 2: Judgment Day* and *The Terminator* starred by *Arnold Schwarzenegger*.

IRIT PUC QUIRA PageRank	actor terminator 🔍	IRIT PUC QUIRA InfoRank	actor terminator 🔍
Actor	**Work: terminator**	**Actor**	**Work: terminator**
Jim Cummings	The Turtle Terminator	Arnold Schwarzenegger	Terminator 2: Judgment Day
Maurice LaMarche	Terminator Tomato from Tomorrow	Arnold Schwarzenegger	The Terminator

Fig. 2. Result for query $K = \{actor, terminator\}$ in PageRank and InfoRank.

The HITS Hubs fails in all queries that refer to a person (e.g. *Denzel Washington*) since instances of class Person do not have outgoing edges. Furthermore, the Degree Decoupled (DD) PageRank fails because it penalizes instances with a high degree, whereas many important instances (e.g. *Star Wars*) have a high degree.

To summarize, InfoRank achieves the best MAP result in Coffman's IMDb Benchmark queries, since it successfully finds a balance between degree and informativeness. Furthermore, Table 5 indicates that these type of centrality measures, based on the Power Iteration method, can be computed in a feasible time.

Finally, we used 25 queries from QALD-2 (https://github.com/ag-sc/QALD) to evaluate the impact of InfoRank in a keyword search system over MusicBrainz. InfoRank achieved a MAP of 0.80 and PageRank a MAP of 0.75. For instance, PageRank gives a priority to music albums that have a higher number of tracks, since more tracks imply more links. However, we argue that the number of tracks is not necessarily related to the importance of an album.

6 Conclusions and Future Work

In this paper, we addressed two problems: (1) how to define importance measures in RDF graphs; (2) how to use these measures to help compute and rank answers of keyword queries over RDF graphs. To solve these problems, we proposed a novel family of measures, called InfoRank, and a keyword search system, called *QUIRA*, for RDF graphs. *QUIRA* uses the proposed importance measures: to narrow the retrieved pieces of information; to choose the best paths to connect the resources (nodes) in a graph; and to rank the retrieved answers. We concluded with experiments that show that the proposed solution improves the quality of results in popular keyword search benchmarks.

As future work, we plan to use InfoRank to improve Entity Linking and Entity Summarization solutions, to evaluate *QUIRA* with larger schemas, and to test ranking functions that take advantage of domain knowledge.

Acknowledgments. This work was partly funded by CAPES under grant 88881.134081/2016-01, by CNPq under grants 153908/2015-7, 302303/2017-0 and by FAPERJ under grant E-26-202.818/2017.

References

1. Agarwal, A., et al.: Learning to rank networked entities. In: Proceedings 12th ACM SIGKDD International Conference on Knowledge Discovery and Data Mining - KDD 2006, pp. 14–23 (2006)
2. Agrawal, S., et al.: DBXplorer: a system for keyword-based search over relational databases. In: Proceedings 18th International Conference Data Engineering, pp. 5–16 (2002)
3. Balmin, A., et al.: ObjectRank: authority-based keyword search in databases. In: Proceedings 13th International Conference on Very Large Data Bases - Volume 30, pp. 564–575 (2004)

4. Bast, H., et al.: Semantic Search on Text and Knowledge Bases. Foundation and Trends® in Information Retrieval, vol. 10, no. 2–3, pp. 119–271 (2016)
5. Bhalotia, G., et al.: Keyword searching and browsing in databases using BANKS. In: Proceedings 18th International Conference on Data Engineering, pp. 431–440. IEEE Computer Society (2002)
6. Brin, S., Page, L.: The anatomy of a large-scale hypertextual Web search engine. Comput. Netw. ISDN Syst. **30**(1–7), 107–117 (1998)
7. Chirita, P.A., et al.: Beagle ++: semantically enhanced searching and ranking on the desktop. In: The Semantic Web: Research and Applications - ESWC 2006, pp. 348–362 (2006)
8. Coffman, J., Weaver, A.C.: A framework for evaluating database keyword search strategies. In: Proceedings 19th ACM International Conference on Information and Knowledge Management, pp. 729–738 (2010)
9. De Oliveira, P., et al.: Ranking Candidate Networks of relations to improve keyword search over relational databases. In: Proceedings 31st International Conference on Data Engineering, pp. 399–410 (2015)
10. Ding, L., et al.: Swoogle: a search and metadata engine for the semantic web. In: Proceedings 13th ACM Conference on Information and Knowledge Management - CIKM 2004, pp. 652–659 (2004)
11. Elbassuoni, S., Blanco, R.: Keyword search over RDF graphs. In: Proceedings 20th ACM International Conference on Information and Knowledge Management - CIKM 2011, pp. 237–242 (2011)
12. Franz, T., Schultz, A., Sizov, S., Staab, S.: TripleRank: ranking semantic web data by tensor decomposition. In: Bernstein, A., et al. (eds.) ISWC 2009. LNCS, vol. 5823, pp. 213–228. Springer, Heidelberg (2009). https://doi.org/10.1007/978-3-642-04930-9_14
13. García, G.M., et al.: RDF Keyword-based query technology meets a real-world data set. In: Proceedings 20th International Conference on Extending Database Technology (EDBT), pp. 656–667 (2017)
14. Graves, A., et al.: A method to rank nodes in an RDF graph. In: Proceedings 7th International Semantic Web Conference, pp. 84–85 (2008)
15. Harth, A., Kinsella, S., Decker, S.: Using naming authority to rank data and ontologies for web search. In: Bernstein, A., et al. (eds.) ISWC 2009. LNCS, vol. 5823, pp. 277–292. Springer, Heidelberg (2009). https://doi.org/10.1007/978-3-642-04930-9_18
16. He, H., et al.: BLINKS: ranked keyword searches on graphs. In: Proceedings 2007 ACM International Conference on Management of Data - SIGMOD 2007, pp. 305–316 (2007)
17. Hiemstra, D.: Information retrieval models. In: Information Retrieval: Searching in the 21st Century, pp. 1–17 (2009)
18. Hogan, A., et al.: ReConRank: a scalable ranking method for semantic web data with context. In: Proceedings 2nd Workshop on Scalable Semantic Web Knowledge Base System (2006)
19. Hristidis, V., Papakonstantinou, Y.: Discover: keyword search in relational databases. In: Proceedings 28th International Conference on Very Large Databases, pp. 670–681. Elsevier (2002)
20. Izquierdo, Y.T., García, G.M., Menendez, E.S., Casanova, M.A., Dartayre, F., Levy, C.H.: QUIOW: a keyword-based query processing tool for RDF datasets and relational databases. In: Hartmann, S., Ma, H., Hameurlain, A., Pernul, G., Wagner, R.R. (eds.) DEXA 2018. LNCS, vol. 11030, pp. 259–269. Springer, Cham (2018). https://doi.org/10.1007/978-3-319-98812-2_22
21. Kasneci, G., et al.: NAGA: searching and ranking knowledge. In: Proceedings 2008 IEEE 24th International Conference on Data Engineering, pp. 953–962 (2008)

158   E. S. Menendez et al.

22. Kim, J.H., et al.: PageRank revisited: on the relationship between node degrees and node significances in different applications. In: Proceedings 5th International Workshop on Querying Graph Structured Data at EDBT/ICDT, pp. 1–8 (2016)

23. Kleinberg, J.M.: Authoritative sources in a hyperlinked environment. J. ACM **46**(5), 604–632 (1999)

24. Komamizu, T., Okumura, S., Amagasa, T., Kitagawa, H.: FORK: feedback-aware objectrank-based keyword search over linked data. In: Sung, W.K., et al. (eds.) Information Retrieval Technology AIRS 2017. Lecture Notes in Computer Science, vol. 10648, pp. 58–70. Springer, Cham (2017). https://doi.org/10.1007/978-3-319-70145-5_5

25. Marx, E., et al.: DBtrends: exploring query logs for ranking RDF data. In: Proceedings 12th International ACM Conference on Semantic Systems, pp. 9–16 (2016)

26. Mirizzi, R., Ragone, A., Di Noia, T., Di Sciascio, E.: Ranking the linked data: the case of DBpedia. In: Benatallah, B., Casati, F., Kappel, G., Rossi, G. (eds.) ICWE 2010. LNCS, vol. 6189, pp. 337–354. Springer, Heidelberg (2010). https://doi.org/10.1007/978-3-642-13911-6_23

27. Nie, Z., et al.: Object-level ranking. In: Proceedings 14th International Conference on World Wide Web - WWW 2005, pp. 567–674 (2005)

28. Oren, E., et al.: Sindice.com: a document-oriented lookup index for open linked data. Int. J. Metadata Semant. Ontol. **3**(1), 37–52 (2008)

29. Park, H., et al.: A link-based ranking algorithm for semantic web resources. J. Database Manag. **22**(1), 1–25 (2011)

30. Roa-Valverde, A.J., Sicilia, M.-A.: A survey of approaches for ranking on the web of data. Inf. Retr. **17**(4), 295–325 (2014)

31. Tran, T., et al.: Top-k exploration of query candidates for efficient keyword search on graph-shaped (RDF) data. In: Proceedings 25th International Conference on Data Engineering, pp. 405–416 (2009)

32. Turpin, A., Scholer, F.: User performance versus precision measures for simple search tasks. In: Proceedings 29th Annual International ACM SIGIR Conference on Research and Development in Information Retrieval, pp. 11–18 (2006)

33. Wei, W., et al.: Rational Research model for ranking semantic entities. Inf. Sci. **181**(13), 2823–2840 (2011)

34. Yu, J.X., et al.: Keyword Search in Databases. Morgan & Claypool, San Francisco (2010)

35. Yumusak, S., et al.: A short survey of linked data ranking. In: Proceedings 2014 ACM Southeast Regional Conference on - ACM SE 2014, pp. 1–4 (2014)

36. Zenz, G., et al.: From keywords to semantic queries - Incremental query construction on the semantic web. Web Semant. Sci. Serv. Agents W.W.W. **7**(3), 166–176 (2009)

37. Zhou, Q., Wang, C., Xiong, M., Wang, H., Yu, Y.: SPARK: adapting keyword query to semantic search. In: Aberer, K., et al. (eds.) ASWC/ISWC -2007. LNCS, vol. 4825, pp. 694–707. Springer, Heidelberg (2007). https://doi.org/10.1007/978-3-540-76298-0_50

Querying in a Workload-Aware Triplestore Based on NoSQL Databases

Luiz Henrique Zambom Santana$^{(\boxtimes)}$ and Ronaldo dos Santos Mello

Universidade Federal de Santa Catarina, Florianópolis, Brazil
luiz.santana@posgrad.ufsc.br, r.mello@ufsc.br

Abstract. RDF and SPARQL are increasingly used in a broad range of information management scenarios (*e.g.*, governments, corporations, and startups). Scalable SPARQL querying has been the main issue for virtually all the recent RDF triplestores. This paper presents *WA-RDF*, a middleware that addresses workload-adaptive management of large RDF graphs. Our middleware not only employs all the most used NoSQL data models but also provides a novel RDF data partitioning approach based on a fragmentation strategy that maps RDF data into multiple NoSQL databases. This workload-aware partitioning scheme provides, in turn, efficient processing of SPARQL queries over these NoSQL databases. Our experimental evaluation shows that the solution is promising, outperforming three recent baselines.

Keywords: RDF · SPARQL · NoSQL · Workload · Triplestore

1 Introduction

In the last decade, RDF, the standardized data model that, along with other technologies, like RDFS, and SPARQL, grounds the vision of the Semantic Web, was affected by a wide range of data management problems. The main reason for that is the current scale of Big Data intensive applications, which generates very large datasets and need to efficiently store massive RDF graphs that goes beyond the processing capacities of existing RDF storage systems operating on a single node. This scenario includes innovations in the frontier of Semantic Web research fields. For example, semantic technologies can enhance the storage of moving object trajectories [6], generating huge datasets about traffic, people behaviour and citizen routine. The scale of this kind of domain raises the need for new triplestores that can, for instance, take advantage of NoSQL databases to store and access large volumes of RDF data.

This paper presents *WA-RDF*, a triplestore composed of a middleware and multiple NoSQL databases. Our middleware includes a novel RDF data partitioning approach with a fragmentation strategy that maps pieces of an RDF graph into NoSQL databases with different data models. We consider a workload-aware partitioning approach based on the ideas from *Estocada* [1] to develop a

© Springer Nature Switzerland AG 2019
S. Hartmann et al. (Eds.): DEXA 2019, LNCS 11707, pp. 159–173, 2019.
https://doi.org/10.1007/978-3-030-27618-8_12

multiformat RDF storage that takes into account the query workload to decide which NoSQL data model is the best fit for each incoming RDF fragment.

The main contributions of this paper are: *(i)* a workload-aware RDF data partitioning approach based on the current graph structure and, mainly, on the typical application queries; *(ii)* a query processing mechanism that takes advantage of the partitioning approach to define efficient query planning to access RDF data; *(iii)* a set of experiments that evaluate our solution against three baselines (*Rainbow* [2], *ScalaRDF* [4] and S2RDF [8]) by considering the NoSQL databases MongoDB and Neo4J. Our strong point is the ability to process queries over large RDF graphs stored on multiple NoSQL database servers with a subtle amount data joining cost. The experimental evaluation shows that our middleware scales well.

The rest of the paper is organized as follows. Section 2 contains the background and related work. Sections 3 and 4 detail the *WA-RDF* approach. Section 5 reports the experimental evaluation and Sect. 6 concludes the paper.

2   Background and Related Work

The most important pillars of this work are the Semantic Web and the NoSQL databases movement.

Currently, the Semantic Web is defined mainly in terms of well-established standards for expressing shared meaning, defined by WWW Consortium (W3C)[1], like *Resource Description Framework (RDF)* and *the Simple Protocol and RDF Query Language (SPARQL)*. RDF is expressed by triples that define a relationship between two resources. RDF triples can be modeled as graphs, where the resources, called *subject* and *object*, are vertexes, and the relationship, called *predicate*, is a directed edge from the subject to the object. SPARQL is a query language for searching and retrieving RDF information. The most important part of a SPARQL query is the *triple pattern*, which defines the RDF subject, predicate and object to be searched. Moreover, sets of triple patterns define *Basic Graph Patterns (BGP)*, being each BGP a function that transforms the RDF datasets into the answer of a SPARQL query in the form of RDF triples. Traditionally, SPARQL queries can be categorized into *star*, *chain* and *complex* queries [8]. These query shapes depend on the location of the variables in the triple patterns, which can heavily influence the query performance [8].

There are many works that employs NoSQL systems for scalable RDF data management [5]. Among the recent works, we highlight *Rainbow* [2] (a polyglot NoSQL-based triplestore), *ScalaRDF* [4] (an in-memory solution) and *S2RDF* [8] (a scalable query processor). *Rainbow* is a distributed triplestore that uses the *HBase* columnar database and the *Redis* key-value database (K/V) as distributed storages to speed up query processing. Based on a previous analysis of the dataset and the expected workload, it decides on which NoSQL database the RDF data will be maintained. *ScalaRDF* introduces a distributed in-memory

---

[1] https://www.w3.org/.

triple store that uses *Redis* as a fault-tolerant and distributed RDF store. Additionally, *S2RDF* proposes a *Spark*-based SPARQL query processor that offers very fast response time for star queries by extending the vertical partitioning. Their partition scheme uses the *Apache Parquet*[2] columnar format to store the triples excluding unnecessary data from query processing. In order to reduce the intermediate results, S2RDF maintains statistics about the size of the dataset tables and places the subqueries corresponding to the smallest tables at the beginning of a joining in order to reduce the intermediate result size.

WA-RDF represents an advance on the state-of-the-art in sense that it is the first triplestore that considers the typical workload to decide which is the best NoSQL database to store an RDF triple.

3   WA-RDF

WA-RDF is a workload-aware middleware for storing and querying RDF data in multiple NoSQL database nodes. Its inspiration comes from *Estocada*, which argues that a mixed-model layer, relying on a set of diverse and heterogeneous data stores, can provide performance advantages for the applications using this layer. However, *Estocada* is neither a workload-aware approach nor a storage solution for RDF data. Another idea we borrowed from *Estocada* is a *fragment-based* storage that is entirely transparent to the client applications. It means that the data flow in *WA-RDF* is most of the time in the format of fragments. Figure 1 gives an overview of the *WA-RDF* architecture.

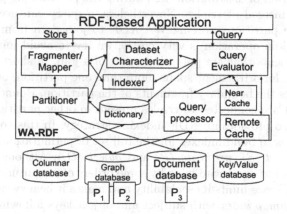

Fig. 1. WA-RDF architecture

An *RDF-based application* issues *store* or *query* requests to *WA-RDF*, which is normally deployed into multiple dedicated physical nodes. When an *RDF-based Application* submits a *store* request for a triple to the *Fragmenter/Mapper*

---

[2] https://parquet.apache.org/.

component, *WA-RDF* expands this triple to a fragment F_{RDFi} and maps F_{RDFi} to the target NoSQL database(s). This process is performed by the *Dataset Characterizer*, which is the main component of our middleware. During a triple storage, it decides on translating F_{RDFi} to a NoSQL document or graph database (or both) according to the usual query workload, and indexes it with the aid of the *Indexer* component. Once F_{RDFi} is created, the *Partitioner* registers this fragment into the *Dictionary* repository - supported by a NoSQL columnar database - and stores it in the NoSQL databases.

When an *RDF-based Application* submits a SPARQL *query* request, the *Query Evaluator* component decomposes this query into subqueries and reports to the *Dataset Characterizer* about them. In the following, the *Query Evaluator* verifies, with the aid of the *Dictionary*, the partitions on which the triples for the query are potentially located. Based on this information, it checks which triples are available in the *Near Cache* (a data structure in the main memory of the server) and the *Remote Cache* (a remote NoSQL key/value database), and sends the SPARQL subqueries for the missing triples to the *Query Processor* component that, in turn, translates them to graph and/or document NoSQL database queries. Finally, the *Query Processor* sends back the query results to the *Query Evaluator* that translates them back to RDF triples with the aid of the *Dictionary*, and returns the result to the *RDF-based Application*.

The main purpose of *WA-RDF* is to store large RDF graphs. In such a scenario, the number of RDF triples can easily surpass the performance capacity (*e.g.*, disk, memory, CPU) of a single server. When it occurs, *WA-RDF* distributes the RDF fragments among potentially many NoSQL nodes. A fragment is our smallest grain of distribution, *i.e.*, during the partitioning process we deal with fragments instead of triples. Nevertheless, a query can eventually access data in multiple partitions, forcing *WA-RDF* to join data from different partitions. Since a join operation is very costly, we try to avoid join processes by replicating fragments that are potentially part of a join. In short, whenever the typical workload for a fragment spans more than one partition, our *partitioning* scheme replicates the boundary fragments of the partition. Boundary fragments have triples that are connected to triples present in other partitions.

WA-RDF also provides an RDF indexing strategy. In this context, a traditional approach is to build indexes for the full set of permutations of each triple component (*subject (S), predicate (P)* and *object (O)*). Although this method has been designed to accelerate joins by some orders of magnitude, the overhead with large index space limits its scalability and makes it heavyweight. Hence, we developed a *hashmap index* with subject and object keys following the patterns *S-PO* and *O-PS*. In *WA-RDF*, the *Indexer* component is responsible to manage these indexes. It is accessed in two situations: *(i)* during the fragment creation; and *(ii)* to process queries with one triple pattern.

WA-RDF is an evolution of *Rendezvous* [7]. In this version, all the NoSQL databases are employed. Also, as stated before, a graph database replaced the columnar database for triple storage, and the dictionary uses now a columnar database as the main storage.

4 The Workload-Aware Approach

A *workload-aware* approach is the cornerstone of *WA-RDF*. Based on it, *WA-RDF* decides where to place each triple, which influences mapping, partitioning and querying strategies. In order to be aware of the typical workload, *WA-RDF* registers information about the *triple patterns* of each incoming SPARQL query. We consider triple patterns because they determine BGPs that define the query shape (star, chain or complex). For instance, in the SPARQL query SELECT ?x WHERE { ?x p1 B }, the triple pattern is ?x p1 B. WA-RDF registers historical information about the queries into two *hashmaps*, as shown in the example of Figs. 2(iii) and 3(iii) for the RDF graph (i) of both figures. One hashmap registers all the chain-shaped queries indexed by the predicate, and the other one all the star-shaped queries indexed by the subject. For example, Fig. 2 (iii) shows that a typical star query around C containing the triple patterns C p6 G, C p9 I and C p8 H, and Fig. 3(iii) shows a chain query starting on *p1* containing the triple patterns A p1 B, B p5 C *and* C p6 ?.

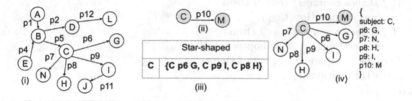

Fig. 2. Star fragmentation strategy

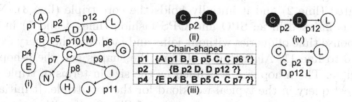

Fig. 3. Chain fragmentation strategy

4.1 Storage: Fragmentation and Partitioning

When a new RDF triple $t_{new} = (s, p, o)$ is inserted through *WA-RDF*, the hashmaps are checked to decide if t_{new} is more frequent on star or chain-shaped queries. Algorithm 1 shows the workload-based triple storage procedure. The input parameter is t_{new}, and it generates an RDF fragment f that is stored in one or more partitions. In Figs. 2(ii) and 3(ii), for instance, we have two new triples C p10 M and C p2 D, respectively. An RDF fragment represents an expansion of t_{new} (called *core triple*) with all of its neighbors according to a *n-hop replication horizon* managed by *WA-RDF*. The n-hop is used to avoid frequent joins by wisely expanding the core triple to include its neighbors until a maximal distance n. The value n is calculated as the mode of the number of triple patterns in the queries related to t_{new}. For star queries, it is the most frequent diameter of the queries. For chain queries, it is the most frequent length of the queries. For example, the diameter mode for the triple C p10 M is 1 according to the frequent star-shaped query in the index of Fig. 2(iii).

Algorithm 1: Workload-based triple storage

Input: Triple t_{new}

```
1  if !exists(t_new) then
2      f = new Fragment;
3      f.core = t_new;
4      indexSPO.put(t_new.s, t_new);
5      indexOPS.put(t_new.o, t_new);
6      f.shapes = getShapes(t_new);
7      hop = 1;
8      if f.shapes.contains('chain') then
9          hop = chainHop(t_new);
10     if f.shapes.contains('star') then
11         if(starHop(t) > hop) hop = starHop(t_new);
12     f.triples = expand(t_new, hop, f.shapes);
13     writeToPartitions(f);
14 end
```

Back to Algorithm 1, if t_{new} does not exists (line 1), a new RDF fragment f is generated (line 2) and it initially holds the core triple (line 3). Next, the core triple is indexed in an SPO and OPS fashion (lines 4 and 5) in order to reduce response time of queries without joins and facilitate the query expansion. From line 6 to line 11, Algorithm 1 obtains the shapes and the n-hop size for the core triple. The n-hop size is defined as the size in terms of triple patterns of the biggest query in the typical workload for the core triple. It initially finds the shapes and registers them in the fragment f (line 6). If neither the predicate nor the subject exist in the chain and star hashmaps, respectively (no shape is found), it defaults to a star-shaped query with one triple n-hop size (hop = 1) (line 7). Otherwise, it determines the hop based on the found shapes (lines 8 to 11). In line 12, t_{new} is expanded to the n-hop size. $f.triples$ is an array with up to 2 positions: one for the chain fragment and another one for the star fragment. In the example of Fig. 2, the new triple C p10 M is expanded to the RDF fragment in the left of Fig. 2(iv), and the new triple C p2 D to the RDF fragment in the top of Fig. 3(iv).

Formally, an *RDF Fragment* is a set $F_{RDFi} = \{t_{RDF}\}$ of RDF triples $t_{RDF} = (s, p, o)$ whose content may overlap with other fragment F_{RDFj}. After the document or graph fragment is created, *WA-RDF* distributes it among potentially NoSQL nodes (line 13). A NoSQL node can store one or more partitions. We discuss RDF data partitioning further on in this section.

It is important to observe here that a core triple can generate two RDF fragments (graph and document). It happens when the subject of this core triple is in the star hashmap and its predicate is in the chain hashmap at the same time. If an RDF fragment is translated to a document fragment, we have a mapping to a JSON document and it is stored into a NoSQL document database. If an RDF fragment is translated to a graph fragment, we have a mapping to a NoSQL graph database.

A *document fragment* is a tuple $f_{df} = (k_d, A)$ where $f_{df}.k_d$ is the JSON document key and $f_{df}.A = \{(k_\alpha : v)\}$ is a set of attributes, being k_α the attribute key and v a value whose domain can be atomic, a list, a set or a tuple. In short, the core triple t_{core} in the RDF fragment F_{RDFi} is mapped to a document whose key is $t_{core}.s$, and each outgoing predicate from the subject becomes a document attribute with a key $t_{core}.p$. If F_{RDFi} is 1-hop, the attribute value of each outgoing predicate is the object $t_{core}.o$ reached from it. Otherwise, the predicate value is an inner document that maintains the target object as the inner document key, and its outgoing predicates as attributes. If any of these outgoing predicates is, in turn, an n-hop, n > 1, the generation of other inner documents proceeds recursively. Figure 2(iv) illustrates an RDF fragment (left) and its corresponding document fragment (right).

A *graph fragment* is a triple $f_{gf} = (s_g f, T, o_g f)$ where $s_g f$ is a vertex representing the first subject of a chain, $o_g f$ is a vertex representing the last object of a chain, and $T = \{t_n\}$ denotes an edge that holds a set of triples as property, i.e., the intermediary triples between $s_g f$ and $o_g f$, including the object of the first triple and the subject of the last triple. A graph fragment summarizes a chain of triples by transforming this chain into a triple where the subject of the first triple and the object of the last triple are mapped to two vertexes, and the edge between these two vertexes is created with a property that maintains all the triples of the chain. In Fig. 3(iv) we see an RDF fragment (top) and a graph fragment obtained from it (bottom).

We now explain the partitioning strategy of *WA-RDF*. Given the RDF graph of Fig. 4 (the resulting graph after the storage of the triples C p10 M and C p2 D into the graph of Figs. 2(i) and 3(i)), the fragments are stored in document partitions (for instance, P_1) and/or in graph partitions (for instance, P_2 and P_3). In *WA-RDF*, a fragment is the finest unit for a partition. As defined in the following, a partition is a set of fragments stored into the same physical NoSQL node, and a fragment can be replicated in multiple partitions.

An *RDF Partition* P_m of an RDF graph G, such that $G \subseteq P_1 \cup P_2 \cup ...P_n$, is a set of RDF fragments $P_m = \{F_{RDFi}\}$, being not required that $P_m \cap P_t = \emptyset$, for m $\neq$ t. Also, given $SP = \{P_1, P_2, ..., P_n\}$ the set of RDF partitions, the *partition boundary* B_{P_i} of a partition $P_i \subset SP$ is the set of RDF fragments $B_{P_i} = Fb_{P_1} \cup Fb_{P_2}... \cup Fb_{P_n}$, where $Fb_{P_k} \subset P_k$ for any k. Each $Fb_{P_i} \in B_{P_i}$ has one or more RDF triples $t_i F_{P_i} = (s_i, p_i, o_i)$ where $o_i = s_j$, being s_j the subject of any other triple $t_j F_{P_j}$ of a partition P_j where $t_j F_{P_j} = (s_j, p_j, o_j)$.

The *Dictionary* shown in Fig. 4 registers each fragment location. It holds three *hashsets* for each partition to keep track of the RDF elements stored in

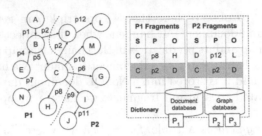

Fig. 4. Fragment partitioning

each partition (represented in the tables *P1 Fragments and P2 Fragments*), so during a query request we can avoid accessing unnecessary partitions that cannot answer this query. If a *WA-RDF* node manages more than one partition of a NoSQL database type, in face of a new core triple we have to decide which is the best partition to store its fragments. For doing so, *WA-RDF* finds out the typical workload for the triples that belong to the fragment generated by the core triple. With this information, we can query the partition sets in the *Dictionary* to verify in which partition this fragment can be more useful (this is represented by the line 13 of Algorithm 1) in sense that joins outside the fragment can be answered within a single partition. In Fig. 4, the size n = 1 for boundary replication repeats the fragment with core triple C p10 D in partitions *P1* and *P2*.

Algorithm 2 presents an overview of the query planning and partition processes. The input is the set of *triple patterns* from the query and the output is the result set *R*. If the query has only one triple pattern, the result is retrieved from SPO and OPS indexes (lines 1 and 2). Otherwise, Algorithm 2 looks for the shapes of the query to define its execution plan. Firstly (lines 4 to 6), *WA-RDF* loads the triple patterns into two multilevel hash tables *mhtSPO* and *mhtOPS* in order to speedup the further steps. Then, it looks for S-S star shapes (lines 11 to 14), O-O star shapes (lines 15 to 18) and chains (lines 20 to 26). The star shapes are identified when a subject has more than 2 entries in the *mhtSPO* (line 11), or an object have more than 2 entries on the *mhtOPS* (line 15). In this case, it expands the star shape with all the entries from the multilevel hash tables, registers the results in the star hashmap and add it to the query execution plan stored into the set *stars* that will be later translated to the document database query language (line 28). The triple patterns that do not define star shapes are expanded to chains (line 20). If the expanded chain has size 1 (*i.e.*, the triple pattern itself), the indexes are accessed to get the result triples (line 22 and 23). Otherwise, the expanded chain is registered in the chain hashmap and added to the query execution plan stored into the set *chains*, which is later translated to the graph database query language (line 29). Finally, with the aid of the *Dictionary*, after the *stars* and *chains* sets are processed by the document and graph databases, the algorithm returns the result set R (line 30).

Algorithm 2: Workload-based triple querying

Input: SPARQL query triple patterns = $\{tp_1, tp_2, ..., tp_n\}$, where
$\quad\quad tp_i = (s_i, p_i, o_i)$
Output: Result set R = $\{t_1, t_2, ..., t_m\}$

```
 1  if n == 1 then
 2      R.add(getFromIndex(tp₁));
 3  else
 4      for i = 1 to n do
 5          mhtSPO.put(sᵢ, tpᵢ);
 6          mhtOPS.put(oᵢ, tpᵢ);
 7      end
 8      stars = {};
 9      chains = {};
10      for i = 1 to n do
11          if mhtSPO.get(sᵢ).size() > 2 then
12              expandedStar = expandSubject(mhtSPO.get(sᵢ));
13              register(expandedStar, 'star', expandedStar.hop);
14              stars.add(expandedStar);
15          else if mhtOPS.get(oᵢ).size() > 2 then
16              expandedStar = expandObject(mhtOPS.get(oᵢ));
17              register(expandedStar, 'star', expandedStar.hop);
18              stars.add(expandedStar);
19          else
20              expandedChain = expandChain(tpᵢ);
21              if expandedChain.horizon==1 then
22                  R.add(indexSPO.get(sᵢ));
23                  R.add(indexOPS.get(oᵢ));
24              else
25                  register(expandedChain, 'chain', expandedChain.hop);
26                  chains.add(expandedChain);
27      end
28      R.add(readFromDocument(stars));
29      R.add(readFromGraph(chains));
30  return R;
```

4.2 Query Processing

From a performance point of view, the most important task accomplished by *WA-RDF* is the query processing. The queries analyzed by the *Query Evaluator* component are processed by the *Query Processor* component, which determines the best way to read data from the NoSQL databases.

The *Query Processor* usually has many options to process a query. Even so, its main strategy is to foster the early execution of triples with low selectivity to reduce the number of intermediate results and, consequently, to boost the query performance. Our work focuses on selectivity estimation of single BGPs based on statistics of the queried data.

Suppose, for example, the query Q in the following. It could be decomposed into BGPs that define star queries where ?x and ?y are the star shape centers, or BGPs that define chain queries that starts in ?x, follows to ?y and then goes to other nodes. However, during the query processing we have to decide if the process first execute one of the star or chain queries, as there are dependencies between the queries.

Q: SELECT ?x WHERE { ?x p1 ?y . ?x p2 ?z .
?x p3 ?w . ?y p5 ?k . ?y p6 G . ?k p7 ?l . ?l p8 H . ?l p8 J }

Suppose, for example, that the star-shaped BGP ?x p1 ?y . ?x p2 ?z . ?x p3 ?w potentially returns 100 triples and the chain-shaped BGP ?x p1 ?y . ?y p5 ?k . ?k p7 1 . ?l p8 J returns only 10 triples. In this case, we would process first the chain-shaped BGP.

In short, the selectivity estimation is the number of triples that is returned for each BGP. This number depends on the shape of the query. For star shapes, it is calculated by the number of times that the center of the star (subject or object) is present in the *Dictionary*. For chain shapes, it is calculated as how many times the predicates of the chain are presented in the chain.

The selectivity is the input for the query translation processes accomplished by the *Query Processor* component into the target databases. The *star* queries (O-O or S-S joins) are converted to queries over NoSQL document databases. For instance, the *star* queries $Q1$ (O-O) and $Q2$ (S-S) in the following are converted to the access methods $D1$ and $D2$, respectively (MongoDB NoSQL database syntax). The $exists function of MongoDB filters the JSON documents that have all the predicates of each query. In $D2$, we also filter by the subject M.

Q1: SELECT ?x WHERE {x? p5 y? . x? p2 z? .}
Q2: SELECT ?x WHERE {x? p9 y? . M p10 y? .}

D1: db.partition1.find({p5:{$exists:true},
p2:{$exists:true}}})
D2: db.partition1.find({p9:{$exists:true},
subject:M}})

The *chain* queries are converted to queries over NoSQL graph databases. For example, given the query $Q3$ in the following, with O-S joins, *WA-RDF* translates it to the set of query $G1$ according to the Cypher[3] query language of the Neo4J NoSQL database.

Q3: SELECT ?x WHERE {x? p1 y?. y? p2 z?. z? p3 w?.}

G1: MATCH (f:Fragment)
WHERE ANY(item IN f.p WHERE item = p1 OR
item = p2 OR item = p3)
RETURN p

---

[3] https://neo4j.com/developer/cypher-query-language/.

The processing of joins occurs when a query as a whole cannot be executed on a single partition. In this case, it needs to be decomposed into a set of subqueries, being each subquery evaluated separately and joined at the *WA-RDF* node.

For example, if we consider the graph of Fig. 4, the query *Q4* in the following is not able to be completed only querying the partitions *P1* or *P2* alone. In this case, the *Query Processor* divides it into subqueries *SQ5* and *SQ6*, issues it to the partitions *P1* and *P2*, respectively, and joins the result sets by matching the predicate p9 (the connection between *P1* and *P2*).

Q4: SELECT ?x WHERE {x? p1 y?. y? p5 z?.
z? p9 w?. w? p11 J.}
SQ5: SELECT ?x WHERE {x? p1 y?. y? p5 z?.
z? p9 w?.}
SQ6: SELECT ?x WHERE {z? p9 w?. w? p11 J.}

As explained before, a complex query is a combination of the *star* and *chain* patterns, potentially connected by simple queries. Query *Q5* in the following is an example, where the BGP x? p1 y? . y? p2 z? . z? p3 w? is a *chain* pattern, the BGP z? p5 ?k is a simple query, and the BGP k? p6 G . k? p7 I . k? p8 H is a *star* pattern. In this case, the decomposition process works as follows: *(i)* it first sorts the triple patterns by subject and object; *(ii)* if it is identified a subset with two or more patterns with the same subject or object, it is considered a *star* subquery, like the subquery *P1* in the following. Then, chains are identified in the remaining query patterns, *i.e.*, *(iii)* for each triple pattern, we navigate from object to subject creating chains, and we pick up the longest chain and consider this a *chain* subquery, like subquery *P2*.

Q5: SELECT ?x WHERE { x? p1 y? . y? p2 z? .
z? p3 w? . z? p5 ?k . k? p6 G . k? p7 I . k? p8 H }
P1: {k? p6 G . k? p7 I . k? p8 H }
P2: {x? p1 y? . y? p2 z? . z? p3 w?}
P3: {z? p5 ?k}

We repeat step *(iii)* until there are no more chains, or there are only simple patterns, like the subquery *P3*. Each *star* and *chain* subquery is processed separately, and the join of the results (along with the simple patterns) is performed at the *WA-RDF* node. In case of ambiguity, *i.e.*, a pattern that is presented in more than one query type, we consider the following priority: *(1)* subject-based *star* query; *(2)* object-based *star* query; *(3)* the longest *chain* query; and *(4)* simple queries. The star queries are processed with high priority for two reasons. star queries are most common, and the MongoDB translation permits that we query mostly the document keys, what lets queries over documents much faster when compared to queries over graphs.

5   Experimental Evaluation

This section presents an evaluation of the proposed approach. The considered dataset comes from the *Lehigh University Benchmark (LUBM)* [3], which features an ontology for an *University* domain, synthetic RDF data, and 14 extensional queries representing a variety of properties. In our experiments, we generate a dataset with 4000 universities. The dataset size is around 100 GB and contains around 500 million triples. Regarding query complexity, we have 12 queries with joins, all of them having at least one star join, and 6 of them also having at least one chain join.

We ran experiments for data insertion and data querying to evaluate the performance and scalability of *WA-RDF*. *WA-RDF* was developed using Apache Jena version 3.2.0 with Java 1.8, and we use MongoDB 3.4.3 and Neo4J 3.2.5 as the document and graph NoSQL databases, respectively, on considering their maturity as representatives of these NoSQL data models. All the nodes are Amazon m3.xlarge spot instances[4] with 7.5 GB of memory and 1×32 SSD capacity. For all the experiments, the nodes represent the number of MongoDB + Neo4J servers, always with half of each database. We also create one partition for each server, and the *WA-RDF* servers were installed alone in each node. All the queries were issued from a server in the same network, so the latency between the client and *WA-RDF* was inexpressive.

We reproduce the query processing strategies of *Rainbow* and *ScalaRDF* because we could not find the implementation of these baselines in public repositories. To test *S2RDF*, we use the version found in GitHub[5], with small changes in the source code so we could use LUBM. The machines we use to run *Rainbow*, *ScalaRDF* and *S2RDF* are similar to the m3.xlarge of *WA-RDF*. We considered only one processing server for *Rainbow* and *ScalaRDF*, and we deployed an Apache Spark cluster with one master and 3 workers for *S2RDF* (the same size of our *WA-RDF* installation). The baselines were chosen because they hold different strategies: Rainbow also applies multiple databases by using Redis as a cache, ScalaRDF use a native storage along with Redis, and S2RDF uses Apache Spark.

Table 1 details the ingestion response time for three different triples. LUBM is a synthetic benchmark based on the educational domain, creating a model and data simulating a university with students, courses and professors. We first ran the queries *Q1* to *Q5* to provide workload information to *WA-RDF* and, in the following, we inserted the fragments *F1* to *F3*. The queries and the fragments are available at Appendix A. *F1* presents the insertion of a university. As shown in Table 1, the fragmentation is faster and only MongoDB was used. *F2* presents the insertion of a Department in the University of *F1*. During *F2* processing, the fragmentation phase is slower because the triples are expanded to include the University and, as the relation ub:subOrganizationOf is part of the chain in query *Q5*, it is added to Neo4J. *F3* inserts a professor that is also a chair of the department inserted before. It generates fragments for MongoDB and Neo4J so the fragmentation and partitioning tasks are slower than the other ones.

---

[4] https://aws.amazon.com/ec2/instance-types/.
[5] https://github.com/mxhdev/S2RDF_BSBM.

Table 1. Detailed ingestion time (ms)

Work	F1	F2	F3
WA-RDF - Parsing	9	12	13
WA-RDF - Fragmenting	13	19	23
WA-RDF - Indexing	5	6	4
WA-RDF - Partitioning	24	39	41
WA-RDF - Inserting MongoDB	102	-	204
WA-RDF - Inserting Neo4J	-	300	320
WA-RDF total	153	356	605
Rainbow	201	209	198
ScalaRDF	233	253	208
S2RDF	129	197	291

Table 2. Detailed query response time (ms)

Work	Q1	Q2	Q3	Q4	Q5
WA-RDF - Parsing	10	13	20	21	19
WA-RDF - Index access	13	14	11	18	15
WA-RDF - Decomposition	-	20	35	33	54
WA-RDF - MongoDB	-	70	102	123	132
WA-RDF - Neo4J	-	-	-	-	302
WA-RDF - Result set creation	5	20	30	40	60
WA-RDF total	28	144	199	233	582
Rainbow	33	162	203	594	1022
ScalaRDF	34	190	182	602	892
S2RDF	27	98	182	493	921

Table 2 details the querying response time for five different triples. The queries used here were proposed by Guo et al. [3]. For sake of simplicity, we discuss only a simple, a star, a chain and two complex queries, instead of all the queries available in LUBM. $Q1$ is the most basic query, and it is solved directed by the *WA-RDF* OPS index. $Q2$ is a small star-shape query around X that causes an access to MongoDB. $Q3$ is composed of two stars connected by Y. It takes more time to generate the result set because some triples have to be cleaned. $Q4$ is a big star composed of five BGPs. However, it is very fast to be processed by *WA-RDF* because we can solve it with only one MongoDB access. $Q5$ is a complex query that is decomposed into two stars and a chain (`?X ub:memberOf ?Z . ?Z ub:subOrganizationOf ?Y . ?Y rdf:type ub:University.`). It touches Neo4J and avoids multiple calls to MongoDB. As shown in Table 2, *WA-RDF* is specially interesting for complex queries like $Q4$ and $Q5$.

6   Conclusion

This paper presents *WA-RDF*, a workload-aware RDF partitioning and querying approach for RDF data stored into NoSQL databases. We based it on a middleware that can, according to the typical shape of SPARQL queries, define RDF fragments and store them into the document and graph NoSQL databases. Our experiments show that *WA-RDF* outperformed three recent baselines in terms of large queries (*Q4* and *Q5*). For most of the other ones, we ran under the average of the baselines executions. However, there is still room for improvements regarding data ingestion time and storage size.

In general, *WA-RDF* is a contribution to the problem of efficient management of RDF data persisted into NoSQL databases. To the best of our knowledge, this is the first work that deals with RDF data fragmentation, partitioning and efficient query processing (including optimization issues to deal with intermediate results) for massive RDF graphs stored in multiple NoSQL databases. Even so, we have some future works in mind. First of all, we are considering the development of an algorithm for triples compression. The lack of this feature lets *WA-RDF* uses exponentially more storage space as the n-hop horizon grows. Moreover, we intend to consider update and delete operations and cluster capabilities in the *WA-RDF* server. With these improvements, we aim at comparing it again with the related work. Finally, we intend to evaluate *WA-RDF* against other benchmarks, like the *Waterloo SPARQL Diversity Test Suite (WatDiv)*.

A   Fragments and Queries

F1 - Insert a university:
University0.edu rdf:type ub:University

F2 - Insert a department for the university:
Department0.University0.edu rdf:type ub:Department
Department0.University0.edu ub:subOrganizationOf University0.edu

F3 - Insert a professor for the department:
Professor0 rdf:type ub:Professor
Professor0 rdf:type ub:Chair
Professor0 ub:worksFor Department0.University0.edu

Q1 - SELECT ?X WHERE {?X rdf:type ub:UndergraduateStudent}

Q2 - SELECT ?X WHERE {?X rdf:type ub:GraduateStudent . ?X ub:takesCourse Department0.University0.edu GraduateCourse0}

Q3 - SELECT ?X, ?Y WHERE {?X rdf:type ub:Chair . ?Y rdf:type ub:Department . ?X ub:worksFor ?Y . ?Y ub:subOrganizationOf University0.edu}

Q4 - SELECT ?X, ?Y1, ?Y2, ?Y3 WHERE {?X rdf:type ub:Professor . ?X ub:worksFor Department0.University0.edu . ?X ub:name ?Y1 . ?X ub:emailAddress ?Y2 . ?X ub:telephone ?Y3}

Q5 - SELECT ?X, ?Y, ?Z WHERE {?X rdf:type ub:GraduateStudent .?Y rdf:type ub:University .?Z rdf:type ub:Department .?X ub:memberOf ?Z .?Z ub:subOrganizationOf ?Y . ?X ub:undergraduateDegreeFrom ?Y}

References

1. Bugiotti, F., Bursztyn, D., Diego, U.C.S., Ileana, I.: Invisible glue: scalable self-tuning multi-stores. In: CIDR 2015 (2015)
2. Gu, R., Hu, W., Huang, Y.: Rainbow: a distributed and hierarchical RDF triple store with dynamic scalability. In: Proceedings - 2014 IEEE International Conference on Big Data, IEEE Big Data 2014, pp. 561–566 (2015). https://doi.org/10.1109/BigData.2014.7004274
3. Guo, Y., Pan, Z., Heflin, J.: LUBM: a benchmark for OWL knowledge base systems. Web Semant. Sci. Serv. Agents WWW 3(2), 158–182 (2005)
4. Hu, C., Wang, X., Yang, R., Wo, T.: ScalaRDF: a distributed, elastic and scalable in-memory RDF triple store (2016)
5. Ma, Z., Capretz, M.A.M., Yan, L.: Storing massive Resource Description Framework (RDF) data: a survey. Knowl. Eng. Rev. 31(04), 391–413 (2016). https://doi.org/10.1017/S0269888916000217, http://www.journals.cambridge.org/abstract_S0269888916000217
6. Mello, R.D.S., et al.: Master: a multiple aspect view on trajectories. Trans. GIS (2019)
7. Santana, M.: Workload-aware RDF partitioning and SPARQL query caching for massive RDF graphs stored in NoSQL databases. In: Brazilian Symposium on Databases (SBBD), pp. 1–7. SBC (2017)
8. Schätzle, A., Przyjaciel-Zablocki, M., Skilevic, S., Lausen, G.: S2RDF: RDF querying with SPARQL on Spark. Proc. VLDB Endowment 9(10), 804–815 (2016)

Reverse Partitioning for SPARQL Queries: Principles and Performance Analysis

Jorge Galicia[1(✉)], Amin Mesmoudi[1,2], Ladjel Bellatreche[1],
and Carlos Ordonez[3]

[1] LIAS/ISAE-ENSMA, Chasseneuil-du-Poitou, France
{jorge.galicia,bellatreche}@ensma.fr
[2] Université de Poitiers, Poitiers, France
amin.mesmoudi@univ-poitiers.fr
[3] University of Houston, Houston, USA

Abstract. RDF and SPARQL have been widely adopted for modeling
and querying Web objects as facts in the Semantic Web. The amount
of data stored in RDF format has grown significantly pushing RDF processing systems to implement efficient query processing techniques in
parallel and distributed architectures. In such environments, the data
partitioning is a pre-condition for query performance. Traditionally, the
graph-based RDF systems store the data using adjacency lists formed
by a vertex and its outgoing edges. Nevertheless, for a certain type
of queries, considering entities and their ongoing edges may speed up
their execution. This point motivates us to present a new partitioning
technique (called reverse partitioning) dedicated to graph-based triple
stores that is complementary to traditional ones. In this paper, we first
detail its main principles by illustrating its functioning. Secondly, the
best classes of queries for which reverse partitioning gives better performance are discussed. Finally, we report on intensive experiments using
large RDF datasets that show significant performance improvements for
certain queries in a graph-based triple store and in a relational-based
system.

Keywords: RDF · Partitioning · Distributed computing

1 Introduction

The Semantic Web strives for a worthwhile integration of the data published
on the Web to be exchanged and reused in a variety of applications, communities and scenarios. Accordingly the W3C promotes standard data formats and
exchange protocols, most fundamentally the **R**esource **D**escription **F**ramework
(RDF) and SPARQL [11] as its query language. RDF has been widely adopted
for modeling web objects as facts in the semantic web representing data as a

© Springer Nature Switzerland AG 2019
S. Hartmann et al. (Eds.): DEXA 2019, LNCS 11707, pp. 174–183, 2019.
https://doi.org/10.1007/978-3-030-27618-8_13

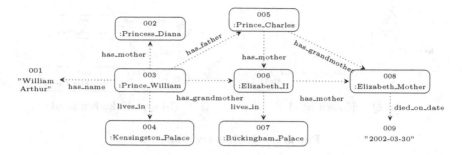

Fig. 1. RDF example graph G

collection of triples of the form $<subject, property, object>$. A collection of RDF triples form an RDF graph as the one shown in Fig. 1.

With the advent of low-cost distributed architectures and the need to scale to process datasets with several millions of triples, the number of research projects on distributed RDF systems[1] has significantly increased. Indeed, distributed computing raises other challenges such as data distribution and execution skewness that are less relevant in centralized architectures. In distributed engines, a correct data placement strategy is a pre-condition to balance the loads and optimize the performance of the processing system. In this context, many algorithms have been proposed for specific platforms, applications and constraints.

Most of distributed RDF processing systems are based on the relational model. These approaches map triples to relations and apply partitioning strategies used in relational databases (e.g. hashing functions, vertical partitions). In our work, we focus in other kind of systems storing the data as graphs, without a relational database layer. We are interested in systems persisting the data as adjacency lists. This storage model is embraced in the gStoreD [8] system and also in systems built on top of key-value stores (e.g. Trinity.RDF [12]). In this representation, each node (generally the subject) is stored together with its *outgoing* edges and *1-hop* neighbors. This paper explores adjacency lists storing each node and its *ingoing* edges. We name our strategy *reverse partitioning* and we show that this representation is useful for queries with specific shapes. Then, we propose and compare three allocation strategies in a distributed RDF system.

The contributions of this paper are: *(i)* The introduction of the *reverse partitioning* main principles firstly by means of a motivating example that is used in the formalization part to clarify the main concepts, *(ii)* An experimental study performed in a graph-based parallel RDF engine to evaluate our complimentary partitioning solution, and *(iii)* The comparison of distinct physical storing strategies simulating different partitioning schemas in a relational-based system.

The organization of the paper is as follows. In the next section (Sect. 2) we provide a motivating example to clarify our reasoning. In Sect. 3 we describe and

---

[1] We use the term distributed RDF systems to denote both parallel and distributed architectures.

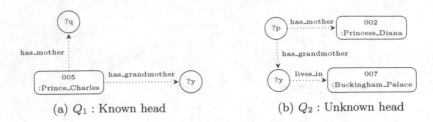

(a) Q_1 : Known head   (b) Q_2 : Unknown head

Fig. 2. SPARQL query graphs

formalize our partitioning approach. Section 4 shows our experimental results. Section 5 gives the study of related work and we conclude and give future perspectives in Sect. 6.

2   Motivating Example

Let us consider the RDF graph G of Fig. 1 stored in an adjacency list as shown in Fig. 3a. Each element of the list is called an entity class depicting a vertex and its *outgoing* edges. Generally, the entity labels (*eLabel* in Fig. 3a) are indexed to improve the performance of queries seeking for a specific subject. Consequently, conventional adjacency lists are adept to answer linear and star queries in which the subject or *head* is known as it is the case of Q_1 in of Fig. 2a. However, in many cases the query is not selective on the subject and instead its properties are given to identify the subject vertex (e.g. Q_2 in Fig. 2b). In these types of queries, the index mentioned previously on subject labels cannot be used to prune based on a known subject, bearing a full scan of the adjacency list to solve the SPARQL query.

Queries on which the head of the outgoing edge is unknown (e.g. Q_2 in Fig. 2b) are very frequent when exploring RDF graphs to obtain meaningful information. A vertex is described by its properties, therefore if a node or a set of vertices are to be identified, their properties should be clearly stated in the query. An efficient searching process in the adjacency list should be able to prune irrelevant results and avoid a full scan of the list when possible. We propose the creation of a *reverse* adjacency list (illustrated in Fig. 3b) that stores the graph and groups its vertices in terms of its *ongoing* edges.

eID	eLabel	adjList
003	:Prince_William	(has_mother, :Princess_Diana), (has_father, :Prince_Charles), (has_grandmother, :Elizabeth_II), (lives_in, y:Kensington_palace), (has_name, "William Arthur")
008	:Elizabeth_Mother	(died_on_date, "2002-03-30")
006	:Elizabeth_II	(has_mother, :Elizabeth_Mother), (lives_in, Buckingham_Palace)
005	:Prince_Charles	(has_mother, :Elizabeth_II), (has_grandmother, :Elizabeth_Mother)

eID	eLabel	adjList
006	x:Elizabeth_II	(has_grandmother, x:Prince_William), (has_mother, x:Prince_Charles)
008	x:Elizabeth_Mother	(has_grandmother, x:Prince_Charles), (has_mother, x:Elizabeth_II)
001	"William Arthur"	(has_name, x:Prince_William)
007	y:Buckingham_Palace	(lives_in, x:Elizabeth_II)
...	...	...

(a) Regular Adjacency List for G (b) Reverse Adjacency List for G

Fig. 3. Adjacency Lists for G

3 Our Approach

In this section we propose the *Reverse Partitioning* strategy which formalizes the intuition presented in Sect. 2.

3.1 Preliminaries

As we have previously mentioned, graph-based triple store engines represent the data on disk using an adjacency list. Each row of the list represents the subject and its *outgoing* edges. For example, x:Prince_Charles → {(has_mother, x:Elizabeth_II), (has_grandmother,x:Elizabeth_Mother)} depicts the entity Prince Charles. The Prince Charles's entity is *described* by its properties and objects. Each row of the adjacency list is named a *forward entity*.

Definition 2 Forward Entity: *A forward entity denoted as* $\overrightarrow{E}$ *is the quadruple* $< V_R, L_R, \mathcal{F}(V_R), L_{\mathcal{F}(V_R)} >$. $\overrightarrow{E}$ *is a subgraph of G where* V_R, L_R *are the root and label respectively, and* $\mathcal{F}(V_R) = \{< v_r, v'_r > | \exists < v_r, v'_r > \in E\}$ *(i.e. the set of all out-going edges from* v_R *and* v_R's *one-hop neighbors in G) as well as the binding labels* $L_{\mathcal{F}(V_R)}$.

The forward entities are the base partitioning unit of systems like EAGRE [13] for example. This partitioning strategy is ideal for star-shaped queries, especially when the head of the query is known and an efficient index is created on the adjacency list keys. However, when the head of the query is not known, the entire adjacency list (of size n) must be read to find the query matches.

Definition 3 Backward Entity: *A backward entity denoted as* $\overleftarrow{E}$ *is the quadruple* $< V_R, L_R, \mathcal{B}(V_R), L_{\mathcal{B}(V_R)} >$. $\overleftarrow{E}$ *is a subgraph of G where* V_R, L_R *are the root and label respectively, and* $\mathcal{B}(V_R) = \{< v'_r, v_r > | \exists < v'_r, v_r > \in E\}$ *(i.e. the set of all in-going edges from* v_R *and* v_R's *one-hop neighbors in G) as well as the binding labels* $L_{\mathcal{B}(V_R)}$.

Backward entities are ideal to solve queries in which the head of the query is unknown. Similarly to the Forward Entities, we assume that the adjacency list is efficiently indexed. In this case, a graph matching is easily found exploring the index (we assumed an $O(1)$ cost).

3.2 Partition Algorithm

In this section we define the partitioning algorithm used to distribute the data among the nodes of a distributed/parallel system using Forward or Backward entities as the distribution units. We represent the number of nodes as P. We consider the following partitioning strategies.

Hashing Strategies: These methods apply a hashing function on the node's label L_R of $\overrightarrow{E}$ or $\overleftarrow{E}$. The hashing value modulo the number of computer nodes (P) returns the site to which the adjacency list's row is assigned. The risk of applying this method is that since the connectivity between entities is not considered, two entities (backward or inward) that are highly connected may be found in two distinct sites making the join operation between them very costly.

Min-Cut Algorithms: In response to the drawback of hashing methods, graph partitioning methods have been applied to this problem. EAGRE [13] for example used the min-cut strategy to distribute forward entities. The first step of this strategy consists in mapping the forward/backward entities to a weighted graph that is partitioned with robust heuristics (e.g. METIS [6]). The METIS heuristic, for example, takes the number of partitions as a parameter; in our case, the number of partitions equals the number of sites. Other works like [4], have also explored scalable graph partitioning algorithms on massive graphs. To reduce the number of nodes to be partitioned, forward and backward entities are grouped according to their predicates (entity classes).

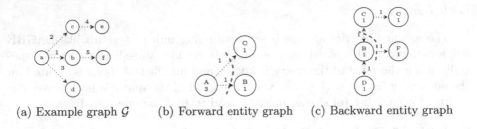

(a) Example graph $\mathcal{G}$ (b) Forward entity graph   (c) Backward entity graph

Fig. 4. Partition models, $P = 2$

Definition 4 Entity Class: $\mathcal{E}_C$ *is a set containing only either* $\overrightarrow{E}$ *or* $\overleftarrow{E}$. *Two entities belong to the same entity class set **iff** they share the same (or almost the same according to a threshold) set of edge labels* $L_{\mathcal{F}(V_R)}$ *or* $L_{\mathcal{B}(V_R)}$.

Let the functions $nodes(\mathcal{E}_C), edges(\mathcal{E}_C)$ returning the set of nodes V_R and edges E belonging to all entities in $\mathcal{E}_C$ respectively.

Definition 5 Compressed Entity Graph: *A compressed entity graph denoted as* $\mathcal{C}(G) = <V_c, w_{V_c}, C(E), w_{C(E)}>$ *is a **weighted graph** where* $V_C = \{v_c | v_c$ *is an entity class* $\mathcal{E}_C\}$, w_{V_c} *is the node weight equal to the number of triples contained in* $\mathcal{E}_C$, $C(E) = \{< v_c, v_c' > | \exists < v_r, v_r' > \in edges(v_c)$ *where* $v_r \in nodes(v_c)$ *and* $v_r' \in nodes(v_c')\}$, *and the weight* $w_{C(E)}$ *indicates the number of exchanged tuples.*

Definition 6 Reverse Partitioning: *The reverse partitioning algorithm consists in applying a partitioning heuristic to the compressed entity graph $\mathcal{C}(G)$ obtained checking the relationships between the backward entities in the RDF graph.*

An example of both, forward and backward entity graphs are shown in Fig. 4. In Fig. 4b, the weights of the nodes correspond to the number of triples in the forward entity, and the weighted edges correspond to the number of triples exchanged between entities. A graph partitioning heuristic creates partitions that are balanced according to the node's weights and that cut the least amount of weighted edges. The *Reverse Partitioning* heuristic is shown on Fig. 4c.

4   Experimental Evaluation

In this section we evaluate and compare the performance of the *Reverse Partitioning* strategy in different scenarios. The first scenario, detailed in Sect. 4.2, compares the reverse partitioning strategy with two physical storage approaches applied by two state of the art systems. The scenario in Sect. 4.3 evaluates the performance of the reverse partitioning strategy in a distributed graph-based system.

4.1   Experimental Setup

- **Hardware:** The scenario described in Sect. 4.2 was performed on a Dell Tower Precision 3620 running Windows 10. This computer features an Intel(R) Core(TM) i7-7700 CPU @ 3.60 GHz processor, 16 GB of main memory and 2 TB of hard disk. The experiments on a distributed graph-based triple store were performed on a 5 machine cluster (i.e. $P = 5$) connected by a 10 Gbps Ethernet switch. The cluster runs a 64-bit Linux and each site has a 8 GB RAM, a processor Intel(R) Xeon(R) Gold 5118 CPU @ 2.30 GHz and 100 GB of hard disk.
- **Software:** The reverse partitioning core module is implemented in Scala and runs in Spark 2.12.2. The translation module from SPARQL to SQL was implemented in Java and the data were stored on PostreSQL 11. The distributed version of gStore [8] is the graph-based triple store used to test partitioning configurations on a cluster.
- **Datasets and queries:** We tested our approach with the WatDiv framework for datasets of 1, 10 and 20 million triples. More details are found on Table 1. For each of these datasets we generated 80 queries (20 of each query type).

4.2   Experiments in a Single-Node Relational Database System

We stored RDF datasets into a relational database using three different strategies: (i) single big table of three columns (subject, predicate, object) similar to

Table 1. Experimental datasets M: millions, #S #O: number of distinct subjects and objects

Dataset	Size (GB)	#S	#P	#O	#Backward Entities
Watdiv1M	0.148	52,505	86	105,492	222
Watdiv10M	1.54	521,585	87	1,003,136	587
Watdiv20M	3.28	1,042,785	87	2,473,723	641

RDF-3X's strategy [7], (ii) vertical partitioning (one table per predicate) similar to the strategy applied by SW-Store [1] and (iii) applying our reverse partitioning strategy gathering the data by incoming edges. We evaluated on each schema the execution time of queries with different forms[2]. The results are shown in Fig. 5. Creating vertical partitions on the predicates gives the most performant execution times for the majority of queries considering that there was not an intense intermediary indexing strategy as it is the case for RDF-3X. The major drawback of the vertical partitioning strategy is that the data are not well distributed in terms of volume. The Reverse Partitioning strategy performs almost as good as the vertical partitioning, especially when the dataset size is bigger and exploring a single table becomes more costly. Reverse partitioning has a very important overhead for queries with patterns in which the subject and object are unknown.

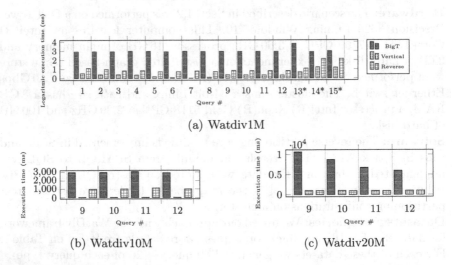

(a) Watdiv1M

(b) Watdiv10M   (c) Watdiv20M

Fig. 5. Performance of partitioning configurations in relational based system

---

[2] The tested queries are available in: bit.ly/2VCi6tL.

4.3   Experiments in a Distributed Graph-Based Triple Store

We stored the dataset of 20 million triples in the gStoreD [8] system that allows to choose among different partitioning strategies. The selected partitioning configurations were: (1) simple hashing on the subject, (2) min-cut algorithm applied to an entity graph and (3) reverse partitioning strategy.

We configured gStoreD to create the adjacency lists on the triple's objects. At query runtime, 7 complex queries did not send any result for both the in-going and the out-going configurations, 13 queries (11 linear and 2 snowflake) did not send a result either by the ongoing or the outgoing configuration. Our final SPARQL query set is composed then of 60 queries (9 linear, 13 complex, 18 snowflake and 20 stars).

Data Distribution: Our results show that the technique that is more efficient in terms of data skew is hashing the data on the subject that distributes the data almost evenly. Our reverse partitioning strategy sends almost 29.4% of the data to one machine but distributes nearly evenly in the four other sites. The min-cut algorithm on the outgoing edges entities has two sites with 28.7% and 27.3% of the data, and a site with only 12.5% being the one with the worst performance in terms of data skewness.

Storage Overhead: Considering that our *Reverse Partitioning* strategy creates an adjacency list for the node and its in-going edges, the number of individual entities stored on the list is greater than the number of entities stored in an adjacency list of the node and its outgoing edges. Therefore, the V*-Tree[3] index size is larger. The sizes of the hashing, mincut and reverse strategies are 1345, 1246 and 1568 MB respectively. In average compared to the other strategies, the Reverse Partitioning creates an index 21% larger but that benefits in a much greater percentage some queries.

Query Performance: In general, the *Reverse Partitioning* strategy improves the performance to solve SPARQL queries considerably. The majority of star queries try to find the head based on the value of its properties, following what was illustrated in the motivating example of Sect. 2, an inverse adjacency list will provide a much better performance as proven by our experiments in Fig. 6b. The 4th and 18th star queries of Fig. 6b are both queries having contrarily to the majority the variable not located in the center of the star, degrading the performance of a *Reverse Partitioning*. With the snowflake queries we confirmed our intuition that queries having the variable in the center, benefit greatly from a reverse partitioning strategy.

If the workload of the system is composed only of very complex queries, the reverse partitioning strategy is not the best option. As shown in Fig. 6d, the performance of the system is not significantly improved, the cost of storing a

---

[3] bit-based B-Tree index on the subjects and predicates used by gStoreD.

much greater index is not compensated based on the reported performance. We can represent complex queries as a union of star queries on which the variables are located on both, the center of star queries, and its on its properties.

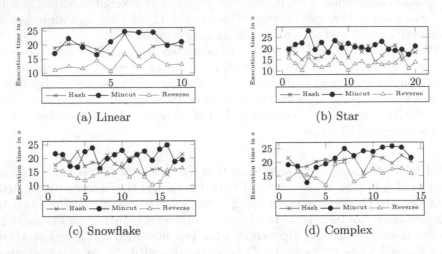

(a) Linear

(b) Star

(c) Snowflake

(d) Complex

Fig. 6. Individual query results

5   Related Work

Most of distributed RDF processing systems are dependent on a single partitioning strategy. This strategy relies on how the data are physically stored on the disk or main memory and also on whether the system is built on top of a distributed computing platform. A few works have explored RDF partitioning, [2] for example, proposes a strategyusing the query workload. We classify the existing systems in three categories:

- *Cloud-based:* The data distribution is performed by the cloud platform on which the system is built on. For example SHARD [9] and PigSparql [10].
- *Specialized systems:* This category considers systems specifically built to process RDF. We considered two sub-categories of these systems based on their processing model: *(i) Partitioned-query based:* At runtime a SPARQL query is decomposed into several subqueries such that each subquery is solved locally on a site and the results are finally aggregated (e.g. TriAD [5]), *(ii) Partial query evaluation:* contrary of partitioned-query based systems, each site receives the full SPARQL query and executes it on the local RDF graph fragment to parallelise the execution (e.g. gStoreD [8]).
- *P2P systems:* Distributed RDF systems in Peer-to-Peer networks. The system 3rdf [3], for instance, is built on top of the 3nuts (p2p network).

6   Conclusions

In this paper we proposed a novel partitioning strategy for graph-based RDF distributed systems. Our partitioning method, named *reverse partitioning*, defines first an adjacency list based on the *in-going* edges of each node to store the data. Secondly, the entries in the adjacency list having similar in-going edges are grouped together and the relations between them are represented in an undirected weighted graph that is partitioned using graph partitioning heuristics. Experiments confirmed that our partitioning strategy is effective to solve Linear and Star queries for which the unknown parameters are located in the center of the star query. *Subject hash-based* and the *min-cut based* partitioning strategies are still more performant to solve a majority of snowflake and complex queries. Our partitioning strategy is therefore complimentary to the ones already proposed in the literature.

As future perspectives, we consider furthering research in a system that considering replication to enhance performance and fault-tolerance. Besides, we acknowledge exploring algorithms to manage highly skewed vertices. Defining which properties allow breaking groups into smaller pieces is a promising hint.

References

1. Abadi, D.J., Marcus, A., Madden, S., Hollenbach, K.: SW-Store: a vertically partitioned DBMS for semantic web data management. VLDB J. **18**(2), 385–406 (2009)
2. Al-Ghezi, A.I.A., Wiese, L.: Adaptive workload-based partitioning and replication for RDF graphs. In: 29th International Conference, DEXA, pp. 250–258 (2018)
3. Ali, L., Janson, T., Lausen, G.: 3rdf: storing and querying RDF data on top of the 3nuts overlay network. In: DEXA, International Workshops, pp. 257–261 (2011)
4. Cabrera, W., Ordonez, C.: Scalable parallel graph algorithms with matrix-vector multiplication evaluated with queries. Distrib. Parallel Databases **35**, 335–362 (2017)
5. Gurajada, S., Seufert, S., Miliaraki, I., Theobald, M.: TriAD: a distributed shared-nothing RDF engine based on asynchronous message passing. In: SIGMOD, Snowbird, UT, USA, 22–27 June, pp. 289–300 (2014)
6. Karypis, G., Kumar, V.: A fast and high quality multilevel scheme for partitioning irregular graphs. SIAM J. Sci. Comput. **20**(1), 359–392 (1998)
7. Neumann, T., Weikum, G.: The RDF-3X engine for scalable management of RDF data. VLDB J. **19**(1), 91–113 (2010)
8. Peng, P., Zou, L., Özsu, M.T., Chen, L., Zhao, D.: Processing SPARQL queries over distributed RDF graphs. VLDB J. **25**(2), 243–268 (2016)
9. Rohloff, K., Schantz, R.E.: Clause-iteration with mapreduce to scalably query datagraphs in the SHARD graph-store. In: DIDC 2011, pp. 35–44 (2011)
10. Schätzle, A., Przyjaciel-Zablocki, M., Lausen, G.: PigSPARQL: mapping SPARQL to pig latin. In: Proceedings of SWIM, p. 4 (2011)
11. W3C: RDF 1.1 concepts and abstract syntax (2014). https://www.w3.org/TR/rdf11-concepts/, https://www.w3.org/TR/rdf-sparql-query/
12. Zeng, K., Yang, J., Wang, H., Shao, B., Wang, Z.: A distributed graph engine for web scale RDF data. PVLDB **6**(4), 265–276 (2013)
13. Zhang, X., Chen, L., Tong, Y., Wang, M.: EAGRE: towards scalable I/O efficient SPARQL query evaluation on the cloud. In: 29th ICDE, pp. 565–576 (2013)

PFed: Recommending Plausible Federated SPARQL Queries

Florian Hacques, Hala Skaf-Molli[✉], Pascal Molli, and Sara E. L. Hassad

LS2N, University of Nantes, Nantes, France
{Hala.Skaf,Pascal.Molli,Sara.elhassad}@univ-nantes.fr
Florian.Hacques@etu.univ-nantes.fr

Abstract. Federated SPARQL queries allow to query multiple inter-linked datasets hosted by remote SPARQL endpoints. However, finding federated queries over a growing number of datasets is challenging. In this paper, we propose PFED, an approach to recommend plausible federated queries based on real query logs of different datasets. The problem is not to find similar federated queries, but plausible complementary queries over different datasets. Starting with a real SPARQL query from a given log, PFED stretches the query with real queries from different logs. To prune the research space, PFED proposes semantic summary to prune the query logs. Experimental results with real logs of DBpedia and SWDF demonstrate that PFED is able to prune drastically the logs and recommend plausible federated queries.

Keywords: Semantic web · Federated SPARQL query · Plausible · Joinable

1 Introduction

Following the Linked Open Data cloud (LOD) principles many datasets have been published. Federated SPARQL query engines [1,15] have been developed to query multiple interlinked datasets hosted by remote SPARQL endpoints. However, finding federated queries over a growing number of datasets is challenging. This requires to fully understand the datasets and find potential joins among them. In this paper, we propose PFED, an original approach to recommend federated queries for end-users. Instead of using datasets to recommend federated queries, PFED recommends federated queries using query logs of different SPARQL endpoints. This is not a classical recommendation problem. In recommender systems [2], the problem is to recommend resources (or items) for users based on similar ones already seen by the users. In PFED, we start with a SPARQL query from a given log and we stretch this query with real queries from other existing query logs. The main advantage of using real logs rather than using datasets is to produce *plausible* federated queries, i.e. queries that generated by combining real queries. This is useful, especially for data portal owners who can recommend federated queries for end-users. Imagine a data portal such

© Springer Nature Switzerland AG 2019
S. Hartmann et al. (Eds.): DEXA 2019, LNCS 11707, pp. 184–197, 2019.
https://doi.org/10.1007/978-3-030-27618-8_14

as Sage[1], or LodLaundromat[2] hosting thousands of linked datasets. The portal owner can see that some users are looking for information about "United Kingdom" in DBpedia, others are looking for conferences in SWDF dataset. Using PFED, the portal owner can suggest to extended conferences with information about country.

To illustrate, consider queries extracted from real SPARQL query logs of SWDF (SWDF 2012) and DBpedia (DBpedia 3.5.1)[3] presented in Fig. 1.

```
Q1S: SELECT * WHERE {
        ?inst rdf:type ?dClass .
        ?inst foaf:based_near ?place
}#results:5025

Q3S: SELECT * WHERE {
        {?paper swrc:author ?author}
        UNION {?paper foaf:maker ?author}
        OPTIONAL {?paper swrc:abstract ?abstract}
}#results:21649
```

(a) SWDF query log

```
Q1D:SELECT * WHERE {
        ?country rdfs:label "United Kingdom"@en .
        ?country dbp:capital ?capital .
        ?capital geo:lat ?lat .
        ?capital geo:long ?long
}#results: 4

Q3D: SELECT * WHERE {
        {dbpedia:Paris ?property ?hasValue}
        UNION
        {?isValueOf ?property dbpedia:Paris}
}#results:41482
```

(b) DBpedia query log

Fig. 1. SPARQL queries from the logs of SWDF and DBpedia

Consider the $Q1S$ from the log of SWDF, this query can be extended with the query $Q1D$ from the log of DBpedia. The result is the SPARQL 1.1 federated Query $Q1S1D$ given in Fig. 2. $Q1S1D$ is generated by joining the variable ?place of the query $Q1S$, i.e. the object of the predicate foaf:based_near with the variable ?country of the query $Q1D$, i.e. the subject of the predicates rdfs:label and dbpedia2:capital. The joined variable ?country has been renamed by ?place, in the generated query $Q1S1D$. The execution of this query over a federation of SWDF and DBpedia produces 1388 results.

The generated query $Q1S1D$ can be recommended as a plausible federated query. In the same way, we can generate a more complex federated query such as the query $Q2S2D$ shown in Fig. 2b. $Q2S2D$ is obtained by extending the query $Q2S$ from the log of SWDF with the query $Q2D$ from the log of DBpedia. The joining variable ?sameAs is renamed as ?person in $Q2S2D$.

Recommending plausible federated queries is challenging because the size of logs. The log of DBpedia contains 217 812 queries, and the log of SWDF contains 64 030 queries [12]. To overcome this problem, we propose a semantic summary that allows to reduce drastically the size of logs by excluding non joinable queries. The main contributions of the paper are:

- a new semantic summary for pruning query logs.
- an algorithm to exclude non joinable queries from logs.

---

[1] http://sage.univ-nantes.fr.

[2] http://lodlaundromat.org/.

[3] All information about logs, and prefixes are available at the project site: https://github.com/GDD-Nantes/PFed.

```
SELECT * WHERE {
  SERVICE <http://swdf-2012>
  { ?inst rdf:type ?dClass .
    ?inst foaf:based_near ?place
  SERVICE <http://dbpedia-3.5.1>
  { ?place rdfs:label "United Kingdom"@en .
    ?place dbp:capital ?capital .
    ?capital geo:lat ?lat .
    ?capital geo:long ?long }}}
#results = 1388
```

(a) $Q1S \bowtie Q1D$

```
SELECT * WHERE {
  SERVICE <http://swdf-2012> {
  swc:tim-finin rdf:type foaf:Person
    {swc:tim-finin foaf:name ?name1}
  UNION
    {swc:tim-finin rdfs:label ?name1}
  OPTIONAL
    {swc:tim-finin foaf:mbox_sha1sum ?mbox_sha1sum}
  OPTIONAL
    {swc:tim-finin foaf:homepage ?homepage}
  OPTIONAL
    {swc:tim-finin foaf:page ?page}
  OPTIONAL
    {swc:tim-finin owl:sameAs ?person
    SERVICE <http://dbpedia-3.5.1> {
      ?person skos:subject ?subject .
      ?person dbo:birthDate ?birth .
      ?person foaf:name ?name2 .
      ?person rdfs:comment ?description
      FILTER (lang(?description) = "en")}}}
  OPTIONAL
    {swc:tim-finin rdfs:seeAlso ?seeAlso}
}}
#results = 178
```

(b) $Q2S \bowtie Q2D$

Fig. 2. Plausible federated query generated from logs of SWDF and DBpedia in Fig. 1

– an algorithm for generating plausible federated queries using the pruned logs.
– an experimentation using real queries logs of SWDF 2012 and DBpedia 3.5.1.

This paper is organized as follows. Section 2 summarizes related works. Section 3 details PFED approach and algorithms. Section 4 presents our experimental results. Finally, conclusions and future work are outlined in Sect. 5.

2   Related Work

Many efforts have been done to automatically generate SPARQL queries, either for individual dataset [4,12] or multiple datasets as Splodge [7] and Fed-Bench [14]. Federated queries benchmarks have been proposed for evaluating the performance of federated query engine. Existing benchmark rely either on hand-crafted queries or on automatically generated ones.

FedBench [14] rely on hand-crafted queries. The datasets of FedBench are real datasets preselected from the Linked Data Cloud, e.g. Life Science, Cross domain. FedBench is commonly used for the evaluation of federated query engines. FedBench is not designed to recommend plausible federated queries over a federation of SPARQL endpoints. LargeRDFBench [11] attempts to generate more realistic federated queries. The benchmark comprises a total of 32 queries for SPARQL endpoint federation. Queries are ranging from simple queries extracted from FedBench queries and large data queries created by the authors with the help of the expert domain. As FedBench, LargeRDFBench are designed for preselected datasets and queries are designed for specific domains and cannot be used for automatic generation of realistic federated queries.

Splodge [7] proposes heuristics for automatic query generation. Splodge generates only conjunctive queries of triple patterns, i.e., Basic Graph Patterns (BGP) with bound predicate, unbound subject and unbound object. Other

SPARQL operators such as FILTER, OPTIONAL are not considered. However, recent analytical study of large SPARQL query logs [6] shows that 74.83% of studied queries have JOIN, FILTER and OPTIONAL and only 7.49% have JOIN alone (conjunctive queries). Consequently, the queries of Splodge cannot reflect the reality. Feta [8] is a federated query tracker that computes Basic Graph Patterns from a federated log. It supposes the existence of a federated query log. In this work, we want to build and recommend federated queries rather than analyzing federated query logs.

Existing approaches of automatic generation of federated queries do not reflect reality and hand-crafted federated queries are designed for specific datasets with the purpose to stress the performance of a federated query engine. Benchmarks are not designed for recommending plausible federated queries.

3   Generation of Plausible Federated Queries

Intuitively, for generating a plausible federated query over n datasets, we propose to start by combing (joining) the query logs log_1 and log_2 of two datasets d_1 and d_2, respectively. Then, we generate new federated queries by joining the resulting queries and the log log_3 of the dataset d_3. We repeat the same process iteratively until processing the n query logs.

In the following, for simplicity, we restrict our discussion to the case of two real query logs. Given two queries Q_1 and Q_2 belong to different query logs, we want to build a plausible federated query FQ. We call FQ a plausible federated query because it is composed of two real queries. Our intuition is FQ is more likely to be a real query than a synthetic one.

3.1   Datasets Capabilities

We can distinguish different type of join combinations: subject-subject or object-subject leading to different query structures star-shaped, path-shaped, or hybrid queries [14]. To find joinable predicates, one can rely on the Vocabulary Of Interlinked Datasets VoID [3]. This vocabulary describes metadata about RDF datasets and the *linkset*. A *linkset* is a collection of RDF links between two datasets[4]. An RDF link is an RDF triple whose subject and object are described in different datasets. This corresponds to the joinable predicates in the example of the Fig. 2. However, we cannot use VoID to detect joinable predicates because a large number of RDF datasets do not provide VoID [16], only 13.65% of datasets[5] (77/564) present a VoID description.

Another solution is to use the capabilities of data sources as defined in Hibiscus [13] to check the *possible* existence of matching. According to [13], the data summary of a source $d \in D$ is the set $CA(d)$ of all *capabilities* of that source. In Hibiscus, this summary is used to remove endpoints during the source selection during federated query processing.

---

[4] https://www.w3.org/TR/void.
[5] http://sparqles.ai.wu.ac.at/.

```
[] a ds:Service ;
ds:url <http://swdf-2012> ;
ds:capability [
    ds:predicate foaf:based_near ;
    ds:sbjAuthority <http://data.semanticweb.org> ;
    ds:objAuthority <http://dbpedia.org>,
    <http://www.w3.org>, <http://sws.geonames.org>,
    <http://data.semanticweb.org> ; ] ;
ds:capability [
    ds:predicate owl:sameAs ;
    ds:sbjAuthority <http://data.semanticweb.org> ;
    ds:objAuthority <http://dbpedia.org>, ...] ;
ds:capability [
    ds:predicate swc:hasLocation ;
    ds:sbjAuthority <http://data.semanticweb.org>;
    ds:objAuthority <http://data.semanticweb.org>,
    <http://dbpedia.org> ; ] ;
ds:capability [
    ds:predicate swrc:author ;
    ds:sbjAuthority <http://data.semanticweb.org> ;
    ds:objAuthority <http://data.semanticweb.org> ; ] ;
ds:capability [
    ds:predicate foaf:maker ;
    ds:sbjAuthority <http://data.semanticweb.org> ;
    ds:objAuthority <http://data.semanticweb.org> ; ] ;
ds:capability [
    ds:predicate swrc:abstract ;
    ds:sbjAuthority <http://data.semanticweb.org> ; ] ;
ds:capability [
    ds:predicate skos:prefLabel ;
    ds:sbjAuthority <http://dbpedia.org>, ...] ;
```

(a) SWDF data summary

```
[] a ds:Service ;
ds:url <http://dbpedia-3.5.1> ;
ds:capability [
    ds:predicate dbpedia2:capital ;
    ds:sbjAuthority <http://dbpedia.org> ;
    ds:objAuthority <http://dbpedia.org> ; ] ;
ds:capability [
    ds:predicate dbo:birthDate ;
    ds:sbjAuthority <http://dbpedia.org> ; ] ;
ds:capability [
    ds:predicate rdfs:comment ;
    ds:sbjAuthority <http://dbpedia.org> ; ] ;
ds:capability [
    ds:predicate foaf:name ;
    ds:sbjAuthority <http://dbpedia.org> ; ] ;
ds:capability [
    ds:predicate dbo:abstract ;
    ds:sbjAuthority <http://dbpedia.org> ; ] ;
ds:capability [
    ds:predicate dbo:thumbnail ;
    ds:sbjAuthority <http://dbpedia.org> ;
    ds:objAuthority <http://upload.wikimedia.org> ; ] ;
ds:capability [
    ds:predicate foaf:depiction ;
    ds:sbjAuthority <http://dbpedia.org> ;
    ds:objAuthority <http://upload.wikimedia.org> ; ] ;
ds:capability [
    ds:predicate dbpedia2:party ;
    ds:sbjAuthority <http://dbpedia.org> ;
    ds:objAuthority <http://dbpedia.org>,
    <http://www.xat.org> ; ] ;
```

(b) DBpedia data summary

```
[] a ds:Service ;
ds:url <http://swdf-2012> ;
ds:capability [
    predicate: foaf:based_near ;
    sbjClasses: foaf:Person, ...] ;
    objClasses: dbo:Country, dbo:Place,
    dbo:PopulatedPlace, ...] ;
ds:capability [
    ds:predicate: owl:sameAs ;
    objClasses: dbo:Person, dbo:Scientist, ...] ;
ds:capability [
    ds:predicate: skos:prefLabel ;
    sbjClasses: foaf:Organization, foaf:Person,
    skos:Concept, swc:WorkshopEvent ; ] ;
```

(c) SWDF classes summary

```
[] a ds:Service ;
ds:url <http://dbpedia-3.5.1> ;
ds:capability [
    ds:predicate: dbpedia2:capital ;
    sbjClasses: dbo:Country, dbo:Place,
    dbo:PopulatedPlace, ...] ;
    objClasses: dbo:City, dbo:Place,
    dbo:PopulatedPlace, ...] ;
ds:capability [
    ds:predicate: dbo:birthDate ;
    sbjClasses: dbo:Person,dbo:Scientist, ...] ;
ds:capability [
    ds:predicate: dbo:abstract ;
    sbjClasses: foaf:Person, dbo:Ship, ...] ;
```

(d) DBpedia classes summary

Fig. 3. Sample of authorities and classes summaries of logs of SWDF and DBpedia

Definition 1 (Authority Capability). *Given a source d, an authority capability is a triple (p, SA(d,p), OA(d,p)), which contains (1) a predicate p in d, (2) the set SA(d,p) of all distinct subject authorities of p in d and (3) the set OA(d,p) of all distinct object authorities of p in d.*

The total number of capabilities of a source is equal to the number of distinct predicates in it. The definition of the authorities of a subject or an object relies on the analysis of the Unified Resource Identifier (URI) syntax. The URI syntax consists of a hierarchical sequence of

```
SELECT DISTINCT ?type
WHERE {
      ?s foaf:based_near ?o .
      FILTER isURI(?s)
      ?s rdf:type ?type
}
```

```
SELECT DISTINCT ?type WHERE {
  SERVICE<http://swdf-2012> {
    ?s foaf:based_near ?o .
    FILTER isURI(?s)
    FILTER regex(STR(?s), 'http://dbpedia.org')
    SERVICE<http://dbpedia3.5.1> {
      ?s rdf:type ?type
  }}
}
```

```
SELECT DISTINCT ?type
WHERE {
      ?s foaf:based_near ?o .
      FILTER isURI(?o)
      ?o rdf:type ?type
}
```

```
SELECT DISTINCT ?type WHERE {
  SERVICE<http://swdf-2012> {
    ?s foaf:based_near ?o .
    FILTER isURI(?o)
    FILTER regex(STR(?o), 'http://dbpedia.org')
    SERVICE<http://dbpedia3.5.1> {
      ?o rdf:type ?type
  }
}
```

(a) Retrieving types of *foaf:based_near* from SWFD

(b) Retrieving types for *foaf:based_near* from DBpedia

Fig. 4. Class Capability for *foaf:based_near* predicate in SWDF

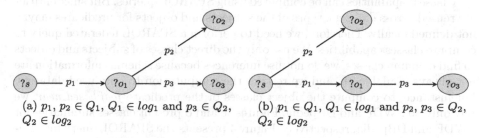

(a) $p_1, p_2 \in Q_1$, $Q_1 \in log_1$ and $p_3 \in Q_2$, $Q_2 \in log_2$

(b) $p_1 \in Q_1$, $Q_1 \in log_1$ and $p_2, p_3 \in Q_2$, $Q_2 \in log_2$

Fig. 5. Possible structures for hybrid federated queries

components referred to as the *scheme, authority, path, query,* and fragment[6]. For example, the uri <http://dbpedia.org/ontology/Plant> contains a schema "http", an authority "dbpedia.org" and a path "ontology/Plant". To compute the set of capabilities for a source, the first two components (path, authority) are combined as the *authority* of the URI. Figure 3 presents a sample of the summary of SWDF 2012 and DBpedia 3.5.1. For instance, in Fig. 3a, the first capability of SWDF data source is the predicate *foaf:based_near*, its subject authority is <http://data.semanticweb.org> and its object authorities are <http://dbpedia.org>, <http://www.w3.org>, <http://sws.geonames.org>, and <http://data.semanticweb.org>.

Authority summary allows to prune the query logs only if many predicates have different subjects or objects authority. However, this not always the case, especially for the subject authority. For instance, the majority of subjects of DBpedia have the authority <http://dbpedia.org>, only six predicates out of 39672 predicates of DBpedia 3.5.1 do not have <http://dbpedia.org> as a subject authority. Therefore, if a query Q_1 in SWDF query log is joinable with a query Q_2 in DBpedia query log on the subject authority

---

[6] URI Syntax Components: https://tools.ietf.org/pdf/rfc3986.pdf.

<http://dbpedia.org>, then Q_1 will be joinable with a large number of queries in the log of DBpedia. Therefore, for query logs of SWDF and DBpedia, authority summary will prune mostly queries with unbounded predicates.

To further prune the log, we define new data summary that considers semantic of subjects and objects for finding joinable predicates. Intuitively, a subject or an object from one dataset could be joinable with a subject or object from another dataset, if they share some common types. More precisely, we define a new summary called *Class summary*. A class summary is a set of classes capabilities.

Definition 2 (Class Capability). *Given a source d, a class capability is a triple (p, SC(d,p), OC(d,p)), which contains (1) a predicate p in d, (2) the set SC(d,p) of all distinct subject classes of p in d and (3) the set OA(d,p) of all distinct object classes of p in d.*

Classes capabilities can be computed using SPARQL queries. But since entities are reused across datasets, types of the subjects and objects for predicates maybe not defined locally. Therefore, we need to perform a SPARQL federated query to compute classes capabilities. We use only the direct classes of subjects and objects to find common classes, we do not use inferences because schemas information are not always available [9], and we restrict the computation to only used datasets. For instance, to compute the object classes of the predicate *foaf:based_near*, we rely only on SWDF and DBpedia. Figures 3c and d present classes summaries for SWDF and DBpedia, respectively. Figure 4 presents the SPARQL query for computing the Class Capability of *foaf:based_near* predicate in SWDF.

3.2   Pruning Query Logs

Based on authorities summaries and classes summaries, we can prune the logs of corresponding datasets by retaining only joinable queries.

Definition 3 (Joinable Queries). *Let D be a set of distinct data sources, $d_1, d_2 \in D$. Let log_1 and log_2 are the real query log of d_1 and d_2, respectively. For two queries $Q_1 \in log_1$ and $Q_2 \in log_2$ with $tp_1 = (s_1, p_1, o_1) \in Q_1$ and $tp_2 = (s_2, p_2, o_2) \in Q_2$, we say that Q_1 and Q_2 are joinable if p_1 and p_2 have a predicate joinable path or predicate joinable star.*

Definition 4 (Predicate Joinable Path). *joinablePath(p_1, p_2) = true, if $OA(d_1, p_1) \cap SA(d_2, p_2) \neq \emptyset$ and $OC(d_1, p_1) \cap SC(d_2, p_2) \neq \emptyset$.*

Definition 5 (Predicate Joinable Star). *joinableStar(p_1, p_2) = true, if $SA(d_1, p_1) \cap SA(d_2, p_2) \neq \emptyset$ and $SC(d_1, p_1) \cap SC(d_2, p_2) \neq \emptyset$.*

The hybrid join pattern is built as a mix of a path join pattern and a star join pattern. Figure 5 presents possible structures of hybrid federated queries. The query generated in Fig. 5a is built from the path query of $p_1 \in Q_1$, $Q_1 \in log_1$ and $p_3 \in Q_2$, $Q_2 \in log_2$. The query generated in Fig. 5b built from the star query of $p_1 \in Q_1$, $Q_1 \in log_1$ and $p_2 \in Q_2$, $Q_2 \in log_2$.

Algorithm 1: Joinable predicates

Input: $AS1, CS1, AS2, CS2$ ▷ Authorities and classes summaries for the two datasets

Output: $JPred$ ▷ Set of joinable predicates

1 **Function** JoinPred($AS1, CS1, AS2, CS2$):

2    $JPred \longleftarrow \varnothing$;

3   **foreach** $cap_1 \in AS_1$ **do**

4   **foreach** $cap_2 \in AS_2$ **do**

5   **if** $AS1.objAuthority(cap_1) \cap AS2.sbjAuthority(cap_2) \neq \varnothing$ **then**

6   **if** $CS1.objClasses(cap_1) \cap CS2.sbjClasses(cap_2) \neq \varnothing$ **then**

7 | $JPred \longleftarrow JPred \cup (cap_1.predicate, cap_2.predicate)$;

8   **end**

9   **end**

10   **end**

11   **end**

12   **return** $JPred$;

13 **End Function**

The objective now is to prune query logs and conserve only joinable queries. First, the Algorithm 1 uses summaries to conserve predicate joinable (predicate joinable path), then the Algorithm 2 excludes non joinable queries from logs. For logs in Fig. 3, the Algorithm 1 keeps the couple ($foaf:based\_near, dbpedia2:capital$) because they share http://dbpedia.org as object and subject authority, respectively, and they share *Country, Place* and *PopulatedPlace* as object and subject classes, respectively.

To compute predicate joinable star, we only need to modify conditions in lines 5–6 of the Algorithm 1 to compare subjects parts of both capabilities. With this modification, the algorithm will keep the couple ($skos:prefLabel, dbo:abstract$) as they share same authorities and classes as subjects. The Algorithm 1 can be iteratively called to compute predicate joinable path or star for more than two datasets.

We use the result of the Algorithm 1 to exclude non joinable queries as shown in the Algorithm 2. After the execution of the Algorithm 2 for joinable path, $Q1S$ of SWDF and $Q2D$ of DBpedia will be preserved, because they have the joinable predicates ($foaf:based\_near, dbpedia2:capital$) as shown previously. We exclude $Q3S$ because it cannot be joined with any query from dbpedia, *i.e.* no predicate in DBpedia has <http://data.semanticweb.org> as subject authority. We also eliminate $Q3D$ because the capability of unbound predicate is undefined.

3.3 Building Plausible Federated Queries

We rely on the results of the Algorithm 2 to build plausible federated queries. For sake of simplification, we start by illustrating the generating of minimal federated queries PFED_{min}. A minimal federated contains one triple from log_1 and one triple from log_2.

Algorithm 2: Joinable queries

Input: $log_1, log_2, JPred$ ▷ Logs of both dataset and the set of corresponding joinable predicates
Output: $feds$ ▷ Set of federated queries
1 **Function** GenFed($log_1, log_2, JPred$):
2 │ $feds \longleftarrow \varnothing$;
3 │   **foreach** $Q_1 \in log_1$ **do**
4 │   │   **foreach** $Q_2 \in log_2$ **do**
5 │   │   │   **if** $\exists (p_1, p_2) | p_1 \in Q_1, p_2 \in Q_2 \wedge (p_1, p_2) \in JPred$ **then**
6 │   │   │   │ $feds \longleftarrow feds \cup (Q_1, Q_2)$
7 │   │   │   **end**
8 │   │   **end**
9 │   **end**
10 │   **return** $feds$;
11 **End Function**

```
SELECT * WHERE {
  SERVICE <http://swdf-2012>
    {?obj foaf:based_near ?place .
  SERVICE <http://dbpedia-3.5.1>
    {?place dbpedia2: capital ?capital }}}
```

(a) Path query by joining a triple pattern from $Q1S$ and a triple pattern from $Q1D$

```
SELECT * WHERE {
  SERVICE <http://swdf-2012>
    {?x rdfs:prefLabel ?o1 .
  SERVICE <http://dbpedia-3.5.1>
    {?x dbo:thumbnail ?o2 }}}
```

(b) Star query by joining a triple from $QS4$ and a triple pattern from $Q4D$

Fig. 6. Minimal federated queries generated from pruned logs of SWDF and DBpedia in Fig. 3

In order to construct a path (star) join, we substitute the object (subject) of p_1 and the subject of p_2 by the same value as given in the Table 1.

Figure 6a presents a minimal path-shaped federated query between *foaf:based_near* $\in Q1S$ and *dbpedia2:capital* $\in Q1D$ in Fig. 3. Figure 6b presents a minimal star-shaped federated query between *skos:prefLabel* $\in Q4S$ and *dbo:thumbnail* $\in Q4D$.

PFED_{min} are not required to generate plausible federated queries. But they can help to reduce the number of potential joinable predicates by only keeping PFED_{min} producing results. They can also be used to navigate through datasets.

The construction of Q_{PFED} is tricky, if the original queries contain OPTIONAL operator. We have to construct only correct plausible federated query. A plausible federated query is *correct* if it is well designed [10] and service-safeness [5].

Definition 6 (Well Designed [10]). *A graph pattern P is well designed if for every occurrence of a sub-pattern P′ = (P1 OPT P2) of P and for every variable ?X occurring in P, the following condition holds:*

if ?X occurs both inside P2 and outside P′, then it also occurs in P1.

Table 1. All substitution values possible to create path join. $?x$, $?y$ are variables and a, b are constants (URIs or literals)

tp_1 object	$tp2$ subject	Substitution value
$?x$	$?y$	$?x$
$?x$	a	a
a	$?x$	a
a	b	null

```
SELECT * WHERE {
    ?s1  p1  ?o1 .
    ?s1  p2  ?o2   }
```

(a) Q_1 from log
endpoint1

```
SELECT * WHERE {
    ?s3  p3  ?o3
    OPTIONAL { ?o2 p4 ?o3 }}
```

(b) Q_2 from log endpoint2

```
SELECT * WHERE {
    SERVICE <endpoint1>
    { ?s1  p1  ?o1 .
      ?s1  p2  ?o2
    SERVICE <endpoint2>                :P'
    { ?s3  p3  ?o3                     :P1
      OPTIONAL { ?o2 p4 ?o3 }}}}       :P2
```

(c) $Q_1 \bowtie Q_2$

Fig. 7. A non well designed federated query

The federated query in Fig. 7c is not well designed because the variable $?o2$ occurs in P2 and outside the P' (i.e. clause SERVICE <dataset2>), but it not occurs in P1.

The service-safeness provides condition that ensures that a SPARQL query containing SERVICE operator can be safely evaluated. Our generated queries ensure service-safeness because each SERVICE clause has only bounded service, i.e., during the construction the URI of the SPARQL endpoints are known.

The main issue is to build well designed queries to avoid cartesian products as illustrated in Fig. 7c. If Q_2 does not have a mapping for $?o2$, a result will still produced. To avoid this problem, we define the following strategy:

- If Q_1 and Q_2 are conjunctive queries (a.k.a BGPs) then $Q_{\text{PFED}} = Q_1 \bowtie Q_2$, Q_{PFED} is a simple concatenation of queries ($Q_1 . Q_2$), as in Fig. 2, $Q1S1D = Q1S \bowtie Q1D$.
- If Q_1 contains binary operators like UNION or OPTIONAL, we distinct two cases:
  - If a joinable predicate is outside binary clauses of Q_1, we add Q_2 in the BGP part of Q_1.
  - If a joinable predicate of Q_1 is inside the UNION or OPTIONAL clauses, we append Q_2 inside this clause after the substitution of the join variables (subject or object of the triple) according to Table 1.

– If a joinable predicate of Q_2 is inside an OPTIONAL clause, we make sure to not generate non well designed queries like query shown in Fig. 7c.

4 Evaluation

The objective of the evaluation is to answer empirically the following questions: Do authorities summaries prune non joinable predicates? Do classes summaries prune further non joinable predicates? Does PFED able to generate plausible federated queries?

All data, codes, and generated query are available at the project web page[7].

Table 2. Real datasets and real logs

Dataset	|triples|	|dataset predicates|	|original log|	|SELECT queries|	|log predicates|
SWDF	242 256	170	64 030	37 592	201
DBPedia	232 542 405	39 672	217 812	127 812	247

4.1 Experimental Setup

Dataset and Queries: We use SWDF 2012 and DBPedia 3.5.1 datasets and clean queries of Feasible[8]. We use only SELECT queries to construct plausible federated queries. Table 2 reports statistics about the datasets and query logs. It is strange that the query log of SWDF contains more predicates than the original dataset hosted at the SPARQL endpoint. Some queries in the logs use predicates that are not defined in the dataset. As they appear inside OPTIONAL or UNION, they do not stop queries from returning results. Using DBpedia to generate plausible federated queries is challenging because DBpedia dataset has a high number of predicates and the log of DBpedia has a high number of queries.

4.2 Experimental Results

Do authorities summaries prune non joinable predicates? Table 3 presents the results of pruning using authorities summaries. As we can see, the reduction is 62.75% for SWDF query log for path-shaped queries (all path refers to path from SWDF to DBpedia) and by 42.82% for star-shaped. The reduction is only 2.15% for DBpedia log for both path-shaped queries and star-shaped generation. This reduction is not significant because most of predicates in DBPedia has the authority <http://dbpedia.org>.

Table 3. Logs pruning using authorities summaries

Dataset	path-shaped			star-shaped		
	|predicate joinable|	|pruned log|	% reduce	|predicate joinable|	|pruned log|	% reduce
SWDF	6	14 003	62.75	3	21 495	42.82
DBPedia	230	125 078	2.15	229	125 070	2.15

---

[7] https://github.com/GDD-Nantes/PFed.
[8] https://github.com/dice-group/feasible.

Table 4. Logs pruning using authorities and classes summaries

	path-shaped			star-shaped		
Dataset	|predicate joinable |	|pruned log|	% reduce	|predicate joinable |	|pruned log|	% reduce
SWDF	3	9 355	75.12	3	21 495	42.82
DBPedia	139	36 522	71.42	83	36 449	71.48

Do classes summaries prune further non joinable predicates? We now use our classes summaries on top of authorities summaries. We observe in Table 4 that the sizes of logs are reduced. The reduction is impressive for DBpedia, it is about 72%. Therefore, classes summaries are affective for pruning non joinable queries.

We observe also an important reduction in the number of minimal federated queries $PFED_{min}$ (Table 5). This reduction is important as each $PFED_{min}$ contributes to many federated queries.

Table 5. Number of $PFED_{min}$ generated using authorities and classes summaries

With authorities		With authorities and classes	
path-shaped	star-shaped	path-shaped	star-shaped
1 146	687	352	432

Does PFED *Generate Plausible Federated Queries?* Due to the size of the pruned logs, we can generate a large number of plausible federated queries. In our experimentation, we focus on the generation of path-shaped between *foaf:based_near* from SWDF and *dbpedia2:capital* from DBPedia. The pruned SWDF query log contains 2 866 queries that contains *foaf:based_near*. Many of these queries have the same structure but with different literals and variables. Therefore, instead of producing $2866 \times 14 = 40124$ queries where 14 is the number of queries that contains *dbpedia2:capital* in pruned DBpedia log, we define patterns for *foaf:based_near* queries. We differentiate 9 patterns for *foaf:based_near* queries and we generate 24 queries. All generated queries are executed correctly and 19 of these queries have non empty results set (see Table 6).

We generate star-shaped plausible federated queries based on *skos:prefLable* from SWDF and *dbpedia:thumbnail* from DBPedia (see Table 6). The 42 generated queries are executed correctly and 28 of these queries produce results.

Table 6. PFED path and star, $p_1 \in$ SWDF and $p_2 \in$ DBPedia

	p_1	$\|p_1\|$	p_2	$\|p_2\|$	|PFED|	|with result|	%
PFED path	foaf:based_near	9	dbpedia2:capital	5	24	19	79.17
PFED star	skos:prefLabel	3	dbo:thumbnail	14	42	14	33.33

5   Conclusion and Future Work

We presented PFED an approach for automatic generation of plausible federated queries based on real query logs. PFED starts by pruning the logs to exclude non joinable queries using data summaries. The first one is based on the authorities and the second is based on the type of subjects and objects of predicates. Experimentations with real query logs of SWDF and DBpedia demonstrate that PFED is able to prune considerably the logs and generate plausible federated queries.

As future work, we would like to experiment PFED with more real query logs and produce plausible federated queries over a large number of SPARQL endpoints. Finally, we plan to extend PFED with statistical information to generate only queries that return results.

Acknowledgement. This work is part of the multidisciplinary project SEDELA, funded by CominLabs, that brings together three laboratories: LS2N, CREAD and Lab-STICC.

References

1. Acosta, M., Vidal, M.-E., Lampo, T., Castillo, J., Ruckhaus, E.: ANAPSID: an adaptive query processing engine for SPARQL endpoints. In: Aroyo, L., et al. (eds.) ISWC 2011. LNCS, vol. 7031, pp. 18–34. Springer, Heidelberg (2011). https://doi.org/10.1007/978-3-642-25073-6_2
2. Adomavicius, G., Tuzhilin, A.: Toward the next generation of recommender systems: a survey of the state-of-the-art. IEEE Trans. Knowl. Data Eng. **17**(6), 734–749 (2005)
3. Alexander, K., Cyganiak, R., Hausenblas, M., Zhao, J.: Describing linked datasets. In: LDOW (2009)
4. Aluç, G., Hartig, O., Özsu, M.T., Daudjee, K.: Diversified stress testing of RDF data management systems. In: The International Semantic Web Conference, pp. 197–212 (2014)
5. Arenas, M., Pérez, J.: Federation and navigation in SPARQL 1.1. In: Eiter, T., Krennwallner, T. (eds.) Reasoning Web 2012. LNCS, vol. 7487, pp. 78–111. Springer, Heidelberg (2012). https://doi.org/10.1007/978-3-642-33158-9_3
6. Bonifati, A., Martens, W., Timm, T.: An analytical study of large SPARQL query logs. PVLDB **11**(2), 149–161 (2017). http://www.vldb.org/pvldb/vol11/p149-bonifati.pdf
7. Görlitz, O., Thimm, M., Staab, S.: SPLODGE: systematic generation of SPARQL benchmark queries for linked open data. ISWC 2012. LNCS, vol. 7649, pp. 116–132. Springer, Heidelberg (2012). https://doi.org/10.1007/978-3-642-35176-1_8
8. Nassopoulos, G., Serrano-Alvarado, P., Molli, P., Desmontils, E.: FETA: Federated QuEry TrAcking for Linked Data. In: Hartmann, S., Ma, H. (eds.) DEXA 2016. LNCS, vol. 9828, pp. 303–312. Springer, Cham (2016). https://doi.org/10.1007/978-3-319-44406-2_24
9. Neumann, T., Moerkotte, G.: Characteristic sets: accurate cardinality estimation for RDF queries with multiple joins. In: 2011 IEEE 27th International Conference on Data Engineering (ICDE), pp. 984–994. IEEE (2011)

10. Pérez, J., Arenas, M., Gutierrez, C.: Semantics and complexity of SPARQL. In: Cruz, I., et al. (eds.) ISWC 2006. LNCS, vol. 4273, pp. 30–43. Springer, Heidelberg (2006). https://doi.org/10.1007/11926078_3
11. Saleem, M., Hasnainb, A., Ngonga Ngomo, A.C.: LargeRDFBench: A billion triples benchmark for sparql endpoint federation. J. Web Semant. (JWS) (2017). https://svn.aksw.org/papers/2017/LargeRDFBench_JWS/public.pdf
12. Saleem, M., Mehmood, Q., Ngonga Ngomo, A.-C.: FEASIBLE: a feature-based SPARQL benchmark generation framework. In: Arenas, M., et al. (eds.) ISWC 2015. LNCS, vol. 9366, pp. 52–69. Springer, Cham (2015). https://doi.org/10.1007/978-3-319-25007-6_4
13. Saleem, M., Ngonga Ngomo, A.-C.: HiBISCuS: hypergraph-based source selection for SPARQL endpoint federation. In: Presutti, V., d'Amato, C., Gandon, F., d'Aquin, M., Staab, S., Tordai, A. (eds.) ESWC 2014. LNCS, vol. 8465, pp. 176–191. Springer, Cham (2014). https://doi.org/10.1007/978-3-319-07443-6_13
14. Schmidt, M., Görlitz, O., Haase, P., Ladwig, G., Schwarte, A., Tran, T.: Fed-Bench: a benchmark suite for federated semantic data query processing. In: International Semantic Web Conference, pp. 585–600 (2011). https://doi.org/10.1007/978-3-642-25073-6_37
15. Schwarte, A., Haase, P., Hose, K., Schenkel, R., Schmidt, M.: FedX: optimization techniques for federated query processing on linked data. In: Aroyo, L., et al. (eds.) ISWC 2011. LNCS, vol. 7031, pp. 601–616. Springer, Heidelberg (2011). https://doi.org/10.1007/978-3-642-25073-6_38
16. Vandenbussche, P.Y., Umbrich, J., Matteis, L., Hogan, A., Buil-Aranda, C.: SPARQLES: monitoring public SPARQL endpoints. Seman. Web 8(6), 1049–1065 (2017)

Representing and Reasoning About Precise and Imprecise Time Points and Intervals in Semantic Web: Dealing with Dates and Time Clocks

Nassira Achich[1(⊠)], Fatma Ghorbel[1,2], Fayçal Hamdi[2],
Elisabeth Metais[2], and Faiez Gargouri[1]

[1] MIRACL Laboratory, University of Sfax, Sfax, Tunisia
achichnassira@gmail.com, fatmaghorbel6@gmail.com,
faiez.gargouri@isims.usf.tn
[2] CEDRIC Laboratory, Conservatoire National des Arts et Métiers (CNAM),
Paris, France
{faycal.hamdi,metais}@cnam.fr

Abstract. Temporal data may be precise or imprecise. Representing and reasoning about these kinds of data in ontology still needs to be addressed. A significant number of approaches exist. However, they handle only precise temporal data and lack imprecise ones. In this paper, we propose a crisp-based approach for representing and reasoning about temporal data in term of quantitative (i.e., time points that can be dates and clocks, and time intervals) as well as qualitative relations (e.g., "*before*") in ontology. It aims to support not only precise time points and intervals, but also imprecise ones e.g., "The journey starts *by the beginning of June* and ends *by mid-June*". It relies only on crisp exiting Semantic Web standards and it is modeled in crisp ontology. Our approach is based on three blocks. (*i*) We extend the 4D-fluents approach with new crisp ontological components to represent the mentioned precise and imprecise temporal data. (*ii*) We extend the Allen's interval algebra to reason about imprecise time intervals. Compared to related work, our extension is entirely based on crisp set theory. The resulting interval relations preserve many of the desirable properties of the original algebra. We adapt these relations to allow relating a time interval and a time point, and two time points; where time points and intervals may be both precise or both imprecise. All proposed relations can be used for temporal reasoning by means of transitivity tables. (*iii*) We propose an OWL 2 ontology based on our extensions. It proposes a set of SWRL rules to infer the proposed qualitative temporal relations. A prototype based on this ontology is implemented. We apply our approach to the Travel ontology.

Keywords: Precise and imprecise temporal data ·
Temporal representation and reasoning · Crisp ontology · 4D-fluent approach ·
Allen's interval algebra

© Springer Nature Switzerland AG 2019
S. Hartmann et al. (Eds.): DEXA 2019, LNCS 11707, pp. 198–208, 2019.
https://doi.org/10.1007/978-3-030-27618-8_15

1   Introduction

Temporal data given by people are often imprecise. For instance, if they give the information "Holidays start by the end of May", an imprecise measure is introduced. Indeed, "by the end of May" could be May 28th, 29th, 30th or 31th. This paper focuses on representing and reasoning about precise and imprecise temporal data in ontology.

In the Semantic Web field, many approaches have been proposed to represent and reason about precise temporal data. However, most of them handle only time intervals and associated qualitative relations i.e., they are not intended to handle time points and qualitative relations between a time interval and a time point or two time points. Besides, to the best of our knowledge, there is no approach devoted to handle imprecise temporal data in ontology.

In our previous work [11], we have proposed a fuzzy-based approach for representing and reasoning about imprecise time intervals in ontology. It is entirely based on fuzzy set theory and dates and time clocks are not considered. In this paper, we propose a crisp-based approach for representing and reasoning about concepts evolving in time in ontology. Quantitative temporal data, i.e., time points (that can be dates or time clocks) and intervals, as well as qualitative ones (e.g., "before") are taken into consideration. Our approach supports not only precise time points and intervals, but also imprecise ones, such as "The journey starts by June 5th and finishes by the end of July". We adopt only crisp existing Semantic Web standards. It is modeled in crisp ontology. Our approach is based on three facets: (i) Representing precise and imprecise temporal data in ontology by extending the 4D-fluents approach [25] with new crisp ontological components. (ii) Reasoning about precise and imprecise temporal data by extending the Allen's interval algebra [1]. This latter proposes 13 temporal relations between precise time intervals. However, it is not designed to handle imprecise time intervals. Moreover, it is not intended to relate a time interval and a time point or even two time points. We extend this algebra to propose qualitative temporal relations between imprecise time intervals. Compared to related work, these relations are defined based on crisp set theory. Properties of reflexivity/irreflexivity, symmetry/asymmetry and transitivity are preserved. We adapt the resulting interval relations to propose temporal relations between a time interval and a time point, and two time points that may be precise or imprecise. All temporal relations that we propose can be used for temporal reasoning by means of transitivity tables. (iii) Proposing an OWL 2 ontology based on our extensions. It infers the proposed qualitative temporal relations via a set of SWRL rules.

This paper is organized as follows. Preliminaries and related work are reviewed in Sect. 2. We introduce, respectively, our approaches for representing and reasoning about precise and imprecise temporal data in Sects. 3 and 4. In Sect. 5, we present our ontology implemented based on our extensions. Section 6 presents some experimentations. Section 7 summarizes our contributions and gives some future directions.

2 Preliminaries and Related Work

Related work in the fields of temporal data representation in Semantic Web and Allen's interval algebra are discussed in this section.

2.1 Representing Temporal Data in the Semantic Web Field

There is a need for representing temporal data in ontology. However, representing ontology languages such as OWL and RDF provide a minimal support. They are all based on binary relations that simply connect two instances without adding any temporal data. There is a significant number of approaches for representing temporal data in ontology. We classify them into two categories: (i) approaches which extend the OWL or RDF syntax to incorporate temporal data through defining new OWL or RDF operators and semantics; and (ii) approaches which are implemented directly using OWL or RDF to represent temporal data without extending their syntax.

One of the approaches that belongs to the first category is the Temporal Description Logics [3]. It extends the standard description logics with new temporal semantics such as "until". This approach does not suffer from data redundancy and retain decidability. However, extending OWL or RDF, which is a tedious task, makes it an avoidable solution. Another approach is the Concrete Domains [18]. It requires introducing additional data types and operators to OWL. Several implementations based on this approach have been proposed, such as OWL-MeT [8] and TL-OWL [16]. Temporal RDF [12], which also belongs to the first category, uses only RDF triples. It does not have all the expressiveness of OWL. It cannot express qualitative relations. In [15], the authors present a comprehensive framework to incorporate temporal reasoning into RDF.

The second category includes: Versioning [17], Reification [6], N-ary Relations [20], 4D-Fluents and Named Graphs [23] approaches. Versioning is described as the ability to handle changes in ontology by creating and managing different variants of it. All the versions are independent from each other which require exhaustive searches in the all versions. Reification is a technique for representing N-ary relations when only binary relations are allowed. A new object is created whenever a temporal relation has to be represented. N-ary Relations proposes to represent an N-ary relation as two properties each related with a new object. It maintains property semantics. These approaches suffer from data redundancy. Unlike them, 4D-fluents approach which represents time intervals and their evolution in OWL, minimizes the problem of data redundancy, as the changes occur only in the temporal parts and concepts varying in time are represented as 4-dimensional objects with the 4th dimension being the temporal data. The Named Graphs approach represents each time interval by exactly one named graph, where all triples belonging share the same validity period. Reasoning and querying are supported in [2, 5, 13, 14, 21].

All the reviewed approaches handle only precise temporal data and neglect imprecise ones. They are not intended to handle time points and qualitative temporal relations between a time interval and a time point or even two time points. Based on this study, we choose to extend the 4D-fluents approach to represent precise and imprecise quantitative temporal data and associated qualitative temporal relations in crisp ontology. Our choice is based on that we need an approach which relies on existing OWL constructs. Therefore, we exclude the Temporal Description Logic, Concrete Domain and Temporal RDF approaches. We also exclude the Named Graphs approach as it does not support OWL and it is not a W3C compliant solution. Compared to the Reification, N-ary Relations and Versioning approaches, the 4D-fluents approach minimizes data redundancy as the changes occur on the temporal parts and keep the static part unchanged.

2.2   Allen's Interval Algebra: Definition and Extensions

13 qualitative relations between precise time intervals are proposed by Allen. Their definitions are expressed in Table 1. A characteristic of Allen's algebra is that we can deduce new relations through the composition of other ones. For instance, "Before(A, B)" and "Equals(B, C)" gives "Before(A, C)". Allen's interval algebra is not dedicated to represent imprecise time intervals. Furthermore, it does not relate neither a time point and a time interval nor two time points. Several approaches have been extended this algebra. Some of them propose temporal relations between precise time intervals [4, 7, 9] and other ones propose temporal relations between imprecise time intervals [10, 19, 22]. However, these extensions are based on theories related to imperfect data and cannot be supported in the context of crisp ontology.

Table 1. Allen's relations between two precise time intervals A = [A^+, A^-] and B = [B^-, B^+]

Relation(A, B)	Relations between interval bounds	Illustration	Inverse(B, A)
Before	$A^+ < B^-$		After
Meets	$A^+ = B^-$		Met-by
Overlaps	$(A^- < B^-) \wedge (A^+ > B^-) \wedge (A^+ < B^+)$		Overlapped-
Starts	$(A^- = B^-) \wedge (A^+ < B^+)$		Started-by
During	$(A^- > B^-) \wedge (A^+ < B^+)$		Contains
Ends	$(A^- > B^-) \wedge (A^+ = B^+)$		Ended-by
Equals	$(A^- = B^-) \wedge (A^+ = B^+)$		Equals

3   Our Crisp-Based Approach to Representing Precise and Imprecise Temporal Data in Ontology

We extend the 4D-fluents approach to represent precise and imprecise quantitative temporal data as well as associated qualitative temporal relations in ontology.

3.1   Quantitative Temporal Data Representation

We extend the 4D-fluents approach to represent: (i) precise and imprecise time points and (ii) imprecise time intervals. Some of the introduced components are already defined in OWL-Time[1] ontology. Some others that we define, do not exist.

Representing Precise and Imprecise Time Points. We introduce a class "TimePoint" and an object property "TsTimePoint". The latter relates an instance of "TimeSlice" and an instance of "TimePoint". To express the dates and time clocks, we use the class named "time:DateTimeDescription" defined in OWL-Time ontology.

We present precise time points (dates and time clocks). For the dates, let D, Mo and Y be, respectively, precise day, month and year. We use three datatype properties from OWL-Time named "time:day", "time:month" and "time:year" to relate, respectively, the "Date" class and D, Mo and Y. For instance, if we have "The journey begins in *June 05th 2019*", "*June 05th 2019*" is represented as a precise date. "time:day" has the range "*05th*", "time:month" has the range "*June*" and "time:year" has the range "*2019*". Similarly, for the time clocks, let S, Mi and H be, respectively, precise seconds, minutes and hours. We use three datatype properties from OWL-Time named "time: second", "time:minute" and "time:hour", to connect, respectively, the "Clock"class with S, Mi and H. For instance, if we have "The breakfast in the hotel starts at *07:30:00*", "*07:30:00*" is represented as a precise time clock. "time:second" has the range "*00*", "time:minute" has the range "*30*" and "time:hour" has the range "*07*".

We present imprecise time points (dates and time clocks). For the dates, let D, Mo and Y be, respectively, imprecise day, month and year. We represent them by disjunctive ascending sets $\{D^{(1)}...D^{(d)}\}$, $\{Mo^{(1)}...Mo^{(mo)}\}$ and $\{Y^{(1)}...Y^{(y)}\}$. As an example "The price of the train tickets was much cheaper *during the seventies*", "*during the seventies*" is represented as the disjunctive ascending set $\{1970 ... 1979\}$. We define for each of D, Mo and Y, respectively, two datatype properties: "HasDayFrom" and "HasDayTo", "HasMonthFrom" and "HasMonthTo", "HasYearFrom" and "HasYearTo". They are all connected to the "Date" class. For instance, "The journey begins *by the June 05th, 2019*", "*by the June 05th, 2019*" is represented as an imprecise date since the day part is imprecise. "HasDayFrom" has the range "*03rd*" and "HasDayTo" has the range "*07th*". Similarly, for the time clocks, let S, Mi and H be, respectively, imprecise seconds, minutes and hours, represented by disjunctive ascending sets $\{S^{(1)}...S^{(s)}\}$, $\{Mi^{(1)}... Mi^{(mi)}\}$ and $\{H^{(1)}... H^{(h)}\}$. As an example "We should finish the breakfast *at most at 10 o'clock*", "*at most at 10 o'clock*" is represented as the disjunctive ascending set $\{07 ... 10\}$. We define for each of S, Mi and H, respectively, two datatype properties: "HasSecondsFrom" and "HasSecondsTo", "HasMinutesFrom" and "HasMinutesTo", "HasHoursFrom" and "HasHoursTo". They are all connected to the "Clock" class. For instance, if we have "We leave the hotel *after lunch before 5 pm*. Lunch time is between 12 am and 02 pm.", "*after lunch before 05 pm*" is represented as an imprecise time clock. "HasHoursFrom" and "HasHoursTo" have the ranges "*02 pm*" and "*05 pm*".

---

[1] https://www.w3.org/TR/owl-time/.

Representing Imprecise Time Intervals. An imprecise time interval has beginning and ending bounds. We represent them using instances of "TimePoint". In our extension, "time:hasBeginning" and "time:hasEnd" are object properties defining the beginning and the ending bounds of the interval. They relate an instance of "TimeInterval" (domain) and an instance of "TimePoint" (range). For example "We will visit the national park *at 14:30* and we will leave *in the evening*. The national park closes at 22:00" We represent the time of the park closure "*22:00*" as a precise time clock. We represent the duration of the visit as an imprecise time interval. The beginning bound is represented as a precise time clock. The property "time: hasBeginning" represents the range "*14:30*". The ending bound is represented as an imprecise time clock. It could be between 19:00 and 22:00. The property "time: hasEnd" models the range "*until the evening*". The properties "hasHoursFrom" and "hasHoursTo" has the ranges "*19*" and "*22*".

3.2 Qualitative Temporal Data Representation

Four temporal relations may exist between time points and time intervals: Point-Point, Interval-Point, Point-Interval and Interval-Interval relations. Hence we assign four crisp object properties. The property "RelationPoints" connects two instances of the "TimePoint" class to represent Point-Point relations. "RelationIntervalPoint" property connects an instance of the "TimeInterval" (domain) class and an instance of the "TimePoint" class (range) to represent Interval-Point relation. "RelationPointInterval" property connects an instance of the "TimePoint" (domain) class and an instance of the "TimeInterval" class (range) to represent Point-Interval relation. "RelationIntervals" connects two instances of the "TimeInterval" class to represent Interval-Interval relations. Figure 1 represents our 4D-fluents approach extension.

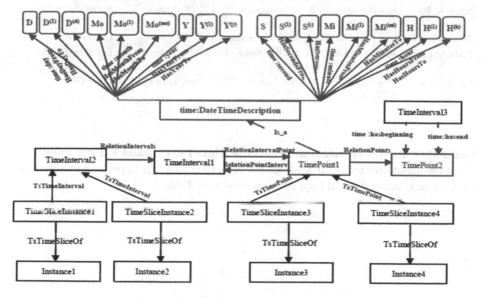

Fig. 1. Our 4D-fluents approach extension

4   Our Approach to Reasoning About Precise and Imprecise Temporal Data

Our approach consists of extending the Allen's interval algebra to: *(i)* reason about precise and imprecise quantitative temporal data to infer qualitative temporal relations and *(ii)* to reason about the qualitative temporal relations to infer new ones.

4.1   Qualitative Temporal Relations

We define temporal relations in a crisp way. At the beginning, we propose qualitative temporal relations between imprecise time intervals. Then, we adapt these relations to relate a time interval and a time point or two time points.

Qualitative Temporal Relations between Time Intervals. The proposed temporal relations are based on orderings between the time points contained in the intervals. They may be expressed using time point comparators like the ones proposed in Vilain and Kautz's Algebra [24]. When considering precise time intervals, our approach reduces to Allen's interval algebra. We redefine the 13 Allen's relations to propose temporal relations between imprecise time intervals as shown in Table 2.

Table 2. Temporal relations between the imprecise time intervals A and B.

Relation(A, B)	Definition	Inverse(B, A)
Before(A, B)	Precedes $(A^{+(E)}, B^{-(1)})$	After(B, A)
Meets(A, B)	Min (Same $(A^{+(1)}, B^{-(1)}) \wedge$ Same $(A^{+(E)}, B^{-(B)}))$	Met-by(B, A)
Overlaps(A, B)	Min (Precedes $(A^{-(B)}, B^{-(1)}) \wedge$ Precedes $(B^{-(B)}, A^{+(1)}) \wedge$ Precedes$(A^{+(E)}, B^{+(1)}))$	Overlapped-by(B, A)
Starts(A, B)	Min (Same $(A^{-(1)}, B^{-(1)}) \wedge$ Same $(A^{-(B)}, B^{-(B)}) \wedge$ Precedes $(A^{+(E)}, B^{+(1)}))$	Started-by(B, A)
During(A, B)	Min (Precedes $(B^{-(B)}, A^{-(1)}) \wedge$ Precedes $(A^{+(E)}, B^{+(1)}))$	Contains(B, A)
Ends(A, B)	Min (Precedes $(B^{-(B)}, A^{-(1)}) \wedge$ Same $(A^{+(1)}, B^{+(1)}) \wedge$ Same $(A^{+(E)}, B^{+(E)}))$	Ended-by(B, A)
Equals(A, B)	Min (Same $(A^{-(1)}, B^{-(1)}) \wedge$ Same $(A^{-(B)}, B^{-(B)}) \wedge$ Same $(A^{+(1)}, B^{+(1)}) \wedge$ Same $(A^{+(E)}, B^{+(E)}))$	Equals(B, A)

Qualitative Temporal Relations between a Time Interval and a Time Point. We adapt the qualitative temporal relations between time intervals to propose relations between a time interval and a time point as shown in Table 3.

Table 3. Temporal relations between a time interval A and a point P

Relation (P, A)	Definition	Inverse (A, P)
Temporal relations between a precise time interval A and a precise time point P		
Before(P, A)	Precedes(P, A^-)	After(A, P)
After(P, A)	Precedes(A^+, P)	Before(A, P)
Starts(P, A)	Same(P, A^-)	Started-by(A, P)
During(P, A)	Precedes(A^-, P) $\wedge$ Precedes(P, A^+)	Contains(A, P)
Ends(P, A)	Same(P, A^+)	Ended-by(A, P)
Temporal relations between an imprecise time interval A and an imprecise time point P		
Before(P, A)	Precedes($P^{(P)}$, $A^{-(1)}$)	After(A, P)
After(P, A)	Precedes($A^{+(E)}$, $P^{(1)}$)	Before(A, P)
Starts(P, A)	Same($P^{(1)}$, $A^{-(1)}$) $\wedge$ Same($P^{(P)}$, $A^{-(B)}$)	Started-by(A, P)
During(P, A)	Precedes($A^{-(E)}$, $P^{(1)}$) $\wedge$ Precedes($P^{(P)}$, $A^{+(1)}$)	Contains(A, P)
Ends(P, A)	Same($P^{(1)}$, $A^{+(1)}$) $\wedge$ Same($P^{(P)}$, $A^{+(E)}$)	Ended-by(A, P)

Qualitative Temporal Relations between Time Points. We adapt the qualitative temporal relations between time intervals to propose relations between time points, as shown in Table 4. For example, let N and R be two imprecise time points; represented respectively using the disjunctive ascending sets {Just before lunch … One hour before lunch} and {Just after lunch … Just before dinner}. We conclude that: After (R, N) = 1.

Table 4. Temporal relations between time points N and R.

Relation(N, R)	Definition	Inverse(R, N)
Temporal relations between precise time points N and R		
Before(N, R)	Precedes(N, R)	After(R, N)
Equals(N, R)	Same(N, R)	Equals(R, N)
Temporal relations between imprecise time points N and R		
Before(N, R)	Precedes($N^{(P)}$, $R^{(1)}$)	After(R, N)
Equals(N, R)	Same($N^{(1)}$, $R^{(1)}$) $\wedge$ Same($N^{(P)}$, $R^{(L)}$)	Equals(R, N)

4.2 Transitivity

The Allen's transitivity table lets us obtain from R1(A, B) and R2(B, C) that R3(A, C) holds, where A – [A^-, A^+], B = [B^-, B^+] and C = [C^-, C^+] are precise time intervals and R1, R2 and R3 are Allen's relations. Based on Table 1, we deduce from "During (A, B)" and "Meet(B, C)" that "Before(A, C)" holds. "During(A, B)" means "Precedes ($B^{-(B)}$, $A^{-(1)}$) $\wedge$ Precedes($A^{+(E)}$, $B^{+(1)}$)", and "Meet(B, C)" means "Same($B^{+(1)}$, $C^{-(1)}$) $\wedge$ Same($B^{+(E)}$, $C^{-(B)}$)". Considering precise relations, our transitivity table coincides with the Allen's one. We introduce three transitivity tables[2] to reason about the qualitative temporal relations between a time interval and a time point and time point relations.

5   Our Ontology for Representing and Reasoning About Precise and Imprecise Temporal Data in OWL2

Based on our extensions of the 4D-fluents approach and Allen's interval algebra, we implement our OWL 2 temporal ontology. We instantiate the crisp object properties {"RelationIntervals", "RelationIntervalPoint", "RelationPointInterval" and "Rela- tionPoints"} based on our Allen's extension. Our ontology proposes a set of SWRL rules to infer missing qualitative temporal relations. For each temporal relation, we associate an SWRL rule to deduce it from the quantitative temporal data given by the user. Based on the transitivity tables, we associate an SWRL rule for each transitivity relation.

6   Experimentations

To validate our approach, we introduce a prototype based on our proposed ontology.

6.1   Our Ontology-Based Prototype

Our ontology-based[3] prototype offers user interfaces to explore our approach. It is implemented based on JAVA language. It uses JENA API[4] and SPARQL-DL API[5] for managing and querying crisp ontology. First, the user instantiates our ontology. After each new temporal data input, the "Qualitative Temporal Data Inference" component is automatically executed to infer missing data. It is based on the proposed SWRL rules. The third component allows users to query our ontology via SPARQL queries.

6.2   Application to the Travel Ontology

We apply our work to the Travel ontology[6]. It needs to be extended to represent and reason about precise and imprecise temporal data (we merge our temporal ontology and the Travel ontology). For example: "The journey starts by *the end of the school year*. The school year ends *by the end of June*. The journey lasts 7 days. The program of the journey contains two main activities. The first one lasts 3 days and the second lasts 2 days". Let P = {June 27th...June 30th} be an imprecise time point which represents the end of the school year. Let A = [A⁻, A⁺] and B = [B⁻, B⁺] be two imprecise time intervals representing the durations of both activities. Assume that A⁻ = {June 29th...

---

[2] https://cedric.cnam.fr/~hamdif/upload/DEXA19/Transitivity_Tables.pdf.

[3] http://cedric.cnam.fr/isid/ontologies/files/CrispTimeOnto.html.

[4] https://jena.apache.org/.

[5] http://www.derivo.de/en/resources/sparql-dl-api.html.

[6] https://protege.stanford.edu/junitOntologies/testset/travel.owl.

June 30th}, A^+ = {01^{st} July ... July 2nd}, B^- = {July 02nd... July 03rd} and B^+= {July 03rd...July 04th}. The associated qualitative temporal relations "BeforeIntervals" and "AfterIntervals" are inferred based on the second component of our prototype.

7 Conclusion and Future Directions

In this paper, we introduce our crisp-based approach for representing and reasoning about temporal data in terms of quantitative and qualitative relations in ontology. It supports precise and imprecise time points and imprecise time intervals in ontology. It is based only on existing crisp Semantic Web standards.

Our approach is based on three contributions. The first one is about extending the 4D-fluents approach with crisp components to represent precise and imprecise time points, precisely dates and time clocks, and imprecise time intervals. The second contribution consists of extending the Allen's interval algebra in a crisp way to reason about precise and imprecise temporal data. It preserves reflexivity/irreflexivity, symmetry/asymmetry and transitivity. We introduce four transitivity tables to reason about the resulting temporal relations. The third contribution is about creating an ontology based on our extensions. A prototype is created to explore our approach. Our Allen's interval algebra extension can be applied to other research fields such as databases. Our approach can be implemented with crisp standards and researchers are not obliged to learn technologies related to fuzzy ontology. Thus, it is suitable for marketed products.

We plan to extend our work to handle other imperfections such as the uncertainty.

References

1. Allen, J.: Maintaining knowledge about temporal intervals. Commun. ACM **26**, 832–843 (1983)
2. Anagnostopoulos, E., Batsakis, S., Petrakis, E.: CHRONOS: a reasoning engine for qualitative temporal information in OWL. Procedia Comput. Sci. **22**, 70–77 (2013)
3. Artale, A., Franconi, E.: A survey of temporal extensions of description logics. Ann. Math. Artif. Intell. **30**, 70–77 (2000)
4. Badaloni, S., Giacomin, M.: The algebra IAfuz: a framework for qualitative fuzzy temporal reasoning. Artif. Intell. **170**(10), 872–908 (2006)
5. Batsakis, S., Petrakis, E.G.M.: SOWL: a framework for handling spatio-temporal information in OWL 2.0. In: Bassiliades, N., Governatori, G., Paschke, A. (eds) Rule-Based Reasoning, Programming, and Applications. RuleML 2011. LNCS, vol. 6826, pp. 242–249. Springer, Heidelberg (2011). https://doi.org/10.1007/978-3-642-22546-8_19
6. Buneman, P., Kostylev, E.: Annotation algebras for RDFS. In: Workshop on the Role of Semantic Web in Provenance Management (2010)
7. Dubois, D., Prade, H.: Processing fuzzy temporal knowledge. IEEE Trans. Syst. Man Cybern. **19**(4), 729–744 (1989)
8. Ermolayev, V., et al.: An agent-oriented model of a dynamic engineering design process. In: Kolp, M., Bresciani, P., Henderson-Sellers, B., Winikoff, M. (eds.) AOIS -2005. LNCS (LNAI), vol. 3529, pp. 168–183. Springer, Heidelberg (2006). https://doi.org/10.1007/11916291_12

9. Freksa, C.: Temporal reasoning based on semi-intervals. Artif. Intell. **54**, 199–227 (1992)
10. Gammoudi, A., Hadjali, A.: Fuzz-TIME: an intelligent system for managing fuzzy temporal information. Intell. Comput. Cybern. **10**, 200–222 (2017)
11. Ghorbel, F., Hamdi, F., Métais, E., Ellouze, N., Gargouri, F.: A fuzzy-based approach for representing and reasoning on imprecise time intervals in fuzzy-owl 2 ontology. In: Silberztein, M., Atigui, F., Kornyshova, E., Métais, E., Meziane, F. (eds.) NLDB 2018. LNCS, vol. 10859, pp. 167–178. Springer, Cham (2018). https://doi.org/10.1007/978-3-319-91947-8_17
12. Gutierrez, C., Hurtado, C., Vaisman, A.: Temporal RDF. In: Gómez-Pérez, A., Euzenat, J. (eds.) ESWC 2005. LNCS, vol. 3532, pp. 93–107. Springer, Heidelberg (2005). https://doi.org/10.1007/11431053_7
13. Harbelot, B.A.: Continuum: a spatiotemporal data model to represent and qualify filiation relationships. In: ACM SIGSPATIAL International Workshop, pp. 76–85 (2013)
14. Herradi, N.: A Semantic Representation of Time Intervals in OWL2. KEOD, pp. 1–8 (2017)
15. Hurtado, C., Vaisman, A.: Reasoning with temporal constraints in RDF. In: Alferes, J.J., Bailey, J., May, W., Schwertel, U. (eds.) PPSWR 2006. LNCS, vol. 4187, pp. 164–178. Springer, Heidelberg (2006). https://doi.org/10.1007/11853107_12
16. Kim, S.-K., Song, M.-Y., Kim, C., Yea, S.-J., Jang, H.C., Lee, K.-C.: Temporal ontology language for representing and reasoning interval-based temporal knowledge. In: Domingue, J., Anutariya, C. (eds.) ASWC 2008. LNCS, vol. 5367, pp. 31–45. Springer, Heidelberg (2008). https://doi.org/10.1007/978-3-540-89704-0_3
17. Klein, M.C.: Ontology versioning on the semantic web. In: Semantic Web Working Symposium, Stanford University, pp. 75–91. California (2001)
18. Lutz, C.: Description logics with concrete domains. In: Advances in Modal Logic, pp. 265–296 (2003)
19. Nagypál, G., Motik, B.: A fuzzy model for representing uncertain, subjective, and vague temporal knowledge in ontologies. In: Meersman, R., Tari, Z., Schmidt, D.C. (eds.) OTM 2003. LNCS, vol. 2888, pp. 906–923. Springer, Heidelberg (2003). https://doi.org/10.1007/978-3-540-39964-3_57
20. Noy, N.R.: Defining N-Ary Relations on the Semantic-Web. W3C Working Group (2006)
21. O'Connor, M.J., Das, A.K.: A method for representing and querying temporal information in OWL. In: Fred, A., Filipe, J., Gamboa, H. (eds.) BIOSTEC 2010. CCIS, vol. 127, pp. 97–110. Springer, Heidelberg (2011). https://doi.org/10.1007/978-3-642-18472-7_8
22. Sadeghi, K.M.: Uncertain interval algebra via fuzzy/probabilistic modeling. In: IEEE International Conference on Fuzzy Systems, pp. 591–598 (2014)
23. Tappolet, J., Bernstein, A.: Applied temporal RDF: efficient temporal querying of RDF data with SPARQL. In: Aroyo, L., et al. (eds.) ESWC 2009. LNCS, vol. 5554, pp. 308–322. Springer, Heidelberg (2009). https://doi.org/10.1007/978-3-642-02121-3_25
24. Vilain, M.B.: Constraint propagation algorithms for temporal reasoning. In: National Conference on Artificial Intelligence, pp. 377–382. Philadelphia (1986)
25. Welty, C., Fikes, R.: A reusable ontology for fluents in OWL. In: FOIS, pp. 226–336 (2006)

Information Processing

Context-Aware Multi-criteria Recommendation Based on Spectral Graph Partitioning

Rim Dridi[1,2(✉)], Lynda Tamine[3], and Yahya Slimani[2,4]

[1] ENSI, University of Manouba, Tunis, Tunisia
`rim.dridi@ensi-uma.tn`
[2] LISI Laboratory, INSAT, University of Carthage, Tunis, Tunisia
`yahya.slimani@fst.rnu.tn`
[3] IRIT, University of Toulouse 3, Toulouse, France
`tamine@irit.fr`
[4] ISAMM, University of Manouba, Tunis, Tunisia

Abstract. Both multi-criteria recommendation and context-aware recommendation are well addressed in previous research but separately in most of existing work. In this paper, we aim to contribute to the under-explored research problem which consists in tailoring the multi-criteria rating predictions to users involved in specific contexts. We investigate the application of simultaneous clustering based on the application of a spectral partitioning graph method over situational contexts in the one hand and criteria in the other hand. Besides, we conjecture that even with similar criteria-related ratings, the importance of criteria might differ among users. This idea leads us to use prioritized aggregation operators as means of multi-criteria rating aggregations. Our experimental results on a real-world dataset show the effectiveness of our approach.

Keywords: Recommender system · Multi-criteria · Context

1 Introduction

The key problem of recommendation is designing the utility function that measures the usefulness of items to target users. Traditionally, recommender systems are based on a single-criterion utility function. Some studies have begun employing multi-criteria recommender systems (MCRS) [1,10,12] that model a user's utility of an item as a vector of ratings along several criteria.

Yet, previous recommenders have highlighted the impact of context dimensions (e.g., time, location, *etc.*) on user's judgments. In this respect, several researches have been devoted to context-aware recommender systems (CARS) [2].

However, most of previous CARS still consider single item ratings while either the item criteria and their strength might evolve while context evolves.

In our work, we attempt to contribute to this under-explored research area. Specifically, we explore the idea of clustering situational recommendations

© Springer Nature Switzerland AG 2019
S. Hartmann et al. (Eds.): DEXA 2019, LNCS 11707, pp. 211–221, 2019.
https://doi.org/10.1007/978-3-030-27618-8_16

embedding users providing similar criteria ratings to target items under similar contexts. Our assumption is that users in similar contextual situations tend to have similar interests for similar criteria. Following this assumption, we consider the joint clustering of two types of entities, where both contextual situations and criteria are simultaneously assigned to clusters. Then, users' predicted criteria ratings from the co-clusters are aggregated based on their personal preferences. We formulate the recommendation problem in terms of two sub-problems: (1) Criteria rating prediction: we transform the first sub-problem to a bipartite graph partitioning problem that we solve using the well known spectral co-clustering [5]. Then, we exploit the obtained co-clusters with a rating prediction algorithm for predicting criteria ratings. (2) Overall rating prediction: the key issue within this second sub-problem is the design of an appropriate aggregation of the criteria ratings resulting from the co-clusters. Therefore, we explore the use of two prioritized aggregation operators [3,4], where the criteria weights are computed on the basis of their priority order in accordance with the users' interests.

2   Related Work

2.1   Multi-criteria Recommender Systems

One of the popular efficient MCRS approaches is the aggregation-based one [12,14] which builds an aggregation function f (Eq. 1) that represents the relationship between the overall rating r_0 and the criteria ratings $(r_1, .., r_N)$:

$$r_0 = f(r_1, .., r_N) \tag{1}$$

In [7,10], a linear aggregation function was applied to predict the overall rating using criteria preferences. In [12], Zheng used criteria chains for multi-criteria rating predictions and conditional aggregations by viewing the criteria predictions as contexts. These criteria ratings are predicted and employed in the chain, which might lead to an accumulated loss while predicting the global rating.

2.2   Context-Aware Recommender Systems

The first category of work in this area, considers context in a single-criterion based recommendation framework. For example, in [2], a context-aware matrix factorization (CAMF) was proposed for item rating prediction.

Unlikely, the second category of work which is closest to ours, explores the exploitation of context information in addition to multi-criteria ratings to provide more accurate predictions [9,14]. Li et al. [9] defined a 4-order tensor recommendation space, where the contextual information and the multi-criteria ratings are considered besides the users and items. This tensor was then reduced by using the relevant context to find the closest neighbors based on the multi-linear singular value decomposition. Recently, Zheng [14] integrated context information into four MCRS baselines. The independent and dependent methods were used for the multi-criteria rating predictions step, and the linear and conditional aggregation methods for the rating aggregations step.

Beside the differences in the used prediction methods, what basically differentiates our proposal is considering that criteria are both item and user-dependent.

3 Context-Aware Multi-criteria Recommendation Framework

3.1 Basic Notation

User's Situational Context. A user's situational context refers to the situation characterized by a user involved in a specific surrounding context. We represent distinct pairs (user, context) as distinct contextual situations. Let users set, noted Us is represented by $Us = \{u_1, .., u_k\}$, where k is the total number of users, and contexts Co are represented by $Co = \{co_1, .., co_l\}$, where l is the total number of contexts. A contextual situation is built up as an entity noted s_{ij}, represented by a contextual situation that implicitly refers to the pair user u_i in context co_j. For care of the simplicity of the notations, s_{ij} is noted as s_i where i is in the range $1..m$ leading the whole set of situations noted as $S = \{s_1, .., s_m\}$.

Criteria. The criteria set contains rated item aspects involving in situational contexts. The set of entities referring to rated item criteria is noted $C = \{c_1, .., c_n\}$, where n is the number of criteria considered for rating an item.

Situational Bipartite Graph. A situational bipartite graph is a triple $G = (S, C, E)$ where S, C are the two vertex sets and E is the set of edges that connect nodes from vertex S to vertex C such as $(E = <s_i, c_j> \mid s_i \in S, c_j \in C)$.

3.2 Situational Bipartite Graph Co-clustering

We focus on extending the conventional rating prediction process using a co-clustering method to find sub-groups of contextually similar users and criteria that these users are interested in. Our driving hypothesis is the following:

H: "Users in similar contextual situations tend to have similar interests for similar criteria".

To solve the partitioning problem, we employ the popular spectral co-clustering algorithm [5] which approximates the normalized cut of the bipartite graph to find co-clusters. An approximate solution to the optimal normalized cut may be found via the decomposition of the normalized $m \times n$ rating matrix R as follows: $R_n = D_1^{-1/2} R D_2^{-1/2}$, where D_1 is the diagonal matrix with entry i equal to $\sum_j R_{ij}$ and D_2 is the diagonal matrix with entry j equal to $\sum_i R_{ij}$. Then, the singular value decomposition of the resulting matrix $R_n = U \Sigma V^\top$ provides the desired partitions of the rows and columns of R. U is an $m \times m$

matrix, Σ is an $m \times n$ diagonal matrix, and V^T is the transpose of an $n \times n$ matrix. The columns of U and V are called the left and right singular vectors respectively. A subset of the left singular vectors will give the users' situational contexts partitions, and a subset of the right singular vectors will give the criteria partitions. Later, the singular vectors are used to build the matrix Z.

$$Z = D_1^{-1/2} U D_2^{-1/2} V$$

Finally, the resulting matrix Z is decomposed using k-means++ to obtain the desired co-clusters to be used as input to the prediction process detailed below.

3.3   Rating Prediction Algorithm

Criteria Rating Predictions. The Algorithm 1 aims to provide, as an output, the criteria predicted ratings for each co-cluster of situational contexts and criteria. As stated in the algorithm, for each co-$cluster_k$, we can extract a rating sub-matrix $R_k \in \mathbb{R}^{m_k \times n_k}$ from the original rating matrix $R \in \mathbb{R}^{m \times n}$, m_k and n_k denote respectively the number of users' situational contexts and criteria in co-$cluster_k$. Then, we use the Matrix Factorization (MF) [8] as the rating prediction algorithm on each obtained sub-matrix R_k due to its efficiency and scalability. In line 2, the algorithm calls the $MatrixFactorization$ function. This routine applies the MF algorithm where we assume there are F hidden factors, which capture users' situational contexts features and criteria features to model users' preferences. Matrix factorization algorithm works by decomposing the $m_k \times n_k$ rating sub-matrix R_k into the product of two lower dimensionality matrices. Users' situational contexts are represented by a $m_k \times F$ matrix called P, where each row of P would represent the strength of the associations between a user's situational context and the features. In order to relate users' situational context with criteria, the latter are also represented by a matrix called Q, where each row of Q would represent the strength of the associations between a criterion and the features. P and Q are learned using stochastic gradient descent method by minimizing the rating prediction errors. The predicted preference $\hat{r}_{ij}$ of a user's situational context s_i for a criterion c_j can be computed as follows:

$$\hat{r}_{ij} = p_i q_j{}^T \tag{2}$$

Overall Rating Prediction. We make the first attempt to apply "Scoring" and "And" prioritized aggregation operators [3,4] for overall rating prediction. The criteria weights depend on users' preference order of criteria extracted on the basis of their expressed criteria ratings. Besides, regarding the problem of contextual recommendation at hand, we conjecture that the criteria strength also varies in accordance with users' contexts. Hence, the prioritized operators allow flexible personalization of the overall rating prediction by considering the criteria weights based on users' criteria preferences under different contexts.

Algorithm 1: Criteria Rating Prediction for each Co-cluster

Input: Rating matrix with multicriteria: $R \in \mathbb{R}^{m \times n}$, the number of co-clusters: L, and the number of factors: F.

begin

    for *each co-cluster* $k \in \{1, .., L\}$ **do**

1   R_k=ExtractSubmatrix $(R, co\text{-}cluster_k)$

2   P_k, Q_k=MatrixFactorization(R_k, F)

        for *each* $i \in P_k$ **do**

            for *each* $j \in Q_k$ **do**

                for *each* $t \in \{1, .., F\}$ **do**

3                    $\hat{r}_{ij} = p_{i,t} \times q_{j,t}$

Output: Criteria predicted ratings

The importance weight computation of a criterion c_i, with $i \neq 1$, depends on users' preference order of criteria, and depends also on both the weight associated to criterion c_{i-1}, and the preference of c_{i-1}. The user preference ordering of the considered criteria is based on computing an average score for each criterion in accordance with the users expressed criteria ratings. More formally, let $C = \{c_1, ..., c_N\}$ be a set of ordered criteria, where c_1 presents the most preferred criterion and c_N is the least one. We indicate by w_p the importance weight of the criterion $c_p \in C$ for a given item and user's context. The weights associated with the ordered criteria are computed as follows:

- The weight associated with the most important criterion c_1 is set to be 1.
- The weights of the other criteria c_p for $p \in [2, N]$, are computed as follows:

$$w_p = w_{p-1}.r_{p-1} \tag{3}$$

r_{p-1} denotes the preference rating given by a user on criterion c_{p-1} of an item. We define in the following a new way in which the function f (See Eq. 1) is defined according to the mentioned prioritized aggregation operators.

- **Prioritized "Scoring" operator** (F_s): This operator calculates the overall item rating r_0 from several criteria evaluations, where the weight associated with each criterion depends both on the weights and on the preferences of the most important criteria. The higher the satisfaction degree of a more important criterion, the more the satisfaction degree of a less important criterion impacts the overall rating. F_s is defined as: $F_s : [0,1]^N \longrightarrow [0, N]$

$$r_0 = F_s(r_1, .., r_N) = \sum_{p=1}^{N} w_p.r_p \tag{4}$$

For example, let us consider that a user is looking for an hotel. His choice depends on two criteria $c_1 = $ "comfort" and $c_2 = $ "inexpensiveness" with $c_1 > c_2$. An hotel with a "comfort" degree of 1 and an "inexpensiveness" degree of 0 would have an overall rating of 1.

- **Prioritized "And" operator** (F_a): This operator models a situation where the overall rating r_0 strongly depends on the importance of the least satisfied criterion. If it is the most important criterion, the value of the least satisfied criterion is considered as the overall rating merely. F_a is defined as follows: $F_a : [0, 1]^N \longrightarrow [0, 1]$

$$r_0 = F_a(r_1, .., r_N) = \min_{p \in [1, N]} (\{r_p\}^{w_p}) \qquad (5)$$

Let us come back again to the previous example. $c_1 =$ "comfort" and $c_2 =$ "inexpensiveness" with $c_1 > c_2$. Here, an hotel with a "comfort" degree of 1 and a "inexpensiveness" degree of 0 would have an overall rating of 0. So in this case, the under-satisfaction of the inexpensiveness criterion cannot be compensated by the satisfaction of the "comfort" criterion.

4   Experimental Evaluation

4.1   Experimental Settings

The only suitable dataset with respect to our evaluation purpose is TripAdvisor data [6] since: (1) user's context is available based on a contextual dimension which refers to the *season*. This contextual dimension is derived from the trip date expressed in months in the dataset (e.g., March, April and May are the spring season months). (2) Users' ratings of seven individual criteria, plus one overall rating are provided. The used criteria are: *value for the money, quality of rooms, the hotel location, cleanliness of the hotel, experience of check-in, overall quality of service and business services.* There are a total of 22.130 ratings given by 1502 users on 14.300 hotels. The bipartite graph modeling is built upon $m = 3916$ users situational contexts connected to $n = 7$ criteria.

We measure the performance by mean absolute error (MAE) on this dataset by adopting a training-testing methodology for both parameter tuning and evaluation. For this purpose, we fixed a splitting ratio of training/test of 80/20. For comparison, we used a single rating approach (BiasMF [8]), multi-criteria rating approaches (Agg [1], CluAllCrit [10], CIC [12], CCA [12], CCC [12]) and a context-aware rating approach (CAMF [2]).

4.2   Research Hypothesis Validation

To validate our research hypothesis **H** (See Sect. 3.2), we perform a statistical analysis to determine the strength of the relationships between contextually similar users according to their criteria importance. More precisely, we run a correlation analysis on all the users providing criteria preferences of similar items in similar context situations from the real-world TripAdvisor dataset. First, we compute the importance of each criterion for each user to identify users preferred criteria according to their contexts [11]. Having computed the users criteria importance, we examine the strength of the relationship between these users with respect to their criteria importance through the computation of the Spearman's rank correlation coefficient. To interpret the strength of the obtained correlation coefficient values, we use the rule of thumb (See Fig. 1). We can clearly see from Fig. 1, the high percentage of the very strongly correlated users in similar situations. This result shows that the majority of contextually similar users achieve a fairly strong positive correlation coefficient with respect to their interests for similar criteria which represents a good agreement between contextually similar users on criteria importance order. Hence, we could conjecture that the more similar the users contexts, the more these users tend to have similar criteria importance which provides a strong support for our research hypothesis **H**.

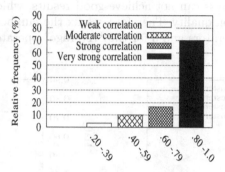

Fig. 1. Distribution of the correlation measures between users' criteria importance in similar contexts

4.3   Evaluation of the Prioritized Aggregation Operators

We begin by tuning the latent factor number F which is one of the important parameters for matrix factorization. As shown in Fig. 2(a), we can observe, when F is equal to 12 the MAE of our proposal using "And" operator declines to the lowest in cluster 2 and cluster 3. So, we come to a conclusion that F = 12 is a better choice for both cluster 2 and cluster 3. For cluster 1 and 4, the MAE of "And" operator model shows a good prediction accuracy when F = 10. While the prediction accuracy of "Scoring" operator model in all clusters improves as the number of latent factors reaches 10 (Fig. 2(b)). Then, we assess in this experimental scenario, the effectiveness of the "Scoring" and "And" prioritized operators for improving the overall rating prediction in comparison with the standard "Average" operator. Particularly, to evaluate the joint effect of the aggregation operators and the number of co-clusters on rating prediction accuracy, we experiment different numbers of co-clusters ranging from 2 to 10. From Fig. 3, we can observe that the "Scoring" operator (resp. the "And" operator) achieves an average improvement of 19.9% (resp. 14.6%) over the "Average" aggregation operator for a number of co-clusters ranging from 5–8. This result confirms the effectiveness of the prioritized combination of the considered criteria in the co-clusters, which allows flexible personalization of the overall prediction results according to users' preferences. The "Scoring" operator is the best performing operator in these comparisons due to the appropriateness of the importance order of relevant criteria in accordance with users' contexts. Fig. 3 also reveals that the prediction accuracy is affected by the number of co-clusters. We can observe that the accuracy slightly increases as the co-clusters number increases from 2 to 4 since the information within each co-cluster is more tied to users. However, when the co-clusters number continues to increase, the prediction accuracy tends to be steady. This observation could be explained by the fact that increasing the number of co-clusters would lead to divide the rating matrix into several more small sub-matrices. Yet, the criteria rating prediction using the MF algorithm requires a sufficient volume data to provide accurate predictions. Thus, under a reasonable threshold of data provided by the co-clusters, the criteria aggregation process can not achieve good results, which have a downside effect on the prediction quality. Therefore, we fix the number of co-clusters to 4 for the prioritized operators and 3 for the "Average" operator.

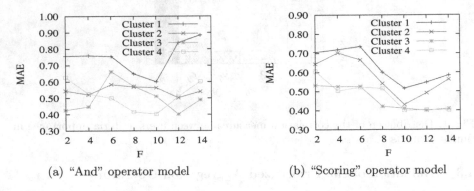

(a) "And" operator model   (b) "Scoring" operator model

Fig. 2. F variation on the prioritized operators

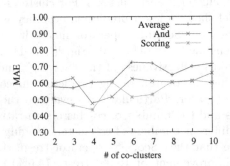

Fig. 3. Effectiveness of prioritized operators

4.4   Comparison Effectiveness Evaluation with Baselines

The multi-criteria baselines results are reported from the published corresponding research papers referenced in Table 1 using their optimal parameters and using the same dataset we used in our experiments. While the results of the other categories of baselines are obtained from the toolkit CARSKit [13].

In Table 1, *IR Scoring* and *IR And* indicate the improving rate achieved using the "Scoring" and the "And" operators respectively. According to Table 1, our proposed approach is able to outperform the baselines by achieving higher prediction accuracy. More precisely, our model based "Scoring" operator allows achieving a considerable improvement of +72.1%, +72.9% and +62.4% over Agg, CIC and CCA models respectively. The same trend of improvement holds for the model based on the "And" operator. These results could be explained by the fact that the multi-criteria Agg, CIC and CCA models use either a traditional way for predicting multi-criteria ratings, a linear aggregation, or both which may decrease prediction accuracy. The multi-criteria algorithm based on clustering (CluAllCrit) which uses a linear aggregation degrades the prediction results compared with other multi-criteria algorithms. Therefore, our model allows a huge improvement over CluAllCrit (+482.4% by the "Scoring" operator and +434.7% by the "And" operator), this may be because the problem with the automatic criteria coefficients obtained by the linear aggregation function. Even when employing a clustering technique to enhance prediction results, using such coefficients in the aggregation process may generate many rating prediction results with negative values or outside of the [1..5] scale. Comparing with the CCC model, which considers criteria dependency to predict the criteria ratings and uses conditional aggregations, there is a little difference in the accuracy results between this latter model and ours. These results reveal that there might exist complementary criteria affecting the user's choice for choosing an item. Meanwhile, using a conditional aggregation may not always be a good choice, since CIC model which uses a conditional aggregation performs worse than CCA model which uses a linear function.

For the contextual baseline, CAMF works better than the majority of baselines but still outperformed by our model (+46.2% using the "Scoring" operator and +34.2% using the "And" operator); this may be because it does not take extra information such as multi-criteria ratings.

Overall, our results indicate that particularly in situations where different criteria ratings are available, it can be advantageous to consider the criteria strength with respect to user's context. This explanation is corroborated by cross-comparing the results obtained using the prioritized operators in the one hand versus the average aggregation and the CAMF approach on the other hand. We can see that the MAE decreased from 0.639 to 0.570 when leveraging context and decreased more to less 0.480 when additionally applying the prioritized operators.

Table 1. Comparison results for the rating prediction task

Category	Algorithms	MAE	IR Scoring	IR And
Traditional single	BiasMF [8]	0.894	+104.5%	+87.8%
Multi-criteria rating approaches	Agg [1]	0.752	+72.1%	+57.9%
	CIC [12]	0.756	+72.9%	+58.8%
	CCA [12]	0.710	+62.4%	+49.2%
	CCC [12]	0.460	+5.3%	−3.5%
	CluAllCrit [10]	2.545	+482.4%	+434.7%
Context-aware rating approach	CAMF [2]	0.639	+46.2%	+34.2%
Our model	Average	**0.570**	-	-
	Scoring	**0.437**	-	-
	And	**0.476**	-	-

5 Conclusion

In this paper, we have proposed a context-aware recommendation approach that relies on multi-criteria rating predictions. The key characteristics of the proposed approach consist in jointly clustering users involved in contextual situations while rating items with respect to multiple facets. For this purpose, we used the spectral graph partitioning method. The obtained co-clusters provide partial user's item ratings that are aggregated using prioritized aggregation operators which allow tailoring the criteria strengths to the user's preferences.

The experiments shows that: (1) the prioritized operators outperform basic average aggregation but that improvement is achieved only with a limited number of co-clusters and that (2) the co-clusters of contextual situations and criteria provide relevant signals about the users' perceptions about item aspects.

In the future, we plan to evaluate our recommendation framework on other datasets allowing a multi-dimensional-based context evaluation. Within this line of work, we will support our model with an in-depth analysis of the users' ratings on item aspects in various contexts and study the correlation between them. This analysis would give insight into the relevance of extending the bipartite graph to deal with different context nodes and the usefulness of filtering relevant interactions between contexts and item criteria before applying the aggregation.

References

1. Adomavicius, G., Kwon, Y.: New recommendation techniques for multicriteria rating systems. IEEE Intell. Syst. **22**(3), 48–55 (2007)
2. Baltrunas, L., Ludwig, B., Ricci, F.: Matrix factorization techniques for context aware recommendation. In: RecSys 2011, New York, USA, pp. 301–304 (2011)
3. da Costa Pereira, C., Dragoni, M., Pasi, G.: Multidimensional relevance: a new aggregation criterion. In: Boughanem, M., Berrut, C., Mothe, J., Soule-Dupuy, C. (eds.) ECIR 2009. LNCS, vol. 5478, pp. 264–275. Springer, Heidelberg (2009). https://doi.org/10.1007/978-3-642-00958-7_25

4. da Costa Pereira, C., Dragoni, M., Pasi, G.: A prioritized "and" aggregation operator for multidimensional relevance assessment. In: Serra, R., Cucchiara, R. (eds.) AI*IA 2009. LNCS (LNAI), vol. 5883, pp. 72–81. Springer, Heidelberg (2009). https://doi.org/10.1007/978-3-642-10291-2_8

5. Dhillon, I.S.: Co-clustering documents and words using bipartite spectral graph partitioning. In: Proceedings of the 7th ACM SIGKDD International Conference on Knowledge Discovery and Data Mining, pp. 269–274. KDD, New York (2001)

6. Jannach, D., Zanker, M., Fuchs, M.: Leveraging multi-criteria customer feedback for satisfaction analysis and improved recommendations. J. IT Tourism **14**(2), 119–149 (2014)

7. Jhalani, T., Kant, V., Dwivedi, P.: A linear regression approach to multi-criteria recommender system. In: Tan, Y., Shi, Y. (eds.) Data Mining and Big Data, pp. 235–243. Springer, Cham (2016). https://doi.org/10.1007/978-3-319-40973-3_23

8. Koren, Y., Bell, R., Volinsky, C.: Matrix factorization techniques for recommender systems. Computer **42**(8), 30–37 (2009)

9. Li, Q., Wang, C., Geng, G.: Improving personalized services in mobile commerce by a novel multicriteria rating approach. In: WWW, NY, USA, pp. 1235–1236 (2008)

10. Liu, L., Mehandjiev, N., Xu, D.L.: Multi-criteria service recommendation based on user criteria preferences. In: RecSys 2011, pp. 77–84. ACM, New York (2011)

11. Sreepada, R.S., Patra, B.K., Hernando, A.: Multi-criteria recommendations through preference learning. In: CODS, pp. 1:1–1:11. ACM, New York (2017)

12. Zheng, Y.: Criteria chains: A novel multi-criteria recommendation approach. In: IUI 2017, New York, NY, USA, pp. 29–33 (2017)

13. Zheng, Y., Mobasher, B., Burke, R.D.: CARSKIT: A java-based context-aware recommendation engine. In: ICDMW, Atlantic City, USA, 14–17, pp. 1668–1671 (2015)

14. Zheng, Y., Shekhar, S., Anna Jose, A., Kumar, S.: Integrating context-awareness and multi-criteria decision making in educational learning. In: SAC, ACM (2019)

SilverChunk: An Efficient In-Memory Parallel Graph Processing System

Tianqi Zheng[1,2] , Zhibin Zhang[1], and Xueqi Cheng[1,2]

[1] CAS Key Laboratory of Network Data Science and Technology,
Institute of Computing Technology, Chinese Academy of Sciences, Beijing, China
zhengtianqi@ict.ac.cn
[2] University of Chinese Academy of Sciences, Beijing, China

Abstract. One of the main constructs of graph processing is the two-level nested loop structure. Parallelizing nested loops is notoriously unfriendly to both CPU and memory access when dealing with real graph data due to its skewed distribution. To address this problem, we present `SilverChunk`, a high performance graph processing system. `SilverChunk` builds edge chunks of equal size from original graphs and unfolds nested loops statically in pull-based executions (`VR-Chunk`) and dynamically in push-based executions (`D-Chunk`). `VR-Chunk` slices the entire graph into several chunks. A virtual vertex is generated pointing to the first half of each sliced edge list so that no edge list lives in more than one chunk. `D-Chunk` builds its chunk list via binary searching over the prefix degree sum array of the active vertices. Each chunk has a local buffer for conflict-free maintenance of the next frontier. By changing the units of scheduling from edges to chunks, `SilverChunk` achieves better CPU and memory utilization. `SilverChunk` provides a high level programming interface combined with multiple optimization techniques to help developing efficient graph processing applications. Our evaluation results reveal that `SilverChunk` outperforms state-of-the-art shared-memory graph processing systems by up to 4×, including Gemini, Grazelle, etc. Moreover, it has lower memory overheads and nearly zero pre-processing time.

Keywords: Graph processing · Parallel scheduling · Chunking

1 Introduction

1.1 Background

Graphs are commonly used to represent interactions between real world entities. Graph analytics are algorithms that extract information from a graph, which are widely used in social network analytics, transportation, ad and e-commerce recommendation systems. As a result, a large number of graph processing systems are proposed to facilitate graph analytics. Recently there is a rising interest of building multi-core shared memory graph processing systems on a single machine because (1) distributed graph systems incur a lot of communication overheads;

© Springer Nature Switzerland AG 2019
S. Hartmann et al. (Eds.): DEXA 2019, LNCS 11707, pp. 222–236, 2019.
https://doi.org/10.1007/978-3-030-27618-8_17

(2) real world graphs, e.g., Twitter's follower graph, despite its billions of edges, can still fit into main memory; and (3) memory capacity and bandwidth are increasing and will keep increasing in the near future. These systems [5,6,8–14] process a big graph in main memory of a single high-end server with large RAM space. They provide high level interfaces for programming simplicity and aim at full utilization of all CPU and memory resources without manual tweaking. For example, Ligra [9] provides two simple primitives, EdgeMap and VertexMap, for iterating over edges and vertices respectively in parallel. These simple primitives can be applied to various graph algorithms which operate on a subset of vertices during each iteration.

1.2 Problems

Parallel graph processing is nontrivial due to complex data dependencies in graphs, however, it is essential for efficient graph analytics. In this paper we discuss two problems of building an high-performance in-memory graph processing system.

Preliminaries. In-memory graph processing systems often organize outgoing edges in the Compressed Sparse Row (CSR) format and incoming ones in the Compressed Sparse Column (CSC) format, as shown in Fig. 1. A frontier is a subset of the vertices which are active in the current iteration, as shown in Fig. 2. Graph algorithms visit the destination vertices of the active edges and apply an algorithm-specific function to propagate the value from each edge's source to its destination. This operation is repeated until the current frontier is empty or user defined condition is met. We refer to this process as frontier-based computing.

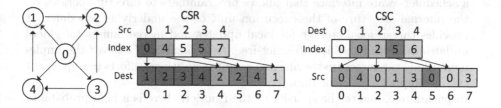

Fig. 1. Compressed Sparse Row/Column format

The frontier structure may be implemented either as a bitmap (dense format) or as an array directly storing the vertex IDs (sparse format). Which one is better depends on the density of the frontier. Frontier-based computing can have two different execution modes, namely push and pull. Both modes contains a two-level nested loop. In push mode, frontiers are used in the outer loop and updates are propagated from active vertices to their neighbors, while in pull mode, the outer loop is the entire vertex list and each vertex receives updates from its

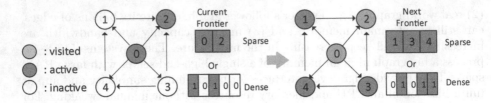

Fig. 2. Frontiers in a simple BFS algorithm

in-bound edges by checking if the source vertex is inside the current frontier or not. There are active researches [1,7] studying whether to push or pull. The basic principle is to push when the frontier is sparse and to pull if otherwise. As a result, graph processing engines like Ligra [9] automatically switches between these two execution modes based on the density of the current frontier.

Problems. We discuss the following two problems:

- In both execution modes, the outer loop is parallelized in order to leverage the multiple cores of modern processor chips. Unfortunately, due to the power-law nature of real world social graphs, only a small fraction of vertices has a significant large number of neighbors while a major fraction of vertices has relatively few neighbors. As a result, parallelizing only the outer loop is insufficient as it can lead to significant load imbalance. One naive approach is to use traditional parallel schedulers such as Cilk [2] or OpenMP [3] to parallelize the inner loop. However, this approach can lead to numerous conflicting writes and scheduling overhead which completely negates the benefits of the pull execution mode. Grazelle [5] solves this problem by introducing a scheduler-aware interface that allows programmers to directly operates on the internal structure of the execution unit of the underlying scheduler. It provides thread local storage for local updates and merge buffers for global updates in order to achieve conflict-free parallelization. However the implementation is architecture-specific and requires additional efforts to implement even a simple graph algorithm.
- In push mode, due to the sparsity of the frontier, there is a high probability that the next frontier will also be sparse, hence building the next frontier as a sparse array instead of a bitmap is more efficient. However, building sparse frontiers in parallel is nontrivial. Ligra [9] does this by first allocating a scratch buffer that is large enough to hold all possible vertices in the next frontier, and then computing an offset array via parallel prefix summing over the active vertices' degrees in the current frontier. When a vertex successfully updates one of its neighbor, Ligra puts the neighbor into the scratch place pointed by its corresponding offset and atomically adds one to the offset. Finally it gathers all the valid vertices inside the scratch buffer into the next frontier. This process is both CPU and memory unfriendly. It scatters the vertices in the scratch buffer with random writes and relies on atomic instructions to synchronize the updates of the offset values.

1.3   Our Solutions and Contributions

To address these problems, we present `SilverChunk`, a graph processing system that enables balanced execution of parallel nested loop and conflict-free frontier maintenance. `SilverChunk` consists of two different chunking schemes, namely `VR-Chunk` for pull mode and `D-Chunk` for push mode. It also provides a high level programming interface with additional optimizations. The main contributions of our work are summarized as follows:

- **VR-Chunk.** We show that our `VR-Chunk` solves the first problem in a clean way. Instead of tuning the parallel scheduler, we change the scheduling unit directly from vertices to chunks. `VR-Chunk` splits the edge list statically into small chunks and generates additional virtual vertices to ensure conflict-free updates.
- **D-Chunk.** To tackle the second problem, we propose `D-Chunk`, a dynamic chunking scheme that applies to sparse frontiers. Since the vertices in a sparse frontier is discrete in memory, we build a list of virtual chunks that contains the information to help iterate over the edge list one piece of at a time. A virtual chunk provides a scratch space to aggregate vertices for the next iteration, which alleviates concurrent conflicts when building sparse frontiers.
- **Hybrid Polymorphic Interface and Optimizations.** We propose a new programming interface addressing different execution modes and graph algorithm properties for further optimizations. We design a new execution mode: `AllPull` mode, which optimizes the execution when the current frontiers are very dense.
- **Extensive Experiments.** We carry out extensive experiments using both large-scale real-world graphs and synthetic graphs to validate the performance of `SilverChunk`. Our experiments look into the key performance factors to all in-memory systems including the pre-processing time, the computational time and the effectiveness of main memory utilization. The results reveal that `SilverChunk` outperforms the state-of-the-art graph processing systems in most test cases by up to 4×.

The rest of this paper is organized as follows. Section 2 describes the main constructs of `SilverChunk`. Section 3 shows the high level programming interface and additional optimizations. Section 4 contains experimental results. Finally, Sect. 5 discusses the related works and Sect. 6 gives the concluding remarks.

2   Constructs

The main constructs of `SilverChunk` are the two chunking schemes: `VR-Chunk` and `D-Chunk`. Both schemes output similar chunk structures which are used to iterate over the input graphs. As a result, we unfold the nested loop into one flat loop which is efficient for parallel scheduling.

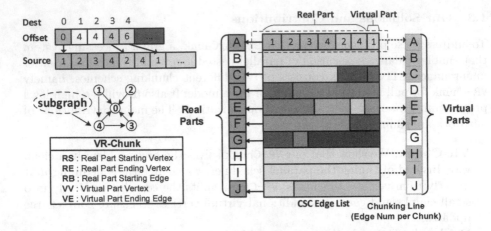

Fig. 3. VR-Chunk

2.1 VR-Chunk

In pull-based execution, we always iterate over the entire edge list to pull updates from the active vertices, thus the chunking scheme is static. Figure 3 shows how chunks are built from the original CSC array. Due to the dense feature of the frontier in pull mode, we assume that every edge requires the same amount of computation. Hence we slice the edge list into several chunks with equal number of edges, and assign each thread the same number of chunks to process.

Each chunk only needs to maintain five data fields: the starting and the ending destination vertices, the first edge, the virtual vertex and the last edge. The first two fields are obvious. As VR-Chunk might break the edge list, we need to maintain the first edge at each boundary. These fields form the real part of a chunk. The interesting one is the virtual vertex field, which stores the virtual vertex's ID, referring to the virtual part of a chunk. A different approach of dealing skewed distribution would be directly slicing the giant vertices into small virtual vertices. However, it cannot generate balanced chunks with respect to the edge number. VR-Chunk always slices giant vertices if its neighbor size is greater than the chunk size. Virtual vertices are used as delegates to the real vertices so that each vertex is assigned to exact one chunk. Virtual vertices are appended at the end of the vertex array to enlarge the vertex space so that the application data such as the PageRank value array gets transparently expanded too. Therefore, every application data gets a dedicated merge buffer which is appended at the end and there is no need to explicitly maintain a separate one.

2.2 D-Chunk

In push-based execution, since the active frontier is known only at runtime, VR-Chunk cannot be applied directly. Also the push execution always incurs random writes, synchronization is unavoidable. However, we can still benefit from chunking because it allows the destination vertices be collected in a conflict-free manner, therefore improving the sparse frontier's maintenance.

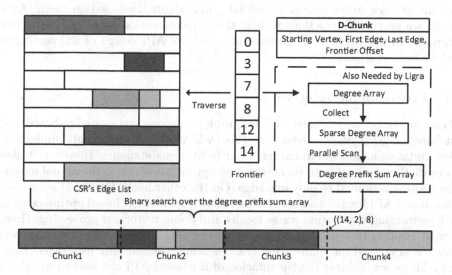

Fig. 4. D-Chunk

To build a chunk list dynamically in push-based execution, we extend the sparse frontier construction process used in Ligra [9], which requires calculating the prefix sum of the degree array. Figure 4 shows the building process of D-Chunk. An astute reader might notice that we need to rebuild the chunk list every time when entering push mode. This might sound problematic but actually building a chunk list for sparse frontier is very fast. Since we already have the prefix sum of the vertices' degrees in the current frontier during the original construction process, the running time of building the chunk list is proportional to the logarithm of the frontier's size. The additional work that D-Chunk does is a binary search to generate chunks with equal number of edges.

Each chunk only needs to maintain four data fields: the starting source vertex, the first and last edges, and the frontier offset. The first three data fields are used together with the current sparse frontier to iterate over the active edges. The frontier offset is a variable that helps collecting the vertices into the next sparse frontier. Since it is local to each chunk and there is no inter-chunk parallelism, the collecting process is conflict-free. Moreover, it generates sequential writes for each chunk. Hence the frontier maintenance is both CPU and memory friendly. Note that by using chunking in push mode, we can reuse the parallel scheduler in pull mode, which leads to better thread locality too. The actual scheduler is a simple thread pool implemented using a user-space thread barrier. Each thread is bound to a unique CPU core and the scheduler does round-robin work-stealing over the chunk list.

3   Implementations and Optimizations

Both VR-Chunk and D-Chunk are computational efficient but may require some amount of work to implement an actual graph algorithm based on them. As a result, we provide abstractions to hide the implementation details of the chunk internals. In this section we discuss the high level API design of SilverChunk and its optimizations.

3.1   Programming Interface

There are two commonly used APIs for graph processing systems: edge-based and list-based. Ligra [9] uses an edge-based API which allows users to only implement edge updating logic without caring about frontier maintenance. However, it also prevents the application to do customized optimizations since the actual execution context is limited to only one edge. On the other hand, Gemini [15] exposes a list-based API for the end users which allows application based optimizations, such as merging application states locally and doing vectorized processing. However, it requires the end users to maintain the next frontier in application code which is nontrivial for sparse frontiers. Therefore Gemini only uses dense frontiers. Moreover, a direct implementation of list-based API can lead to workload imbalance due to the skewed distribution of a input graph.

   As a result, we adopt these two API styles into SilverChunk and propose a hybrid interface. For push mode, we use the edge-based API similar to Ligra. The main reason is, since we are already doing random writes in push mode, there is little chance for a list-based API to provide further optimizations. Instead, we can hide the nontrivial frontier maintenance from the end users. An actual implementation of graph algorithms in push mode is instantiated as a push operator. A push operator accepts a source vertex and a destination vertex. It requires synchronization when updating to the destination vertex. A push operator can return a boolean value indicating whether the destination vertex should be put into the next frontier. It can also return nothing so that any sane compilers will get rid of unnecessary instructions of the frontier maintenance.

   For pull mode, we use the list-based API similar to Gemini. Thanks to our VR-Chunk scheme, giant vertices are already sliced, so workload balance is guaranteed. The running instance is called the pull operator. A pull operator accepts the starting and ending pointers of a source edge list, a real destination vertex and a destination vertex that might be real or virtual. Every update is guaranteed to be conflict-free when the pull operator is executed in parallel. The destination vertex is equal to the real destination vertex unless the vertex has its source edge list sliced by VR-Chunk. In that case, it is equal to the corresponding virtual vertex. In additional, pull mode also requires a pull reduce operator to be specified so that at the end of each iteration, all virtual vertices' states are merged to their corresponding real ones.

   Listing 1 shows a vanilla implementation of the PageRank algorithm using the SilverChunk's API. The graph argument contains the input graph data and is able to run a graph algorithm. The `Algorithm` class is instantiated with the aforementioned three operators, written as C++ lambdas.

```
void PageRankFunction::run(Graph & graph) {
  ... // initialization code
  Algorithm algo(
    [&](UInt32 s, UInt32 d) {// push
      atomicAdd(pr_new[d], pr[s]); },
    [&](UInt32* b, UInt32* e, UInt32 rd, UInt32 d) {// pull
      Float y = 0;
      while (b < e) y += pr[*b++];
      pr_new[d] = y; },
    [&](UInt32 rd, UInt32 d) {// pull reduce
      atomicAdd(pr_new[rd], pr_new[d]); }
  );
  while (!finish) {
    graph.run(algo);
    ... /* other related code */ }}
```

Listing 1. Page Rank Implementation

3.2 Optimizations

In the previous section we briefly described the polymorphism of the `push` operator, which enables optimizations when returning nothing. We call algorithms having this kind of operators `Immutable` since the frontier does not change after each iteration. We also identify other properties of graph algorithms for potential optimizations, as shown in Table 1. When all vertices are activated, the code path of propagating updates can be further optimized by removing unnecessary checks. We refer to this execution mode as `AllPull`.

Table 1. Algorithm properties

Algorithm	Immutable	Bypassable	Idempotent
PageRank	✓		
BFS		✓	✓
Components			✓
BellmanFord			✓

An algorithm is `Bypassable` if every vertex is supposed to be activated only once. An example is the simple breadth first search algorithm which finds any one traversing tree from the starting vertex. As shown in Listing 2, the `Algorithm` class accepts a `Bypassable` flag that checks if a vertex is already activated and can be bypassed for any further updates. When `Bypassable` is specified, the frontier maintenance does not interact with the application, hence it can be optimized statically. Note that the `pull reduce` operator is not needed in this algorithm.

```
void SimpleBFSFunction::run(Graph & graph) {
  ... // initialization code
  Algorithm algo(
    [&](UInt32 s, UInt32 d) {// push
      parent[d] = s; },
    [&](UInt32* b, UInt32* e, UInt32 rd, UInt32 d) {// pull
      while (b < e)
        if (graph.isActive(*b)) { parent[rd] = *b; return; }},
    Bypassable());
  while (!finish) {
    graph.run(algo);
    ... /* other related code */ }}
```

Listing 2. Simple BFS Implementation

An algorithm is `Idempotent` if algorithm correctness is not affected by propagating updates from inactive vertices to their neighbors. An example is the
label propagation algorithm for computing connected components. As shown
in Listing 3, the `Algorithm` class accepts a `Idempotent` threshold that switches
to `AllPull` execution when current frontier's density is greater than the threshold. The reason of specializing this property is because when frontiers are near
full, `AllPull` is faster than normal pull mode.

```
void LabelPropagationFunction::run(Graph & graph) {
  ... // initialization code
  Algorithm algo(
    [&](UInt32 s, UInt32 d) {// push
      return writeMin(id[d], id[s]); },
    [&](UInt32* b, UInt32* e, UInt32 rd, UInt32 d) {// pull
      UInt32 m = MAX_UINT32;
      while (b < e) if (graph.isActive(*b)) m = min(m, *b);
      if (m < id[rd]) { id[d] = m; return true; }
      return false; },
    [&](UInt32 rd, UInt32 d) {// pull reduce
      writeMin(id[rd], id[d]); },
    Idempotent(0.5));
  while (!finish) {
    graph.run(algo);
    ... /* other related code */ }}
```

Listing 3. Label Propagation Implementation

4   Experiments

In this section, we evaluate `SilverChunk`'s performance using a physical server
with four applications (PageRank, BFS, WCC and BellmanFord) and five
datasets (RMat24, RMat27, Twitter, Powerlaw and USARoad). The physical
server contains two Intel Xeon E5-2640v4 CPUs with 128 GB memory. We synthesized graphs using the R-MAT generator, following the same configuration used by

the graph500 benchmark. The synthetic power-law graph (PowerLaw) with fixed power-law constant 2.0 was generated using the tool in PowerGraph [4], which randomly samples the degree of each vertex from a Zipf distribution and then adds edges. We also use two types of real-world datasets, a social network graph (twitter-2010[1]) and a geometric graph (USARoad[2]). All graphs are unweighted except USARoad. To provide a weighted input for the SSSP algorithm, we add a random edge weight in the range [1, 100] to each edge. Following Table 2 shows the basic information of used datasets .

Table 2. Data set

Dataset	Vertex Num	Edge Num	Avg Deg	Max Indeg	Max Outdeg	Size (CSV)
RMat24	16M	0.3B	16.0	18.0K	17.3K	4.0 GB
RMat27	134M	2.1B	15.8	0.90M	0.86M	34 GB
Twitter	42M	1.5B	35.3	0.77M	3.0M	25 GB
Powerlaw	10M	0.1B	9.2	10	2.1M	1.4 GB
USARoad	23M	58M	2.4	9	9	1.3 GB

We compare SilverChunk to a number of different in-memory graph engines. Primarily, we compare SilverChunk with Ligra [9], Polymer [13], Gemini [15], Grazelle [5] and Galois [8] as these systems achieves state-of-the-art performance on a single-machine environment using in-memory storage. We run these systems with four graph algorithms on five different data sets using two different configuration of one commodity machine (Dell PowerEdge R730xd). We run iterative algorithms like Pagerank (PR) as well as traversal algorithms such as Bellman-Ford (BF) algorithm on these engines. This allows a comparison on how well a graph engine can handle different kinds of graph algorithms with different graph data distributions. The detailed information of the evaluated graph algorithms are as follow:

PageRank (PR) computes the rank of each vertex based on the ranks of its neighbors. We use the synchronous, pull-based PageRank in all cases and apply the division elimination optimization to all applications except Grazelle.

Breadth-first search (BFS) traverses an unweighted graph by visiting the sibling vertices before visiting the child vertices. The source is vertex one for this test.

Connected components (CC) calculates a maximal set of vertices that are reachable from each other for a directed graph. All systems adopt label propagation algorithm except Galois, which provides a topology-driven algorithm based on a concurrent union-find data structure.

---

[1] http://law.di.unimi.it/datasets.php.
[2] http://www.dis.uniroma1.it/challenge9/.

Table 3. Running times (in seconds) of algorithms over various data sets

System	Data set	PR (5 iterations)		BFS		CC		SSSP	
		one cpu	two cpus	one cpu	two cpus	one cpu	two cpus	one cpu	two cpus
SilverChunk	R-Mat24	1.35	0.84	0.13	0.10	0.79	0.48	3.50	2.49
	R-Mat27	9.52	5.86	0.62	0.42	5.56	2.69	7.66	4.63
	Twitter	4.55	2.64	0.41	0.30	4.49	2.35	7.54	4.59
	Powerlaw	0.34	0.21	0.13	0.10	0.53	0.27	0.93	0.64
	US Road	0.36	0.23	0.55	0.80	23.18	15.01	117.29	70.19
Ligra	R-Mat24	2.78	1.81	0.23	0.20	1.93	1.03	3.84	2.51
	R-Mat27	19.13	14.80	1.07	1.06	13.32	7.62	7.84	5.08
	Twitter	9.18	6.69	0.68	0.61	10.97	6.75	7.65	5.03
	Powerlaw	0.94	0.72	0.18	0.12	1.44	0.98	1.26	0.93
	US Road	0.88	0.65	1.46	1.57	62.42	40.12	169.23	87.26
Polymer	R-Mat24	4.71	1.88	0.26	0.22	1.64	0.80	4.23	2.53
	R-Mat27	43.98	19.08	1.36	1.04	13.82	6.58	9.48	4.91
	Twitter	28.82	12.02	0.79	0.65	16.51	8.61	7.69	5.15
	Powerlaw	1.54	0.71	0.18	0.20	1.58	1.02	1.29	0.73
	US Road	0.61	0.52	1.21	1.25	82.94	45.59	258.03	180.71
Gemini	R-Mat24	1.52	0.85	0.18	0.14	3.12	1.35	7.06	3.55
	R-Mat27	9.64	6.14	0.86	0.76	18.28	8.77	16.21	8.14
	Twitter	4.88	2.56	0.56	0.74	19.06	9.84	12.66	6.51
	Powerlaw	0.46	0.41	0.15	0.23	1.25	0.54	1.39	0.72
	US Road	0.61	0.31	20.42	21.64	176.23	123.24	533.04	379.54
Grazelle	R-Mat24	2.18	1.42	0.14	0.13	1.02	0.63	No Impl	
	R-Mat27	13.30	9.05	0.69	0.70	7.67	4.36		
	Twitter	6.27	3.81	0.54	0.44	6.27	4.47		
	Powerlaw	0.45	0.33	0.14	0.13	0.88	0.43		
	US Road	0.39	0.22	2.91	1.85	26.23	15.66		
Galois	R-Mat24	5.09	2.72	0.61	0.32	1.04	0.64	4.40	4.18
	R-Mat27	36.93	20.48	4.14	2.41	7.87	4.90	7.40	4.48
	Twitter	10.47	6.12	2.19	1.55	3.86	2.48	90.92	69.74
	Powerlaw	1.71	0.86	0.38	0.22	0.51	0.35	3.88	4.27
	US Road	4.40	2.15	0.33	0.30	0.81	0.45	0.90	1.02

Fastest time is denoted as underline. Second fastest time is denoted as underwave.

Single-source shortest-paths (SSSP) computes the distance of the shortest path from a given source vertex to other vertices. The source is vertex one for this test. All systems implement SSSP based on the Bellman-Ford algorithm with synchronously data-driven scheduling, while Galois uses a data-driven and asynchronously scheduled delta-stepping algorithm.

4.1 Graph Algorithm Test

Table 3 gives a complete runtime comparison. Of all the test cases, we report the execution time of their five runs. For PageRank algorithm, `SilverChunk` achieves optimal performance against other systems using only one CPU. Gemini and Grazellel are the second best. With two CPUs enabled, systems like Polymer, Gemini and Grazelle scales better than `SilverChunk`, however `SilverChunk` still holds three best results out of five. On the other hand, the graph traversal algorithms, including BFS, CC and SSSP, are not sensitive to the memory accesses of NUMA systems, since they have much fewer active vertices in each iteration, resulting in fewer memory accesses. Therefore, `SilverChunk` outperforms all other systems except Galois, which either adopts different algorithms for the problem or uses specialized scheduler for asynchronous execution. In most test cases, `SilverChunk` takes a leading position, except the `USRoad` graph. For high-diameter graphs like `USRoad`, the asynchronous scheduling and special implementations in Galois are able to exploit more parallelism for the graph traversal algorithms, such as CC and SSSP. In general, our graph chunking technique achieves 99% of CPU usage without any dynamic coordination in pull mode. It also gives consistent load balance in push mode.

4.2   VR-Chunk Test

As can be seen from Fig. 5, compared to other systems, `VR-Chunk` does not introduce pre-processing overheads, while still achieves the best performance. Figure 6 compares the running time of the PageRank algorithm on the twitter graph with three different implementations: Cilk [2], `VR-Chunk` and `VR-Chunk` with work-stealing. The static execution of `VR-Chunk` already excels the Cilk scheduler. Adding a simple chunk-based work-stealing mechanism gives another 10% performance gain.

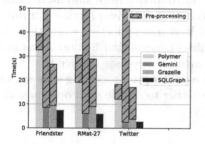

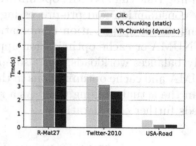

Fig. 5. Comparision among different systems

Fig. 6. Comparision with hand-written code

4.3   AllPull Test

We test different thresholds of `AllPull` execution combined with adaptive Push-Pull switching. Figure 7 shows the test result of running the Connected Components algorithm. With `AllPull` mode enabled, we get 30% performance gain. All three different data set achieve the best running time when the threshold is between 0.3 and 0.5. Therefore it can serve as a proper reference value for optimizing idempotent algorithms.

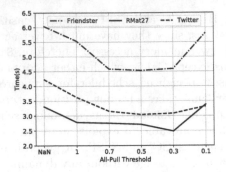

Fig. 7. Connected components execution time with different **AllPull** thresholds

4.4 NUMA and Cache Optimization Test

Since NUMA based engine Polymer [13] does not reveal proper performance, and cache based engine Cagra [14] does not open source their code, we implement both optimization schemes in order to complete our testing. We also combine NUMA and cache optimizations along with the optimizations used in SilverChunk. As can be seen from Table 4, both NUMA or cache optimizations can effectively improve the performance. The last column lists the memory consumption with values related to the lowest one. Cache optimization gives better running time than NUMA optimization but it introduces a huge amount of memory consumption and pre-processing time. SilverChunk gives further improvements in all optimization combinations, and it is more effective when there is no NUMA or cache optimization applied, which suggests that SilverChunk not only balances workloads, but also optimizes memory accesses. Notice that both NUMA and cache optimizations in this test have their pre-processing time longer than the actual running time. As a result, Whether to enable such optimization needs further considerations.

Table 4. PageRank (5 iters) over Twitter-2010

	Nested loop	VR-Chunk	Pre-processing	Peak memory
No NUMA, No Cache	3.13 s	2.64 s	0 s	1.0
NUMA, No Cache	2.28 s	2.08 s	3.83 s	1.05
No NUMA, Cache	1.91 s	1.68 s	6.52 s	1.56
NUMA, Cache	1.67 s	1.55 s	11.84 s	1.64

5 Related Works

The field of single machine graph processing in main memory has seen efforts in both parallel scheduling and graph partitioning. Ligra [9] proposes an EdgeMap interface to hide the inner loop parallelism, however it does not solve the actual workload imbalance issue. Grazelle [5] adopts a schedule-aware to achieve workload balance which however makes graph applications hard to implement. Polymer [13], Gemini [15] and Grazelle [5] are exponents in NUMA optimizations. They partition graph into subgraphs for each NUMA node, trying to reduce remote memory access. However it takes more time in pre-processing and its effectiveness is related to the graph data distribution and the actual running modes. For sparse frontiers, pre-partitioned graphs are less effective. Systems like GRACE [12] and Cagra [14] partition the input graph even further, at the CPU cache level. Cagra manually partitions the graph in order to make sure one batch of concurrent workload would end up only reading data from CPU's LLC. However, this adds a lot of complexity to the initialization process, and similar to NUMA-aware partitioning, it barely helps when the frontiers are sparse. Graph-Grind [10] uses partition-based optimization only when the frontier's density exceeds certain threshold, which is 50% in their experiments, while still keeps the vanilla CSR/CSC formats for sparse and medium-dense frontiers. However, they add one additional copy of the graph data to store the partitioned graph, resulting in 50% more memory consumption.

6 Conclusion

We present SilverChunk, an efficient in-memory parallel graph processing system running on a single machine. SilverChunk solves the workload imbalance issue of frontier-based computing by unfolding the nested loop into a flat loop over a chunk list. We extend the chunking scheme to support both pull and push modes and provide a unified high level API for implementing graph applications. In addition, we address new optimization opportunities based on different execution modes and algorithm properties, and use a policy based API to automatically apply the corresponding optimizations. Currently SilverChunk cannot handle graphs too big to fit into main memory. We plan to extend the ideas presented in this paper to external memory and distributed environment in near future.

Acknowledgements. We thank anonymous reviewers whose comments helped improve and clarify this manuscript. This work is funded by the National Key Research and Development Program of China.

References

1. Besta, M., Podstawski, M., Groner, L., Solomonik, E., Hoefler, T.: To push or to pull: on reducing communication and synchronization in graph computations. In: Proceedings of the 26th International Symposium on High-Performance Parallel and Distributed Computing, pp. 93–104. HPDC 2017, ACM, New York (2017). https://doi.org/10.1145/3078597.3078616

2. Blumofe, R.D., Joerg, C.F., Kuszmaul, B.C., Leiserson, C.E., Randall, K.H., Zhou, Y.: Cilk: an efficient multithreaded runtime system. In: Proceedings of the Fifth ACM SIGPLAN Symposium on Principles and Practice of Parallel Programming, PPOPP 1995, pp. 207–216. ACM, New York (1995). https://doi.org/10.1145/209936.209958

3. Dagum, L., Menon, R.: OpenMP: an industry standard api for shared-memory programming. IEEE Comput. Sci. Eng. **5**(1), 46–55 (1998). https://doi.org/10.1109/99.660313

4. Gonzalez, J.E., Low, Y., Gu, H., Bickson, D., Guestrin, C.: PowerGraph: distributed graph-parallel computation on natural graphs. In: Presented as part of the 10th USENIX Symposium on Operating Systems Design and Implementation (OSDI 2012), pp. 17–30. USENIX, Hollywood (2012)

5. Grossman, S., Litz, H., Kozyrakis, C.: Making pull-based graph processing performant, pp. 246–260. ACM Press (2018). https://doi.org/10.1145/3178487.3178506

6. Hong, S., Chafi, H., Sedlar, E., Olukotun, K.: Green-Marl: a DSL for easy and efficient graph analysis. In: Proceedings of the Seventeenth International Conference on Architectural Support for Programming Languages and Operating Systems, ASPLOS XVII, pp. 349–362. ACM, New York (2012). https://doi.org/10.1145/2150976.2151013

7. Malicevic, J., Lepers, B., Zwaenepoel, W.: Everything you always wanted to know about multicore graph processing but were afraid to ask. In: 2017 USENIX Annual Technical Conference (USENIX ATC 2017), pp. 631–643. USENIX Association, Santa Clara (2017)

8. Pingali, K., et al.: The tao of parallelism in algorithms. In: Proceedings of the 32Nd ACM SIGPLAN Conference on Programming Language Design and Implementation, PLDI 2011, pp. 12–25. ACM, New York (2011). https://doi.org/10.1145/1993498.1993501

9. Shun, J., Blelloch, G.E.: Ligra: a lightweight graph processing framework for shared memory, p. 135. ACM Press (2013). https://doi.org/10.1145/2442516.2442530

10. Sun, J., Vandierendonck, H., Nikolopoulos, D.S.: GraphGrind: addressing load imbalance of graph partitioning. In: Proceedings of the International Conference on Supercomputing, ICS 2017, pp. 16:1–16:10. ACM, New York (2017). https://doi.org/10.1145/3079079.3079097

11. Sundaram, N., et al.: GraphMat: high performance graph analytics made productive. Proc. VLDB Endowment **8**(11), 1214–1225 (2015). https://doi.org/10.14778/2809974.2809983

12. Wang, G., Xie, W., Demers, A.J., Gehrke, J.: Asynchronous large-scale graph processing made easy. In: CIDR, vol. 13, pp. 3–6 (2013)

13. Zhang, K., Chen, R., Chen, H.: NUMA-aware graph-structured analytics, pp. 183–193. ACM Press (2015). https://doi.org/10.1145/2688500.2688507

14. Zhang, Y., Kiriansky, V., Mendis, C., Amarasinghe, S., Zaharia, M.: Making caches work for graph analytics. In: 2017 IEEE International Conference on Big Data (Big Data), pp. 293–302, December 2017. https://doi.org/10.1109/BigData.2017.8257937

15. Zhu, X., Chen, W., Zheng, W., Ma, X.: Gemini: a computation-centric distributed graph processing system. In: OSDI 2016, pp. 301–316. USENIX Association (2016)

A Modular Approach for Efficient
Simple Question Answering
Over Knowledge Base

Happy Buzaaba[1(✉)] and Toshiyuki Amagasa[2]

[1] Graduate School of Systems and Information Engineering,
University of Tsukuba, Tsukuba, Japan
happy-b@kde.cs.tsukuba.ac.jp
[2] Center for Computational Sciences, University of Tsukuba, Tsukuba, Japan
amagasa@cs.tsukuba.ac.jp

Abstract. In this work, we propose an approach for efficient question answering (QA) of simple queries over a knowledge base (KB), whereby a single triple consisting of (subject, predicate, object) is retrieved from a KB for a given natural language query. In fact, most recent state-of-the-art methods exploit complex end-to-end neural network approaches to achieve higher precision while making it difficult to perform detailed analysis of the performance and suffering from long execution time when training the networks. To this problem, we decompose the simple QA task in a three step-pipeline: entity detection, entity linking and relation prediction. More precisely, our proposed approach is quite simple but performs reasonably well compared to previous complex approaches. We introduce a novel index that relies on the relation type to filter out subject entities from the candidate list so that the object entity with the highest score becomes the answer to the question. Furthermore, due to its simplicity, our approach can significantly reduce the training time compared to other comparative approaches. The experiment on the SimpleQuestions data set finds that basic LSTMs, GRUs, and non-neural network techniques achieve reasonable performance while providing an opportunity to understand the problem structure.

Keywords: Question answering · Knowledge base

1 Introduction

Large scale knowledge bases like Freebase [2], consist of a large pool of information with real-world entities as nodes and their relations as edges. Each directed edge, along with its head entity and tail entity, constitute a triple, i.e., (head entity, relation, tail entity), which is also named as a fact. Because of their large volume and complex data structures, it is difficult for non-technical users to access the substantial and valuable knowledge in them. To bridge this gap, Question answering over knowledge base (KB) aims at automatically translating

© Springer Nature Switzerland AG 2019
S. Hartmann et al. (Eds.): DEXA 2019, LNCS 11707, pp. 237–246, 2019.
https://doi.org/10.1007/978-3-030-27618-8_18

the end users' natural language questions into structured queries and returning entities from the KB as answers. For example, given a question "where was Barack Obama born?", simple Question answering over KB aims at identifying its corresponding triple, i.e., *(Barack Obama, people/person/place/of/birth, Honolulu)*. Recent developments in deep learning have triggered a line of work that have attracted many researchers to investigate more end-to-end approaches together with complex neural network architectures that performs well on a variety of natural language processing tasks like opinion extraction [10], sentence classification [11] etc. It is, however, difficult to perform detailed analysis of the performance in the end-to-end setting and they also suffer from long execution time when training the networks.

We focus on simple factoid questions based on the simpleQuestions benchmark [3], in which answering a question requires the extraction of a single fact from the knowledge base. We decompose the simple question-answering task in three different components: (1) Entity detection where standard Recurrent Neural Network (RNN), and Conditional Random Field (CRF) are applied to identify entities in the question, (2) We then link the identified entities to there corresponding nodes in the KB using an inverted index to generate a candidate list with their respective score and (3) Relation prediction where a question is classified as one of the relation types in the KB, we apply standard (RNN) plus standard Convolutional Neural Network (CNN) to do this.

In our work we make the following contributions; (i) Propose a simple yet effective approach, our approach is faster/efficient to train the network and performs reasonably well compared to previous complex approaches of Bodes et al. [3], Golub and He [8], Lukovinikov et al. [14], that apply end-to-end neural network on a similar task of simple question answering. (ii) Introduce a novel index that relies on the relation type to filter out subject entities from the candidate list so that the object entity with the highest score becomes the answer to the question.

2   Related Work

For several years, research has been conducted on question answering by directly parsing the natural language question into a structured query using semantic parsing [12], more recent work includes designing knowledge specific logical representation and grammar parsing [1]. In his work, Bodes et al. [3], proposed the single-relation factoid question answering. This work introduces a **Simple-Questions** data set which has 108,442 questions built on Freebase and proposes a memory network to solve the simple question answering task. This prompted a line of work that have led researchers like Golub and He [8], Lukovinikov et al. [14], Dai et al. [6] to apply even more complex neural network architectures to address this problem. Golub and He [8] proposed a character-level attention-based encoder-decoder model, Lukovinikov et al. [14] applies a hierarchical word-level and character-level question encoder to train a neural network in an end-to-end manner. Dai et al. [6] proposes a conditional probabilistic framework using BIGRUs to infer the target relation first and then the target subject associated with the candidate relations. Yin et al. [19] used character-level convolutional Neural Network for entity linking and a separate word-level convolutional Neural Network with attentive max-pooling that models the relationship between the predicate and question pattern more effectively. Yu et al. [20] applied a residual hierarchical BILSTM that performs hierarchical matching between questions and knowledge base relations for relation prediction, the results were then combined with the entity linking output. The above deep learning approaches, exploit increasingly complex techniques. Our work builds on a related work by Ture and Jojic [18] which argues that baseline methods when fully explored can equally produce competitive results. His work applies simple recurrent neural network and urges that taking advantage of the problem structure yields accurate and efficient results compared to complex neural network methods.

With the goal of enhancing the performance of the simple question answering system using baseline methods, we examine the necessity of complex models for the simple question answering task as applied by previous related work, and we do this by exploring the performance of baseline methods both standard neural network and non neural network techniques that perform reasonably well on a similar task.

3   Proposed Approach

The task of question answering over knowledge base for simple questions can be formally represented as follows; Let $G = \{(s_i, p_i, o_i)\}$ be the knowledge base representing a set of triples where s_i represents a subject entity, p_i a predicate also denoted as a relation, o_i an object entity. Given a simple natural language question q represented as a sequence of words, $q = \{w_1, w_2, ..., w_T\}$, the simple question answering task is to find a triple $(\hat{s}, \hat{p}, \hat{o}) \in G$ such that $\hat{o}$ is the intended answer to the question.

We therefore formulate this task to finding the right subject $\hat{s}$ and predicate or relation $\hat{p}$ referred to in the question q that characterizes a set of triples in the knowledge base G that contain the answer $\hat{o}$ to the question.

3.1   Entity Detection

To identify the entity in the question we formulate this as a sequence labeling problem where each word or token is tagged as entity or non-entity; I: entity and 0:non-entity. We apply both neural network and non-neural network methods to this task.

In Fig. 1, Each question word/token is represented with a word embedding, the input word representation is then combined with the hidden layer representation from the previous time step using either BiLSTM [9] or BiGRU [4] standard RNNs which then applies a non-linear transformation to compute the hidden layer representation at the current time step. The final hidden representation at the current time step is then projected to the output dimensional space and normalized into a probability distribution via a softmax layer.

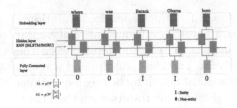

Fig. 1. RNN architecture for entity detection

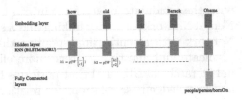

Fig. 2. RNN architecture for relation prediction.

Gated recurrent units (GRU) Fig. 3, are commonly used due to their ability to process longer sequences. As we read the sentence from left to right, the GRU is going to have a new memory variable called the memory cell $C^{<t>}$, so that when the network gets further into the sentence it can still remember the subject of the sentence. At time step t the GRU will output an activation function equivalent to the memory cell at that time step. The current memory cell $C^{<t>}$ at time step t is computed by interpolating between the previous hidden state $C^{<t-1>}$ at previous time step and the candidate state $\hat{C}^{<t>}$ at the current time step; $C^{<t>} = \Gamma_u \odot \hat{C}^{<t>} + (1 - \Gamma_u) \odot C^{<t-1>}$. With Γ_u as the update vector and $\odot$ the element wise vector product. For interpolation, the update gate which determines how much of the previous state is leaked into the current state Γ_u is computed using the current input $X^{<t>}$ and the previous state $C^{<t-1>}$ as $\Gamma_u = \sigma(W_u[C^{<t-1>}, X^{<t>}] + b_u)$. Where W_u and b_u are parameter metrics to be learned during training and σ the Sigmoid activation function $\sigma(x) = \frac{1}{1+e^{-x}}$ applied element wise to the vector entries. The update gate can decide to forget the previous state altogether or copy the previous state and ignore the current input. The candidate memory cell/hidden state at the current time step $\hat{C}^{<t>}$ is computed based on the current input $X^{<t>}$ and the previous hidden state $C^{<t-1>}$ give by $\hat{C}^{<t>} = tanh(W_c[\Gamma_r \odot C^{<t-1>}, X^{<t>}] + b_c)$. W_c and b_c are parameter metrics, $tanh$ the hyperbolic tangent activation function and Γ_r is the reset gate which determines the parts of the previous state ignored in computation of the candidate state and it is computed as $\Gamma_r = \sigma(W_r[C^{<t-1>}, X^{<t>} + b_r])$ with W_r and b_r the parameters.

We also apply **Conditional random field (crf)** to sequence labeling [13] to compare the entity detection performance with RNN. CRF, represents the probability of a hidden state sequence given some observations. Given $\mathbf{x}$; input sequence, $x = (x_1, x_2, ..., x_m)$ and $s = (s_1, s_2, ..., s_m)$ the output states (crf tags), the conditional probability cp; is given by $c_p = p(s_1, s_2, ..., s_m|x_1, x_2, ..., x_m)$. We define a feature map as $\Phi(x_1, x_2, ..., x_m, s_1, s_2, ..., s_m) \in R^d$. This feature map, maps x paired with s to d: dimensional feature vector. The probability is therefore modeled as a log linear $p(s|x, w) = \frac{exp(w.\phi(x,s))}{\sum_{s'} exp(w.\phi(x,s'))}$ with parameter vector $w \in R^d$. s'; ranges over all possible outputs. We estimate the parameter vector w by assuming that we have a set of n labeled samples $\{(x^i, s^i)\}^n$, $i = 1$. The regularized log likelihood is given by $\mathcal{L}(w) = \sum_{i=1}^{n} \log p(s^i|x^i, w) - \frac{\lambda 2}{2}||w||_2^2 - \lambda 1||w||_1$, where $\frac{\lambda 2}{2}||w||_2^2$ and $\lambda_1||w||_1$ forces w to be small in the respective

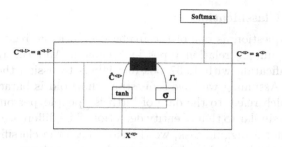

Fig. 3. Schematic representation of GRU

norm. The parameter vector w^* is estimated as $w^* = argmax(w \in R^d \mathcal{L}(w))$. If we estimate w^* the parameter vector, we can then find the most likely tag a sentence s^* for a sentence x by $s^* = argmax_s p(s|x : w^*)$.

We train crf using Stanford Named Entity Recognizer (NER) [7], a tool that labels word sequences in the sentence into four classes; person, organization, location and non-entity. This tool extracts features such as current/previous/next word, POS tag, character n-gram etc, and trains a crf model.

We tagged the question into four classes, the first three classes of (person, organization, and location) were tagged as entity. So we ultimately have two tags in our experiment of entity and non-entity and trained the stanford NER on the training set and labeled the test set questions.

3.2 Entity Linking

The generated candidate entities are then linked to the actual knowledge base node. We use 2M Freebase subset as our knowledge base. For linking the extracted entity to the actual knowledge base node, we build different indexes using dictionaries in python. First, a names index which maps all entity machine identifiers (MID's) in the Freebase subset to their names in the Freebase names file [6]. Second, the inverted index to map any entity n-gram to all nodes in the knowledge base, this association of entity n-grams to nodes in the knowledge base is computed by term frequency inverse document frequency (tf-idf). For example; assuming a node referring "Barack Obama" exists in the knowledge base, the tf-ifd weights would be computed by; $\mathcal{I}(\text{"Barack"}) \rightarrow \{node: e_i, score: tf\text{-}idf(\text{"Barack"}, \text{"Barack Obama"})\}$ and $\mathcal{I}(\text{"Barack Obama"}) \rightarrow \{node: e_i, score: tf\text{-}idf(\text{"Barack"}, \text{"Barack Obama"})\}$ We perform this for every n-gram of every entity node in the KB. We are able to generate a list of candidate entities with their associated scores. Once we have a list of candidate entities, we use each candidate node as a starting point to reach candidate answers. We limit our search to a single hop for the purpose of the current approach and retrieve all nodes that are reachable from the candidate where the path from is consistent with the predicted relation.

3.3   Relation Classification

We classify the question as one of the freebase knowledge base relation types. There are 1,837 unique relation types in Freebase, And the task is to do a large scale classification with 1,837 possible labels to assign the relation type to the question. Assuming we have a question "how old is barack obama", the relation type which refers to the date of birth is "people/person/bornOn". We examine a model similar to that of entity detection. The difference is that relation classification is not a tagging task, we therefore base the classification decision on the output of the last hidden layer for prediction as shown in Fig. 3.

We also apply **Convolutional Neural Networks CNN** for relation classification, similar to Kim et al. [11]. We modify the multi-channel model described in his paper to a single static channel instead, and apply the same model to our task of relation classification. We adopt CNNs because of their ability to do extract local features by sliding filters over the word embeddings.

The sentence is represented by concatenating words and padding where necessary as follows; $x_{1:n} = x_1 \oplus x_2 \oplus .... \oplus x_n$, and we use convolutional filters to generate new features from a window of words as represented by $c_i = f(W.x_{i:i+h-1} + b)$. We apply the filter to each of the possible window of words in the sentence to produce a feature map represented by $c = [c_1, c_2, \ldots, c_{n-h+1}]$. We then use the max-over-time pool over the filter to take the maximum value as a feature corresponding to this particular filter. The idea is to capture the most important feature, which is basically one with the highest value for each feature map. And finally these features are passed on to the fully connected softmax layer whose output is the probability distribution over labels. We are able to come up with a relation type for each of the questions in the data set from the relation prediction step.

3.4   End to End Question Answering

After generating the candidate entities with their respective scores, and the relation types in the previous steps, we come up with all possible (entity, relation) pairs, from which we believe we can get an answer to the question. The next step is to filter out those entity nodes that do not seem to have an answer to the question. For example if we had a question *"how old is Barack Obama"* which has a relation type *"people/person/bornOn"*, to do filtering, we look at the relation type, and all those candidate entity nodes with a different relation type leading to another node are filtered out. Only those with a relation type leading to another node similar to the one generated are kept and the entity node with a high score in the remaining candidate list has an object entity node which is the answer to the question.

4   Experiment and Results

4.1   Experimental Setup

We experiment on the SimpleQuestions benchmark [3], with Freebase as the knowledge base [2]. We experiment on a 2M Freebase subset to be able to compare with previous work that applied a similar subset as the knowledge base. In freebase knowledge base, entities are connected by predefined predicates connecting from the subject to the object. A triple (Subject, Predicate, Object) denoted as *(S, P, O)* which describe a fact for example *(Barack Obama, people/person.bornOn, 8/4/1961)* refers to the fact that Barack Obama's date of birth is 8th, *April 1961*. We use the training, validation and test splits of 75,910, 10,845 and 21,687 questions respectively as provided by the dataset in our experiment. We initialize the model word embeddings with a 300-dimensional pre-trained vectors provided by Glove [15]. The pre-trained word embeddings implicitly integrate word semantics inferred from large text corpus based on the distributional hypothesis [17]. The pre-trained word embeddings allows to find better matches between words in the question and subject labels or relation URI's. It also allows to handle unseen words during training when it comes to testing.

We compute precision, recall and F1 for every sequence tags against the ground truth for evaluation in entity detection. We evaluate recall for top results at k *(R@k)* for both entity linking and relation prediction to see if the correct answer appears in the top k results. The final prediction is marked as correct if both entity and relation match the ground truth in end-to-end evaluation. We follow Bodes et al. [3] to do this and we measure accuracy which is equivalent to R@1.

We also use the Stanford Named Entity Tagger (NER) [7] a tool for labeling word sequences in the sentence to conduct the experiment with conditional random field (crf).

4.2 Results

In this subsection, we present our results on the SimpleQuestions task and we begin with the results on each individual component.

Entity Detection: Table 1 shows each models' results on the task of entity detection. We evaluate the precision, recall and F1-score on the token span level. This means that the predicted entity token span exactly matches the ground truth (a true positive span). The results reveal that RNN (LSTM & GRU) perform better with F1-score of 92.5% for the GRU. It can also be noticed that the crf result of 90.2% is comparable.

Entity Linking: Table 2 shows the performance of each model on the entity linking task. The CRF entity linking results accuracy is comparable to both LSTM and GRU. Although the crf may have performed slightly lower than the LSTM and GRU on entity detection, the bottleneck is entity linking because there are more entities in the knowledge graph with the same label which makes it difficult to identify the correct entity.

Table 1. Entity detection results for a given model.

Model	Split	P (%)	R (%)	F1 (%)
LSTM	Val	91.89	92.87	92.26
LSTM	Test	91.08	91.21	91.53
GRU	Val	92.56	93	92.78
GRU	Test	92.09	92.92	92.5
CRF	Val	90.71	89.92	90.36
CRF	Test	90.72	89.8	90.2

Table 2. Entity linking results for a given model.

Model	Split	R@1	R@5	R@10	R@20
LSTM	Val	0.679	0.827	0.863	0.889
LSTM	Test	0.662	0.811	0.849	0.876
GRU	Val	0.676	0.825	0.86	0.885
GRU	Test	0.661	0.808	0.848	0.876
CRF	Val	0.663	0.809	0.845	0.871
CRF	Test	0.649	0.796	0.834	0.861

Table 3. Relation prediction results for a given model.

Model	Split	P (%)	R@3	R@5
GRU	Val	82.22	93.75	95.93
GRU	Test	81.59	93.68	95.76
LSTM	Val	81.76	93.73	95.85
LSTM	Test	81.28	93.66	95.47
CNN	Val	82.88	93.75	95.86
CNN	Test	81.92	93.68	95.64

Table 4. End to end combination of entity detection and relation prediction models.

Entity	Relation	Accuracy
BiLSTM	**BiGRU**	**74.64**
BiLSTM	CNN	74.63
BiLSTM	BiLSTM	74.59
BiGRU	BiGRU	74.54
BiGRU	CNN	73.92
CRF	CNN	73.42
CRF	**BiGRU**	**73.39**
CRF	BiLSTM	73.34
Model	Description	Accuracy
Yin et al. 2016	Max-pooling	76.4
Dai et al. 2016	Probabilistic	75.7
Lukovinikov 2017	Neural embedding	71.2
Golub and He 2016	Character-based	70.9
Bodes et al. 2015	Memory network	62.7

Relation Prediction: For the task of predicting the relation type of the question, the relation or predicate is given in the data-set. We conduct a large scale classification with 1,837 possible labels to assign a relation type to the question. In Table 3 we can see that on precision, CNN out performs both RNN's (LSTM and GRUs). We however see that both RNN and CNN retrieval results (R@3) are essentially similar but RNN better at (R@5).

Table 4 shows end-to-end results for various combinations of entity detection and relation prediction on test set. The best model combination which achieves 74.64% accuracy is the BiLSTM for entity detection and BiGRU for relation prediction. When we replace BiLSTM with CRF for entity detection, the accuracy decrease by only 1.25 this shows that non neural network baselines can still perform well. Despite the immense contribution of neural networks to the meaningful improvements in the state of the art on the simple questions data set, our results suggest that the improvements directly attributed to complex neural networks are modest than previous researchers may have led the readers to believe. We also compare our results with existing state-of-art complex models to examine the necessity of model complexity on this task. Our results outperform the complex neural network models of Bodes et al.'s [3] memory network, Golub and He's [8] attention-enhanced encoder-decoder framework and Lukovinikov [14] complex character and word-level encoding. Our model is however comparable to Dai et al. [6] and Yin et al. [19] which apply a separately trained segmentation. Our best accuracy is less than 2 points away from the next highest reported result in the literature. It is important to pay attention when interpreting the results due to non-determinism associated with training neural networks that can yield differences in accuracy Reimers and Gurevych [16]. It was also demonstrated that for answer selection in question answering, issues ranging from software versions, can significantly impact the accuracy Crane [5].

4.3 Training Time

Our observation is that the proposed method is quicker and efficient to train. Training each of the component of entity detection and relation prediction for 50 epochs, using our PC, takes approximately 8 h. While it takes close to 6 days to train for example Lukovinikov et al. [14] for the same number of epochs on a similar PC.

5 Conclusion and Future Work

In this work we explore simple yet effective approach for simple question answering. We decompose the simple question answering task in sub-problems of entity detection, entity linking and relation prediction and solve each separately using simple baseline methods. Our results show that there is need to adequately examine simple baselines and take advantage of the problem structure before rushing to sophisticated deep learning techniques at least for the simple question answering task.

References

1. Berant, J., Chou, A., Frostig, R., Liang, P.: Semantic parsing on freebase from question-answer pairs. In: Proceedings of the 2013 Conference on Empirical Methods in Natural Language Processing, pp. 1533–1544 (2013)
2. Bollacker, K., Evans, C., Paritosh, P., Sturge, T., Taylor, J.: Freebase: a collaboratively created graph database for structuring human knowledge. In: Proceedings of the 2008 ACM SIGMOD International Conference on Management of Data, pp. 1247–1250. ACM (2008)
3. Bordes, A., Usunier, N., Chopra, S., Weston, J.: Large-scale simple question answering with memory networks. arXiv preprint arXiv:1506.02075 (2015)
4. Cho, K., et al.: Learning phrase representations using RNN encoder-decoder for statistical machine translation. arXiv preprint arXiv:1406.1078 (2014)

5. Crane, M.: Questionable answers in question answering research: reproducibility and variability of published results. Trans. Assoc. Comput. Linguist. **6**, 241–252 (2018)
6. Dai, Z., Li, L., Xu, W.: CFO: Conditional focused neural question answering with large-scale knowledge bases. arXiv preprint arXiv:1606.01994 (2016)
7. Finkel, J.R., Grenager, T., Manning, C.: Incorporating non-local information into information extraction systems by gibbs sampling. In: Proceedings of the 43rd Annual Meeting on Association for Computational Linguistics, pp. 363–370. Association for Computational Linguistics (2005)
8. Golub, D., He, X.: Character-level question answering with attention. arXiv preprint arXiv:1604.00727 (2016)
9. Graves, A., Schmidhuber, J.: Framewise phoneme classification with bidirectional LSTM and other neural network architectures. Neural Netw. **18**(5–6), 602–610 (2005)
10. Irsoy, O., Cardie, C.: Opinion mining with deep recurrent neural networks. In: Proceedings of the 2014 Conference on Empirical Methods in Natural Language Processing (EMNLP), pp. 720–728 (2014)
11. Kim, Y.: Convolutional neural networks for sentence classification. arXiv preprint arXiv:1408.5882 (2014)
12. Kwiatkowski, T., Choi, E., Artzi, Y., Zettlemoyer, L.: Scaling semantic parsers with on-the-fly ontology matching. In: Proceedings of the 2013 Conference on Empirical Methods in Natural Language Processing, pp. 1545–1556 (2013)
13. Lafferty, J., McCallum, A., Pereira, F.C.: Conditional random fields: Probabilistic models for segmenting and labeling sequence data (2001)
14. Lukovnikov, D., Fischer, A., Lehmann, J., Auer, S.: Neural network-based question answering over knowledge graphs on word and character level. In: Proceedings of the 26th International Conference on World Wide Web, pp. 1211–1220. International World Wide Web Conferences Steering Committee (2017)
15. Pennington, J., Socher, R., Manning, C.: Glove: global vectors for word representation. In: Proceedings of the 2014 Conference on Empirical Methods in Natural Language Processing (EMNLP), pp. 1532–1543 (2014)
16. Reimers, N., Gurevych, I.: Reporting score distributions makes a difference: Performance study of LSTM-networks for sequence tagging. arXiv preprint arXiv:1707.09861 (2017)
17. Sahlgren, M.: The distributional hypothesis. Ital. J. Disabil. Stud. **20**, 33–53 (2008)
18. Ture, F., Jojic, O.: No need to pay attention: Simple recurrent neural networks work!(for answering "simple" questions). arXiv preprint arXiv:1606.05029 (2016)
19. Yin, W., Yu, M., Xiang, B., Zhou, B., Schütze, H.: Simple question answering by attentive convolutional neural network. arXiv preprint arXiv:1606.03391 (2016)
20. Yu, M., Yin, W., Hasan, K.S., Santos, C.D., Xiang, B., Zhou, B.: Improved neural relation detection for knowledge base question answering. arXiv preprint arXiv:1704.06194 (2017)

Scalable Machine Learning in the R Language Using a Summarization Matrix

Siva Uday Sampreeth Chebolu[✉], Carlos Ordonez, and Sikder Tahsin Al-Amin

Department of Computer Science, University of Houston, Houston, TX 77204, USA
sivauday.sampreeth8@gmail.com

Abstract. Big data analytics generally rely on parallel processing in large computer clusters. However, this approach is not always the best. CPUs speed and RAM capacity keep growing, making small computers faster and more attractive to the analyst. Machine Learning (ML) models are generally computed on a data set, aggregating, transforming and filtering big data, which is orders of magnitude smaller than raw data. Users prefer "easy" high-level languages like R and Python, which accomplish complex analytic tasks with a few lines of code, but they present memory and speed limitations. Finally, data summarization has been a fundamental technique in data mining that has great promise with big data. With that motivation in mind, we adapt the Γ (Gamma) summarization matrix, previously used in parallel DBMSs, to work in the R language. Γ is significantly smaller than the data set, but captures fundamental statistical properties. Γ works well for a remarkably wide spectrum of ML models, including supervised and unsupervised models, assuming dimensions (variables) are either dependent or independent. An extensive experimental evaluation proves models on summarized data sets are accurate and their computation is significantly faster than R built-in functions. Moreover, experiments illustrate our R solution is faster and less resource hungry than competing parallel systems including a parallel DBMS and Spark.

1 Introduction

Machine Learning has become popular and gained a lot of demand in the present world with the availability of abundant data and abundant processing power. There are a lot of tools and technologies like Python, R, Scala, Java, C# and many more which compute these machine learning models. However, data sets can be so large that they do not fit in the main memory. For these types of data, Hadoop stack or distributed systems or DBMSs like Vertica, SciDB is a popular choice to compute the Machine Learning models [10,15]. Contrary to the popular belief, we propose that the size of the cleaned data set, rather than its raw counterpart should dictate the data processing platform to be used. Data cleaning strips off a lot of unwanted and inaccurate data. As a result, the size of the data set is significantly reduced and with it, the need to use a heavyweight

© Springer Nature Switzerland AG 2019
S. Hartmann et al. (Eds.): DEXA 2019, LNCS 11707, pp. 247–262, 2019.
https://doi.org/10.1007/978-3-030-27618-8_19

data processing platform like Hadoop. Therefore, with a refined data set, data processing can be limited to a single system environment like, in our case, R.

With a vast package ecosystem coupled with extensive developer support, R ticks all the right boxes when it comes to being a data analytics platform [7]. However, R has few shortcomings of which memory management, speed and efficiency are the most noticeable. While parallelism in R can be achieved by using packages like *parallel*, the shortcoming becomes evident with an increasing number of cores. The language design sometimes poses a great problem when working with large data sets since the data has to be stored in physical memory. With the dedicated physical memory, R cannot scale to deal with data sets larger than the proportion of memory allocated to it and is forced to crash in such cases. So, the physical memory limitation clearly outweighs the need to address the issue of parallelization in R. In an attempt to address the above limitation in R, we used the summarization technique in the first Phase of our approach. But again, summarization technique can be used only for those models which accept Gramian Matrix product like Linear Regression (LR), Principal Component Analysis (PCA), Naïve Bayes (NB), K-means (KM) and few others. Furthermore, we built upon the parallel database systems algorithm in [10] to make it work in a serial scalable manner in R. Here, we implemented the models from [10] and also explored new models like Naïve Bayes and K-means which require a new gamma matrix, Diagonal Gamma, instead of the old ones stated in [10]. The environment does not crash even for large data sets, works independent of the physical memory allocated to the R environment and gives as accurate results as the existing packages that compute the above models in R.

2   Definitions

This is a reference section which introduces definitions of input data sets and models from mathematical perspective, R runtime and RCpp package. Each subsection can be skipped by a reader familiar with the material.

2.1   Mathematical Definitions

First, we define the inputs given to the models. The most obvious one is the input data set, interpreted as a matrix, which is defined to be a set of n column-vectors. All the models take a $d \times n$ matrix as input. Let the input data set be defined as X, which is considered to have n points, where each point is a vector in R. Therefore, we can see X as a wide rectangular matrix. In the case of Linear Regression (LR) and Principal Component Analysis (PCA), we take an extra dimension (output variable Y) resulting a change in the dimensions of X to $(d+1) \times n$, which we call $\mathbf{X}$. We use $i = 1...n$ and $j = 1...d$ as matrix subscripts. We augment X with an extra row of n 1s and call that as matrix $\mathbf{Z}$ $((d+2) \times n)$ for mathematical convenience. Column-vectors and column-oriented matrices are used for mathematical convenience because they allow simpler equations.

We use Θ to represent a statistical model in general. That is, Θ can be a LR or PCA model as well as any of the clustering and classification models such as Naïve Bayes (NB) and K-means (KM). PCA is an unsupervised model to reduce dimensionality. LR is a fundamental supervised model, whose solution helps in understanding and building other linear models. Naïve Bayes is another classic supervised model, whose solution assigns a numerical value between 0 and 1 to each class label denoting the probability of data belonging to a specific class. K-means is a clustering algorithm whose goal is to find k similar groups in the data. The algorithm works iteratively to assign each data point to one of k groups based on the features that are provided. Data points are clustered based on feature similarity. Therefore, for each model, $\Theta = \{list\ of\ matrices/vectors\}$, as follows. For LR: $\Theta = \beta$ where β is the vector or regression coefficients; for PCA: $\Theta = U, D$ where U are the eigen vectors and D contains the squared eigenvalues obtained from SVD; for NB: $\Theta = \{\pi, \mu, \sigma\}$, where π is the vector of k class priors, μ is a set of k mean vectors and σ are k diagonal matrices with standard deviations; and for KM: $\Theta = \{W, C, R\}$, where W is a vector of k (number of clusters) weights, C is a set of k centroid vectors and R is a set of k variance matrices.

2.2 R Runtime and RCpp Package

R is a dynamic language for statistical computing that combines lazy functional features and object-oriented programming [6, 12]. In R, vectors are stored as one contiguous block, matrices are 2-dimensional arrays of real numbers, which are stored as one block in column major order dynamically allocated, Lists are the most general ones and can have elements of diverse data types, including atomic data types and nested data structures. R uses a dynamic interpreter and also it utilizes C language for matrix and data frame operations and LAPACK library for linear algebra and numerical methods. When R functions are called, the R run-time creates nested variable environments, which are dynamically scoped.

The advantage of the RCpp package is its memory management. We can pass values to and from R and RCpp. When we pass the values, only the reference gets passed to the other side but not the actual value. So, memory consumption is very efficient and the runtime is the same. We can even pass matrices, lists, vectors and similar data to RCpp and return any of those from RCpp.

3 Theory and Algorithm

We present our main technical contribution in this section. First, we propose our main algorithm and then we discuss it in details. Then we discuss the implementation of our algorithm in R and RCpp. Finally, we give the run time complexity of our algorithm.

3.1 Algorithm

Our main algorithm consists of two steps:

1. Phase 1: Compute summarization matrix: one matrix Γ or k matrices Γ_j.
2. Phase 2: Compute model Θ based on Gamma martix (matrices).

In phase 1, first, we review the Gamma matrix (Non-Diagonal Gamma) and the statistics in it which was proposed in [10]. Matrix Γ (Gamma), defined below, is a fundamental matrix which contains a complete, accurate and sufficient summary. Then we describe the design and implementation of our main technical contribution, the Diagonal Gamma matrix. Both Non-Diagonal Gamma and Diagonal Gamma provides summarization for a different set of models which are presented in phase 2. For PCA and LR, we need one full Γ matrix assuming element off-diagonal is not zero. And for NB and KM, we need k matrices Γ_j (k classes, or k clusters respectively), where each Γ_j is "diagonal" meaning we assume Q is diagonal where off-diagonal elements are assumed to be zero. We discuss both phases in details in the following sections.

3.2 Phase 1: Computing Summarization Matrices

First we review the sufficient statistics for X which are integrated to form the Non-Diagonal Gamma Matrix, which are:

$$n = |X|, \tag{1}$$

$$L = \sum_{i=1}^{n} x_i, \tag{2}$$

$$Q = XX^T = \sum_{i=1}^{n} x_i \cdot x_i^T \tag{3}$$

Here, X is the data set, n counts total number of points in the data set, L is a linear sum of x_i and Q is a sum of vector outer product where x_i is multiplied by itself, i.e., Q is simply the "quadratic" sum of x_i. As defined earlier in Sect. 2.1, X is $d \times n$, $\mathbf{Z}$ has $(d+2)$ rows and n columns, where row $[0]$ are 1s and row $[d+1]$ is Y. Hence, z_i can be defined as $z_i = [1, x_i, y_i]$. Then the $\mathbf{Z}$ matrix becomes:

$$\mathbf{Z} = \begin{bmatrix} 1 & 1 & ... & 1 \\ x_1 & x_2 & ... & x_n \\ y_1 & y_2 & ... & y_n \end{bmatrix} \tag{4}$$

Matrix Γ (Gamma), which is defined below, is a fundamental Gamma matrix which contains a complete, definite, and sufficient summary of X to efficiently compute models like LR and PCA that have been previously defined. We define a complementary Gamma matrix, Diagonal Gamma, in Sect. 3.2 for models assuming variable independence, like Naïve Bayes and K-means.

$$\Gamma = \begin{bmatrix} n & L^T & 1^T \cdot Y^T \\ L & Q & XY^T \\ Y \cdot 1 & YX^T & YY^T \end{bmatrix} = \begin{bmatrix} n & \sum x_i^T & \sum y_i \\ \sum x_i & \sum x_i x_i^T & \sum x_i y_i \\ \sum y_i & \sum y_i x_i^T & \sum y_i^2 \end{bmatrix} \quad (5)$$

Here, Γ which can be computed in two ways from [10]. Alternative (1) is matrix-matrix multiplications i.e. $\mathbf{ZZ}^T$; Alternative (2) is sum of vector outer products i.e. $\sum_i z_i \cdot z_i^T$. So, $\Gamma = \mathbf{ZZ}^T = \sum_{i=1}^n z_i \cdot z_i^T$. That is, the square of matrix $\mathbf{Z}$ gives us Γ, which is significantly smaller than X. In general, if $d << n$, Γ comfortably fits in main memory.

Diagonal Q Matrix Assuming Dimensions Are Independent: From [10], it is clear that Non-Diagonal Gamma matrix, despite being iterative algorithms, avoids reading the entire data sets at every iteration. But that approach cannot be applied on models like Naïve Bayes (NB) or K-means (KM) which require more than one summarization matrix and may also require to read the entire data set more than once. For example, Naïve Bayes requires k summarization matrices for a given data set, where k being the number of unique class labels in the data set and K-means requires k matrices for summarization of a data set with k as the number of clusters given by the user, i.e., one for each cluster. Furthermore, these models do not require the complete computation of the Non-Diagonal Gamma as described in Sect. 3.2. The reason behind that is, the LR and PCA are computed in rotated space whereas in NB and KM we assume that the dimensions are independent, making Gamma diagonal. Due to this reason, we introduce another matrix, Diagonal Gamma, which helps to compute these models. Here, we do not require the Y parameter for Naive Bayes and K-means as used in LR and PCA. The major difference between the two forms of Gamma is we do not require parameters off the diagonal in Diagonal Gamma matrix as in Non-Diagonal Gamma matrix. So, we need only a few parameters out of the whole Non-Diagonal Gamma, namely, n, L, L^T, Q. That is, we require only a few sub-matrices from Non-Diagonal Gamma, which can be visualized as:

$$\Gamma_{diag} = \begin{bmatrix} n & L^T & 0 \\ L & Q & 0 \\ 0 & 0 & 0 \end{bmatrix}, where\ Q = \begin{bmatrix} Q_{11} & 0 & 0....... & 0 \\ 0 & Q_{22} & 0....... & 0 \\ 0 & 0 & Q_{33}..... & 0 \\ 0 & 0 & 0........ & Q_{dd} \end{bmatrix} \quad (6)$$

Furthermore, if we see the above sub-matrix, we observe that if we compute the terms in the lower triangle, we can get the whole sub-matrix just by copying the L to L^T, i.e., we need to compute the terms in the lower triangle and copy it to the upper triangle. This is the major change in definition of the Non-Diagonal Gamma to that of the Diagonal Gamma. Also, in Non-Diagonal Gamma, the Q is computed completely. On the other hand, in Diagonal Gamma, the Q is diagonal. From which we came up with the name of the matrices as Diagonal and Non-Diagonal Gamma. So, Q is diagonal or non-diagonal but not Γ.

3.3   Phase 2: Computing Models

Models are computed using the two versions of Gamma. One is with one Non-Diagonal Gamma Matrix and another one is k-Diagonal Gamma Matrices. Both of them were introduced previously.

Models Based on One Non-Diagonal Gamma:

Linear Regression (LR): From [10], the standard definition of LR is given as $Y = \beta^T \mathbf{X} + \epsilon$, where β is the column vector of regression coefficients and ϵ represents the Gaussian error. $\mathbf{X}$ is a $(d+1) \times n$ augmented matrix where we have X with a row of n 1s. β can be defined as $\hat{\beta} = (\mathbf{X}\mathbf{X}^T)^{-1}\mathbf{X}Y^T$. From the discussed Non-Diagonal Gamma, we can rewrite this equation as

$$\hat{\beta} = Q^{-1}(\mathbf{X}Y^T) \tag{7}$$

Principal Component Analysis (PCA): PCA is mainly implemented on a data set to reduce noise and redundancy of dimensions. PCA can be computed on the covariance matrix (V), or the correlation matrix (ρ), of the data set from [4]. This model require two parameters. First is U, which is a set of d orthogonal vectors, principal components of the data set, ordered in decreasing order by their variance. Second is the diagonal matrix D^2 which contains the squared eigen values. From [10], we can compute ρ, the correlation matrix, from the two parameters, D and U as $\rho = UD^2U^T = (UD^2U^T)^T$. We can also compute the covariance matrix as $V = Q/n - LL^T/n^2$. Then we compute PCA by using Eigen decomposition of the ρ, which is a symmetric matrix factorization. That is, we compute PCA from the correlation matrix by solving Singular Value Decomposition (SVD) on it. Also, we express ρ in terms of the sufficient statistics to compute SVD as follows:

$$\rho_{ab} = \frac{(nQ_{ab} - L_a L_b)}{(\sqrt{nQ_{aa} - L_a^2}\sqrt{nQ_{bb} - L_b^2})} \tag{8}$$

Models Based on k Diagonal Gammas:

Naïve Bayes (NB): The input for this model is a data set X and the output is a Naïve Bayes classification model which contains C (mean per dimension), R (variance per dimension), and W (prior per class). First, we take the data set X as input in chunks of fixed size. In each chunk, we split the data based on number of classes in the data set. We compute one gamma for each part of the chunk and at last add up these Γ matrices with respect to the classes and arrive at a final list of Γ matrices one for each class. We focus on $k = 2$ classes for NB. Then finally we have Γ_0 for class 0 and Γ_1 for class 1. We extract N_g, L_g, Q_g as defined in Sect. 3.2, from this final list of Γs. So, we arrive at lists of N_g, L_g, Q_g from where we compute π, μ and σ per dimension per item in the list separately like:

$$\pi_g = \frac{N_g}{n}, \tag{9}$$

$$\mu_g = \frac{L_g}{N_g}, \tag{10}$$

$$\sigma_g = \frac{Q_g}{N_g} - diag[\frac{L_g L_g^T}{N_g^2}] \tag{11}$$

Here, $N_g = |Xg|$ and we take the diagonal of $L \cdot L^T$ and Q, which can be manipulated as a 1-dimensional array instead of a 2D array. These are the 3 parameters included in the Naïve Bayes model. Now, we can predict class labels for new data using this model. For the prediction, for each point in the input data, we compute a probability value per class using the model parameters and assign the class with maximum probability. We compute the probability using,
$$P_{xi_{class}} = (1/\sqrt{2\pi\sigma_{g_j}^2})e^{(-0.5(x_i - \mu_{x_i})^2/\sigma_{g_j}^2)}.$$

K-means (KM): The input for this model is a data set X and the number of clusters (k) and the output is three matrices C, R, W, containing the means, the variances and the weights respectively for each cluster of X. For K-means with k clusters, we have list of matrices as $\Gamma_1, \Gamma_2, .., \Gamma_k$, where $k \geq 2$. Following definitions from Sect. 3.2, we introduce similar model parameters X_j, N_j, L_j, Q_j as the subset of X which belong to cluster j, the total number of points per cluster ($|X_j|$), the sum of points in a cluster ($\sum_{\forall x_i \in X_j} x_i$) and the sum of squared points in each cluster ($\sum_{\forall x_i \in X_j} x_i x_i^t$) respectively. From these statistics, we compute C_j, R_j, W_j as:

$$C_j = \frac{L_j}{N_j}, \tag{12}$$

$$R_j = \frac{Q_j}{N_j} - diag[\frac{L_j L_j^t}{N_j^2}], \tag{13}$$

$$W_j = \frac{N_j}{n} \tag{14}$$

Here $N_j = |Xj|$ and we take diagonal of $L \cdot L^T$ and Q, which can be treated as vectors instead of a matrix. The algorithm iterates executing two steps starting from random initialization until cluster centroids become stable.

Step 1 determines the closest cluster for each point and adds the point to it. K-means uses Euclidean distance to determine the closest centroid to each point x_i which is defined as $d(x_i, C_j) = (x_i - C_j)^t (x_i - C_j)$.

Step 2 updates all centroids C_j by computing the mean vector of points belonging to cluster j. The cluster weights W_j and diagonal covariance matrices R_j are also updated based on the new centroids. The quality of a clustering solution is measured by the average quantization error $q(C)$, defined in [8] (also known as distortion and squared reconstruction error). Lower is the value of $q(C)$, better is the quality of clustering. $q(C) = \frac{1}{n}\sum_{i=1}^n d(x_i, C_j)$, where $x_i \in X_j$.

The K-means algorithm stops when centroids change by a marginal fraction in consecutive iterations which is measured by the quantization error.

With decreasing $q(C)$ at each iteration, K-means is theoretically guaranteed to converge, yet a threshold is set on the number of iterations to avoid excessively long runs.

3.4   Computing Gamma Matrix and Machine Learning Models in R

We discuss how Γ is computed exploiting RCpp and how the models are computed in R itself. Depending on the models, we choose between the Non-Diagonal or the Diagonal Gamma matrix to compute at first.

Phase 1: This part is computed exploiting RCpp package. From Sect. 3.1, phase 1 takes are of computing the sum, $\sum_i z_i z_i^T$. The main idea is to evaluate this equation in C++ code instead of R code, following the same UDF idea presented in [10].

First, we take the input data set (X) and split that into chunks of equal size. Chunks are a subset of X so that chunk fits in RAM and it has many points. If there are M chunks, then X is partitioned into $X_1, X_2, .., X_M$ chunks, where each chunk X_I (uppercase i) fits in RAM. Regarding chunks most libraries in R use data frames and therefore it is sort of a table, not a matrix. It seems the conversion from data frame to matrix is done somewhere. We read text files because they are the most common. However, our program would be more efficient with binary files.

If the model to be computed is LR or PCA, we compute the Non-Diagonal Gamma based on the type of data set (whether it is dense or sparse) for each chunk. So, we have a list of Γs. If the model is Naïve Bayes, we compute the Diagonal Gamma, one for each class label for every chunk. If the model K-means, for the first iteration and first chunk, we initialize the k cluster centroids randomly and for successive iterations, we initialize the k cluster centroids with that of the first chunk. Then, we assign a cluster number to each data point and compute the Diagonal Gamma, one for each cluster in every chunk. Hence, we have a list of list of k Γs. Since Γ is additive, we can add all the intermediate Γs to obtain a final Γ. This is straightforward for LR and PCA. But for Naïve Bayes and K-means, since we have list of list of Γs, we need to add the Γs corresponding to a given class/cluster respectively such that we arrive at a final Γ which is a list of matrices representing each class/cluster.

Phase 2: In this part we compute each model (θ). While Phase 1 is basically exploiting RCpp, Phase 2 uses R itself "as is" (we use R existing functions and operators). After obtaining the final Γ, we use Non-Diagonal Gamma to compute LR and PCA and Diagonal Gamma to compute Naïve Bayes and K-means using the mathematical equations discussed previously. Since the models LR, PCA, and NB do not need to converge to a best solution like K-means, that will be the end of Phase 2 for them. On the other hand, K-means is not trivial to compute as it needs to converge to a best solution by the reduction of the quantization error to a minimum value. So, we need to repeat the

Phase 1 and Phase 2 iteratively in order to achieve this. Every time we read the data set, we take the cluster centroids from the previous pass, which improves the accuracy of the model. This process terminates when there is no change in the clusters formed from previous iteration. In summary, for LR, PCA, and Naïve Bayes, we read each and every point in the data set only once but for K-means, we read the data set multiple times until a best solution is achieved. It is beyond the scope of this paper to justify why Γ eliminates the need to read X multiple times in LR and PCA, but not in KM.

Here, the input data set X, intermediate computations and output model, everything is a matrix. In summary, the Γs are computed in Cpp exploiting RCpp package and the models are computed in R itself. To compute LR and PCA, we are forced to call R routines. But for NB and KM, we can compute it ourselves, helped by the fact that diagonal Q simplifies computations in addition to efficiency.

3.5 Time and Space Complexity Analysis

From [10], it is clear that the time complexity for the Phase 1 of the Non-Diagonal Gamma with dense data is $O(d^2n)$ and sparse data is $O(k^2n)$, assuming k entries in x_i are non-zero on an average. In Phase 2, we compute the machine learning models based on the Gamma from Phase 1. So, time for Phase 2 does not depend on n and is $\Omega(d^3)$, which for a dense matrix may approach $O(d^4)$, when the number of iterations in the factorization numerical method is proportional to d. This Non-Diagonal Gamma is used by models like LR and PCA.

A separate Gamma matrix, Diagonal Gamma, is used owing to the fact that a major set of the traditional Non-Diagonal Gamma has little-to-no utility for models like Naïve Bayes and K-means. Time complexity of Diagonal Gamma computation is $O(dn)$ as we compute only L and diagonal of Q of the whole Non-Diagonal matrix. This time complexity applies for all the models utilizing the Diagonal Gamma except K-means. The time complexity of K-means would be $O(kdn)$, where k is the number of clusters.

When we come to the space complexity, space required by Non-Diagonal Gamma matrix in main memory with dense representation is $O(d^2)$. However, it is $O(kd)$ for K-means and $O(d)$ for Naïve Bayes. In short, we can state that Diagonal Γ consumes much less memory than full Γ. However, Diagonal Gamma does not mean faster algorithms since KM requires multiple iterations.

4   Experimental Evaluation

We present an experimental evaluation of our R package and the machine learning models based on the Γ matrix. First, we show the models computed by our R package are accurate, down to almost zero error. Second, we compare the times from our package in R with those times obtained in three alternatives: a columnar DBMS (Vertica [5]), well-known R functions computing each model and the popular Hadoop stack system, Spark.

4.1 Experimental Setup

Hardware and Software: The system and software configuration used for the experiments is a four core 2.83 GHz system with Linux Ubuntu as operating system with 4 GB physical memory and 294 GiB storage space.

Data Sets: The data sets which are used for the experiments are described in Table 1. All the data sets are taken from the UCI Machine Learning repository. We also include the information about the models which utilize these data sets. We replicated each of the data sets in order to get various combinations of n and d without altering statistical properties of the data. The first one was sampled and replicated to get combinations of $d = (9, 91)$ and $n = (0.5M, 1M, 10M)$, second was replicated to get the combinations of $d = 30$ and $n = (0.2M, 1M, 10M, 100M)$ and the third one is replicated to get $d = 4$ and $n = (0.1M, 1M, 10M, 100M)$.

Table 1. Base data sets description

Data set	d	n	Description	Used for model
CreditCard	30	285K	Predict if there is raise in credit line	Naïve Bayes
YearPredictionMSD	90	515K	Predict if there is rain or not	LR and PCA
Iris	3	150	To distinguish the flower species	K-means

4.2 Accuracy Evaluation

Table 2 below shows the results of the experiments that were performed using the two forms of Gamma. We compared the accuracy of model computations of our package with similar packages in R, which is a popular language and environment for statistical computing. We implemented four models in our package, namely, LR, PCA, Naïve Bayes and K-means. For each model, we have a different way of measuring the accuracy with the common underlying metric being Relative Error. From Table 2, we understand that the results from the functions of our package are almost an exact match with the output given by the currently existing best packages in R.

For LR, we get an intercept and a β per attribute as an output for the model computed by Gamma matrix. This is similar to the output given by $lm()$, the preferred default routine in R for LR, for the same input data set. We then compute the absolute differences among all the respective values of intercept and βs, from which we compute the Relative differences. Finally, we report the maximum of the relative differences among the intercept and the βs in Table 2.

For PCA, we get a diagonal matrix, D, of Eigen values and two ortho-normal matrices, S and V, which are Eigen vectors of the given input matrix. Unlike other models, we do not compute PCA completely in Cpp as it gives inaccurate results. Rather we use pure R routines to compute SVD of the correlation matrix generated from the Gamma matrix. The values in D depict the relative importance of each column in S and V matrices. So, we imply on the point that, for the computation of relative error, we take the values from D whose value is greater than 1. We first find the absolute differences among the pairs of corresponding values from the output of the Gamma matrices and that of the default R routines, from which we compute the relative differences. We report the maximum of these relative differences in Table 2.

In Naïve Bayes, we build a model to predict the class labels for the test data set. For that, we compute two separate Naïve Bayes models on the given input training data set using the default R routine and the aforementioned Gamma functions. Consequently, we compute the prediction accuracy by finding the degree to which the predictions made by the functions of our package conforms to that from standard routine in R.

For K-means, we group the input data into k clusters, where k is pre-defined by the user. We compute K-means with both the default R routine and the previously discussed Gamma functions. The output from both the techniques have three vectors, namely, Centers, Radii and Weights. We take the weight vectors, sorted in decreasing order, from both the models and obtain the respective absolute errors. We use this absolute error to compute the relative errors with respect to the weight vector of the model computed from the default R routine. We report the maximum value of relative error in Table 2.

Table 2. Accuracy of models on respective data sets.

Model	Maximum relative error	Data set used
LR	5.89E-10	YearPredictionMSD
PCA	4.75E-13	YearPredictionMSD
Naïve Bayes	0	CreditCard
K-means	4.7E-2	Iris

4.3   Time Performance Evaluation and Benchmarking

We compare the performance of the models in our proposed package with the currently available best packages in R to compute the respective models, a similar implementation done in Vertica, which is a very fast columnar database [5] and also popular for big data analytics nowadays [1] and Spark which is the best representative from the Hadoop world. Since Naïve Bayes and K-means are new models that we explored in our research, there are no prior implementations of these in Vertica. So, we made the comparisons with Vertica for LR and PCA only

Tables 3 and 4 compares the time to compute PCA and LR on YearPrediction data set with Vertuca, R and Spark. We can see that as the as the size of the data set increases, the inbuilt R packages crash. One of the main reasons can be attributed to the fact that it tries to load the whole data set into main memory, eventually resulting in untimely aborts of the program. However, our package overcome this problem by not loading the entire data set into the memory, instead breaking the data set into chunks according to allocated memory. Also, though Vertica and Spark are able to compute the models even for large data sets, they perform slower than our package in R. As n grows, the time complexity of our method for LR and PCA is shown in Fig. 1.

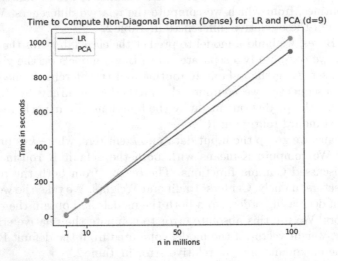

Fig. 1. Time complexity to compute LR and PCA as n grows.

Table 5 compares the time to compute Naïve Bayes model in our package with the one given by R. We see that R crashes for large values of n which is not the case with our package. From Table 6, although the current packages in R scale well for small data sets, they result in untimely aborts for large data sets. As the size of data set increases, the performance of our package improves greatly. Spark, on the other hand, is able to compute the models though it is much slower than our package.

Table 3. Time to compute PCA on YearPrediction data set (Dense) (in secs)

n	d	R+ $\Gamma_{non-diag}$ (dense)	R+ $\Gamma_{non-diag}$ (sparse)	Vertica	R	Spark
0.5M	91	22	33	46	336	67
1M	91	66	80	115	575	130
10M	91	726	800	1290	Crashed	1074
1M	9	9	9	10	21	31
10M	9	91	75	110	205	286
100M	9	1018	1020	1560	Crashed	1780

4.4 Strengths and Weaknesses

Even though this model works efficiently for data sets with rows in the order of millions, it does not work as intended with the billion or higher rowed counterparts. This issue is magnified with the K-means algorithm as it requires multiple reads of the data set before returning the final clusters. Notwithstanding the long execution times, it still gives accurate results in contrast to the existing packages that result in untimely session aborts. As we see in the experimental results of K-means, the existing most efficient package for K-means model in R is aborted for a data set with five million rows or higher. In a similar manner, even for Naïve Bayes, the most efficient package in R is aborted when a data set with ten million rows is given as input while our solution returned accurate results within a reasonable amount of time.

Table 4. Time to compute LR on YearPrediction Data set (Dense) (in secs)

n	d	R+ $\Gamma_{non-diag}$ (dense)	R+ $\Gamma_{non-diag}$ (Sparse)	Vertica	R	Spark
0.5M	91	22	36	46	276	67
1M	91	74	74	115	630	130
10M	91	720	828	1290	Crashed	1074
1M	9	6	6	10	24	31
10M	9	91	69	110	285	286
100M	9	941	928	1560	Crashed	1780

Table 5. Time to compute Naïve Bayes on Credit card data set (Dense) (in secs)

n	d	R+Γ_{diag}	R
0.2M	30	7	51
1M	30	40	158
10M	30	399	Crashed
100M	30	1132	Crashed

Table 6. Time to compute K-means on Iris data set (Dense) (in secs)

n	d	R+ Γ_{diag}	R	Spark
150	4	0	0	3.2
0.1M	4	6	0	7.5
1M	4	65	6	43.3
5M	4	380	Crashed	1370
10M	4	756	Crashed	3012

Our solution adapts to the local machine and customizes the chunk size with respect to the available physical memory. The main drawback is that R cannot be easily parallelized unlike the Hadoop stack or other parallel systems to completely utilize the cores available in a system thus resulting in a decreased performance.

5 Related Work

There are many techniques to improve the performance of the models PCA, Naïve Bayes and K-means few of which are [14], which used decomposition of

Classes via Clustering to improve Naïve Bayes, [3,13], which used the triangle inequality and collaboration of compressed sensing theory and K-SVD approach to accelerate K-means, [8,15], which did Fast PCA computation in a DBMS with Aggregate UDFs and LAPACK and improved performance on MapReduce environment. If we observe carefully, LR, Naïve Bayes or PCA does not require any initialization unlike the K-means model which require the number of clusters and their respective centroids to be initialized. If the initialization is bad, we never converge at a solution.

Summarization of scalable Machine Learning algorithms was done in a parallel manner in [10]. The authors of the [10] exploited HP Vertica's parallelization feature, similar to [11], to perform summarization on multiple systems simultaneously. We adapted the algorithms in [10] and implemented them such that they are serial, scalable and are 99% accurate in R. We made use of the chunking ability in R to read the infinite amount of input data which also makes the process faster. We removed the use of database system completely which is the main component in [10]. In [9], Naïve Bayes is computed inside the database with pure SQL queries. We adapted model computation from [9] and implemented it in R. We compared our work with the most efficient ones in R and have shown that our package is faster and reliable than the former. Alternatively, there is another technology, Microsoft R Open, which is also designed to include an updated R engine (R 3.2.2), new fuzzy matching algorithms, the ability to write to databases via ODBC, and a streamlined install experience. This can also be used to obtain some optimization in building the models. Computers, nowadays have more physical memory, more computing power. So, using a single system, our solution is better for millions of records with all the 4 models. The algorithms programmed in R and C++ are presented in [2].

6 Conclusions

We introduced a powerful summarization matrix to compute fundamental ML models in two phases: Phase 1 to compute one summarization matrix or multiple summarization matrices and Phase 2 to update model parameters based on summarization, where Phase 1 is I/O intensive and Phase 2 is CPU bound. Based on our summarization matrix we developed an R package capable of computing ML models with high accuracy, high speed, and no main memory limitations. Specifically, our R package computes LR, PCA, NB and KM models in one pass over the input data set, except for KM which requires iterative processing. The main memory limitation is solved by reading the data set in small blocks (relative to available RAM) and incrementally updating summarization (with either one summarization matrix or multiple summarization matrices). High speed is achieved by computing the summarization matrix in high-performance C++ code, compiled and linked to run inside the R runtime. We introduced several variants on the Gamma matrix to work with sparse data sets, diagonal and non-diagonal variance matrices, as well as supervised and unsupervised models. That is, we cover a wide spectrum of data sets and ML models, thereby offering wide applicability. We presented interesting experiments to evaluate accuracy and time performance. We show our summarization matrix produces practically the same model, with negligible error, compared to standard R functions. On the other hand, we show our R algorithms are much faster than R built-in functions, removing main memory limitations, but preserving the ease of use. Extensive benchmarks show our package is faster than competing parallel systems: a parallel DBMS and the popular Spark system. In short, our R package opens the possibility of analyzing large data sets on an average personal computer.

Even though our research proves we can get better performance and scalable computing in the R language beyond RAM limits with single-threaded processing, there are many opportunities for future research. We need to explore non-linear ML models, like logistic regression and Support Vector Machines. We need to explore mechanisms to parallelize summarization inside the R runtime, via parallel C or C++ code running on multicore CPUs. Our approach has the promise to be applied in other high-level languages including Python, Matlab, and Javascript, being Python our first target. Given extensive past research work on parallel processing on big data it is worth investigating a data set size threshold to move processing from a single machine to a parallel cluster.

Acknowledgements. The second author would like to thank the guidance of Simon Urbanek, from ATT Labs, to understand the R language runtime source code.

References

1. Al-Amin, S.T., Ordonez, C., Bellatreche, L.: Big data analytics: exploring graphs with optimized SQL queries. In: Elloumi, M., et al. (eds.) DEXA 2018. CCIS, vol. 903, pp. 88–100. Springer, Cham (2018). https://doi.org/10.1007/978-3-319-99133-7_7
2. Chebolu, S.U.S.: A General Summarization Matrix for Scalable Machine Learning Model Computation in the R Language. Master's thesis, University of Houston (2019)
3. Elkan, C.: Using the triangle inequality to accelerate k-means. In: Machine Learning International Conference, vol. 20, p. 147 (2003)
4. Hastie, T., Tibshirani, R., Friedman, J.: The Elements of Statistical Learning, 1st edn. Springer, New York (2001). https://doi.org/10.1007/978-0-387-84858-7
5. Lamb, A., et al.: The vertica analytic database: C-store 7 years later. Proc. VLDB Endow. **5**(12), 1790–1801 (2012)
6. Morandat, F., Hill, B., Osvald, L., Vitek, J.: Evaluating the design of the R language. In: Noble, J. (ed.) ECOOP 2012. LNCS, vol. 7313, pp. 104–131. Springer, Heidelberg (2012). https://doi.org/10.1007/978-3-642-31057-7_6
7. Ordonez, C., Johnson, T., Urbanek, S., Shkapenyuk, V., Srivastava, D.: Integrating the R language runtime system with a data stream warehouse. In: Benslimane, D., Damiani, E., Grosky, W.I., Hameurlain, A., Sheth, A., Wagner, R.R. (eds.) DEXA 2017. LNCS, vol. 10439, pp. 217–231. Springer, Cham (2017). https://doi.org/10.1007/978-3-319-64471-4_18
8. Ordonez, C., Omiecinski, E.: Efficient disk-based K-means clustering for relational databases. IEEE Trans. Knowl. Data Eng. (TKDE) **16**(8), 909–921 (2004)

9. Ordonez, C., Pitchaimalai, S.: Bayesian classifiers programmed in SQL. IEEE Trans. Knowl. Data Eng. (TKDE) **22**(1), 139–144 (2010)
10. Ordonez, C., Zhang, Y., Cabrera, W.: The Gamma matrix to summarize dense and sparse data sets for big data analytics. IEEE Trans. Knowl. Data Eng. (TKDE) **28**(7), 1906–1918 (2016)
11. Raychev, V., Musuvathi, M., Mytkowicz, T.: Parallelizing user-defined aggregations using symbolic execution. In: Proceedings of the 25th Symposium on Operating Systems Principles, pp. 153–167. ACM (2015)
12. Stadler, L., Welc, A., Humer, C., Jordan, M.: Optimizing R language execution via aggressive speculation. In: Proceedings of the 12th Symposium on Dynamic Languages, DLS 2016, pp. 84–95 (2016)
13. Ueda, N., Nakano, R., Ghahramani, Z., Hinton, G.: SMEM algorithm for mixture models. Neural Comput. **12**(9), 2109–2128 (2000)
14. Vilalta, R., Rish, I.: A decomposition of classes via clustering to explain and improve naive bayes. In: Lavrač, N., Gamberger, D., Blockeel, H., Todorovski, L. (eds.) ECML 2003. LNCS (LNAI), vol. 2837, pp. 444–455. Springer, Heidelberg (2003). https://doi.org/10.1007/978-3-540-39857-8_40
15. Zhang, Y., Ordonez, C., Cabrera, W.: Big data analytics integrating a parallel columnar DBMS and the R language. In: Proceedings of IEEE CCGrid Conference (2016)

ML-PipeDebugger: A Debugging Tool for Data Processing Pipelines

Felix Kossak$^{(\boxtimes)}$ and Michael Zwick

Software Competence Center Hagenberg, Hagenberg im Mühlkreis, Austria
{felix.kossak,michael.zwick}@scch.at
https://www.scch.at

Abstract. Data pre-processing for data analysis usually requires a considerable number of interdependent steps, many of which are liable to errors or to introduce unwanted biases. Such errors can lead to cases where predictions for similar data instances differ unexpectedly much. An important question is then to find out where in the data processing pipeline the deviation was caused. We present a tool that can help identify critical data processing steps, allowing to "debug" or improve data pre-processing and model generation. More generally, the tool gives a view of how different data instances behave in relation to each other throughout a pipeline. The task to identify critical steps turns out to be rather complex, mostly because features of different types and ranges have to be compared, because required statistical measures must be obtained from often small samples, and because time series can be involved.

Keywords: Data analysis · Machine learning · Data pre-processing · Data processing pipeline · Debugging

1 Introduction

"Preparing input for a data mining investigation usually consumes the bulk of the effort invested in the entire data mining process" [10, p. 52]. Probably everyone engaged in data analysis will agree to this sentence; still, little research effort is invested in this area. Due to painful experiences in industrial research projects and lack of explicit tools, we have found it necessary to put considerable efforts into a debugging tool for data pre-processing, which we present here.

Data pre-processing for machine learning and other data analysis applications usually requires a considerable number of interdependent steps, many of which are liable to errors or to introduce unwanted biases. The different pre-processing steps form a so-called pipeline, which can be modelled as a directed graph that

---

The research reported in this paper has been supported by the Austrian Ministry for Transport, Innovation and Technology, the Federal Ministry for Digital and Economic Affairs, and the Province of Upper Austria in the frame of the COMET center SCCH.

S. Hartmann et al. (Eds.): DEXA 2019, LNCS 11707, pp. 263–272, 2019.
https://doi.org/10.1007/978-3-030-27618-8_20

represents the dependencies of generated features on raw input features and other, intermediately generated features. Such a graph can be branched. It could even contain cycles, though our work currently focuses on acyclic graphs.

In machine-learning applications, we have time and again encountered cases where apparently similar data instances lead to considerably deviating predictions. The question then arises where in the data processing pipeline that deviation was caused. This, in turn, requires the possibility to trace particular data instances, and pairs of data instances, through the pipeline in a way that allows to judge, for each feature, how similar or dissimilar particular values are.

This can best be seen in Fig. 1, which was generated by the tool ML-PipeDebugger which we present in this paper. Here, two data instances are followed through a pipeline by depicting their differences, according to some difference measures, at each feature. The features are given on the x-axis. The first few features show low differences, while the last three show high differences, with a considerable jump in between. This can serve as a hint that the 6th feature may be to blame for the deviation of the prediction (last feature).

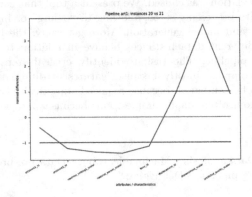

Fig. 1. Example of the development of value differences of two data instances through a pipeline. The first few features show low differences, the last three show high differences.

ML-PipeDebugger (for "Machine-Learning data-processing Pipeline Debugger") is a software prototype whose primary goal is to help identify exactly where in a given pipeline unexpectedly deviating predictions by a machine-learning application originated. More generally, the tool gives an overview of the behaviour of particular (pairs of) data instances throughout data processing.

The task turned out to be surprisingly complex, and this paper is to point out the challenges involved and the solutions we have developed.

The major source of complexity is the need to compare values for features of different types and data distributions and to find comparable difference measures for them. Data types may include scalars as well as time series. How can we say that, say, values (0.21, 0.24) for a scalar feature a are to be considered similar while two time series like those plotted in Fig. 2 are to be judged as considerably different in comparison with the values for a?

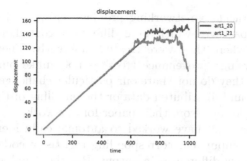

Fig. 2. Two time series to be judged as "considerably deviating".

A related challenge is to obtain significant statistics from often small samples. As we shall see in Sect. 3, we need certain statistics for computing comparable data differences. However, in modern industrial production, batches can be rather small, and often only data from tests are available, with even smaller sample sizes. This situation is even aggravating in the course of tendencies summarised by the slogan "Industry 4.0", with automation enabling arbitrarily small batch sizes (cf. e.g. [7, p. 21]). We thought about how to e.g. combine data from different settings without rendering the resulting statistics irrelevant.

An additional source of complexity is due to time series, which are important in industrial settings: they arise e.g. in cyclic manufacturing processes and can represent such diverse data as pressure curves or trends in quality measures.

In the rest of this paper, we first discuss requirements for the presented tool which we have assembled from concrete needs as well as more general considerations (Sect. 2). Section 3 describes the main challenges for realisation, and how we dealt with them. Section 4 reports on the results we have obtained with the tool so far, and Sect. 5 summarises and discusses ongoing work.

1.1 Related Work

We have failed to find a problem setting or approach sufficiently similar to ours in the literature; we will briefly explain how some similar-looking settings differ from the one tackled by the tool presented here.

A similarity of our work with *delta debugging* is rather superficial. Also delta debugging tries to narrow down the source of a bug from a multitude of possibilities, but it works in a different way and usually also in different settings. To give just one example, Gulzar et al. [6] have proposed a debugging tool for data-intensive scalable computing based on delta debugging; it tries to find *single records* that cause a fault, while we are interested in the relative behaviour of *pairs* of records and in identifying problematic vertices in a pipeline. More generally, typical issues of data provenance are not relevant for our needs. (For an overview of provenance in the context of data processing workflows, see e.g. [4].)

Chen et al. [3] tackle a networking problem which is similar to our goal: Given two similar packages which were differently routed, they want to find the exact location where they were treated differently. They have developed a concept for analysing "provenance trees" – not unlike our data processing pipelines. However, they do not share our particular challenges like comparison of differently typed and distributed data or the handling of time series, and we could not draw any lessons from their paper for our work.

Carbin and Rinard [2] have worked to automatically identify "regions" in complex programme input and corresponding regions of code where small differences can lead to big differences in output. But they analyse input on a byte level, with challenges completely different from those we are concerned with.

Wang et al. [9] have presented "a diagnostic tool for data errors", whose goal is to explain "where and how [...] errors happen in a data generative process". Their approach is based on "finding common properties among erroneous elements" to identify systematic causes of data errors. Their respective extraction of "a hierarchical structure of features" is not relevant for us as we assume a pipeline to be given; however, we may consider their "Bayesian analysis to estimate the *causal likelihood* of a set of features being associated with the causes of the errors" for future work (see Sect. 5).

2 Requirements

We will now list *the most important* requirements for *ML-PipeDebugger*, which were developed in the context of industrial machine-learning projects and supplemented with more general considerations. Some additional requirements will be discussed in Sect. 5. We start with a glossary of important terms.

2.1 Glossary

Data instance. An array of values, for different *data attributes*, pertaining to a single application case.

Data attribute. A variable of a fixed data type which describes one dimension of a *data instance*. We distinguish between *input, intermediate*, and *output attributes*. Note that while we used the more general notion of "feature" in the introduction, we will distinguish between *data attributes* and *data characteristics* (see below) from now on.

Data characteristic. A measure of a *data attribute* which allows for direct comparison between two *data instances*. For each data attribute, multiple characteristics can be defined. An example is the maximum of a time series; the most simple (and common) example is the identity function. The computation of a data characteristic for a particular data attribute may involve values for other data attributes as well – e.g. the value of a time series at a point where another time series attribute is zero.

Data dependency. A data dependency of a *data attribute* a is any other data attribute on whose value the value of a depends.

Pipeline. A data processing pipeline in the form of a directed graph with *data characteristics* as vertices and *data dependencies* of the respective data attributes as edges. Every *data instance* is transformed in every vertex along some path (or multiple paths) until it reaches an output vertex.

Difference measure. A function which takes two values of a *data characteristic* and returns a scalar. For scalars, the canonical difference measure is the absolute value of the arithmetic difference.

Difference statistics. Statistics for *difference measures* over all possible combinations of two *data instances* out of a given set; needed for normalisation.

2.2 Requirements

R1: The Goal of ML-PipeDebugger. ML-PipeDebugger SHALL facilitate the detection of a step in the data processing pipeline of a machine learning application where derived features or predictions for similar input data start diverging considerably. [...]

R8: Visualisation. The relative divergence of values of data characteristics of two given data instances throughout a given pipeline SHALL be visualised in order to facilitate visual detection of steps in the pipeline where derived data for similar instances start diverging.

R25: Time series: Interpolation. It may happen that two time series that are values of the same data attribute [...] have different time points [...]. In such a case, [...] ML-PipeDebugger SHALL use linear interpolation along the time axis to enable comparison of those two time series.

R27: Time series: Comparison. When computing a difference measure for two time series, it SHALL be possible to define a common starting point from which on the time series shall be compared. This starting point may correspond to different time points in each of the time series.

It SHALL be possible to define such a common starting point dynamically (at runtime), based (e.g.) on values of other data attributes.

3 Challenges and Solutions

Identifying a point in a data processing pipeline where data transformation leads to unexpected divergence of instances with similar input values requires to define a notion of similarity for different data types and distributions, but also to make all these different notions of similarity comparable among each other. In this section, we describe the different challenges we have encountered in the context of industrial applications and how we have met them.

3.1 Compare Data Attributes of Different Types and Distributions

In order to be able to compare values from different types or data distributions, the values must be *transformed into the same data type and range*. Ideally, the data of the different data attributes should even have the same distribution. We achieved this with the following basic steps:

- The difference is always a real (float) number (which is natural anyway).
- Differences are normalised in the same way, thus are transformed into similar ranges and distributions.

It may be the case that a data attribute is not relevant for other attributes that depend on it in its raw form but in a transformed form. For instance, a time series may not be relevant in its entirety, but its value at a certain time point may be relevant, or its maximum, etc. However, relevant data transformation modules may be given in black-box or immutable form. Therefore, a user must be able to define "data characteristics" that reconstruct such transformations.

3.2 Difference Measures for Different Data Types

For scalar values, the canonical difference measure is the absolute arithmetical difference. But most of the important data attributes that we have been dealing with are ranging over time series, for which a variety of different similarity measures has been proposed, with research still going on. See [5] for an overview.

Important for the presented tool is that most of the more advanced measures only make sense after transformation of the time series – in particular, compression. Also for compression, various methods have been proposed. In practice, a suitable combination of transformation and difference measure must be found through experimentation. Consequently, users must be able to define difference measures and preceding transformations separately.

3.3 Normalisation

Normalisation of difference measures aims to achieve similar data ranges and distributions for differently typed and ranged data characteristics, under the assumption that the characteristics themselves are following similar distributions.

Normalisation requires statistical measures. Usually we transform the data such that zero represents the arithmetic mean and the standard deviation is 1, but ML-PipeDebugger also provides other standard functions.

Statistics like the mean and the standard deviation must be calculated from a sufficiently large sample of differences. However, in modern industrial production, batch sizes can be fairly small, and we have found that relatively small differences between the batches can already be problematic.

One solution we came up with is to combine data for different but similar batches (or other input parameters) in the following way: Differences are only computed within data for the same batch, but differences from different (yet similar) batches are combined to compute the statistics. This is based on the assumption that while data attribute values for different batches may vary too much, their differences within one batch do not vary significantly from differences within another batch. (This assumption may not always be fulfilled, though.) To this end, ML-PipeDebugger allows users to either provide structured data sets for the computation of difference statistics or to provide criteria for partitioning.

3.4   Additional Pre-processing of Time Series

Certain pre-processing is necessary for time series to make them comparable, and additional, user-defined pre-processing may be desired. The following steps have turned out to be necessary in practice:

- **Interpolation** to align time points;
- **Pruning** (user-defined, possibly data-dependent); and
- Additional pruning of the longer time series to obtain **equal length**.

3.5   Compute and Normalise Differences

For the computation of comparable differences of a particular pair of data instances, we compute the raw difference (including the application of the characteristic function and the transformation of time series) and then apply normalisation. Because the computation of the statistics can be quite time-intensive, this has to be triggered by the user as a separate step. The statistics are then stored on disk and can be used for all normalisations concerning the same pipeline, provided the data fall into the same populations as those used for the statistics.

3.6   Implementation and User Interface

ML-PipeDebugger is implemented in Python 3.6. Scripts are provided for:

- Defining a pipeline (including difference measures, normalisation functions, and pre-processing functions);
- Computing difference statistics;
- Creating difference plots and documents with collections of plots and ancillary information;
- Automatically scanning large data for anomalies (currently in the test phase).

As an output example, Fig. 3 shows part of a document which includes a difference plot (top left), a representation of the pipeline (top right), and comparison plots of all the data characteristics of the pipeline. In the difference plot, the horizontal lines mark the mean difference (in the data used to compute the statistics) and $+/-$ the standard deviation.

4   Practical Results

ML-PipeDebugger has been used with industrial production data which, however, cannot be published. But we have additionally created artificial data, which also allowed us to purposefully construct good test cases. We have strived to render the artificial data close to real data, and results are actually similar. Time series were modelled as splines of different shape types, with separate random jitter. The scenario presented here simulates machine data generated during production and a quality measure for the items produced. Data types include time series and scalars.

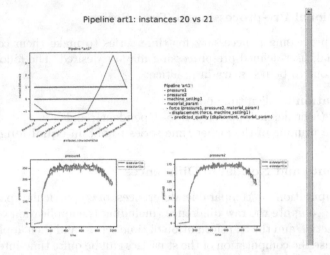

Fig. 3. Example of a difference plot together with a depiction of the pipeline (top right) and comparison plots for all data characteristics (only the first two are shown here).

Apart from a number of similar data instances for the statistics, we also produced, amongst others, a pair of data instances which are intentionally similar with respect to the first couple of data characteristics but dissimilar with respect to the some of the last ones. Figure 4 shows two pairs of time series of these instances. In the first case, the data are to be considered as similar, while in the second case, as considerably diverging.

Figure 1 (see Introduction) shows the respective difference plot. We can see how the relative difference jumps from the 5th data characteristic (more than one standard deviation below the mean) to the 6th (about one standard deviation above the mean) and further to the 7th data characteristic. The differences of the time series were thereby calculated with dynamic time warping (DTW) after simple downsampling. (For a discussion of DTW, see e.g. [1], Sect. 2.6.3.)

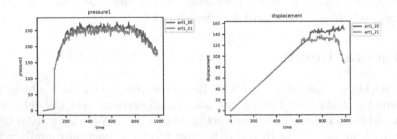

Fig. 4. Example of two similar and two considerably diverging time series.

Results for real industrial data look very similar to these artificially created examples and could actually be used to identify problematic steps in data processing pipelines for real-world machine learning applications.

4.1 Limitations

The most serious problem we have encountered in real-life employment of ML-PipeDebugger is that we often struggle to get enough data for statistics. Only expert knowledge in combination with experimentation can show to what degree data for different batches can be combined for this purpose. Yet we have obtained interesting results, indicating that the algorithms used are not *very* sensitive to the degree of significance of those statistics.

A temporal limitation is given by the need for experimentation and research to find suitable difference measures for particular time series attributes.

Furthermore, the computation of difference statistics can take a long time for real industrial data – in our settings, often more than an hour with a single but relatively powerful workstation. These statistics must be computed separately for every distance measure that is to be tested. The same holds for other custom functions such as "characteristic functions" and pruning functions for time series.

5 Conclusion and Outlook

We have presented ML-PipeDebugger, a software tool for debugging data pre-processing pipelines for data analysis. The importance of such a tool is stressed by the fact that data pre-processing makes up "the bulk of the effort" in data analysis (cf. [10]). We have shown how this tool can visualize the development of relative differences of two data instances as they are processed in a pipeline, which can be used to analyse unexpectedly diverging predictions. The special challenges we have tackled include the comparison of features of different data types and distributions, including time series-type features, and the computation of significant statistics from partially inhomogeneous data.

5.1 Ongoing and Future Work

The ultimate goal of ML-PipeDebugger is to detect suspicious cases like the one shown in Sect. 4 automatically. That is, it should be possible to run a batch process that scans new data instances and flags cases where the pipeline seems to behave suspiciously. In addition to a pipeline and a data source, the user should be able to define thresholds for outlier-detection in terms of a percentage of standard deviation: one below which input attribute values are to be considered as similar, and one above which output attribute values should be considered as considerably diverging. Any pair of data instances for which all input attributes are similar and at least one output attribute is dissimilar in this sense should be flagged by the script, and the "first" intermediate attribute with respect to data dependencies where the same threshold as that for output attributes is exceeded should be highlighted (if it exists). Alternatively, one might give a different threshold for intermediate attributes.

As of the time of submission of this paper, this automatic search is in the test phase. Additionally, a document with difference plot and attribute plots as presented above should be generated automatically for a flagged pair of instances.

Amongst others, such an automated detection of problematic cases could contribute to the challenge of "prioritiz[ing] user attention" to "important" data slices in the course of machine-learning iteration cycles as described in [8, Sect. 3.2.1].

Bayesian analysis similar to that proposed in [9] might be used to refine the flagging of suspicious intermediate attributes. This may be useful if there are several candidates for a suspected cause of divergence.

We are also planning to publish ML-PipeDebugger as a Python package, as well as the source code, after licensing has been accorded with all stakeholders.

References

1. Al-Naymat, G.H.: New methods for mining sequential and time series data. Ph.D. thesis, University of Sydney (2009). https://doi.org/10.1.1.877.2611, http://citeseerx.ist.psu.edu/viewdoc/download?doi=10.1.1.877.2611&rep=rep1&type=pdf
2. Carbin, M., Rinard, M.: Automatically identifying critical input regions and code in applications. In: Proceedings of the 19th International Symposium on Software Testing and Analysis (ISSTA 2010), pp. 37–48 (2010)
3. Chen, A., Wu, Y., Haeberlen, A., Zhou, W., Loo, B.T.: The good, the bad, and the differences: better network diagnostics with differential provenance. In: Proceedings of SIGCOMM 2016, pp. 115–128 (2016). https://doi.org/10.1145/2934872.2934910
4. Fernando, T.: WorkflowDSL: scalable workflow execution with provenance. Master thesis, KTH Royal Institute of Technology, School of Information and Communication Technology, Stockholm, Sweden (2017). http://www.diva-portal.org/smash/get/diva2:1149093/FULLTEXT01.pdf
5. Fu, T.C.: A review on time series data mining. Eng. Appl. Artif. Intell. **24**, 164–181 (2011). https://doi.org/10.1016/j.engappai.2010.09.007
6. Gulzar, M.A., Interlandi, M., Han, X., Li, M., Condie, T., Kim, M.: Automated debugging in data-intensive scalable computing. In: Proceedings of SoCC 2017, pp. 520–534 (2017). https://doi.org/10.1145/3127479.3131624
7. Kagermann, H., Wahlster, W., Helbig, J.: Recommendations for implementing the strategic initiative INDUSTRIE 4.0, April 2013. https://www.din.de/blob/76902/e8cac883f42bf28536e7e8165993f1fd/recommendations-for-implementing-industry-4-0-data.pdf
8. Polyzotis, N., Roy, S., Whang, S.E., Zinkevich, M.: Data lifecycle challenges in production machine learning: a survey. SIGMOD Rec. **47**(2), 17–28 (2018)
9. Wang, X., Dong, X.L., Meliou, A.: Data X-RAy: a diagnostic tool for data errors. In: Proceedings of SIGMOD 2015, pp. 1231–1245 (2015). https://doi.org/10.1145/2723372.2750549
10. Witten, I.H., Frank, E.: Data Mining: Practical Machine Learning Tools and Techniques with Java Implementations, 2nd edn. Morgan Kaufmann, Burlington (2005)

Temporal, Spatial, and High Dimensional Databases

Correlation Set Discovery on Time-Series Data

Daichi Amagata[(✉)] and Takahiro Hara

Osaka University, Osaka, Japan
{amagata.daichi,hara}@ist.osaka-u.ac.jp

Abstract. Time-series data analysis is essential in many modern applications, such as financial markets, sensor networks, and data centers, and correlation discovery is a core technique for the analysis. In this paper, we address a novel problem that computes a k-sized time-series dataset where the minimum Pearson correlation of any two time-series in the set is maximized. This problem discovers a group of time-series, which are highly correlated with each other, from a given time-series dataset without any prior knowledge, thus helps many analytical applications. We show that this problem is NP-hard, and design an approximate heuristic solution that provides a high quality result with fast response time. Extensive experiments on real and synthetic datasets verify the efficiency, effectiveness, and scalability of our solution.

Keywords: Time-series · Correlation set

1 Introduction

In the IoT era, many data can be represented as time-series, i.e., sequences of data points obtained by successive measurements. Time-series analysis is an important task in many applications, such as financial markets [14] and sensor networks [25]. In this paper, we focus on correlation discovery, which is also known to be an important tool for time-series analysis [6,17,20], and address a novel problem of correlation set discovery from a time-series dataset.

For many data mining and discovery tasks, it is interesting to discover an unknown pattern from a given time-series dataset [1,8,11], because such a pattern would be a rule and/or feature of the dataset. In this paper, we consider that a set of time-series, which are highly correlated with each other, indicates a pattern. Because of the large size of the dataset, it is infeasible to obtain the set by visual inspection. Efficient extracting such a set from a given time-series dataset is therefore an interesting problem. Besides, if the obtained set size is still large, it may be hard to analyze, which requires that user can limit the result size. Let $\rho(t, t')$ be the Pearson correlation between two time-series t and t'. Given a result size k, a user-specified threshold θ, and a set of time-series data T, our problem is to compute a set $A \subset T$ such that $|A| = k$, for $\forall t, t' \in A$, $\rho(t, t') \geq \theta$, and the minimum $\rho(t, t')$ is maximized.

© Springer Nature Switzerland AG 2019
S. Hartmann et al. (Eds.): DEXA 2019, LNCS 11707, pp. 275–290, 2019.
https://doi.org/10.1007/978-3-030-27618-8_21

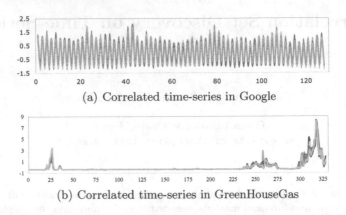

(a) Correlated time-series in Google

(b) Correlated time-series in GreenHouseGas

Fig. 1. Time-series datasets provided by our algorithm where $k = 25$ and $\theta = 0.8$

Our problem can be used in many applications, e.g., pattern (rule) discovery, feature extraction, data exploration, and scientific observation. For example, Fig. 1 illustrates two sets of z-normalized time-series (identified by our algorithm).

Figure 1(a) illustrates a set of 25 time-series in Google dataset (the CPU rate of each machine in Google compute cells) [19]. To achieve high performance computing in data centers, it is important to take into account the correlation of resource utilization (correlated machines should be located in different servers) [9]. By discovering a correlation group (e.g., Fig. 1(a)), administrators can know the machine group that should be divided for performance tuning. This example shows that our problem brings benefits to data center applications.

Environmental analysis is also an application of our problem. It has recently been found that greenhouse gas emissions are spatially correlated (e.g., industrial region) [10]. Investigating how far each emission affects others is also interesting from a scientific viewpoint. Correlation set discovery achieves this by identifying areas where correlated time-series have been observed (e.g., as in Fig. 1(b)). This result is also important for policy makers to establish new environmental protection policies in those areas.

Challenge. In fact, this problem is NP-hard, so exact solutions are impractical, suggesting that approximate heuristic approaches are necessary. To design such a heuristic algorithm, we have to address the following challenges.

(1) *High quality result (effectiveness).* Since the optimal result is not obtained practically, a heuristic algorithm needs to have insights that can be used to discover a data space where time-series in the space are correlated. An intuitive approach is to explore such data spaces in offline pre-processing time. However, thresholds θ are normally different for each user, thus pre-processing for specific thresholds does not make sense.

(2) *Computational efficiency.* In the above applications, users may explore a correlation set with varying k and θ. To enable interactive explorations, an algorithm should provide a high quality result with fast response time.

Contributions. We overcome these non-trivial challenges and propose an efficient greedy algorithm. Our contributions are summarized as follows.

- We address the problem of computing a correlation set on time-series data (Sect. 2). To the best of our knowledge, we are the first to tackle this problem.
- We show that this problem is NP-hard, and propose a heuristic approximate algorithm (Sect. 3). Our greedy algorithm employs locality-sensitive hashing to obtain an approximate result with fast response time. Theoretical analysis shows that the algorithm has linear scalability with respect to $|T|$, l, and k, where T is the set of time-series and l is the time-series length. This result shows a better performance than that of a baseline which employs existing technique and incurs quadratic time w.r.t. $|T|$.
- The results of our experiments using real and synthetic datasets demonstrate the efficiency, effectiveness, and scalability of our solution (Sect. 4).

In addition to the above contents, we discuss related work in Sects. 2 and 5 concludes this paper.

2 Preliminary

2.1 Problem Definition

A time-series t is described as $t = (t[1], t[2], ..., t[l])$, where $t[i]$ is a real value and l is the length of t. We assume that the length of each time-series in a given dataset T is the same [17] and all time-series in T are z-normalized[1] in advance like the real datasets in UCR time-series data archive[2]. (Note that normalizing time-series by z-normalization is currently common assumption to measure time-series similarity and obtain meaningful results [23,24,26].) Let $\|t, t'\|$ be the Euclidean distance between two z-normalized time-series t and t'. The Pearson correlation between t and t', $\rho(t, t')$, is obtained as follows [17].

$$\rho(t, t') = 1 - \frac{\|t, t'\|^2}{2l}$$

We here define correlation set T_θ.

DEFINITION 1 (CORRELATION SET). *Given a threshold θ and a set of time-series data T, a correlation set $T_\theta \subseteq T$ satisfies that $\forall t, t' \in T_\theta$, $\rho(t, t') \geq \theta$.*

---

[1] https://en.wikipedia.org/wiki/Standard_score.
[2] https://www.cs.ucr.edu/~eamonn/time_series_data_2018/.

The idea for selecting the Pearson correlation as similarity measure is twofold. First, its computational cost is linear to l, i.e., $O(l)$. The representative similarity measures for time-series are the Euclidean distance (which corresponds to the Pearson correlation) and dynamic time warping (DTW) [21]. Unfortunately, DTW incurs $O(l^2)$ time to measure the similarity between two time-series. Second, $\rho(t, t') \in [-1, 1]$, thereby specifying θ is not a difficult task (although DTW does not have such a bound).

It is desirable for users to be able to specify a result size k, in order to obtain a reasonable sized correlation set for easy data exploration and pattern discovery. One of the most interesting correlation set T_θ^* is the one of size k that maximizes the minimum Pearson correlation between time-series in the set, which is formally described as:

$$T_\theta^* = \underset{T_\theta \subseteq T, |T_\theta| = k}{\operatorname{argmax}} f(T_\theta) \tag{1}$$

$$f(T_\theta) = \min_{t, t' \in T_\theta} \rho(t, t') \tag{2}$$

where T_θ is a correlation set. If there is no correlation set of size k in a given time-series set T, it is reasonable to provide the correlation set of the largest size, i.e.,

$$T_\theta^* = \underset{T_\theta \subseteq T}{\operatorname{argmax}} |T_\theta|. \tag{3}$$

Ties are broken by selecting the correlation set that maximizes Eq. (2). Now we are ready to define the problem in this paper and its hardness.

DEFINITION 2 (CORRELATION SET DISCOVERY PROBLEM). *Given a set of time-series data T, a result size k, and a threshold θ, this problem is to discover the correlation set A that follows Eq. (1) if there is a correlation set of size k in T. Otherwise, this problem is to discover the correlation set A that follows Eq. (3).*

THEOREM 1 (HARDNESS). *The correlation set discovery problem is NP-hard.*

PROOF. We first assume that there is at least a correlation set of size k in T. We show that our problem corresponds to the k-dispersion problem [18] in this case. The k-dispersion problem is defined as follows: Given a node set $V = \{v_1, v_2, ..., v_{|V|}\}$, this problem is to find a subset V' of V with $|V'| = k$ such that $\min_{v, v' \in V'} dist(v, v')$ is maximized. This problem is shown to be NP-hard. In our problem, each time-series t and the Pearson correlation $\rho(t, t')$ respectively correspond to a node v and $dist(v, v')$. This concludes that computing Eq. (1) is NP-hard. Next, we assume that there is no k-sized correlation set in a given T. In this case, we have to compute Eq. (3). Consider that a time-series t is a node v and if $\rho(t, t') \geq \theta$, there is an edge between v and v'. Now this problem corresponds to finding the maximum clique in a graph, which is also well known to be NP-hard. Theorem 1 therefore holds. □

Due to Theorem 1, it is not feasible to obtain the optimal answer. Hence, we need to design a heuristic algorithm that can efficiently provide an approximate answer set A with high $f(A)$. Note that it is impossible to know in advance whether or not there is a correlation set of size k in a given T. We therefore focus on designing an algorithm that can obtain a result set A *incrementally* to guarantee that A is a correlation set.

2.2 Related Work

The Pearson correlation is a core similarity function, thereby correlation discovery on time-series data has been extensively studied. Literatures [2,6,17,20,27] tackled the problem of discovering (all) correlation pairs. Among them, the most similar to our problem is [17], so we extend the algorithm proposed in [17] for our problem. We compare our algorithm with the extended algorithm, and confirm that computing all correlation pairs does not support efficient correlation set discovery. Our experimental results show that our algorithm significantly outperforms the extended algorithm.

One of other related works is motif discovery. The motif of a given time-series is the most correlated pair of subsequences. Efficient motif discovering algorithms have been proposed for in-memory data [11,16,22] and disk-resident data [15]. Matrix profile project, e.g., [12,17,23], achieves fast motif discovery. However, these works focus only on a single pair, thereby we do not consider their solutions. This discussion also suggests that our problem is different from finding some similar, e.g., kNN, time-series to a given query time-series [7].

3 Proposed Algorithm

This section presents our proposed algorithm Greedy-L. This algorithm employs a novel approach, i.e., greedy heuristic combined with locality sensitive hashing.

3.1 Greedy Heuristic Framework

First, we introduce the framework of the greedy heuristic, and we use the notations in the proof of Theorem 1. Given k and a node set V, this greedy heuristic computes a result set V' as follows.

1. Insert the pair of nodes (v, v') with the maximum distance into V'.
2. Consider an objective function $f(V', v) = \min_{v' \in V'} dist(v, v')$. Insert the node $v \in V \backslash V'$ into V such that v maximizes $f(V', v)$.
3. Iterate the above operation until $|V'|$ becomes k.

This approach can provide an approximate answer in polynomial time, and existing experimental results show that it provides a high quality result in practice [4]. However, straightforward adaptation of this approach to our problem is not efficient. This is because the first operation needs $O(l|T|^2)$ time and each iteration needs $O(kl|T|)$ time, so the straightforward approach incurs $O(l(|T|^2 + k^2|T|))$ time.

3.2   Locality Sensitive Hashing

The above approach incurs quadratic time cost. *We break this quadratic barrier* by optimizing LSH (locality sensitive hashing) usage. We here define LSH.

DEFINITION 3 (LOCALITY-SENSITIVE HASHING). *Given a distance r, an approximate ratio c ($c > 1$), and two probabilities p_1 and p_2 ($p_1 > p_2$), a hash function h is (r, cr, p_1, p_2)-sensitive, if it satisfies the following both conditions:*

- *If $\|t, t'\| \leq r$, then $\Pr[h(t) = h(t')] \geq p_1$;*
- *If $\|t, t'\| \geq cr$, then $\Pr[h(t) = h(t')] \leq p_2$.*

The LSH function commonly used in the Euclidean space is shown below [3].

$$h(t) = \lfloor \frac{a \cdot t + bw}{w} \rfloor \qquad (4)$$

Note that a is a random vector with each dimension independently chosen from the standard normal distribution $\mathcal{N}(0, 1)$, and its length is l. b is a real number randomly chosen from $[0, w)$, and w is a real number that represents the width of h. Recall that we are interested in time-series t and t' satisfying $\rho(t, t') \geq \theta$, thus their hash values should be the same (or very close). To this end, we set

$$w = \sqrt{2l(1 - \theta)}. \qquad (5)$$

Let $\theta_E = \sqrt{2l(1 - \theta)}$, and let $d = \|t, t'\|$. [3] shows that $\Pr[h(t) = h(t')]$ can be obtained as follows:

$$\begin{aligned} p(d) &= \Pr[h(t) = h(t')] \\ &= \int_0^{\theta_E} \frac{1}{d} f_2(\frac{x}{d})(1 - \frac{x}{\theta_E}) dx \\ &= 2norm(\frac{\theta_E}{d}) - 1 - \frac{2}{\sqrt{2\pi}} \frac{d}{\theta_E}(1 - e^{-\frac{\theta_E^2}{2d^2}}) \end{aligned} \qquad (6)$$

where $f_2(z) = \frac{2}{\sqrt{2\pi}} e^{-\frac{z^2}{2}}$ and $norm(\cdot)$ is the cumulative distribution function of a random variable following $\mathcal{N}(0, 1)$. Note that $h(t)$ has the following lemma [5].

LEMMA 1. *The LSH obtained from Eq. (4) is $(\theta_E, c\theta_E, p(\theta_E), p(c\theta_E))$-sensitive.*

Because $h(\cdot)$ provides the same (or similar) hash values if two time-series are very similar, it is intuitive that we do not need to compare two time-series with totally different hash values. However, it is important to note that using a single $h(\cdot)$ cannot avoid unnecessary computation well, because many time-series with far distance (i.e., low Pearson correlation) may have the same hash values. To avoid this, a compound LSH function $G(t) = (h_1(t), h_2(t), ..., h_m(t))$ is employed, where each component of $G(t)$ is $h(t)$ and independently generated [3]. We consider that $G(t)$ is a key of t, and two time-series with high Pearson correlation would have the same or similar keys.

It is important to note that existing studies utilize LSH as indices, i.e., offline processing, but we utilize LSH for online processing to deal with arbitrary θ, see Eq. (5).

3.3 Main Techniques

Assume that each time-series $t \in T$ is assigned its key K and is inserted into the bucket with key K, B_K. One may consider the following simple combination of the greedy heuristic and LSH. We compute a pair of two time-series $\langle t_i, t_j \rangle$, which is firstly added to A, by using LSH. In other words, if we compute the pair with the highest Pearson correlation for $\forall B_K \in B$, where B is the set of buckets, we can obtain $\langle t_i, t_j \rangle$. Then we compute $t^* = \mathrm{argmax}_{t \in T \setminus A} f(A, t)$, where

$$f(A, t) = \min_{t' \in A} \rho(t, t'),$$

by scanning T, and t^* is inserted into A. This operation is iterated until $|A|$ becomes k.

Although this seems to reduce computational cost, it is not sufficient. In each iteration, we compute t^* *based on the intermediate A*, so the pair $\langle t_i, t_j \rangle$, which is firstly added to A, has a large influence on the final quality and size of A. Due to this property, $\langle t_i, t_j \rangle$ has to satisfy the following requirements.

- $\rho(t_i, t_j)$ is high as much as possible: Because $f(A)$, which is described in Eq. (2), has submodularity, i.e., $f(A) \geq f(A \cup \{t\})$, the first pair should have high Pearson correlation. Otherwise, the quality of the final result becomes low.
- $\langle t_i, t_j \rangle$ exists in a large group of time-series which are correlated with each other: This requirement is necessary to provide A such that $|A| = k$.

We below elaborate how to discover such a pair. Assume that each time-series t in T is assigned a key K by $G(t)$. For each bucket $B_K \in B$, we compute the highest Pearson correlation in B_K denoted by ρ_K. Recall that higher ρ_K is better due to the submodularity of $f(A)$. We next consider the size of the *adjacent buckets* which are defined below.

DEFINITION 4 (ADJACENT BUCKET). *Given a set of buckets B and a bucket $B_K \in B$, each bucket $B_{K'}$ which is an adjacent bucket of B_K, satisfies that $|\{i \mid h_i^K = h_i^{K'}\}| = m - 1$ where h_i^K ($h_i^{K'}$) is the i-th hash value of K (K').*

Let B_θ be the set of buckets B_K such that $\rho_K \geq \theta$. We retrieve the adjacent buckets of B_K in B_θ and compute s_K which is the summation of their sizes (the number of time-series in the buckets) and $|B_K|$. More formally,

$$s_K = |B_K| + \sum |B_{K'}|,$$

where $B_{K'}$ is an adjacent bucket of B_K. Recall that time-series in the same bucket or buckets with similar keys tend to be correlated. Therefore, if s_K is large, time-series in B_K would exist in a large group of time-series which are correlated with each other. Based on the above idea, we select the pair of two time-series with the highest Pearson correlation in B_K where $\rho_K \cdot \frac{s_K}{|T|}$ is the maximum among B_θ. (Because $\rho_K \in [\theta, 1]$, s_K has to be normalized and $|T|$ is used to achieve this.) The complexity of this operation is as follows.

LEMMA 2. *We can select the first two time-series with $O(ml|T|)$ time.*

PROOF. Computing $G(\cdot)$ for each time-series needs $O(ml)$ time, thus the hashing incurs $O(ml|T|)$ time. Let β be the number of buckets $\in B$ where $|B_K| \geq 2$, and let n be the average number of time-series in B_K. To obtain $\langle t, t' \rangle$, we need $O(\beta n^2)$. However, by setting a sufficiently large constant as m, n can be very small, so we have $O(\beta n^2) \ll O(ml|T|)$. We can compute the first two time-series by scanning B_θ, and $|B_\theta| \leq |T|$. Then, we can conclude that the time complexity is $O(ml|T|)$. $\qquad\square$

Next, we consider how to efficiently find a time-series which has high Pearson correlation with each time-series in an intermediate result A. Our idea is simple yet effective. Because two time-series with high Pearson correlation share the same or similar key, promising time-series, which can be the next result t^*, exist in the adjacent buckets of the buckets in which the time-series $\in A$ exist. We compute t^* from the set of the buckets denoted by S, and its time complexity is $O(l|S|)$.

LEMMA 3. *We can obtain $t^* = \text{argmax}_{t \in S \setminus A} f(A, t)$ with $O(l|S|)$ time.*

PROOF. Assume that a time-series t is in A and $A = \{t\}$, and for $\forall t_i \in S \setminus \{t\}$, we compute $\rho(t, t_i)$. Assume further that $t^* = t'$ and each time-series t_i caches $f(A, t)$. When we find the next t^*, we can obtain the exact $f(A, t_i)$ of a given $t_i \in S \setminus A$ by comparing $\rho(t', t_i)$ with the cached value, which needs only $O(l)$ time. Thus we can obtain t^* with $O(l|S|)$ time. $\qquad\square$

Besides, a lower-bound of the existing probability of a time-series t, which satisfies that $f(A, t) \geq \theta$, in S is obtained as follows.

LEMMA 4. *We have*

$$\Pr[t \in S, f(A, t) \geq \theta] \geq p(\theta_E)^m + p(\theta_E)^{m-1}(1 - p(\theta_E))m.$$

PROOF. From Eq. (6) and Definition 4. $\qquad\square$

3.4   Algorithm Description

Algorithm 1 details Greedy-L. Greedy-L first obtains the key of each time-series (lines 1–2). Then Greedy-L computes $\langle t, t' \rangle$, where $t, t' \in B_K$ and $\rho(t, t') \cdot \frac{s_K}{|T|}$ is the maximum in B_θ (lines 3–17). The pair $\langle t, t' \rangle$ is inserted into A. Greedy-L retrieves the next result t^* from S which is the union of B_K such that $t, t' \in B_K$ and the adjacent buckets of B_K. Also, for each iteration (lines 20–26), after Greedy-L inserts $t^* = \text{argmax}_{t \in S \setminus A} f(A, t)$ into A, B_K, where $t^* \in B_K$, and its adjacent buckets are inserted into S (line 24). This is repeated until $|A|$ becomes k or Greedy-L identifies that $\nexists t \in S$ such that $f(A, t) \geq \theta$.

Now we show our main result: the time complexity of Greedy-L is linear to each parameter and breaks the quadratic barrier.

Algorithm 1: Greedy-L

---

1 **for** $\forall t \in T$ **do**
2 $\quad\lfloor\ B_K \leftarrow B_K \cup \{t\}$ where $K = (h_1(t), h_2(t), ..., h_m(t))$
3 $B_\theta \leftarrow \varnothing,\ P \leftarrow \varnothing$
4 **for** $\forall B_K \in B$ where $|B_K| \geq 2$ **do**
5 $\quad t_K, t'_K \leftarrow \varnothing,\ \rho_K \leftarrow -1$
6 $\quad$ **for** $\forall t_i \in B_K$ **do**
7 $\quad\quad$ **for** $\forall t_j \in B_K$ **do**
8 $\quad\quad\quad$ **if** $\rho_K < \rho(t_i, t_j)$ **then**
9 $\quad\quad\quad\quad\lfloor\ \rho_K \leftarrow \rho(t_i, t_j),\ \langle t_K, t'_K \rangle \leftarrow \langle t_i, t_j \rangle$

10 $\quad$ **if** $\rho_K \geq \theta$ **then**
11 $\quad\quad$ $B_\theta \leftarrow B_\theta \cup B_K$
12 $\quad\quad$ $P \leftarrow P \cup \langle \rho_K, t_K, t'_K \rangle$

13 $t, t' \leftarrow \varnothing,\ \mu = 0$
14 **for** $\forall B_K \in B_\theta$ **do**
15 $\quad s_K \leftarrow |B_K| + \sum |B_{K'}|$ where $B_{K'} \in B_\theta$ is the nearest bucket of B_K
16 $\quad$ **if** $\rho_K \cdot \frac{s_K}{|T|} > \mu$ **then**
17 $\quad\quad\lfloor\ \mu \leftarrow \rho_K \cdot \frac{s_K}{|T|},\ \langle t, t' \rangle \leftarrow \langle t_K, t'_K \rangle$

18 $A \leftarrow \langle t, t' \rangle$
19 $S \leftarrow B_K \cup B_{K'}^N$ where $t, t' \in B_K$ and $B_{K'}^N$ is the set of the nearest bucket of B_K in B
20 **while** $|A| < k$ **do**
21 $\quad t^* \leftarrow \underset{t \in S \setminus A}{\operatorname{argmax}}\ f(A, t)$
22 $\quad$ **if** $f(A, t^*) \geq \theta$ **then**
23 $\quad\quad$ $A \leftarrow A \cup \{t^*\}$
24 $\quad\quad$ $S \leftarrow S \cup B_K \cup B_{K'}^N$ where $t^* \in B_K$ and $B_{K'}^N$ follows line 19
25 $\quad$ **else**
26 $\quad\quad\lfloor$ **break**

THEOREM 2. *Greedy-L needs $O(ml|T| + kl|S|)$ time to provide A, where $S \subseteq T$.*

PROOF. From Lemma 2, lines 1–17 need $O(ml|T|)$ time. To obtain s_K, we need to find the adjacent buckets of B_K. We cache the value range of each LSH h_i, and z-normalization provides the fact that the range is very small as shown in Fig. 1. Thus s_K is obtained by $O(m)$ time, i.e., lines 14–17 incurs $O(m|B_\theta|)$ time, and $O(m|B_\theta|) \ll O(ml|T|)$. Lines 19, 21, and 24 respectively need $O(l|S|)$ time. As a result, the time complexity of Greedy-L is $O(ml|T| + kl|S|)$. □

Discussion. We exploit the adjacent buckets to effectively select buckets for the candidates of the result. One may consider about employing *near* buckets that share $(m - m')$ LSHs with a given bucket. If we employ this, Greedy-L loses its efficiency significantly due to large increase of S (the number of near buckets of a given bucket is $\binom{m}{m'}$). Besides, specifying an appropriate m' is not trivial. Greedy-L therefore employs the adjacent buckets.

We next show that Greedy-L is a parallel-friendly framework. Recall that each LSH in $G(\cdot)$ is independently generated. This computation can be parallelized. Also, it can be seen that computing ρ_K, s_K, and t^* is parallelized by dividing B, B_θ, and $|S|$ into some pieces.

4   Experiments

We present our empirical study that evaluates the performance of Greedy-L.

4.1   Setting

Datasets. In our experiments, we used two real datasets and a synthetic dataset introduced below.

- GreenHouseGas [13]: This dataset has 46,736 time-series, and each time-series consists of 327 green house gas concentrations.
- Google: This dataset consists of 10,380 time-series (CPU rates of machines in Google compute cells) with length 128.
- Rand: This dataset is generated by a random walk technique. When generating a time-series t, we randomly choose the first value ($t[1]$) in $\{-1, 1\}$. The subsequent value is generated by $t[i+1] = t[i] + \mathcal{N}(0, 1)$ [16]. We set $|T| = 100,000$ and $l = 1,000$ by default.

(We conducted experiments on other datasets but omit their results because they are consistent.) When we use a dataset, all time-series in the dataset are memory-resident.

Algorithms. We evaluated the following algorithms.

- Greedy-M: this is an extended version of [17], which is a state-of-the-art online algorithm to compute all time-series pairs whose Pearson correlation satisfies θ. Greedy-M employs this technique and the greedy heuristic introduced in the beginning of the proposed algorithm section, to compute A.
- Greedy-L: the proposed algorithm in this paper.
- Greedy-L$^-$: this algorithm utilizes LSH only to obtain the first two time-series. The greedy heuristic is also employed in each iteration to update the result set.
- Greedy-L (wobs): this algorithm normally executes the same operations as those in Greedy-L, but selects the first bucket B_K such that ρ_K is the highest among B.

All algorithms were implemented in C++, and all experiments were conducted on a PC with Intel Xeon E5-2687W v4 processors (3.0 GHz) and 512 GB RAM.

Criteria. We measured the average of each metric introduced below. We run the algorithms 50 times for each experiment.

- Running time (efficiency). This metric is defined as the time to provide a correlation set A.
- $F(A) = f(A) \cdot \frac{|A|}{k}$ (effectiveness). Although the above algorithms guarantee that A is a correlation set, they do not guarantee that $|A| = k$. It is unfair to compare algorithms based on $f(A)$, since the algorithms may provide different result size. We therefore normalize $f(A)$ by $\frac{|A|}{k}$.

Recall that, as shown in Theorem 1, the exact answer A^* is not obtained practically, so comparing $F(A)$, where A is provided by our solution, with $F(A^*)$ is impossible.

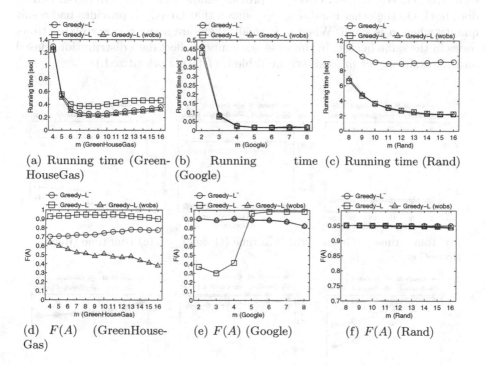

(a) Running time (Green- (b) Running time (c) Running time (Rand)
HouseGas) (Google)

(d) $F(A)$ (GreenHouse- (e) $F(A)$ (Google) (f) $F(A)$ (Rand)
Gas)

Fig. 2. Impact of m

Table 1. Tuning m for each algorithm

Algorithm	GreenHouseGas	Google	Rand
Greedy-L	8	6	13
Greedy-L⁻	13	5	11
Greedy-L (wobs)	6	4	13

4.2   Result

By default, $\theta = 0.8$, $k = 20$ in the cases of GreenHouseGas and Google, and $k = 100$ in the case of Rand.

Varying m. We first tune m (the number of $h(\cdot)$ in the compound LSH function) of Greedy-L$^-$, Greedy-L, and Greedy-L (wobs) for each dataset by using the default parameter setting. Figure 2 illustrates the impact of m. We can see that m affects the performances of the three algorithms. For example, when m is small, there is a large number of time-series in the same bucket, so computing the first two time-series which will be in A needs long time. Figures 2(d) and (f) show that Greedy-L$^-$ and Greedy-L provide stable $F(A)$ (but Greedy-L (wobs) does not). On the other hand, Fig. 2(e) shows that Greedy-L provides bad result quality when m is small. When m is small, there are many non-correlated time-series in the same bucket. In this case, s_K cannot reflect data distribution. Based on the result, we set m as shown in Table 1. ($F(A)$ is prioritized.)

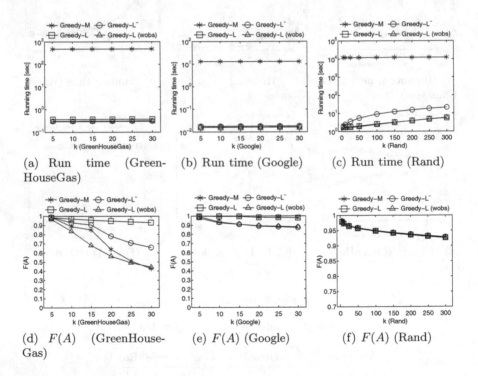

(a) Run time (Green-HouseGas)

(b) Run time (Google)

(c) Run time (Rand)

(d) $F(A)$ (GreenHouse-Gas)

(e) $F(A)$ (Google)

(f) $F(A)$ (Rand)

Fig. 3. Impact of k

Varying k. Figures 3(a) and (b) show that all the algorithms are not affected by k. Since Greedy-M incurs the overhead from computing all pairs Pearson correlation, i.e., $O(l|T|^2)$, it is reasonable. The other algorithms have two main computational overheads: hashing and iteration. When k is small, hashing becomes a dominant factor, thereby the result is obtained. When k is large, on the other hand, the running time of the algorithms except Greedy-M increases as shown in Fig. 3(c). We see that Greedy-L scales better than Greedy-L$^-$, because Greedy-L$^-$ scans the whole dataset in each iteration. Note that Greedy-L runs up to 1,500 times faster than Greedy-M.

Let us focus on result quality, and Figs. 3(d) and (e) show that Greedy-L provides the best result among the four algorithms. (Because Rand has many correlated time-series, the four algorithms provide almost the same result, as shown in Fig. 3(f).) In particular, Greedy-M, Greedy-L$^-$, and Greedy-L (wobs) fail to return a good result in the case of GreenHouseGas. In this dataset, the pair of two time-series with the highest Pearson correlation exists in a very small group. The three algorithm (often) return this set, but Greedy-L can avoid this situation and provides a larger group by exploiting LSH, which verifies the effectiveness of our approach. (Recall that the result obtained by Greedy-L is illustrated in Fig. 1.)

Varying θ. Figure 4 shows the impact of threshold. As shown in Figs. 4(a), (b) and (c), as θ increases, running time of each algorithm decreases. Even in this case, Greedy-M is very slow and the other algorithms keep outperforming Greedy-M significantly.

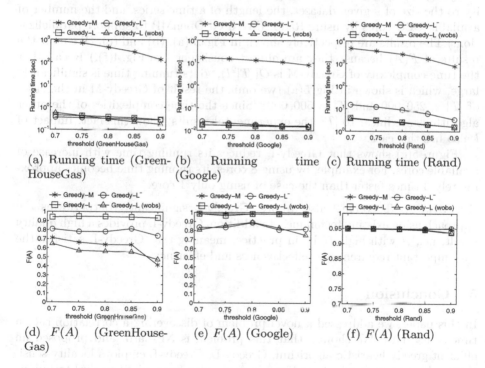

(a) Running time (Green-HouseGas)   (b) Running time (Google)   (c) Running time (Rand)

(d) $F(A)$ (GreenHouse-Gas)   (e) $F(A)$ (Google)   (f) $F(A)$ (Rand)

Fig. 4. Impact of θ

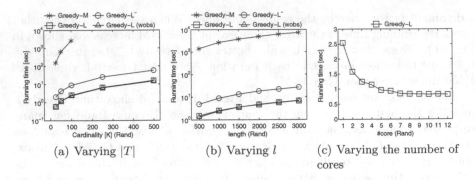

(a) Varying $|T|$ (b) Varying l (c) Varying the number of cores

Fig. 5. Scalability test

Figures 4(d) and (e) show that the Greedy-L (wobs) often returns a worse result than the other algorithms. As well as Greedy-L, Greedy-L (wobs) finds the next result (i.e., t^*) only from a subset of T, and the subset is also dependent on the first two time-series of A. This result implies that ignoring s_K misses identifying a group of time-series, and Greedy-L (wobs) cannot be robust.

Varying $|T|$, l, and the Number of Cores. We also investigate the scalability to the size of a given dataset, the length of a time-series, and the number of available CPU cores by using Rand. (We used OpenMP to support parallelization.) The results are respectively shown in Figs. 5(a), (b) and (c). (We omit the results of $F(A)$ because they are almost consistent like Fig. 4(f).) Recall that the time complexity of Greedy-M is $O(|T|^2l)$, so its running time is significantly large, which is shown in Fig. 5(a) (we omit the result of Greedy-M in the cases of $|T| = 250,000$ and $|T| = 500,000$). Since the time complexities of the other algorithms are linear to $|T|$, the experimental results follow this fact. Impact of l also has this case.

Figure 5(c) shows that Greedy-L reduces its running time with increase of available cores. For example, by using 8 cores, its running time becomes approximately 3 times faster than the case of using only 1 core.

Remark. As Theorem 2 also argues, Greedy-L significantly outperforms the approach using existing techniques. In addition, Greedy-L provides a high quality result, i.e., A with high $f(A)$, in practice, meaning that Greedy-L satisfies the two important requirements, effectiveness and efficiency.

5   Conclusion

In this paper, we addressed a novel problem of discovering a correlation set on time-series data. We showed that this problem is NP-hard, and proposed an efficient greedy heuristic algorithm, Greedy-L. Greedy-L employs locality sensitive hashing to reduce running time. In particular, we devised a novel technique that exploits locality-sensitive hashing to discover a large group of time-series which are correlated with each other. The experimental results demonstrate the efficiency, effectiveness, and scalability.

Acknowledgment. This research is partially supported by JSPS Grant-in-Aid for Scientific Research (A) Grant Number 18H04095, JSPS Grant-in-Aid for Young Scientists (B) Grant Number JP16K16056, and JST CREST Grant Number J181401085.

References

1. Amagata, D., Hara, T.: Mining top-k co-occurrence patterns across multiple streams. TKDE **29**(10), 2249–2262 (2017)
2. Cole, R., Shasha, D., Zhao, X.: Fast window correlations over uncooperative time series. In: KDD, pp. 743–749 (2005)
3. Datar, M., Immorlica, N., Indyk, P., Mirrokni, V.S.: Locality-sensitive hashing scheme based on p-stable distributions. In: SoCG, pp. 253–262 (2004)
4. Drosou, M., Pitoura, E.: Diversity over continuous data. IEEE Data Eng. Bull. **32**(4), 49–56 (2009)
5. Gan, J., Feng, J., Fang, Q., Ng, W.: Locality-sensitive hashing scheme based on dynamic collision counting. In: SIGMOD, pp. 541–552 (2012)
6. Guo, T., Sathe, S., Aberer, K.: Fast distributed correlation discovery over streaming time-series data. In: CIKM, pp. 1161–1170 (2015)
7. Huang, Q., Feng, J., Zhang, Y., Fang, Q., Ng, W.: Query-aware locality-sensitive hashing for approximate nearest neighbor search. PVLDB **9**(1), 1–12 (2015)
8. Kato, S., Amagata, D., Nishio, S., Hara, T.: Monitoring range motif on streaming time-series. In: DEXA, pp. 251–266 (2018)
9. Kim, J., Ruggiero, M., Atienza, D., Lederberger, M.: Correlation-aware virtual machine allocation for energy-efficient datacenters. In: DATE, pp. 1345–1350 (2013)
10. Li, L., Hong, X., Tang, D., Na, M.: GHG emissions, economic growth and urbanization: a spatial approach. Sustainability **8**(5), 462 (2016)
11. Li, Y., Yiu, M.L., Gong, Z., et al.: Quick-motif: an efficient and scalable framework for exact motif discovery. In: ICDE, pp. 579–590 (2015)
12. Linardi, M., Zhu, Y., Palpanas, T., Keogh, E.: Matrix profile x: VALMOD-scalable discovery of variable-length motifs in data series. In: SIGMOD, pp. 1053–1066 (2018)
13. Lucas, D., et al.: Designing optimal greenhouse gas observing networks that consider performance and cost. Geosci. Instrum. Methods Data Syst. **4**(1), 121 (2015)
14. Marti, G., Andler, S., Nielsen, F., Donnat, P.: Clustering financial time series: how long is enough?. In: IJCAI, pp. 2583–2589 (2016)
15. Mueen, A., Keogh, E., Bigdely-Shamlo, N.: Finding time series motifs in disk-resident data. In: ICDM, pp. 367–376 (2009)
16. Mueen, A., Keogh, E., Zhu, Q., Cash, S., Westover, B.: Exact discovery of time series motifs. In: SDM, pp. 473–484 (2009)
17. Mueen, A., Nath, S., Liu, J.: Fast approximate correlation for massive time-series data. In: SIGMOD, pp. 171–182 (2010)
18. Ravi, S.S., Rosenkrantz, D.J, Tayi, G.K.: Facility dispersion problems: heuristics and special cases. In: Dehne, F., Sack, J.-R., Santoro, N. (eds.) WADS 1991. LNCS, vol. 519, pp. 355–366. Springer, Heidelberg (1991). https://doi.org/10.1007/BFb0028275
19. Reiss, C., Wilkes, J., Hellerstein, J.L.: Google cluster-usage traces: format+schema, pp. 1–14. Google Inc., White Paper (2011)
20. Tsytsarau, M., Amer-Yahia, S., Palpanas, T.: Efficient sentiment correlation for large-scale demographics. In: SIGMOD, pp. 253–264 (2013)

21. Vlachos, M., Hadjieleftheriou, M., Gunopulos, D., Keogh, E.: Indexing multidimensional time-series. VLDB J. **15**(1), 1–20 (2006)
22. Yankov, D., Keogh, E., Medina, J., Chiu, B., Zordan, V.: Detecting time series motifs under uniform scaling. In: KDD, pp. 844–853 (2007)
23. Yeh, C.C.M., Kavantzas, N., Keogh, E.: Matrix profile vi: meaningful multidimensional motif discovery. In: ICDM, pp. 565–574 (2017)
24. Yeh, C.C.M., et al.: Matrix profile i: all pairs similarity joins for time series: a unifying view that includes motifs, discords and shapelets. In: ICDM, pp. 1317–1322 (2016)
25. Yi, X., Zheng, Y., Zhang, J., Li, T.: ST-MVL: filling missing values in geo-sensory time series data. In: IJCAI, pp. 2704–2710 (2016)
26. Zhu, Y., et al.: Matrix profile ii: exploiting a novel algorithm and GPUs to break the one hundred million barrier for time series motifs and joins. In: ICDM, pp. 739–748 (2016)
27. Zhu, Y., Shasha, D.: Statstream: statistical monitoring of thousands of data streams in real time. In: VLDB, pp. 358–369 (2002)

Anomaly Subsequence Detection
with Dynamic Local Density
for Time Series

Chunkai Zhang$^{(\boxtimes)}$, Yingyang Chen, and Ao Yin

Department of Computer Science and Technology,
Harbin Institute of Technology, Shenzhen, China
ckzhang812@gmail.com, yingyang_chen@163.com, yinaoyn@126.com

Abstract. Anomaly subsequence detection is to detect inconsistent data, which always contains important information, among time series. Due to the high dimensionality of the time series, traditional anomaly detection often requires a large time overhead; furthermore, even if the dimensionality reduction techniques can improve the efficiency, they will lose some information and suffer from time drift and parameter tuning. In this paper, we propose a new anomaly subsequence detection with Dynamic Local Density Estimation (DLDE) to improve the detection effect without losing the trend information by dynamically dividing the time series using Time Split Tree. In order to avoid the impact of the hash function and the randomness of dynamic time segments, ensemble learning is used. Experimental results on different types of data sets verify that the proposed model outperforms the state-of-art methods, and the accuracy has big improvement.

Keywords: Time series · Anomaly detection · Local Density

1 Introduction

The time series data is stored in the order of the data generation time, and is dynamic and massive. We are interested in finding the abnormal subsequence in complete time series, in other words, anomaly subsequences are inconsistent with the shape of most other subsequences. Anomaly detection for time series is an analysis of inconsistent data with normal data, which always represents an emergency or fault. Itc is applied in many application domains, ranging from financial data [15,19], Electrocardiogram (ECG) data [1,22] to sensor data [8]. For cxample, aualysls of ECG data can timely monitor patients' heart health such as arrhythmia, ventricular atrial hypertrophy, myocardial infarction [13] before diagnosis process. Therefore, timely detection of abnormal data contained in the data is of great significance.

A rich body of literature exist on detecting time series anomalies, however, existing anomaly detection methods [11,17,18,23] still suffer from a lot of problems. Time series is often high-dimensional data, therefore the calculations in

© Springer Nature Switzerland AG 2019
S. Hartmann et al. (Eds.): DEXA 2019, LNCS 11707, pp. 291–305, 2019.
https://doi.org/10.1007/978-3-030-27618-8_22

the original data storage format often require large storage and computational overhead. In recent years, the different time series data representation methods were proposed to achieve the purpose of dimensionality reduction. Discrete Fourier Transformation (DFT) [5] can convert time series of length n into m coefficients by discrete Fourier transform method; Discrete Wavelets Transformation (DWT) [3] is a multi-resolution representation of the data signal but can only be used in time series of integer powers of length 2; and Piecewise Aggregate Approximation (PAA) [6] divides the time series into equal length segments, then takes the average for each segment. As for Symbolic Aggregate Approximation (SAX) [10], it maps the mean of the segments to a symbolic representation based on PAA as other variants, ESAX [11] and SAX-TD [23]. All these methods can reduce the dimensionality but losing information on local time segment. However, there are some problem that the size of the window needs to be set manually, which requires the relevant expert knowledge [22]. And the average in the sliding window will lose some important information. In addition, these methods have not pay much attention to time drift problem, which will get wrong anomaly subsequence if using Euclidean distance, and the details will be discuss in Sect. 2.

We also need to perform anomaly calculations on the representation of time series. The simplest and straightforward method of anomaly subsequence detection is to calculate the similarity between each pair of subsequences by double-loop violence, and treat the most dissimilar subsequences with most other subsequences as abnormal subsequences [7]. In order to improve the efficiency of the brute force algorithm, Keogh et al. proposed HOT SAX [7] to construct an index tree using SAX symbol sequences to optimize the search order of candidate. Li et al. [9] proposed BitClusterDiscord, who used binary representation to approximate the trend information then use K-media clustering and two pruning strategies to reduce the number of similarity calculations. Senin et al. [21] proposed Rare Rule Anomaly to discrete the time series into symbol and derive context-free grammar to discover algorithmic irregularities associated with exceptions. Ren et al. proposed PAPR-RW [17] based on PAPR representation and random walk model [12] to convert time series into similar matrices. All these method use sliding window to split time series into subsequence while set the size of window manually. Once the window setting is not good enough to different kind of data sets, it is easy to detect wrong anomaly subsequence.

In this paper, we propose a novel anomaly subsequence detection of Dynamic Local Density Estimation (DLDE) where TSTree is used to dynamically divide the time series, and hash function to improve the efficiency. In order to avoid the influence of the hash function and the randomness of dynamic time segments, ensemble learning is used in our method. And this algorithm can improve the effect of detection without losing the time series trend information by dynamic segment and has less parameters.

The contribution of this paper can be summarized as follows.

(1) An anomaly detection algorithm is proposed to solve the time drift problem inspired by the idea of DTW. And the detection effect can be improved

without losing the trend information because this algorithm does not com-
press the original time series.

(2) We propose a novel data structure named Time Split Tree (TSTree) and
introduce the three techniques in DLDE, Time Split Tree for time series
randomly division, Hash Table for similarity measurement that the data
points with the same hash value are similar data points, and Ensemble
Learning to ensure the stability of algorithm.

(3) Our algorithm is analyzed with solid theoretical explanation and experimentally verified the effectiveness of the algorithm. DLDE outperforms other
state-of-art algorithms on different types of data sets in accuracy.

The rest of paper is organized as follows. Section 2 sets up the problem def-
initions for anomaly detection in time series. Section 3 proposes the Dynamic
Local Density Estimation algorithm. Experimental results are reported in Sect. 4.
Finally, Sect. 5 concludes the paper.

2 Problem Statement

Dynamic time warping (DTW) is a dynamic programming technique which can
handle nonlinear alignments and local drift time [25] with different length subsequences caused by timeline scaling, amplitude shift and linear drift. Amplitude
shift is ampliotude baseline is different with two similar time series. Timeline
scaling means time series scaling proportionally on the timeline. Linear drift
shows a trend od linear increasing or decreasing for time series. If the corresponding subsequences in two time series do not represent the same meaning,
it is unreasonable to calculate their similarity by means of Euclidean distance.
In order to reduce the time complexity, warping function [16] was proposed as
shown in Fig. 1(a). After adding the optimization width limit, the most similar
data points can be found only within a certain segment.

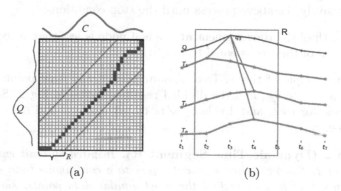

| (a) | (b) |

Fig. 1. The (a) is the DTW calculation matrix with adjustment window R (green
window), C and Q are the two time series. The (b) is the example of DTW calculates
schematics in all data sets. (Color figure online)

The anomaly detection based on DTW needs to calculate the similarity of any two subsequences and the time complexity is $O(mnN^2)$. If adding the search scope limit window R, the calculation process is shown in Fig. 1(a). In Fig. 1(b), suppose we should detect whether the time series Q has anomaly or not, and the other series are $T_1 \sim T_n$, the adjustment window is R. Take the q_5 as an example, finding the minimum distance in the limited R window from T_1 to T_n. From the perspective of anomaly detection, the larger of distance between q_5 and other data points, the more abnormal the point is. In other words, if there is no similar data point in the adjustment window, the test point should be an anomaly. Therefore, inspired by the idea above, we propose a method to quickly evaluate the similarity of subsequence. Based on this method, an anomaly subsequence detection algorithm for dynamic local density estimation is proposed.

3   The Proposed Algorithm

Based on the analysis of dynamic time warping similarity calculation in Sect. 2, we propose a time series anomaly subsequence detection algorithm, Dynamic Local Density Estimation(DLDE) to divide the time series randomly and evaluate the degree of anomaly for data points through dynamic local density of each data point in the subsequence.

3.1   Basic Concept and Definitions

Definition 1 (Time Split Tree (TSTree)). *TSTree randomly divides a time series into several dynamic time segments, each of which is located at the leaf node.*

The process is as follows: there is a time series $\{t_1 \sim t_d\}$, randomly choose time point st as a split point, and divide all time points before st into T_l while others in T_r. Recursively the above process until the stop conditions:

(1) The length of the time segment at the leaf node is less than or equal to 3.
(2) The depth of the tree is equal to $log_2(d)$.

Give an example of the TSTree. Assuming that the time points of the Q time series are t_1 to t_{20}, and the divided result is shown in Fig. 2. Select t_9 as the split node for root, and divide $t_1 \sim t_8$ to left subtree, and $t_9 \sim t_{20}$ to the right subtree.

Definition 2 (Dynamic Time Segment R). *Inspired by limit window R in DTW in Fig. 1, Dynamic time segment refers to a continuous time segment in a subsequence that is used to find the most similar data points, such as $R = \{t_s, t_{s+1}, ..., t_e\}(1 \leq s \leq e)$.*

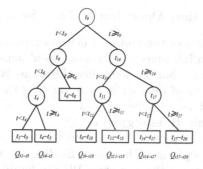

Fig. 2. The structure of TSTree, the circle node represents an internal node, and a rectangle node represents a leaf node.

Definition 3 (Hash Function). *The data set $Q_{t_1 \sim t_d}$ at d time points can be mapped to d hash table $HashTable_{t_1 \sim t_d}$ by hash function (Eq. (1)). If two data points have the same hash function value, the two data points are similar.*

$$hash(p) = \lfloor \frac{p+r}{w} \rfloor \tag{1}$$

where p is the time point, w is the hash function width parameter randomly sampled from the range $[1.0/log_2(N), 1 - 1.0/log_2(N)]$, and r is a parameter randomly selected from the range $[0, w]$.

Definition 4 (Similarity Time Point Set). *Suppose p_r is the value of time point t_r and there is a dynamic time segment $R = t_s, ..., t_e$ and the corresponding dataset $Q_{t_s \sim t_e}$ with $HashTable_{t_s \sim t_e}$. The similarity time point set is calculated as*

$$N(p_r) = \{t_j | t_j \in [t_s, t_e], hash(p_r) \in HashTable_{t_j}\} \tag{2}$$

Definition 5 (True Similarity Relation). *Due to the randomness of the hash function, the set $N(p_r)$ may contains points that are not true similarity relationship with p_r. Therefore, h random hash function are used to find the intersection of $N(p_r)$, which is the true similarity relation set as shown in Eq. (3).*

$$TN(p_r) = N_1(p_r) \cap N_2(p_r) \cap ... \cap N_h(p_r) \tag{3}$$

Definition 6 (Local Density). *Local density $density(Q_{t_j}, q_i)$ refers to the number of similar data points q_i in the data point set Q_{t_j}.*

$$density(Q_{t_j}, q_i) = count\{hash(Q_{k,t_j}) = hash(q_i) | k < N, Q_{k,t_j} \in Q_{t_j}\} \tag{4}$$

Definition 7 (Dynamic Local Density). *Dynamic Local Density refers to evaluating the local density of data points q_i in corresponding dynamic time window*

$$Density(q_i) = \frac{1}{|TN(q_i)|} \sum_{t_j \in TN(q_i)} density(Q_{t_j}, q_i) \tag{5}$$

3.2 Anomaly Detection Algorithm in Time Series

The above section introduces the proposed definition and data structure, in this section, we are going to introduce the dynamic local density estimation, which is the core of the our algorithm. To determine the anomaly of the time series, we evaluate the local density of time series by evaluating the local density of each data point within the dynamic time segment.

(1) **Divide dynamic time segment.**

Dynamic density estimation is to evaluate the local density of data points through dynamic time segments. Therefore, dividing the time series into multiple disjoint time segments is the first step. We randomly construct TSTrees to dynamically divide time series and each leaf contains one segment. The pseudocode are shown in Algorithm 1.

Algorithm 1. Build TSTree (Init_TSTree)

Require:
 Time Series Data Set $Q_{t_1 - t_d}$;
 First Time Point, t_1, The End Time Point, t_d;
 Hight Limit, $hlimit$, Size Limit, $slimit$;
 Current tree height, $height_{cur}$;
Ensure:
 A Time Split Tree, $TSTree$;
1: **if** $t_d\text{-}t_1 \leq slimit$ or $height_{cur} \geq hlimit$ **then**
2:    Return TreeNode($t_1 \widetilde{\ } t_d$);
3: **end if**
4: Randomly select a split time point, st
5: Build Left Tree, Init_TSTree(t_1, $st - 1$, $hlimit$, $slimit$)
6: Build Right Tree, Init_TSTree(st, t_d, $hlimit$, $slimit$)
7: Return $TSTree$;

---

(2) **Build a hash table.**

After dividing the time series into dynamic time segments, we need to use hash function to map data points to hash table in each segment, which can quickly estimate the local density of data points. Suppose the time segment on a leaf node in TSTree is $t_s \sim t_e$. First, h number of hash functions should be generate as $\{hash_1(.), hash_2(.), ..., hash_h(.)\}$ following the Eq. (1). Then, all these hash functions can map leafs to h number of hash tables. Each hash table is a two-dimensional array as Eq. (6), and each element in the hash table is stored in the form of Key-Value, Key ($key_{i,r}$) represents the hash value, and $Value$ ($val_{i,r}$) represents the number of times this hash value appears in the data set. The bigger of $val_{1,s}$, the more data points will be map to $key_{1,s}$ at t_s, and the more likely the corresponding original data point is normal; otherwise, the smaller of $val_{1,s}$, the more likely the original data is anomaly. The width is equal to the length of the time segment contained in the leaf node, and the length of each column may be different.

$$HashTable_j = \begin{bmatrix} (key_{1,s}, val_{1,s}) & \cdots & (key_{1,r}, val_{1,r}) & \cdots & (key_{1,e}, val_{1,e}) \\ \cdots & & \cdots & & \cdots \\ (key_{k,s}, val_{k,s}) & \cdots & (key_{k,r}, val_{k,r}) & \cdots & (key_{k,e}, val_{k,e}) \\ \cdots & & \cdots & & \cdots \\ (key_{x_1,s}, val_{x_1,s}) & \cdots & (key_{x_r,r}, val_{x_r,r}) & \cdots & (key_{x_e,e}, val_{x_e,e}) \end{bmatrix}$$
(6)

The above process uses one hash function to map one leaf node data. In order to calculate the true similarity of the data points on the leaf nodes, h hash tables need to be constructed for each node.

(3) Calculate the dynamic local density of data points

The formula for calculating the dynamic local density of a data point is described in Definition 7. And the Algorithm 2 describes the detailed calculation process after a dynamic time segmentation.

Algorithm 2. Calculate the local density at each time point in the time series.

Require:

Time Point, p_i,corresponding time t_i and TSTree, *tree*;

Ensure:

The local density of p_i, $Density(p_i)$;
1: $Density(p_i) = 0$;
2: Query the leaf node where t_i located;
3: Leaf node t_s contains the start time point of the time period;
4: Leaf node t_e contains the end time point of the time period;
5: $Hash(.) = \{hash_1(.), hash_2(.), ..., hash_h(.)\};$// H hash functions are contained in;
6: $TN(q_i) \leftarrow$ a collection of all time points from t_s to t_e;
7: **for** each $hash_j(.)$ in $Hash(.)$ **do**
8:    $ksy_{i,j} = hash_j(q_i);$// calculate the hash value of p_i;
9:    **for** each $t = t_s$ to t_e **do**
10:        **if** $k_{i,j}$ in $HashTable_{i,t}(.) \rightarrow ksys()$ **then**
11:            $N_j(q_i) \leftarrow t$;
12:        **end if**
13:    **end for**
14:    $TNq_i \leftarrow TN(q_i) \cap N_j(q_i)$;
15: **end for**
16: **for** t in $TN(q_i)$ **do**
17:    **for** each $hash_j(.)$ in Hash(.) **do**
18:        $key_{i,j} = hash_j(q_i)$;
19:        $Density(p_i)+ = HashTable_{j,t} \rightarrow get(key_{i,j})$
20:    **end for**
21: **end for**
22: Return $Density(p_i)$;

---

(4) Calculate the local density of the subsequence

Step 3 completes the dynamic local density estimation of a data point; then the local density of the time series P is estimated as shown in Eq. (7), where d

is the length of time series P and $Density(p_i)$ is calculated by Definition 7. We can see that if the $Density(P)$ value is larger, it indicates that the data points in the time series P are similar to most of the time series data points in the data set, therefore, the time sequence P is more likely to be a normal time series.

$$Density(P) = \frac{1}{d}\sum_{i=1}^{d} Density(p_i) \tag{7}$$

(5) Use Ensemble learning to determine the anomaly

Steps 1 to 4 evaluate the anomaly of each subsequence in the data set by dividing the subsequences into disjoint dynamic time segments once. However, since the data stored by TSTree is randomly segmented, if there is only one TSTree, the algorithm will not get a stable calculation result. Therefore, the idea of using ensemble learning is proposed to construct m TSTrees to form **TSForest**. The score of the subsequence P is calculated by TSForest as the dynamic density mean of m TSTree evaluations, and the formula for calculating the score of the subsequence is as shown in Eq. (8). The smaller the subsequence P is, the more likely subsequence P is an abnormal subsequence.

$$Score(P) = \frac{1}{m}\sum Density(P) \tag{8}$$

Algorithm 3. DLDE anomaly detection algorithm in time series.

Require:

    Time Series P, Subsequence Length, s, Hash Table Number h, TSTree Number m;

Ensure:

    The anomaly score of each subsequence, $Score$;

1: $n \leftarrow$ The length of P;
2: Dividing the time series P into a time series set Q according to the subsequence length s;
3: **for** $i = 1$ to m **do**
4:    Build TSTree;
5:    Initialize h hash functions for each leaf node of TSTree;
6:    Constructing a hash table on each leaf node;
7: **end for**
8: **for** each subsequence in Q **do**
9:    **for** each TSTree in TSForest **do**
10:       Calculating $Density(Q_i)$;
11:    **end for**
12:    $Score \leftarrow Mean(Density(Q_i))$
13: **end for**
14: Return $Score$;

3.3   Analysis

Time Complexity. Suppose the size of time series data set is N, and the length of subsequence is d. The time to build m TSTree needs $O(m * log_2(d))$ and the time complexity of h Hash Table is $O(N * m * d * h)$, therefore, in the detection process, the time complexity is $O(N * m * d * h * log_2(d))$. It is verified in Sect. 4 that m and h can achieve convergence by taking a small constant algorithm.

Space Complexity. DLDE takes advantage of the data structure of the TSTress and the Hash Table. The TSForest composed by m TSTrees and the data in every leaf node needs h Hash Tables to represent. Therefore, the space complexity required by the algorithm is $O(m * h * d * const)$, where $const$ represents the number of hash values.

4   Experimental Evaluation

In this section, the data sets and the evaluation metrics are introduced first. For comparability, we implemented all experiments on our workstation with 2.5 GHz, 64 bits operation system, 4 cores CPU and 16 GB RAM.

4.1   Evaluation Metrics and Experimental Setup

Data Sets: The time series data sets in the experiments are selected from the UCR Time Series Repository [4] and the BIDMC Congestive Heart Failure Database [2]. In UCR, the ECG data and the SENSOR data set are typical time series data sets; MOTION is the sequence data generated by the action, the IMAGE data can extract the time series data. These data sets are described in Table 1. In our experiments, we follow the split subsequences as provided by UCR. For balanced data, we will significantly under-sampling one of two classes to obtain minority (anomaly class). For example, in ECG5000_2_3 we choose class 2 as normal and class 3 as anomaly.

Experimental Setup: We select five anomaly detection algorithms, Relative Density Outlier Score (RDOS) [24], Fast Variance Oulier Angle (FastVOA) [14], Internal [18] and Piecewise Aggregate Pattern Representation (PAPR) [17]. RDOS is the anomaly detection algorithms based on local density, FastVOA is an algorithm based on angle variance. Internal and PAPR are two anomaly detection algorithms based on interval division. The parameter settings of the above comparison algorithm are set according to the reference. For RDOS, the neighbors number will be set to 10. For FastVOA, we will set the hash number to 100. For PAPR, we will set the three parameters $wc = 0.3, wd = 0.4, wr = 0.3$. All these compared algorithms and DLDE are executed for 50 times to get stable results.

4.2   Accuracy

The aim of this experiment is to compare DLDE with other methods in terms of Area Under Curve (AUC). AUC is commonly used for evaluating anomaly

Table 1. The description of UCR time series data sets.

No.	Data sets	Size	Length	Anomaly rate	Type
1	DistalPhalanxOutlineCorrect	876	80	38.47%	Image
2	ECG200	200	96	33.50%	ECG
3	HandOutlines	1370	2709	36.13%	Image
4	Lighting2	121	637	39.66%	Sensor
5	MoteStrain	1272	84	46.14%	Sensor
6	SonyAIBORobotSurfaceII	980	65	38.36%	Sensor
7	ToeSegmentation2	166	343	25.30%	Motion
8	ECG5000_2_3	1863	140	5.15%	ECG
9	ECG5000_2_4	1961	140	9.89%	ECG
10	ECG5000_2_5	1791	140	1.34%	ECG
11	StarLightCurves_2_1	427	1024	35.59%	Sensor
12	DiatomSizeReduction_2_1	132	345	25.75%	Image
13	DiatomSizeReduction_3_1	133	345	25.56%	Image
14	DiatomSizeReduction_4_1	125	345	27.20%	Image

detection algorithm. The experiment results are recorded in Table 2, and the best results are highlighted in bold font. NA indicates that this algorithm cannot be calculated on this data set in the current experimental environment. From this table, we can find that DLDE has better results than other algorithms on the most of all data sets (12/14). It is indicated that DLDE is able to detect anomalies efficiently that other baselines are difficult to detect.

For further analysis of experimental results, the data sets in Table 1 are divided into four parts according to the length, the average of each four parts are the final results of each algorithm on different length of data sets. The RDOS algorithm does not get running results on two data sets which are not be considered in the condition of more than 1000 part. From the experimental results in the Fig. 3(a), we can find that the algorithm DLDE can obtain better experimental results on time series data sets of different lengths. The results are shown in Fig. 3(b), it indicates that DLDE performs better on the first three types of data sets than other algorithms.

To test the impact of different data types on experimental results, the data set in Table 1 is divided into four parts: ECG, MOTION, IMAGE, SENSOR. ECG data and SENSOR data sets are typical time series data sets; MOTION is sequence data generated by actions; IMAGE data can extract time series data. The average of the experimental results of each algorithm on the four data sets is calculated separately as Fig. 3(b). It can be seen from the figure that DLDE performs better than the other algorithms on the first three types of data sets.

Table 2. AUC Performance. The best AUC scores are highlighted in bold.

No.	DLDE	RDOS	FastVOA	Internal	PAPR-RW
1	**0.705**	0.646	0.702	0.632	0.693
2	**0.875**	0.658	0.84	0.609	0.788
3	**0.815**	NA	0.772	0.576	0.734
4	**0.764**	0.611	0.697	0.599	0.653
5	**0.848**	0.529	0.730	0.579	0.724
6	**0.796**	0.52	0.715	0.486	0.571
7	0.736	0.717	0.720	0.671	**0.777**
8	**0.861**	0.669	**0.861**	0.659	0.837
9	**0.735**	0.591	0.716	0.668	0.714
10	0.870	0.904	**0.93**	0.823	0.838
11	**0.966**	NA	0.833	0.769	0.750
12	**1.00**	0.652	0.970	**1.00**	0.998
13	**0.901**	0.774	0.853	0.677	0.761
14	**1.00**	0.768	0.971	**1.00**	**1.00**

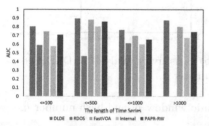

(a) Comparison results of each algorithm on different length time series data.

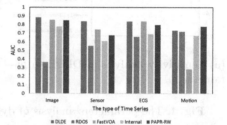

(b) Comparison results of each algorithm on different type of time series data.

Fig. 3. Comparison of various experiments on AUC.

4.3 Parameter Analysis

Dynamic Window m. In the DLDE algorithm, the dynamic time segment window is randomly divided, in order to ensure the stability of the algorithm, we choose to use the idea of ensemble learning. That is randomly divide m times, and the final test result of the algorithm is the average of the number of runs. In this experiment, the sensitivity of the DLDE algorithm to the parameter m will be verified. When m is taken from 1 to 50, the variation of the AUC index on different data sets is recorded. Parameter m is tested under each parameter condition, the average value of the program running 50 times is taken as the final result and recorded in Fig. 4(a). In this figure, it can be noted that the experimental results of DLDE are basically in a stable trend on the data set of all algorithms, that is, the AUC index of the algorithm is basically convergent when m reaches 10. Therefore, this experiment proves that the algorithm does not require a lot of random division of dynamic time segments, and the algorithm has better stability.

Hash Number h. We construct a hash table at the leaf node with h p-stable local-sensitive hash functions [20]. This hash defines a boundary region, and all values in this region have the same hash value. To avoid the instability of random hash functions, we use multiple random hash functions. The intersection of similar sets of data points computed by the generated plurality of hash functions is a final set of similar data points that can more accurately measure similar relationships between data points. It is proved in the Fig. 4(b) that when h is taken from 1 to 108, the algorithm can achieve convergence as long as m and h take a small constant.

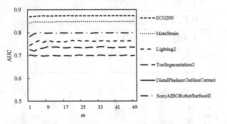

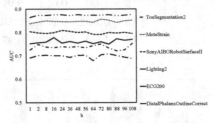

(a) Sensitivity analysis of DLDE to parameter m.

(b) Sensitivity analysis of DLDE to parameter h.

Fig. 4. The parameters analysis of dynamic window m and hash number h.

Computational Time. In order to calculate the consumption time, we selected 8 data sets for testing. We calculate the percentage of calculation time for the five methods in each data set. As can be seen from the Fig. 5, DLDE has a better effect on the short length of the subsequence, and PAPR has a better effect on the long length of the subsequence, and the average performance of other data sets is relatively nearly.

4.4   Performance on ECG Data

In this section, we will demonstrate the effectiveness of our algorithm on ECG data selected from BIDMC Congestive Heart Failure Database. We select two ECG records from this database, $chfdb01\_275$ and $chfdb13\_45590$. These two ECG data contains two ECG signal, and each record contain one anomaly subsequence.

In this experiment, we use the data of one minute length in two data sets as experimental data, and divide the whole time series into 15 sub-time series according to the cycle per second. We will verify the difference between the

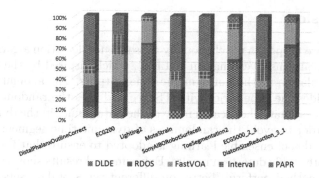

Fig. 5. Comparison results of each algorithm on different length time series data

scores calculated by the method proposed in this paper, and use the line graph to visualize this difference. Since the results calculated by our methods is the density of the subsequence, the score should convert into the abnormal score of the subsequence by using Eq. (9).

$$anomaly\_score(P) = 1 - \sum \frac{p_i - min(P)}{max(P) - min(P)} \qquad (9)$$

These two ECG data are shown in Fig. 6, and the anomaly subsequences are shown in red line. The anomaly scores of each subsequence calculated by DLDE are shown in the dark red line below. It can be clearly found that the higher anomaly score is corresponded to true anomaly subsequence, and other score are around 0.5. We rank the anomaly scores of each subsequence to determine the anomalies in the data, thus avoiding the occurrence of missed detection. Therefore, these results can demonstrate the effective of our algorithm.

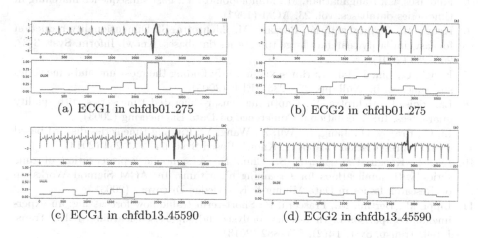

(a) ECG1 in chfdb01_275   (b) ECG2 in chfdb01_275

(c) ECG1 in chfdb13_45590   (d) ECG2 in chfdb13_45590

Fig. 6. The results on ECG data chfdb01_275 and chfdb13_45590

5   Conclusion

In this paper, we propose a novel anomaly subsequence detection algorithm based on dynamic local density estimation (DLDE), which inspired by the idea of the similarity calculation method of dynamic time warping. The anomaly detection algorithm divides the time series with TSTree and uses the random hash function to quickly estimate the local density of the data points in the dynamic time segment. In order to avoid the randomness of dynamic time segments and hash functions, the idea of ensemble learning is adopted to ensure that the algorithm can obtain more stable detection results. Experimental results show that the proposed DLDE method performs better on different types of data sets than other baselines. In the future work, we need to consider whether can set the automatic time segmentation method to reduce the process of algorithm ensemble learning.

Acknowledgment. This study was supported by the Shenzhen Research Council (Grant No. 369 JSGG20170822160842949, JCYJ20170307151518535).

References

1. Argyro, K., George, M., Christophoros, N.: Heartbeat time series classification with support vector machines. IEEE Trans. Inf. Technol. Biomed. Publ. IEEE Eng. Med. Biol. Soc. **13**(4), 512–8 (2009)
2. Baim, D.S., et al.: Survival of patients with severe congestive heart failure treated with oral milrinone. J. Am. Coll. Cardiol. **7**(3), 661–670 (1986)
3. Chan, K.P., Fu, W.C.: Efficient time series matching by wavelets. In: International Conference on Data Engineering (1999)
4. Chen, Y., et al.: The UCR time series classification archive, July 2015. www.cs.ucr.edu/~eamonn/time_series_data/
5. Faloutsos, C., Ranganathan, M., Manolopoulos, Y.: Fast subsequence matching in time-series databases, vol. 23. ACM (1994)
6. Keogh, E., Chakrabarti, K., Pazzani, M., Mehrotra, S.: Dimensionality reduction for fast similarity search in large time series databases. Knowl. Inform. Syst. **3**(3), 263–286 (2001)
7. Keogh, E., Lin, J., Fu, A.: Hot sax: efficiently finding the most unusual time series subsequence. In: Null, pp. 226–233. IEEE (2005)
8. Lazaridis, I., Mehrotra, S.: Capturing sensor-generated time series with quality guarantees. In: International Conference on Data Engineering (2003)
9. Li, G., Bräysy, O., Jiang, L., Wu, Z., Wang, Y.: Finding time series discord based on bit representation clustering. Knowl. Based Syst. **54**, 243–254 (2013)
10. Lin, J., Keogh, E., Lonardi, S., Chiu, B.: A symbolic representation of time series, with implications for streaming algorithms. In: ACM Sigmod Workshop on Research Issues in Data Mining & Knowledge Discovery (2003)
11. Lippi, M., Bertini, M., Frasconi, P.: Short-term traffic flow forecasting: an experimental comparison of time-series analysis and supervised learning. IEEE Trans. Intell. Transp. Syst. **14**(2), 871–882 (2013)
12. Moonesinghe, H.D.K., Tan, P.N.: Outlier detection using random walks. In: IEEE International Conference on Tools with Artificial Intelligence (2006)

13. Ocak, H.: Automatic detection of epileptic seizures in eeg using discrete wavelet transform and approximate entropy. Expert Syst. Appl. **36**(2), 2027–2036 (2009)

14. Pham, N., Pagh, R.: A near-linear time approximation algorithm for angle-based outlier detection in high-dimensional data. In: Proceedings of the 18th ACM SIGKDD International Conference on Knowledge Discovery and Data Mining, pp. 877–885. ACM (2012)

15. Rahmani, A., Afra, S., Zarour, O., Addam, O., Koochakzadeh, N., Kianmehr, K., Alhajj, R., Rokne, J.: Graph-based approach for outlier detection in sequential data and its application on stock market and weather data. Knowl. Based Syst. **61**(1), 89–97 (2014)

16. Rakthanmanon, T., Campana, B., Mueen, A., Batista, G., Keogh, E.: Searching and mining trillions of time series subsequences under dynamic time warping. In: ACM SIGKDD International Conference on Knowledge Discovery & Data Mining (2012)

17. Ren, H., Liu, M., Li, Z., Pedrycz, W.: A piecewise aggregate pattern representation approach for anomaly detection in time series. Knowl. Based Syst. **135**, 29–39 (2017)

18. Ren, H., Liu, M., Liao, X., Li, L., Ye, Z., Li, Z.: Anomaly detection in time series based on interval sets: anomaly detection in time series. IEEE Trans. Electr. Electron. Eng. **13**(9), 757–762 (2018)

19. Ruiz, E.J., Hristidis, V., Castillo, C., Gionis, A., Jaimes, A.: Correlating financial time series with micro-blogging activity. In: ACM International Conference on Web Search & Data Mining (2012)

20. Sathe, S., Aggarwal, C.C.: Subspace outlier detection in linear time with randomized hashing. In: 2016 IEEE 16th International Conference on Data Mining (ICDM), pp. 459–468, December 2016, https://doi.org/10.1109/ICDM.2016.0057

21. Senin, P., et al.: Time series anomaly discovery with grammar-based compression. In: EDBT, pp. 481–492 (2015)

22. Sivaraks, H., Ratanamahatana, C.A.: Robust and accurate anomaly detection in ECG artifacts using time series motif discovery. Comput. Math. Methods Med. **2015**, 453214 (2015)

23. Sun, Y., Li, J., Liu, J., Sun, B., Chow, C.: An improvement of symbolic aggregate approximation distance measure for time series. Neurocomputing **138**(11), 189–198 (2014)

24. Tang, B., He, H.: A local density-based approach for outlier detection. Neurocomputing **241**, 171–180 (2017)

25. Yuan, Y., et al.: Development and application of a modified dynamic time warping algorithm (DTW-S) to analyses of primate brain expression time series. BMC Bioinform. **12**, 347 (2011). https://doi.org/10.1186/1471-2105-12-347

Trajectory Similarity Join for Spatial Temporal Database

Tangpeng Dan[1,2,3], Changyin Luo[1,2,3]([✉]), Yanhong Li[4], and Chenyuan Zhang[1]

[1] School of Computer, Central China Normal University, Wuhan, China
tangpengdan@mails.ccnu.edu.cn, changyinluo@mail.ccnu.edu.cn
[2] Hubei Provincial Key Laboratory of Artificial Intelligence and Smart Learning,
Central China Normal University, Wuhan, China
[3] National Language Resources Monitor and Research Center for Network Media,
Central China Normal University, Wuhan, China
[4] College of Computer Science, South-Central University for Nationalities,
Wuhan, China

Abstract. The trajectory similarity join aims to find similar trajectory pairs from two large collections of trajectories. This join targets applications such as trajectory near-duplicate detection, ridesharing recommendation and so on. Extensive works have been conducted on addressing this join. However, most of them only focus on spatial dimension without combining temporal range together. To address problem, this paper proposes a novel two-level grid index which takes both spatial and temporal range into account when processing spatial-temporal similarity join, and signature based dynamic grid warping (SDGW) approach to evaluate the spatial similarity for trajectory pairs. Some pruning approaches are developed to improve the query processing. In addition, extensive experiments are conducted to verify the efficiency and scalability of our methods.

Keywords: Spatial-temporal database · Two-level grid index · Trajectory similarity join

1 Introduction

With the advancement of GPS-based mobile devices and online map-techniques services (e.g., Google Maps), it is convenient to produce, collect and share trajectories. More and more social network platforms, such as Twitter, Facebook, are supporting trajectory queries and sharing services. Furthermore, a taxi or Uber car generate a trajectory from pick-up point to drop-off point. The availability of massive trajectory data motivates new studies in spatial and temporal data analysis.

Trajectory similarity join, which, given two sets of trajectories and a similarity threshold θ, returns all the trajectory pairs with similarity above θ. By recording a series of key points of a trajectory along the road network at a fixed

© Springer Nature Switzerland AG 2019
S. Hartmann et al. (Eds.): DEXA 2019, LNCS 11707, pp. 306–321, 2019.
https://doi.org/10.1007/978-3-030-27618-8_23

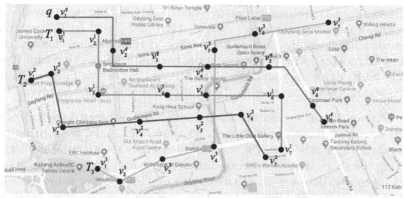

$q \ltimes P_1, (7:30,7:46)>,< P_2, (7:46,7:51)>,< P_3, (7:55,8:06)>,< P_4, (8:09,8:29)>,< P_5, (8:33,8:45)>,< P_6, (8:39,8:52)>,< P_7, (8:46,9:01)>$
$T_1 \ltimes P_1, (7:40,7:49)>,< P_2, (7:44,7:59)>,< P_3, (8:03,8:27)>,< P_4, (8:12,8:35)>,< P_5, (8:39,8:44)>,< P_6, (8:47,8:58)>,< P_7, (9:02,9:16)>$
$T_2 \ltimes P_1, (13:10,13:15)>,< P_2, (13:14,13:22)>,< P_3, (13:25,13:37)>,< P_4, (13:26,13:45)>,< P_5, (13:45,13:56)>,< P_6, (14:03,14:11)>,< P_7, (14:13,02,14:26)>$
$T_3 \ltimes P_1, (7:21,7:32)>,< P_2, (8:49,8:58)>,< P_3, (9:17,9:29)>,< P_4, (9:27,9:40)>,< P_5, (9:42,9:56)>,< P_6, (9:49,10:01)>,< P_7, (10:04,10:13)>$

Fig. 1. Example of spatial temporal join query

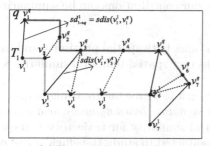

        (a)Point to Point Method   *(b)SDGW Similarity Method*

Fig. 2. Trajectory similarity methods

region, trajectory join can be recorded using in many applications, such as data cleaning [12], taxi recommending system [10], traffic condition analysis [16]. As shown in Fig. 1, in order to reduce the redundancy, similarity join can be used to data cleaning. Given a query trajectory $\{q\}$, we may find two highest ranked similar trajectories $\{T_1, T_2\}$, and only keep the most similar T_1 as the representative trajectory. In the literatures, many studies have been proposed to address the problem of trajectory similarity join [1,12,18,19], and they have their own merits. However, they seldom take both spatial and temporal similarity in a continuous manner.

Example 1. An example depicted in Fig. 1, the query trajectory $Q = \{q\}$ can be regarded as a historical trajectory of a passenger. $P = \{T_1, T_2, T_3\}$ is the set of Uber[1] trajectories. Every trajectory has some sample points associated with the earliest arrival time and the latest arrival time. If the recommendation

---

[1] https://www.uber.com/.

system adopts the traditional trajectory similarity join approaches that do not take temporal factor into account, it will return (q, T_2) and (q, T_1) as a result. Because they are spatially close to each other. If the temporal domain is also taken into consideration in recommendation, because the departure time interval of q is $(7:30, 9:01)$ while that of T_2 is $(13:10, 14:26)$. Thus, T_2 is not matched to q, the system should only return T_1 as a top-1 similarity trajectory.

Furthermore, when evaluating spatial similarity between two trajectories, methods discussed above are inefficient. As shown in Fig. 2a, if adopting point-to-point similarity method [18,19], we need to evaluate their spatial similarity by computing the shortest road network distance for each closest point pairs (e.g., v_1^1 to v_1^q, v_2^1 to v_2^q). However, due to different sampling rates or vehicular speeds, these points may not be aligned well, thus this method consumes much more memory and computation time. In order to address this problem, we propose a Signature based Dynamic Grid Warping (SDGW) method in Sect. 4.3, and evaluate the similarity between T_1 and q by computing the shortest network distance from a sample point on T_1 to the closest signature of the sample point on q, which is more efficient. To summarize, our contributions can be summarized as follows.

- We propose a two-level grid index which takes both spatial proximity and time range into account when processing trajectory spatial-temporal similarity join queries.
- We develop a time-first searching framework to prune unpromising trajectory pairs in an efficient way, and then adopt *signature based dynamic grid warping* (SDGW) approach to evaluate the spatial similarity for trajectory pairs.
- A set of experiments on real data are conducted to study the efficiency of our methods.

2   Related Works

A number of trajectory similarity measurement functions have been proposed, which can be roughly grouped into two types: (1) The spatial based metrics, such as the Closest-Pair Distance (CPD) [13] and the One Way Distance (OWD) [9]. These metrics directly use the Euclidean distance for corresponding sample point pairs to define the similarity. (2) The spatio-temporal metrics, such as the Dynamic Time Warping (DTW) [1,15], the Longest Common Sub Sequence (LCSS) [21], the Sequence Weighted Alignment model (Swale) [11], the Most Similar Trajectory (DISSIM) [5], the model-driven assignment (MA) [16], and the Edit Distance with Projections (EDwP) [14]. Specifically, the One Way Distance (OWD) [9] focuses on shape similarity for trajectories in grid representations. OWD of two grid trajectories is the sum of distances from the grids where one trajectory's sample points reside in to the grids of the other trajectory. The Closest-Pair Distance(CPD) [13] is a variation of Euclidean Distance which was introduced to find closest trajectories for given query in spatial networks. The Dynamic Time Wrapping (DTW) [1,23] distance allows some sample points to repeat in order to achieve the best alignment, i.e., one point in one trajectory can match multiple points in another trajectory. DTW was claimed to be vulnerable to noises since some noise points can introduce large distance between trajectories. However, the authors of [22] argued and experimentally proved that DTW on average is comparative to other similarity measurements on large data sets. The Longest Common Sub Sequence (LCSS) [21] is used to eliminate the effect of noise points. The LCSS method skips points (taking them as noises) if their distance exceeds a matching threshold. Edit Distance with Real Penalty (ERP) [7] uses a threshold θ to quantify a match, and gaps between matched sub-trajectories are assigned penalties to reveal the dissimilarity. As an improvement, Edit Distance on Real Sequence (EDR) [8] combines the strength of DTW and ERP. It handles time shifting and computes distance using a constant reference point.

Trajectory similarity joins can be used such as data cleaning, navigation system, and road planning [9,19,21]. In general, in order to measure spatial temporal similarity between two trajectory sets, people use a time interval threshold to constrain the temporal proximity of two trajectories (in a fixed manner) and can be classified into two categories. Studies in the first category (e.g., [3,6–8]) eliminate trajectory pairs that are temporally further apart than a threshold. We generalize this category of studies and compute temporal similarity by summarizing temporal proximities of sample point pairs from two trajectories in a continuous manner, thus obviating the need for a time threshold. Studies in the other category (e.g., [2,24]) utilize a sliding window (such as "ten minutes from now" or "yesterday between 2 and 3 PM") for all trajectories and eliminate pairs of trajectories with times that fall outside the window. For the remaining pairs, only spatial proximity is considered. Furthermore, unlike existing trajectory similarity joins [2,3,6,24], the trajectory similarity join is applied in spatial networks because in many practical scenarios, objects (e.g., taxies and passengers) move in road networks rather than in Euclidean space. Considering the different sampling rates or different vehicular speeds, the sample points in similar trajectories may not be aligned, a bi-directional mapping similarity (BDS) metric is proposed [12], which allows a sample point of a trajectory to align to the closest location (which may not be a sample point) on the other trajectory, and vice versa. However, their methods may not work well for the scenarios where the timestamp is important.

The most relevant work is proposed in TS-Join [17], a two-phase algorithm was developed based on a divide-and-conquer strategy. The authors proposed an upper bound and a heuristic scheduling strategy to prune the search space effectively. In TS-Join, every sample point on each trajectory has a only timestamp. However, in our work, every sample point has a time period. Therefore, TS-Join may not work well in our setting.

3   Problem Definition

Before formulating the problem of spatial-temporal trajectory similarity, some definitions are given as follows.

Definition 1. Trajectory. *A trajectory T has several sample points, which are arranged in spatial sequence. Each sample point is marked with an estimated arrival time, (e.g., Didi Chuxing has estimated arrival time for each destination) we define T as $T_i = \{v_1^i, v_2^i, v_3^i, \cdots, v_n^i\}$, where v_n^i is a sample point on T_i, $v_n^i = [p_n, (t_{ns}, t_{ne})]$, p_n is the spatial location, (t_{ns}, t_{ne}) indicates the earliest arrival time and the latest arrival time for p_n.*

Note that, for a sample point p_n, we could set its estimated arrival time to t_{ns}, and assign its real arrival time to t_{ne}. Thus, (t_{ns}, t_{ne}) for p_n can be obtained.

Given two trajectories T_i and T_j, a sample point v_n^i on T_i, $Sdis(v_n^i, v_n^j)$ is the shortest road network distance from v_n^i to the sample point v_n^j on T_j. $sd_{i \to j}^n$ is defined as follows.

$$sd_{i \to j}^n = \min_{v_n^i \in T_i, v_n^j \in T_j} \{Sdis(v_n^i, v_n^j)\} \tag{1}$$

Definition 2. Bi-Directional spatial Similarity Function. *Given trajectories T_i and T_j, trajectory-spatial similarity function is defined as follows. We extend Euclidean based trajectory similarity [4] to make it fit into spatial networks.*

$$S_{sim}(T_i, T_j) = \frac{\sum_{k=1}^{|T_i|} e^{-sd_{i \to j}^k}}{|T_i|} + \frac{\sum_{k=1}^{|T_j|} e^{-sd_{j \to i}^k}}{|T_j|} \tag{2}$$

Here, $|T|$ denotes the number of sample points in a trajectory. We use e^{-x} function to measure the similarity for each sample point on trajectory pair. e^{-x} is a monotonically decreasing function, if $x = 0$, $e^{-x} = 1$. If $x > 0$, $e^{-x} < 1$. If the shortest distance from each sample point on T_i to the corresponding point on T_j is 0, that is $sd_{i \to j}^k = 0$, they are completely coincident, the similarity of T_i and T_j reaches maximum. Otherwise, $\frac{\sum_{k=1}^{|T_i|} e^{-sd_{i \to j}^k}}{|T_i|}$ is less than 1, $S_{sim}(T_i, T_j) < 2$.

Similarly, **temporal distance** $td_{i \to j}^n$ denotes the minimum temporal distance between a sample point v_n^i on T_i to T_j, which is defined as Eq. 3, where rt is the threshold time, and $|v_n^i.t_{ne} - v_n^j.t_{ns}| \leq rt$.

$$td_{i \to j}^n = \begin{cases} \frac{\min\{|v_n^i.t_{ne} - v_n^j.t_{ns}|, |v_n^i.t_{ne} - v_n^i.t_{ns}|, |v_n^j.t_{ne} - v_n^j.t_{ns}|\}}{\left|\max\{v_n^i.t_{ne}, v_n^j.t_{ne}\} - \min\{v_n^i.t_{ns}, v_n^j.t_{ns}\}\right|} & if \ |v_n^i.t_{ne} - v_n^j.t_{ns}| \leq rt \\ +\infty & if \ |v_n^i.t_{ne} - v_n^j.t_{ns}| > rt \end{cases} \tag{3}$$

Definition 3. Bi-Directional temporal Similarity Function. *Given trajectories T_i and T_j, trajectory-temporal similarity function is defined as follows.*

$$T_{sim}(T_i, T_j) = \frac{\sum_{k=1}^{|T_i|} e^{-td_{i \to j}^k}}{|T_i|} + \frac{\sum_{k=1}^{|T_j|} e^{-td_{j \to i}^k}}{|T_j|} \tag{4}$$

The value of $T_{sim}(T_i, T_j)$ is in $[0, 2]$. Based on Definitions 2 and 3, we give a linear combination method to combine spatial and temporal similarities as follows.

Definition 4. *Spatial Temporal Similarity Score.*

$$ST_{sim}(T_i, T_j) = \lambda \cdot S_{sim}(T_i, T_j) + (1 - \lambda) \cdot T_{sim}(T_i, T_j) \tag{5}$$

Parameter $\lambda \in [0, 1]$ *controls the relative importance of the spatial and temporal similarities.*

Definition 5. *Spatial Temporal Similarity Joins.* *Given two sets of trajectories* P *and* Q, *a similarity threshold* θ, *find all similar trajectory pairs* $< T_i \in P, T_j \in Q >$ *such that* $ST_{sim}(T_i, T_j) \geq \theta$.

4 Our Solution

4.1 Two-Level Grid Index

In order to efficiently utilize temporal and spatial information to compute trajectory similarity, we build a two-level grid index to organise the trajectories. The first level of the index mainly stores temporal information, and the second level stores spatial information. At first, we describe how to organize the temporal domain in the first level.

Because most movements occur daily, in this work, the value of a temporal domain is set to be within 24 h and the date is not taken into consideration. Considering people always go out and work at daytime, while there is a few activity at night, thus, we partition the temporal domain into m time slots for the daytime, and a larger time slot for the night, each of which corresponds to a leaf node. As shown in Fig. 3, we build up a tree structure in a bottom-up manner.

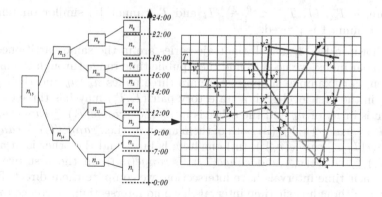

Fig. 3. Two-level grid index

Example 2. Considering T_1, T_2 and T_3 in Fig. 3, suppose their time ranges are $r(T_1) = (9:07, 10:20)$, $r(T_2) = (9:36, 11:10)$ and $r(T_3) = (9:54, 11:52)$, respectively, n_3 is the only choice because $r(T_1) \cup r(T_2) \cup r(T_3) \subseteq r(n_3)$. Suppose there is a trajectory T_4 and its associated time interval is $r(T_4) = (5:35, 8:20)$, then T_4 should be stored in n_{12}. Because the time range of T_4 has already crossed n_1 and n_2 and their parent node is n_{12}.

At the second level, inspired by index methods in [9,20], we adopt the grid-based method to index trajectories.

4.2   Time First Searching Framework

Considering the inefficiency of pruning dissimilarity pairs in spatial domain, we propose a time-first search framework in Algorithm 1. Specifically, we find all the trajectories in every leaf node of the first level of our index, if their time periods are less than the threshold time rt, verify the similarity for each trajectory pair in this node. Otherwise, we prune this trajectory pair based on following lemma:

Lemma 1. *Given a similarity threshold θ and Parameter λ defined in Eq. 5, if $T_{sim}(T_i, T_j) < \frac{\theta - 2\lambda}{1 - \lambda}$, the trajectory pair (T_i, T_j) cannot be similar on temporal domain, we prune it directly.*

Proof. Suppose there is a trajectory pair T_i and T_j having maximum similarity, that is, they are completely coincident, therefore, $sd_{i \to j}^k = 0$ and $e^{-sd_{i \to j}^k} = 1$. Based on this we have:

$$\max\{S_{sim}(T_i, T_j)\} = \max\{\frac{\sum_{k=1}^{|T_i|} e^{-sd_{i \to j}^k}}{|T_i|} + \frac{\sum_{k=1}^{|T_j|} e^{-sd_{j \to i}^k}}{|T_j|}\} = 2$$

Combining with $\lambda \cdot S_{sim}(T_i, T_j) + (1 - \lambda) \cdot T_{sim}(T_i, T_j) \geq \theta$, we have

$$\Rightarrow T_{sim}(T_i, T_j) \geq \frac{\theta - \lambda \cdot S_{sim}(T_i, T_j)}{1 - \lambda} = \frac{\theta - \lambda \cdot 2}{1 - \lambda}$$

Therefore, if $T_{sim}(T_i, T_j) < \frac{\theta - 2\lambda}{1 - \lambda}$, T_i and T_j cannot be similar on temporal domain. Lemma 1 is proved.

As shown in Fig. 3, not all the trajectories are in the same partitioned time period in our index, we need to find their public time range through the merging operation. For example, suppose there are three nodes n_a, n_b and their parent node n_c in the first level-grid, a trajectory pair (T_i, T_j) may has three cases: (i) one item is in n_a or n_b and the other is in n_c (i.e., $range(T_i) \subseteq range(n_a)$ and $range(T_j) \subseteq range(n_c)$); (ii) both of them are in n_c (i.e., $range(T_i) \subseteq range(n_c)$ and $range(T_j) \subseteq range(n_c)$); (iii) one item is in n_a and the other is in n_b (i.e., $range(T_i) \subseteq range(n_a)$ and $range(T_j) \subseteq range(n_b)$). In the first two cases, because their time intervals have intersection, we compute them direct. For the third case, although their time intervals have no intersection, reference to Eq. 3, their time intervals difference is less than the threshold time rt, they may be similar to each other in temporal domain. Therefore, they should be merged into their parent node (lines 7–10).

Algorithm 1. Time-First Searching Framework

Input: two-level grid index T_r, trajectory set T, threshold time rt, λ
Output: $A = \{(T_i, T_j) | ST_{sim}(T_i, T_j) \geq \theta, \forall T_i \in T, \forall T_j \in T\}$

1 we adopt pre-order traversal to search leaf node in T_r;
2 **for** *each leaf node in T_r* **do**
3   **for** *each trajectory pair (T_i, T_j) in node* **do**
4   compute their temporal similarity $T_{sim}(T_i, T_j)$ on Eq.3 and Eq.4
5   **if** $T_{sim}(T_i, T_j) < \frac{\theta - 2 \cdot \lambda}{1 - \lambda}$ **then**
6   prune it based on Lemma 1

7   **if** *The temporal distance between n and n.sibling $\leq rt$* **then**
8   merge n and n.sibling into n.parent;
9   find qualified trajectories in n.parent;
10   spatial similarity search(T, θ, λ);

11   **else if** *The temporal distance between n and n.parent $\leq rt$* **then**
12   find qualified trajectories in n.parent;
13   spatial similarity search(T, θ, λ);

14   **else if** *n.time range $\leq rt$* **then**
15   spatial similarity search(T, θ, λ);

16 **return** A;

4.3 Signature Dynamic Grid Warping

After computed the temporal similarity, it needs to evaluate the spatial similarity for these candidate trajectories. Before introducing SDGW method, we first present how to generate signature for a sample point and trajectory respectively.

Giving a sample point v_n^i and its influence radius rd, the influence zone of v_n^i is defined with v_n^i as the center and rd as the radius. The grids intersected with the influence zone of $v_n^i \in T_i$ are stored in the second-level grid index. Let g denote *grid cell*. For a sample point $v_n^i \in T_i$, its **signature** is defined in Eq. 6.

$$G^r(v_n^i) = \{g | Sdis(v_n^i, g) \leq rd\} \tag{6}$$

$$Sdis(v_n^i, g) = \begin{cases} \min Sdis(v_n^i, l) & v_n^i \notin g \\ 0 & v_n^i \in g \end{cases} \tag{7}$$

where l is a side of a grid cell g.

As shown in Fig. 4(a), the signature for v_n^i is $G^r(v_n^i) = \{g_{12}, g_{22}, g_{32}, g_{13}, g_{23}, g_{33}\}$. If any location at trajectory T_j does not fall in any grid in $G^r(v_n^i)$, so the shortest road network distance from v_n^i to T_j is large than rd, we can infer T_i and T_j can not be similar. Next, we discuss how to check whether the trajectory T_j has a location falling in $G^r(v_n^i)$. For a trajectory T_j, its **signature** is defined as follows.

$$G^t(T_j) = \{g | g \cap T_j \neq \emptyset\} \tag{8}$$

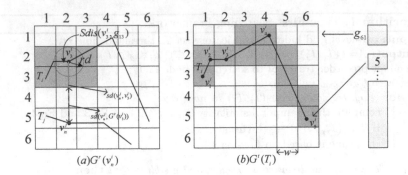

Fig. 4. Signature for a point and a trajectory (Color figure online)

In Fig. 4(b), all the green grids along T_j compose $G^t(T_j)$. We build a sorted list for each cell g to store the vertex-IDs of all the vertices located in it.

Lemma 2. *For each sample point v_n^i on T_i, if $G^r(v_n^i) \cap G^t(T_j) = \emptyset$, the trajectory pair (T_i, T_j) cannot be a similar pair. We prune it directly.*

Proof. If $G^r(v_n^i) \cap G^t(T_j) = \emptyset$, the shortest road network distance from v_n^k to T_j is larger than rd. Based on Definition 2, T_i and T_j cannot be similar. Lemma is proved.

Dynamic Time Warping (DTW) [1,15] is efficient to calculate the distance between sequences whose lengths and/or sampling rates are different. Specifically, DTW is a transformation that allows sequences to be stretched along the time axis to minimize the distance between the sequences. The distance of DTW is calculated by dynamic programming. Inspired by its idea, we propose a novel SDGW method based on signature to retrieve the similarity trajectory pairs on trajectory database.

DTW is feasible of evaluating the similarity of time sequences with different length. Supposed two trajectories $T_i = \{v_1^i, v_2^i, \ldots, v_m^i\}$ and $T_j = \{v_1^j, v_2^j, \ldots, v_n^j\}$, m and n is number of sample points in T_i and T_j respectively. The shortest road network distance is employed to measure similarity between T_i and T_j, and the dynamic programming function is given as follows:

$$DTW(T_i, T_j) = l(m, n) \tag{9}$$

$$l(p, q) = sd(v_p^i, v_q^j) + \min\{l(p-1, q), l(p-1, q-1), l(p, q-1)\} \tag{10}$$

$$l(0,0) = 0, l(p,0) = l(0,q) = \infty$$
$$(p = 1, \ldots, m; q = 1, \ldots, n)$$

where $sd()$ denotes the shortest road network distance between sample point v_p^i and v_q^j. p and q denotes p-th and q-th sample point in T_i and T_j, respectively. $l(m, n)$ is the cumulative distance from $(0, 0)$ to (m, n). However, it is expensive to compute the value of $sd()$. To address this problem, we optimize this computation process as in Eq. 12.

$$SDGW(T_i, T_j) = l(m, n) \tag{11}$$

$$l(p, q) = sd(v_p^i, G^r(v_q^j)) + \min\{l(p-1, q), l(p-1, q-1), l(p, q-1)\} \tag{12}$$

Specifically, $sd(v_p^i, G^r(v_q^j))$ is employed instead of $sd(v_p^i, v_q^j)$ in dynamic programming process. As shown in Fig. 4a, the computation cost of $sd(v_p^i, G^r(v_q^j))$ is much lower than that $sd(v_p^i, v_q^j)$.

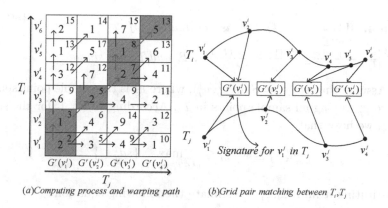

(a)Computing process and warping path   (b)Grid pair matching between T_i, T_j

Fig. 5. Example of SDGW method (Color figure online)

Because of $sd(v_p^i, G^r(v_q^j)) \leq sd(v_p^i, v_q^j)$, so $SDGW(T_i, T_j) \leq DTW(T_i, T_j)$, based on Eq. 2, the similarity computed by SDGW method is the upper bound of $S_{sim}(T_i, T_j)$.

Figure 5a depicts the detail of SDGW, when reaching the last stage, we get $l(m, n) = 13$. If backtracking from last stage, we can obtain the optimal solution. For instance, let focus on $(v_3^i, G^r(v_2^j))$, the number in the center of matrix grid is the shortest road distance, and the number in the upper right corner means the cumulative distance from $(0, 0)$ to $(v_3^i, G^r(v_2^j))$, i.e., 5. Blue grids represent the warping path, and red arrows indicate every stage that we find the sum of the shortest road network distance.

Figure 5b shows the matching process between T_i and T_j. SDGW avoids computing spatial distance point to point, the matching process is transformed to point to signature without synchronism, and is more efficient similarity measurement.

According to Definitions 2 and 3, the computation cost of $T_{sim}(T_i, T_j)$ is much cheaper and easier than that of $S_{sim}(T_i, T_j)$, thus, we compute it first. Suppose $T_{sim}(T_i, T_j)$ is obtained, we propose the following lemmas to prune the dissimilar trajectory pairs.

Lemma 3. *When* $T_{sim}(T_i, T_j)$ *is computed, based on* $sd(v_p^i, G^r(v_q^j))$ *and* $sd(v_q^j, G^r(v_p^i))$ *obtained at each step of SDGW, if* $\dfrac{\sum_p e^{-sd(v_p^i, G^r(v_q^j))}}{|T_i|} + \dfrac{\sum_q e^{-sd(v_q^j, G^r(v_p^i))}}{|T_j|} <$ $\dfrac{\theta - (1-\lambda) \cdot T_{sim}(T_i, T_j)}{\lambda}$, *this trajectory pair cannot be similar, we prune it.*

Proof. If T_i and T_j are similar, we have $S_{sim}(T_i, T_j) \geq \frac{\theta - (1-\lambda) \cdot T_{sim}(T_i, T_j)}{\lambda}$.
Because of $sd(v_p^i, G^r(v_q^j)) \leq sd(v_p^i, v_q^j) = sd_{i \to j}^p$, $\frac{\sum_p e^{-sd(v_p^i, G^r(v_q^j))}}{|T_i|} +$
$\frac{\sum_q e^{-sd(v_q^j, G^r(v_p^i))}}{|T_j|} \geq \frac{\sum_p e^{-sd_{i \to j}^p}}{|T_i|} + \frac{\sum_q e^{-sd_{j \to i}^q}}{|T_j|}$. So, if $\frac{\sum_p e^{-sd(v_p^i, G^r(v_q^j))}}{|T_i|} +$
$\frac{\sum_q e^{-sd(v_q^j, G^r(v_p^i))}}{|T_j|} < \frac{\theta - (1-\lambda) \cdot T_{sim}(T_i, T_j)}{\lambda}$, T_i and T_j cannot be similar. Lemma is
proved.

Lemma 4. *When* $T_{sim}(T_i, T_j)$ *is evaluated, if* $e^{-\max_{v_p^i \in T_i, v_q^j \in T_j} \{sd_{j \to i}^q\}} +$
$\frac{\sum_{v_p^i \in T_i, v_q^j \in T_j} e^{-sd_{j \to i}^q}}{|T_j|} \geq \frac{\theta - (1-\lambda) T_{sim}(T_i, T_j)}{\lambda}$, T_i *and* T_j *are similar.*

Proof. Assume that $sd_{i \to j}^p = Sdis(v_p^i, v_q^j)$, where v_q^j is the sample point spatially
closest to v_p^i among all sample points in T_j. According to Eq. 1, for the sample
point v_q^j, we have that

$$sd_{i \to j}^p \leq sd_{j \to i}^q \leq \max_{v_p^i \in T_i, v_q^j \in T_j} \{sd_{j \to i}^q\}$$

By substituting it into Eq. 2, we estimate the following equation

$$\sum_{v_p^i \in T_i, v_q^j \in T_j} e^{-sd_{i \to j}^p} \geq |T_i| \cdot e^{-\max_{v_p^i \in T_i, v_q^j \in T_j} \{sd_{j \to i}^q\}} \Rightarrow$$

$$S_{sim}(T_i, T_j) \geq e^{-\max_{v_p^i \in T_i, v_q^j \in T_j} \{sd_{j \to i}^q\}} + \frac{\sum_{v_p^i \in T_i, v_q^j \in T_j} e^{-sd_{j \to i}^q}}{|T_j|}$$

(13)

Thus, if $e^{-\max_{v_p^i \in T_i, v_q^j \in T_j} \{sd_{j \to i}^q\}} + \frac{\sum_{v_p^i \in T_i, v_q^j \in T_j} e^{-sd_{j \to i}^q}}{|T_j|} \geq \frac{\theta - (1-\lambda) T_{sim}(T_i, T_j)}{\lambda}$, we
have $S_{sim}(T_i, T_j) \geq \theta$, T_i and T_j are similar. Lemma 4 is proved.

4.4   Trajectory Similarity Search Algorithm

The searching process of trajectory similarity join is shown in Algorithm 2, which
is composed of building (Lines 2 to 4) and refinement (Lines 5, 9, 14).

In order to filter out all the unqualified trajectories, we should first generate
the signature sets for all sample points and trajectories (Lines 3 to 4). After
builded these sets, Lemma 2 can be employed to prune the dissimilar trajectory pairs (Line 6). Then, SDGW method is adopted to compute their spatial
similarity. When $T_{sim}(T_i, T_j)$ is obtained, Lemma 3 is used to further prune
the dissimilar pairs. Next, Lemma 4 can be employed to find a part of similar pairs (Line 11). At last, for the rest candidate trajectory pairs, we compute
$ST_{sim}(T_i, T_j)$ for each of them directly, if its similarity is larger than θ, we add
it into result set A (Line 15).

5   Experiments

5.1   Experimental Settings

In this section, we evaluate the performance of our methods. All of the algorithms are implemented in C++. And the experiments are run on a PC with 3.4 GHz Intel Core I7-6700, 16 GB RAM memory. We use real spatial network, namely New York Road Network (NRN)[2]. It contains more than 95,500 vertices and 260,800 edges. The graphs are stored using adjacency lists. In NRN, we use real taxi trajectory data set from New York (see footnote 2). The trajectories denote the taxi trips, and their average length (number of vertices) is 80. Because every sample point has a timestamp in our setting, in order to construct time period for every sample point, we randomly add or subtract its time stamp within $[0, 15]$ minutes. In experiments, we mainly examine our proposed techniques in filtering and refinement steps. At first, we evaluate candidate set generation techniques in filtering step. Our method is named **Two-level**. Two classic methods are reproduced as our control groups. Specifically, we name TF-matching [17] as **TF-Match**, this method searches the trajectory join with every sample point having a timestamp, we extend it to our time setting. The other one employing

---

Algorithm 2. Trajectory Similarity Search

Input: Two trajectory sets T_i and T_j, similarity threshold θ, maximal
trajectory distance rd, parameter λ
Output: $A = \{(T_i, T_j) \mid ST_{sim}(T_i, T_j) \geq \theta\}$

1 $A \leftarrow \emptyset$;
2 **for** each (T_i, T_j) in Candidate **do**
3   build $G^r(v_n^i)$ and $G^r(v_n^j)$ for all v_n^i and v_n^j on T_i and T_j respectively;
4   build signature set $G^t(T_j)$ for T_j;
5   **if** $G^r(v_n^i) \cap G^t(T_j) = \emptyset$ **then**
6   prune it based on Lemma 2
7   compute $SDGW(T_i, T_j)$ and $SDGW(T_j, T_i)$ by SDGW method;
8   compute temporal similarity $T_{sim}(T_i, T_j)$ by Eq.4
9   **while** $\dfrac{\sum_k e^{-sd(v_k^i, G^r(v_k^j))}}{|T_i|} + \dfrac{\sum_k e^{-sd(v_k^j, G^r(v_k^i))}}{|T_j|} \geq \dfrac{\theta - (1-\lambda) \cdot T_{sim}(T_i, T_j)}{\lambda}$ **do**
10   **if** $e^{-\max_{v_k^i \in T_i, v_k^j \in T_j} \{sd_{j \to i}^k\}} + \dfrac{\sum_k e^{-sd_{j \to i}^k}}{|T_j|} \geq \dfrac{\theta - (1-\lambda) T_{sim}(T_i, T_j)}{\lambda}$ **then**
11              $A.\text{add}(T_i, T_j)$ based on Lemma 3 and Lemma 4;
12   **else**
13   compute $ST_{sim}(T_i, T_j)$ by Eq.5;
14   **if** $ST_{sim}(T_i, T_j) \geq \theta$ **then**
15                  $A.\text{add}(T_i, T_j)$;

16 **return** A;

---

[2] https://lab-work.github.io/data/.

Strain-Join in [12] is called **Strain**, we use this method to do spatial similarity search. We compare the number of searched trajectories and runtime by varying the four parameters: CR threshold θ, influence radius rd, grid width w and the number of candidate trajectories $|P|$. At second, we evaluate the methods in refinement step. The method proposed in Sect. 4.3 is called **SDGW method** and the approach based original DTW is named **DTW method**. We compare their memory cost by varying the four parameters: CR threshold θ, preference parameter λ, influence radius rd, grid width w. Due to space limitation, only parts of experiment results are listed in the following subsection.

5.2   Various Testing

As shown in Fig. 6a, as the threshold increases, the number of searched trajectories grows fewer. The reason is obvious, a lager threshold θ helps us to filter more dissimilarity trajectory pairs. In Fig. 6b, as rd increases, the number of searched trajectories is rising. This is mainly because: (i) a lager rd means more trajectories need to be computed and (ii) influence grid set in Eq. 6 depends on rd, with the increasing of rd, $G^r(v_n^i)$ covers more grids. As in Fig. 6c, the number of the searched trajectories are growing as w is increasing. A larger $|P|$ causes more trajectory pairs to be searched in Fig. 6d. Figure 7a–d show the runtime on NRN dataset when varying four parameters. It is easy to see that the more trajectories we search, the more time we spend. The variation trend on runtime keeps broadly consistent with the number of the searched trajectories in Fig. 6. However, as w and Rd increase, the runtime reduces. This is because larger w leads to fewer grids which saves processing time for constructing signature set, and larger Rd helps reduce the computation cost of spatial similarity.

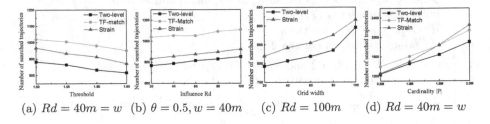

(a) $Rd = 40m = w$ (b) $\theta = 0.5, w = 40m$ (c) $Rd = 100m$ (d) $Rd = 40m = w$

Fig. 6. Evaluating filtering: number of searched trajectories on NRN

The results of evaluating refinement on memory cost are shown in Fig. 8. In each figure, DTW consumes more memory than SDGW. The reason is that, DTW employs the point-to-point method in dynamic programming process to measure the similarity for the trajectory pairs, the computation of the spatial distance between a sample point on T_i to the other trajectory T_j is expensive. However, SDGW computes $sd(v_p^i, G^r(v_q^j))$ instead of $sd(v_p^i, v_q^j)$ in dynamic programming process, its cost is much lower than that $sd(v_p^i, v_q^j)$. As a result, SDGW has better performance than DTW in this experiment.

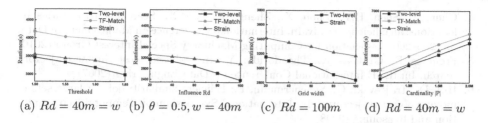

(a) $Rd = 40m = w$ (b) $\theta = 0.5, w = 40m$ (c) $Rd = 100m$ (d) $Rd = 40m = w$

Fig. 7. Evaluating filtering: runtime on NRN

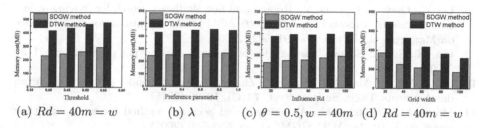

(a) $Rd = 40m = w$ (b) λ (c) $\theta = 0.5, w = 40m$ (d) $Rd = 40m = w$

Fig. 8. Evaluating refinement: memory cost

6 Conclusion

In this paper, we study a novel trajectory similarity join in road networks. To process the trajectory similarity efficiently, a novel index, searching algorithm and pruning method are developed. Experimental results show that our methods can gain good performance. Our future work will study how to extend our methods to various distributed environments.

Acknowledgments. This work is supported in part by Hubei Natural Science Foundation under Grant No. 2017CFB135, and the Fundamental Research Funds for the Central Universities under Grants No. CCNU18QN017, CZZ17003, and Teaching Research Projects NO. JYX17032, and NSFC Grant No. 61309002.

References

1. Assent, I., Wichterich, M., Krieger, R., Kremer, H., Seidl, T.: Anticipatory DTW for efficient similarity search in time series databases. Proc. VLDB **2**(1), 826–837 (2009)
2. Bakalov, P., Hadjieleftheriou, M., Keogh, E.J., Tsotras, V.J.: Efficient trajectory joins using symbolic representation. In: International Conference on Mobile Data Management (2005)
3. Bakalov, P., Tsotras, V.J.: Continuous spatiotemporal trajectory joins. In: Nittel, S., Labrinidis, A., Stefanidis, A. (eds.) GSN 2006. LNCS, vol. 4540, pp. 109–128. Springer, Heidelberg (2008). https://doi.org/10.1007/978-3-540-79996-2_7

4. Chen, Z., Shen, H.T., Zhou, X., Zheng, Y., Xie, X.: Searching trajectories by locations: an efficiency study. In: International Conference on Management of Data, pp. 255–266. Association for Computing Machinery Special Interest Group (2010)
5. Frentzos, E., Gratsias, K., Theodoridis, Y.: Index-based most similar trajectory search. In: IEEE International Conference on Data Engineering (2007)
6. Hui, D., Trajcevski, G., Scheuermann, P.: Efficient similarity join of large sets of moving object trajectories. In: International Symposium on Temporal Representation and Reasoning (2008)
7. Lei, C., Ng, R.: On the marriage of Lp-norms and edit distance. In: Proceedings of the VLDB, pp. 792–803 (2004)
8. Lei, C., Ozsu, M.T., Oria, V.: Robust and fast similarity search for moving object trajectories. In: Proceedings of the 2005 ACM SIGMOD International Conference on Management of Data, pp. 491–502 (2005)
9. Lin, B., Su, J.: Shapes based trajectory queries for moving objects, pp. 21–30 (2005)
10. Lu, J., Wu, D., Mao, M., Wang, W., Zhang, G.: Recommender system application developments. Decis. Support Syst. **74**(C), 12–32 (2015)
11. Morse, M.D., Patel, J.M.: An efficient and accurate method for evaluating time series similarity. In: ACM SIGMOD, pp. 569–580 (2007)
12. Na, T., et al.: Signature-based trajectory similarity join. IEEE Trans. Knowl. Data Eng. **29**(4), 870–883 (2017)
13. Papadias, D., Zhang, J., Mamoulis, N., Tao, Y.: Query processing in spatial network databases. In: Proceedings of the VLDB, vol. 29, pp. 802–813 (2003)
14. Ranu, S., Deepak, P., Telang, A.D., Deshpande, P., Raghavan, S.: Indexing and matching trajectories under inconsistent sampling rates. In: IEEE ICDE, pp. 999–1010 (2015)
15. Sakurai, Y., Yoshikawa, M., Faloutsos, C.: FTW: fast similarity search under the time warping distance. In: Twenty-Fourth ACM SIGMOD-SIGACT-SIGART Symposium on Principles of Database Systems (2005)
16. Sankararaman, S., Agarwal, P.K., Mølhave, T., Pan, J., Boedihardjo, A.P.: Model-driven matching and segmentation of trajectories. In: Proceedings of the 21st ACM SIGSPATIAL International Conference on Advances in Geographic Information Systems, pp. 234–243 (2013)
17. Shang, S., Chen, L., Wei, Z., Jensen, C.S., Zheng, K., Kalnis, P.: Trajectory similarity join in spatial networks. Proc. VLDB **10**(11), 1178–1189 (2017)
18. Shang, S., Ding, R., Yuan, B., Xie, K., Zheng, K., Kalnis, P.: User oriented trajectory search for trip recommendation. In: EDBT, pp. 156–167 (2012)
19. Shang, S., Ding, R., Zheng, K., Jensen, C.S., Kalnis, P., Zhou, X.: Personalized trajectory matching in spatial networks. VLDB J. **23**(3), 449–468 (2014)
20. Vaid, S., Jones, C.B., Joho, H., Sanderson, M.: Spatio-textual indexing for geographical search on the web. In: Bauzer Medeiros, C., Egenhofer, M.J., Bertino, E. (eds.) SSTD 2005. LNCS, vol. 3633, pp. 218–235. Springer, Heidelberg (2005). https://doi.org/10.1007/11535331_13
21. Vlachos, M., Gunopoulos, D., Kollios, G.: Discovering similar multidimensional trajectories. In: IEEE ICDE, pp. 673–684 (2002)
22. Wang, X., Mueen, A., Ding, H., Trajcevski, G., Scheuermann, P., Keogh, E.: Experimental comparison of representation methods and distance measures for time series data. Data Min. Knowl. Disc. **26**(2), 275–309 (2013)

23. Yi, B.K., Jagadish, H.V., Faloutsos, C.: Efficient retrieval of similar time sequences under time warping. In: IEEE ICDE, pp. 201–208 (1998)
24. Yun, C., Patel, J.M.: Design and evaluation of trajectory join algorithms. In: ACM SIGSPATIAL International Conference on Advances in Geographic Information Systems (2009)

Knowledge Discovery

Multiviewpoint-Based Agglomerative Hierarchical Clustering

Yuji Fujiwara and Hisashi Koga[✉]

University of Electro-Communications, Tokyo 182-8585, Japan
koga@sd.is.uec.ac.jp

Abstract. The cosine similarity is a similarity measure useful for document clustering. The cosine similarity between two points is determined by the angle between their corresponding vectors observed from the single reference viewpoint, the origin. Recently, Nguyen *et al.* [6] proposed a new similarity measure called MVS (MultiViewpoint-based Similarity) in which the vectors are observed from multiple viewpoints. They incorporated MVS into some non-hierarchical clustering algorithm and showed that MVS outperforms the original cosine similarity. This paper proposes an agglomerative hierarchical clustering which couples the average-link method with MVS. Despite MVS is more complex than the cosine similarity, our clustering algorithm achieves the same time complexity as the average-link method with the cosine similarity by computing the inter-cluster similarity smartly. Interestingly, our algorithm can be expanded to control the size fairness among clusters. Experimentally in document clustering, our algorithm outputs more accurate clustering results than the average-link method with the cosine similarity almost without lengthening the running time.

Keywords: Hierarchical clustering · Multiview · Similarity measure · Time complexity

1 Introduction

Clustering is a powerful tool for unsupervised data analysis; it provides an insight into the characteristics of the given data by classifying them into several groups. Because of its usefulness, clustering has been used in various application areas such as biology, multimedia analysis and natural language processing. The primary principle of clustering is to make similar data belong to the same cluster. Therefore, the similarity measure between two data is significant for clustering. The cosine similarity is a representative similarity measure for points in a multi-dimensional space and judges the similarity between two points d_i and d_j from the angle between their associated vectors starting from the origin. The cosine similarity is suitable for classifying high-dimensional sparse vectors which emerge in document analysis, market basket analysis, and so on.

Recently, Nguyen *et al.* [6] noticed that the cosine similarity observes all the vectors from the single reference viewpoint, i.e., the origin and proposed a

© Springer Nature Switzerland AG 2019
S. Hartmann et al. (Eds.): DEXA 2019, LNCS 11707, pp. 325–340, 2019.
https://doi.org/10.1007/978-3-030-27618-8_24

new similarity measure using multiple viewpoints chosen from the dataset to be analyzed. They named this similarity as the MultiViewpoint-based Similarity (MVS). Then, they designed a non-hierarchical clustering algorithm exploiting MVS which resembles the k-means and showed that their clustering algorithm outputs more adequate clustering results than the spherical k-means [2] which is a variant of the k-means based on the cosine similarity. Hereafter, we abbreviate the cosine similarity as CS.

Conventionally, clustering methods are categorized into two types: hierarchical and non-hierarchical clustering. The agglomerative hierarchical clustering has an advantage over most non-hierarchical clustering algorithms that the number of clusters need not be specified in advance. On the other hand, its computational complexity tends to be large.

Our main contribution is to incorporate MVS into the agglomerative hierarchical clustering. In particular, we propose an agglomerative hierarchical clustering algorithm which couples the average-link method [9] with MVS. Multiple viewpoints in MVS are expected to slow down the running speed of the average-link method. Nonetheless, our algorithm achieves the same time complexity as the average-link method with CS, supported by our novel means to compute the inter-cluster similarity efficiently. Remarkably, our clustering algorithm can be easily expanded, so that the size fairness among clusters may be controllable. In the task of document clustering, our algorithm outputs more accurate clustering results than the average-link method with CS almost without increasing the running time.

This paper is organized as follows. Section 2 describes (1) MVS proposed in [6] and (2) the average-link method both of which are components of our new algorithm. Then, Sect. 3 presents our algorithm. Section 4 reports the experimental evaluation. Section 5 reviews related works. Section 6 concludes this paper.

2 Preliminaries

2.1 MVS (MultiViewpoint-Based Similarity Measure)

This subsection explains the MVS proposed in [6]. Similarly to CS, MVS treats points on the surface of the unit hypersphere. Let d_i and d_j be such points. The cosine similarity $CS(d_i, d_j)$ between d_i and d_j equals their inner product $d_i^T d_j = (d_i - 0)^T(d_j - 0)$, where 0 is the origin. Thus, CS measures the similarity between d_i and d_j, while specifying the origin as the only reference viewpoint.

By contrast, MVS moves the reference viewpoint to various points in the dataset S to be analyzed, thereby obtaining a similarity value adaptive to S. [6] defines MVS as a similarity measure for two points in the identical cluster. Concretely, MVS between d_i and d_j belonging to the same cluster r is defined as the average of inner products between their associated vectors starting from multiple viewpoints in S outside r. See Eq. (1). In (1), d_h symbolizes the viewpoint and n denotes the number of points in S. Throughout this paper, for a cluster r, we denote the set of members in r by S_r and $n_r = |S_r|$.

$$\text{MVS}(d_i, d_j \mid d_i, d_j \in S_r) = \frac{1}{n - n_r} \sum_{d_h \in S \backslash S_r} (d_i - d_h)^{\mathrm{T}} (d_j - d_h). \tag{1}$$

MVS considers it reasonable that d_i and d_j belong to the same cluster r, if they look similar from the viewpoints outside r. In this way, MVS evaluates the validness to categorize two points in the dataset S into the same cluster based on their similarity value observed not from the origin which is independent of S, but from the viewpoints in S itself.

As $d_h \in S \backslash S_r$ is a unit vector, Eq. (1) is rewritten as Eq. (2), where $C_{S \backslash S_r}$ is the centroid of the points in $S \backslash S_r$.

$$\text{MVS}(d_i, d_j \mid d_i, d_j \in S_r) = d_i^{\mathrm{T}} d_j - d_i^{\mathrm{T}} C_{S \backslash S_r} - d_j^{\mathrm{T}} C_{S \backslash S_r} + 1. \tag{2}$$

From Eq. (2), it holds for two points d_j and d_l that $\text{MVS}(d_i, d_j) > \text{MVS}(d_i, d_l)$ iff $\text{CS}(d_i, d_j) - d_j^{\mathrm{T}} C_{S \backslash S_r} > \text{CS}(d_i, d_l) - d_l^{\mathrm{T}} C_{S \backslash S_r}$. Therefore, even if $\text{CS}(d_i, d_j) < \text{CS}(d_i, d_l)$, $\text{MVS}(d_i, d_j)$ can be greater than $\text{MVS}(d_i, d_l)$, if d_l is close to the outer centroid $C_{S \backslash S_r}$ enough to leave r. Thus, MVS is a similarity measure which adds to CS a new term which evaluates if the two points should group together.

Nguyen *et al.* [6] proposed a non-hierarchical clustering algorithm MVSC based on MVS. MVSC tries to minimize the objective function $F = \sum_{r=1}^{C} n_r \left[\frac{\sum_{d_i, d_j \in S_r} \text{MVS}(d_i, d_j)}{n_r^2} \right]$, where C is the number of clusters. In [6], MVSC outperformed the spherical k-means based on CS [2] in document clustering. Thus, MVS could improve the clustering quality by being adaptive to the dataset S.

On the other hand, MVS has a drawback that it is heavier to compute than CS. Since vector inner products must be computed $(n - n_r)$ times in (1), the time complexity of MVS is as n times large as that of CS in the worst case, if MVS is implemented straightforwardly.

2.2 Average-Link Method

The average-link method is a well-known agglomerative hierarchical clustering algorithm and widely used in many practical applications because of its robustness against outliers. The agglomerative hierarchical clustering initially regards every point in S as a single cluster. Thus, there exist n clusters in the beginning. Then, in the agglomeration step, it repeats merging the most similar cluster pair. This agglomeration step continues until only one cluster remains. Then, this history of cluster merging is outputted as the clustering result. From this history, we can extract the clustering results for any number of clusters.

In the average-link method, the similarity value Sim_{ab} between two clusters a and b is defined as the average similarity value between all the point pairs one of which comes from a and the other of which comes from b, as in Eq. (3).

$$\text{Sim}_{ab} = \frac{1}{n_a n_b} \sum_{d_i \in S_a} \sum_{d_j \in S_b} \text{sim}(d_i, d_j). \tag{3}$$

1: Initialize the similarity matrix A whose size equals $n \times n$
2: **while** the number of cluster > 1 **do**
3:   Find the most similar cluster pair a, b
4:   Merge a and b to form a new cluster c
5:   **for** cluster k other than c **do**
6:     Update the cluster similarity between c and k

Fig. 1. Average-link Method

Figure 1 outlines the average-link method. In the initialization step, each point forms a single cluster and the similarity values between all the points (i.e., clusters) are kept in the similarity matrix A whose size equals $n \times n$. In the agglomeration step, the most similar pair of clusters are found from the matrix A and merged into a new cluster. Suppose that the most similar pair of clusters a and b are merged into a new cluster c. After c is formed, we must update the matrix A to memorize the similarity value Sim_{kc} between c and every other cluster k which is currently registered to A. It is known that Sim_{kc} is represented by a weighted sum of Sim_{ka} and Sim_{kb} in Eq. (4). Since Sim_{ka} and Sim_{kb} have been already stored in A, Sim_{kc} is computable in $O(1)$ time.

$$\text{Sim}_{kc} = \frac{n_a}{n_c}\text{Sim}_{ka} + \frac{n_b}{n_c}\text{Sim}_{kb}. \tag{4}$$

Now, consider the time complexity of the average-link method in Fig. 1, when CS is used. Let m be the dimensionality of vectors. Since the initialization of A computes CS between all the point pairs and it takes $O(m)$ time to calculate an inner product between two m-dimensional vectors, the initialization of A finishes in $O(mn^2)$ time.

As for the agglomeration step, the most similar cluster pair is found in $O(n \log n)$ time at the third line, by using heap structures. Then it costs $O(n)$ time to update A about a new cluster c at the 5th and 6th lines, since the FOR statement there evaluates Eq. (4) less than n times. Because the average-link method performs the cluster merging just $n - 1$ times, its whole time complexity grows $O(mn^2 + (n \log n + n) \times (n - 1)) = O(mn^2 + n^2 \log n)$ together with the initialization of A.

3   Multiviewpoint-Based Hierarchical Clustering

MVS realized more accurate clustering than CS for non-hierarchical clustering [6]. It is natural to expect that MVS is superior to CS also for hierarchical clustering. Hence, this paper designs a clustering algorithm which introduces MVS to the average-link method. As will be explained later, our algorithm has the next two merits.

- Supported by MVS, our algorithm excels to the average-link with CS in terms of classification accuracy. In addition, despite MVS is more complex than CS,

our algorithm theoretically achieves the same time complexity as the average-link method with CS.
– Our clustering algorithm can be easily expanded, so that the size fairness among clusters may be controllable.

Hereafter, our average-link method with MVS is named as MVS-AVE, while the standard average-link method with CS is referred to as CS-AVE.

3.1 Similarity Measure Between Clusters

To apply MVS to the average-link method, it seems at a glance that we have only to replace the function sim(,) with MVS in the definition of inter-cluster similarity in Eq. (3) like Eq. (5) below.

$$\text{Sim}_{ab} = \frac{1}{n_a n_b} \sum_{d_i \in S_a} \sum_{d_j \in S_b} \text{MVS}(d_i, d_j). \tag{5}$$

However, Eq. (5) is not well-defined, because MVS is computed between d_i and d_j which come from different clusters. Recall that MVS in Eq. (1) is originally defined for two points belonging to the same cluster. To solve this problem, we consider a virtual big cluster which includes both the two clusters a and b and formalize $\text{MVS}(d_i, d_j | d_i \in S_a, d_j \in S_b)$ inside the big cluster. See Eq. (6). Note that $S_a \cup S_b$ expresses the members in the virtual cluster and its cardinality equals $n_a + n_b$.

$$\text{MVS}(d_i, d_j | d_i \in S_a, d_j \in S_b) = \frac{1}{n - n_a - n_b} \sum_{d_h \in S \setminus (S_a \cup S_b)} (d_i - d_h)^{\text{T}} (d_j - d_h). \tag{6}$$

3.2 Operations to Be Modified in Average-Link Method

To introduce MVS, operations in the average-link method must be corrected, if they compute the inter-cluster similarity. The average-link method contains two such operations: (I) the initialization of the similarity matrix A and (II) the update of A after merging the most similar cluster pair. First, let us review how large the time complexity of MVS-AVE grows if the above operations are naively implemented.

To initialize A, MVS must be computed between all the point pairs. Thus, MVS is to be computed $O(n^2)$ times. Here, a single MVS computation between two single-point clusters consumes $O(mn)$ time, since the inner product between m-dimensional vectors is calculated just $n - 2$ times in Eq. (6). After all, $O(n^2 \times mn) = O(mn^3)$ time is necessary to initialize A.

Second, consider the update of A after merging the most similar cluster pair into the new cluster c: Unfortunately, MVS-AVE cannot rely on the similarity update formula (4) any more, because Eq. (4) presumes that the similarity does not depend on the cluster membership, which is not the case for MVS. For example, if $d_i \in S_a$ and $d_j \in S_b$, $\text{MVS}(d_i, d_j) = \frac{1}{n - n_a - n_b} \sum_{d_h \in S \setminus (S_a \cup S_b)} (d_i -$

$d_h)^{\mathrm{T}}(d_j - d_h)$. However, if the cluster a is merged with another cluster c at this moment. Then, $\text{MVS}(d_i, d_j)$ changes to $\frac{1}{n-n_a-n_b-n_c}\sum_{d_h \in S \backslash (S_a \cup S_b \cup S_c)}(d_i - d_h)^{\mathrm{T}}(d_j - d_h)$. Thus, $\text{MVS}(d_i, d_j)$ depends on the members of the two clusters holding d_i and d_j.

Without Eq. (4), if we calculate Sim_{kc} according to Eq. (5), the time to compute Sim_{kc} increases up to $O(mn^3)$ in the worst case, since MVS whose time complexity equals $O(m(n - n_k - n_c))$ are to be computed $n_k \times n_c = O(n^2)$ times. Since MVS-AVE merges a pair of clusters just $n - 1$ times and Eq. (5) is evaluated at most n times at each cluster merging, it takes $O(mn^5)$ time for MVS-AVE to update A during the whole running period.

3.3 How to Shrink the Running Time

The previous subsection concluded that, if naively implemented, MVS-AVE runs by far slower than CS-AVE. This subsection explains our novel techniques to accelerate the initialization and the update of the similarity matrix A, which reduce the time complexity of MVS-AVE to $O(mn^2 + n^2 \log n)$ and make it comparable to CS-AVE.

3.3.1 Initialization of Similarity Matrix A

Here, we introduce a technique to initialize A in $O(mn^2)$ time like CS-AVE. When MVS-AVE initializes A, there are n single-point clusters. Then, based on Equations (5) and (6), the similarity between the i-th clusters with d_i and the j-th cluster with d_j is described as Eq. (7). Note that $|d_i| = |d_j| = |d_h| = 1$. Let $D = \sum_{d_i \in S} d_i$ symbolize the vector sum of all the data in S.

$$
\begin{aligned}
\text{Sim}_{ij} = \text{MVS}(d_i, d_j) &= \frac{1}{n-2} \sum_{d_h \in S \backslash (d_i \cup d_j)} (d_i^{\mathrm{T}} d_j - d_i^{\mathrm{T}} d_h - d_j^{\mathrm{T}} d_h + 1) \\
&= \frac{1}{n-2}\{(n-2)d_i^{\mathrm{T}} d_j - d_i^{\mathrm{T}}(D - d_i - d_j) - d_j^{\mathrm{T}}(D - d_i - d_j) + (n-2)\} \\
&= \frac{1}{n-2}(n d_i^{\mathrm{T}} d_j - d_i^{\mathrm{T}} D - d_j^{\mathrm{T}} D + n). \quad (7)
\end{aligned}
$$

If the value of D is known, Eq. (7) is obtained in $O(m)$ time by calculating the inner product between m-dimensional vectors three times. By processing all the elements in the $n \times n$ matrix A in this way, we can initialize A in $O(mn^2)$ time. Although it takes $O(mn)$ time to compute D, we may compute D only once at the very beginning. Thus, MVS-AVE can initialize the matrix A in $O(mn^2 + mn) = O(mn^2)$ time, which is the same time complexity as CS-AVE.

Further Reduction of Inner Product Computation: Though we have shown that both MVS-AVE and CS-AVE initialize A in $O(mn^2)$ time, strictly speaking, MVS-AVE runs as three times slowly as CS-AVE, because the inner product is computed three times in deriving $\text{MVS}(d_i, d_j)$ and once in deriving $\text{CS}(d_i, d_j)$. Because the dimensionality m is enormous in document clustering, the $O(mn^2)$ term often dominates the running time of the average-link method. Thus, it is problematic to neglect this constant-time gap.

Therefore, we devise a novel technique for MVS-AVE which initializes A by computing the inner product just $\binom{n}{2}$ times. Note that CS-AVE also computes the inner product just $\binom{n}{2}$ times to fulfill A whose size is $n \times n$. Therefore, MVS-AVE performs exactly the same number of m-dimensional inner product computations as CS-AVE. The main idea of our technique is to precalculate $d_i^T D$ and $d_j^T D$ in Eq. (7). Our technique executes the 3 next steps in order.

(Step 1): It first computes the cosine similarity between all the point pairs.

(Step 2): Then, for $1 \leq i \leq n$, $d_i^T D$ is calculated by summing up $\mathrm{CS}(d_i, d_j)$ for $1 \leq j \leq n$. That is, $d_i^T D = \sum_{j=1}^{n} \mathrm{CS}(d_i, d_j)$.

(Step 3): For any pair of i and j, $\mathrm{MVS}(d_i, d_j)$ is instantly obtained by subtracting $d_i^T D$ and $d_j^T D$ from $n\mathrm{CS}(d_i, d_j) + n$.

In (Step 2), because $\mathrm{CS}(d_i, d_j)$ is known for any j, $d_i^T D$ is computed in $O(n)$ time. Thus, (Step 2) finishes in $O(n \times n) = O(n^2)$ time which is by far less than $O(mn^2)$. (Step 3) also terminates in $O(n^2)$ time, because each $\mathrm{MVS}(d_i, d_j)$ is computed in $O(1)$ time.

The m-dimensional inner product is computed only in (Step 1) and never computed in (Step 2) nor (Step 3). Since (Step 1) simply computes the cosine similarity between all the point pairs, the inner product is computed just $\binom{n}{2}$ times. Thus, MVS-AVE performs exactly the same number of inner product computations as CS-AVE. The whole time complexity for MVS-AVE to initialize A remains $O(mn^2 + n^2) = O(mn^2)$.

3.3.2   Update of Similarity Matrix A after Merging Clusters

From now on, we first present a new similarity update formula specific to MVS-AVE and then prove that the new formula be efficiently computed.

Consider the situation in which the two clusters a and b are merged into a new cluster c in MVS-AVE's running. Theorem 1 below states that Sim_{kc} between the clusters k and c can be described with Sim_{ka} and Sim_{kb}. Here, D_a and n_a denote the vector sum and the number of vectors in the cluster a.

Theorem 1. *Sim_{kc} is described with using Sim_{ka} and Sim_{kb} as Eq. (8).*

$$\frac{1}{(n_a + n_b)(n - n_k - n_a - n_b)}$$
$$\left\{ n_a(n - n_k - n_a)\mathrm{Sim}_{ka} + n_b(n - n_k - n_b)\mathrm{Sim}_{kb} + 2\left(D_a^T D_b - n_a n_b\right) \right\}. \quad (8)$$

Proof. Due to space limitations, we outline the proof here. Because $\mathrm{MVS}(d_i, d_j | d_i \in S_k, d_j \in S_c) = \frac{1}{n - n_k - n_c} \sum_{d_h \in S \setminus (S_k \cup S_c)} (d_i - d_h)^T(d_j - d_h)$, we have

$$\mathrm{Sim}_{kc} = \frac{1}{n_k n_c(n - n_k - n_c)} \sum_{d_i \in S_k} \sum_{d_j \in S_c} \sum_{d_h \in S \setminus (S_k \cup S_c)} (d_i - d_h)^T(d_j - d_h). \quad (9)$$

In Eq. (9), $\sum_{d_i \in S_k} \sum_{d_j \in S_c} \sum_{d_h \in S \setminus (S_k \cup S_c)} (d_i - d_h)^\mathrm{T} (d_j - d_h)$ is equivalent with

$$\sum_{d_i \in S_k} \left\{ \sum_{d_j \in S_a} \sum_{d_h \in S \setminus (S_k \cup S_a \cup S_b)} (d_i - d_h)^\mathrm{T} (d_j - d_h) \right.$$
$$\left. + \sum_{d_j \in S_b} \sum_{d_h \in S \setminus (S_k \cup S_a \cup S_b)} (d_i - d_h)^\mathrm{T} (d_j - d_h) \right\}. \tag{10}$$

By rewriting the first term of Eq. (10) as $\sum_{d_i \in S_k} \sum_{d_j \in S_a} \sum_{d_h \in S \setminus (S_k \cup S_a)} (d_i - d_h)^\mathrm{T} (d_j - d_h) - \sum_{d_i \in S_k} \sum_{d_j \in S_a} \sum_{d_h \in S_b} (d_i - d_h)^\mathrm{T} (d_j - d_h)$, it can be represented as a term including Sim_{ka} along with the residual. In the same way, the second term of Eq. (10) is represented as a term including Sim_{kb} combined with the residual. These two residuals correspond to $2 \left(D_a^\mathrm{T} D_b - n_a n_b \right)$ in Eq. (8).   □

Given Eq. (8), one may think that MVS-AVE can be easily realized by memorizing the vector sum D_a for each cluster a. Although this approach is surely feasible, it will make MVS-AVE slower than CS-AVE, because it takes $O(m)$ time to evaluate the update formula Eq. (8) owing to the inner product $D_a^\mathrm{T} D_b$. This is in contrast to the fact that CS-AVE computes the update formula Eq. (4) in $O(1)$ time. Since the agglomerative hierarchical clustering evaluates the similarity update formula $O(n^2)$ times during the agglomeration phase, MVS-AVE consumes $O(mn^2)$ time to update A in total, whereas CS-AVE consumes $O(n^2)$ time. Under the condition that both MVS-AVE and CS-AVE initialize A in $O(mn^2)$ time, if MVS-AVE only spends $O(mn^2)$ time to update A, the running time of MVS-AVE will become substantially longer than that of CS-AVE.

To escape from the above undesired situation, we develop a more refined method to update A in $O(n^2)$ time rather than $O(mn^2)$ time. This refined method maintains an $n \times n$ matrix B such that B_{ij} memorizes the inner product value between the vector sum of the i-th cluster and that of the j-th cluster. The matrix B is initialized at no cost because our MVS-AVE has already computed $d_i^\mathrm{T} d_j$ for $1 \le i, j \le n, i \ne j$ in the middle of initializing A.

In the agglomeration step, if a new cluster c is formed by merging two clusters a and b, we must update B by computing $D_c^\mathrm{T} D_k$ for every remaining cluster $k \ne c$. We would emphasize that $D_c^\mathrm{T} D_k$ is notably computed in $O(1)$ time by summing up $D_a^\mathrm{T} D_k$ and $D_b^\mathrm{T} D_k$, because $D_c^\mathrm{T} D_k = \sum_{d_i \in S_c} \sum_{d_j \in S_k} d_i^\mathrm{T} d_j = \sum_{d_i \in (S_a \cup S_b)} \sum_{d_j \in S_k} d_i^\mathrm{T} d_j = \sum_{d_i \in S_a} \sum_{d_j \in S_k} d_i^\mathrm{T} d_j + \sum_{d_i \in S_b} \sum_{d_j \in S_k} d_i^\mathrm{T} d_j = D_a^\mathrm{T} D_k + D_b^\mathrm{T} D_k$. $D_a^\mathrm{T} D_k$ and $D_b^\mathrm{T} D_k$ were saved in B before this cluster merging and taken out in $O(1)$ time. In the same way as A, the matrix B is updated at most n^2 time in MVS-AVE's running. Thus, the total time necessary to maintain B grows $O(n^2)$.

Now that the update formula Eq. (8) is computed in $O(1)$ time by looking up $D_a^\mathrm{T} D_b$ in B, the total time to update A does not go beyond $O(n^2)$.

3.4   Time Complexity of MVS-AVE

Here, let us confirm our MVS-AVE completes in $O(mn^2 + n^2 \log n)$ time in the same way as CS-AVE. First, MVS-AVE initializes the matrix A in $O(mn^2 + n^2)$ time, as explained in Sect. 3.3.1. Next, as for the cluster merging,

- The most similar cluster pair is sought from A in $O(n \log n)$ time per cluster merging, which does not differ from CS-AVE at all. Because MVS-AVE iterates the cluster merging $n - 1$ times, the total time incurred in the agglomeration phase gets $O((n - 1) \times n \log n) = O(n^2 \log n)$.
- The two matrices A and B are both updated in $O(n^2)$ time during the agglomeration phase, as discussed in Sect. 3.3.2.

As a whole, MVS-AVE achieves the time complexity of $O(mn^2 + n^2 + n^2 \log n + n^2) = O(mn^2 + n^2 \log n)$.

3.5   Balancing the Cluster Size

In general, a clustering result is not favored by practical data analysts, if it contains too large clusters and too small clusters simultaneously. The analysts implicitly expect that generated clusters are similar in size. In fact, some clustering algorithms use mechanisms to prevent too small clusters. For example, the famous Normalized Cuts [8] adjusted the definition of graphcut size not to form isolated single-point clusters. The conventional hierarchical agglomerative clustering does not provide any means to prevent too small clusters.

Uniquely and interestingly, our MVS-AVE can be expanded to balance the cluster size. What has to be done is to subtract a constant λ from all the elements in A, when A is initialized, as shown in Eq. (11). Here, λ is a positive real parameter specified by the user. As λ increases, the cluster size is more balanced.

$$\text{Sim}_{ij} = \text{MVS}(d_i, d_j) - \lambda. \tag{11}$$

Why the above mechanism balances the cluster size is explained in the next way. In the update formula specific to MVS-AVE in Eq. (8), the coefficients for Sim_{ka} and Sim_{kb} are $\frac{n_a(n - n_k - n_a)}{(n_a + n_b)(n - n_k - n_a - n_b)}$ and $\frac{n_b(n - n_k - n_b)}{(n_a + n_b)(n - n_k - n_a - n_b)}$ respectively. The point is that the sum of these two coefficients becomes

$$\frac{(n_a + n_b)(n - n_k - n_a - n_b) + 2n_a n_b}{(n_a + n_b)(n - n_k - n_a - n_b)}$$

and strictly greater than 1. Therefore, every time Eq. (8) is called, the term of $-\lambda$ is amplified. As the result, the similarity values in A have a tendency to become lower when they are related to large clusters which have experienced cluster merging many times than to small clusters which have undergone cluster merging few times. Thus, small clusters have more chances to be chosen as the objective of cluster merging than large clusters. In this way, a balanced cluster partition is created.

We remark that the conventional average-link method does not possess this interesting property, because, in the similarity update formula Eq. (4), the sum of coefficients of Sim_{ka} and Sim_{kb} equals $\frac{n_a}{n_c} + \frac{n_b}{n_c} = 1$. Therefore, if we subtract λ from all the elements in A, Eq. (4) does not amplify the term of $-\lambda$.

4   Experimental Evaluation

This section compares our MVS-AVE with CS-AVE in the task of document clustering and shows that our MVS-AVE achieves a higher classification accuracy than CS-AVE almost without increasing the running time.

The experimental platform is a PC (CPU: Intel Core i7 3.6 GHz, OS: Ubuntu 16.04, memory 16 GB). The program codes are compiled with g++ accompanied by the O2 optimization option. In the program codes, we optimize the module to compute vector inner product for high-dimensional sparse vectors having many 0 coordinate values, since the document clustering needs to handle them: Let $x = (x_1, x_2, \cdots, x_m)$ and $y = (y_1, y_2, \cdots, y_m)$ be two m-dimensional vectors. Then, $x^T y = \sum_{i=1}^{m} x_i y_i$. Here, if $x_i = 0$ or $y_i = 0$, $x_i y_i$ trivially does not influence $x^T y$. Therefore, our module accelerates the inner product computation for sparse high-dimensional vectors by skipping the multiplication $x_i y_i$ and the addition of $x_i y_i$ to $\sum_{j=1}^{i-1} x_j y_j$, if either $x_i = 0$ or $y_i = 0$.

4.1   Dataset

Our experiments use the 18 datasets in Table 1 all of which are publicly available at the web site for the CLUTO clustering toolkit [4]. They have been utilized in many previous researches including [6] from which MVS originates.

In Table 1, the symbols c, n, and m represent the class number, the number of objects and the dimensionality of feature vectors in order. m is determined from the word vocabulary, after removing the stop words and discarding too frequent words which appear in more than 99.5% of the documents and too rare words which appear in only one document. After m is fixed, a document is transformed to a feature vector each of whose coordinate values expresses the TF-IDF value of the word responsible for the single dimension. In the last, all the feature vectors are normalized to unit vectors.

4.2   Comparison with CS-AVE

Now, let us compare MVS-AVE with CS-AVE in terms of classification accuracy and running time: The classification accuracy for a clustering algorithm is measured by how consistent the clustering result is with the data partition induced from the ground-truth class labels assigned to the dataset, under the setting that the cluster number equals the the class number c. The consistency degree is measured with NMI (Normalized Mutual Information) defined as $\text{NMI}(X, Y) = \frac{I(X,Y)}{\sqrt{H(X)H(Y)}}$, where X corresponds to the clustering result and Y

Table 1. Dataset for experiments

Dataset	Source	c	n	m
fbis	TREC	17	2,463	2,000
hitech	TREC	6	2,301	13,170
k1a	WebACE	20	2,340	13,859
k1b	WebACE	6	2,340	13,859
la1	TREC	6	3,204	17,273
la2	TREC	6	3,075	15,211
re0	Reuters	13	1,504	2,886
re1	Reuters	25	1,657	3,758
tr31	TREC	7	927	10,127
reviews	TREC	5	4,069	23,220
wap	WebACE	20	1,560	8,440
la12	TREC	6	6,279	21,604
new3	TREC	44	9,558	36,306
sports	TREC	7	8,580	18,324
tr11	TREC	9	414	6,424
tr12	TREC	8	313	5,799
tr23	TREC	6	204	5,831
tr45	TREC	10	690	8,260

corresponds to the ground-truth data partition. $H(X) = -\sum_i P(x_i) \log P(x_i)$ is an entropy of the variable X and $I(X,Y) = \sum_i \sum_j P(x_i, y_j) \log \frac{P(x_i, y_j)}{P(x_i)p(y_j)}$ is the mutual information between X and Y. NMI takes a value in the range from 0 to 1. Intuitively, the NMI reflects the correlation of the clustering result X to the distribution Y of the ground-truth class labels.

The left half of Table 2 summarizes the classification accuracy for MVS-AVE, CS-AVE and BMVS-AVE which will be discussed later in Sect. 4.4 for various datasets. MVS-AVE defeats CS-AVE for 13 out of the 18 datasets. Moreover, the gap of NMI tends to be larger when MVS-AVE exceeds CS-AVE than when CS-AVE defeats MVS-AVE. The average NMI value becomes 0.456 for MVS-AVE and 0.409 for CS-AVE.

Next, the right half of Table 2 presents the running time of MVS-AVE and that of CS-AVE with their ratio, where the running time is the average over three trials. Remarkably, MVS-AVE augments the running time by at most 2.3% as compared with CS-AVE for any dataset.

Thus, MVS-AVE successfully incorporates the high preciseness of MVS into the average-link method with very little overhead.

4.3   Effect of Precalculating $d_i^T D$ and $d_j^T D$

Section 3.3.1 developed a technique to speed up the initialization of the matrix A by precalculating $d_i^T D$ and $d_j^T D$ and reducing the frequency of vector inner product computations to $\frac{1}{3}$. We investigate if this technique helps to shrink the running time by comparing the next two methods to initialize A:

Table 2. Classification accuracy and running time: The bold figures show which of MVS-AVE or CS-AVE is better with regard to classification accuracy.

Dataset	Accuracy			Running time (sec)		
	MVS-AVE	CS-AVE	BMVS-AVE	MVS-AVE	CS-AVE	Ratio ($\frac{\text{MVS-AVE}}{\text{CS-AVE}}$)
fbis	**0.570**	0.561	0.584	21.25	21.13	1.006
hitech	**0.250**	0.059	0.255	17.03	17.02	1.001
k1a	**0.556**	0.550	0.556	21.90	21.74	1.007
k1b	**0.710**	0.666	0.710	17.65	17.52	1.007
la1	**0.383**	0.316	0.376	42.96	42.27	1.016
la2	0.374	**0.390**	0.466	37.14	36.90	1.007
re0	**0.312**	0.296	0.312	4.427	4.394	1.008
re1	0.540	**0.568**	0.540	6.165	6.078	1.014
tr31	**0.670**	0.527	0.670	1.493	1.470	1.016
reviews	**0.410**	0.034	0.420	85.97	85.10	1.010
wap	**0.554**	0.539	0.554	5.676	5.661	1.003
la12	0.367	**0.381**	0.426	291.4	291.3	1.000
new3	**0.535**	0.487	0.535	984.8	979.1	1.006
sports	**0.242**	0.106	0.238	692.8	679.8	1.019
tr11	**0.645**	0.633	0.645	0.180	0.176	1.023
tr12	0.472	**0.523**	0.553	0.096	0.094	1.021
tr23	**0.252**	0.223	0.260	0.045	0.044	1.023
tr45	0.363	**0.495**	0.558	0.663	0.649	1.022
Ave.	**0.456**	0.409	0.481			

- Method 1 which precalculates D, i.e., the vector sum of all the points. Method 1 computes the three inner products $d_i^T d_j$, $d_i^T D$ and $d_j^T D$ to obtain $\text{Sim}_{ij} = \text{MVS}(d_i, d_j)$ in Eq. (7).
- Method 2 which precalculates $d_i^T D$ and $d_j^T D$ after computing the cosine similarity between all the point pairs. Method 2 has only to compute the one inner product $d_i^T d_j$ to obtain $\text{MVS}(d_i, d_j)$.

Table 3 shows the running time necessary for the two methods to initialize A for every dataset. Though we imagined that Method 1 is three times slower than Method 2 due to the frequency of inner product computations, Method 1 is much slower than we expected. For example, for the new3 dataset, Method 1 spent about 40 times longer time than Method 2.

We consider that this result is caused by the optimization to compute the inner product for high-dimensional sparse vectors. Although each d_i is sparse,

$D = \sum_i d_i$ is not sparse any more. Therefore, the optimization technique works effectively in computing $d_i^T d_j$ but not in computing $d_i^T D$ or $d_j^T D$. Thus, it benefits Method 2 much more than Method 1. Obviously, MVS-AVE cannot run as fast as CS-AVE with Method 1. Therefore, precalculating $d_i^T D$ and $d_j^T D$ absolutely helps MVS-AVE to achieve the running speed comparable to CS-AVE.

Table 3. Time necessary to initialize A

Dataset	Running time (sec)		Dataset	Running time (sec)	
	Method 1	Method 2		Method 1	Method 2
fbis	21.2	3.329	reviews	387.2	12.91
hitech	72.61	3.025	wap	22.64	1.180
k1a	77.77	2.750	la12	846.2	24.17
k1b	57.18	2.753	new3	3214	81.59
la1	69.19	5.511	sports	1328	42.87
la2	152.4	5.01	tr11	1.531	0.129
re1	11.33	0.576	tr23	0.376	0.036
tr31	10.60	0.688	tr45	5.132	0.397

4.4 Evaluation of Function to Balance Cluster Size

In the last, we evaluate the function to balance the cluster size in Sect. 3.5. MVS-AVE exploiting this function is referred to as BMVS-AVE (Balanced MVS-AVE). We examine if BMVS-AVE improves the size fairness among clusters against the standard MVS-AVE.

Given a clustering result, we judge the balance of the cluster size from the fairness index $\frac{n^2}{C \sum_{i=1}^{C} n_i^2}$. Here, C is the number of clusters and n_i is the number of points in the i-th cluster ($1 \leq i \leq C$). The fairness index takes a value in the range $[0,1]$ and becomes the maximum value of 1, if the C clusters evenly hold $\frac{n}{C}$ points.

Table 4 shows the fairness index for BMVS-AVE and MVS-AVE. In this experiment, we change the parameter λ in BMVS-AVE in the range from $\frac{2}{n}$ to $\frac{10}{n}$. When $\lambda = \frac{2}{n}$, as compared with MVS-AVE, the fairness index of BMVS-AVE becomes higher for 9 datasets and lower for only one dataset "reviews". For the rest of the datasets, BMVS-AVE outputs the same clustering result as MVS-AVE. As λ increases up to $\frac{10}{n}$, we observed that the fairness index also has a tendency to increase, though it does not augment monotonically. When $\lambda = \frac{10}{n}$, BMVS-AVE attains a higher fairness index than MVS-AVE for 15 datasets. These results show that our function to balance the cluster size works well.

As a side effect, for $\lambda = \frac{2}{n}$, BMVS-AVE yielded an average classification accuracy of 0.481 which is higher than MVS-AVE and outperforms CS-AVE for 17 out of the 18 datasets as summarized in Table 2.

5   Related Works

In the past decade, multi-view clustering has been intensively studied [1,10]. The multi-view clustering treats data characterized by multiple heterogeneous features acquired through distinct sensors. In this context, a view means a sensor or a feature detector. The multi-view clustering aims to ensemble multiple clustering results each of which is for a single view and to get a clustering result more refined than the single-view counterparts.

Table 4. Fairness index: The bold figures indicate that BMVS-AVE has a higher fairness index than MVS-AVE.

Dataset	Fairness index			Dataset	Fairness index		
	MVS-AVE	BMVS-AVE			MVS-AVE	BMVS-AVE	
		$\lambda = 2/n$	$\lambda = 10/n$			$\lambda = 2/n$	$\lambda = 10/n$
fbis	0.299	**0.322**	**0.325**	reviews	0.385	0.382	0.385
hitech	0.483	**0.492**	**0.567**	wap	0.291	0.291	**0.297**
k1a	0.254	0.254	0.251	la12	0.272	**0.373**	**0.372**
k1b	0.343	0.343	**0.467**	new3	0.314	0.314	0.314
la1	0.373	**0.377**	**0.526**	sports	0.189	**0.191**	**0.368**
la2	0.285	**0.384**	**0.395**	tr11	0.507	0.507	**0.510**
re0	0.453	0.453	**0.474**	tr12	0.232	**0.297**	**0.878**
re1	0.258	0.258	**0.356**	tr23	0.465	**0.472**	**0.651**
tr31	0.561	0.561	**0.577**	tr45	0.156	**0.291**	**0.487**

Multiviewpoint-based similarity (MVS) in this paper is very different from the above multi-view clustering. MVS does not care about the sensors. Rather, MVS aims to produce a similarity measure adaptive to the given dataset by migrating the reference point from the origin to the points in the dataset. Speaking of MVS, Yan *et al.* [11] developed a semi-supervised clustering based on MVS. The previous work which is the most related to our work is by Ravoori *et al.* [7] which claimed that they embed MVS into the average-link method. However, [7] does not mention how they incorporated MVS into the average-link method at all. In other words, the contents in Sect. 3 of our paper is completely missing. In fact, the standard average-link method in Fig. 1 in our paper is treated as their proposed method. As a result, neither Eq. (6) which extends MVS to measure the similarity between two points coming from *different clusters* nor the similarity update formula in Eq. (8) is presented. They never discussed the theoretical time complexity nor evaluated the running time experimentally.

Some researches pursued effective similarity measures for document clustering. For example, the pairwise-adaptive similarity [3] picks up some dimensions in the feature vectors and computes the cosine similarity from the chosen dimensions only. Here, the chosen dimensions change, depending on the two points between which the similarity is to be calculated. The similarity measure in [5] penalizes the dimensions for which one of the two points solely takes a non-zero value and assigns a minus constant for such dimensions. These similarity measures enhance the clustering accuracy as compared with CS, though they sacrifice the running speed due to their intricacy.

Our research is novel in that, under the framework of hierarchical clustering, we succeeded in outperforming CS in terms of clustering accuracy without sacrificing the running speed.

6   Conclusion

This paper proposes an agglomerative hierarchical clustering named MVS-AVE which couples the average-link method with the multiviewpoint-based similarity (MVS) [6]. Because MVS is a complex similarity measure, it has some risk of increasing the running time of clustering. Nonetheless, by elaborating the procedures to initialize and update the similarity matrix, our MVS-AVE clustering algorithm achieves the computational time complexity of $O(mn^2 + n^2 \log n)$ which is exactly the same as CS-AVE, the standard average-link algorithm with the cosine similarity. Experimentally, in the task of document clustering, MVS-AVE yields a better classification accuracy than CS-AVE with increasing the running time by at most 2.3%. Interestingly, MVS-AVE can be expanded to control the size fairness among clusters simply by subtracting a constant λ from all the elements in the similarity matrix at the beginning.

Acknowledgments. This work was supported by JSPS KAKENHI Grant Number JP18K11311, 2019.

References

1. Cai, X., Nie, F., Huang, H.: Multi-view k-means clustering on big data. In: International Joint Conference on Artificial Intelligence (2013)
2. Dhillon, I.S., Modha, D.S.: Concept decompositions for large sparse text data using clustering. Mach. Learn. **42**(1–2), 143–175 (2001)
3. D'hondt, J., Vertommen, J., Verhaegen, P.A., Cattrysse, D., Duflou, J.R.: Pairwise-adaptive dissimilarity measure for document clustering. Inf. Sci. **180**(12), 2341–2358 (2010)
4. Karypis, G.: CLUTO - a clustering toolkit. Minnesota University Minneapolis Department of Computer Science, Technical report (2002)
5. Lin, Y., Jiang, J., Lee, S.: A similarity measure for text classification and clustering. IEEE Trans. Knowl. Data Eng. **26**(7), 1575–1590 (2014). https://doi.org/10.1109/TKDE.2013.19
6. Nguyen, D.T., Chen, L., Chan, C.K.: Clustering with multiviewpoint-based similarity measure. IEEE Trans. Knowl. Data Eng. **24**(6), 988–1001 (2012)
7. Ravoori, D.T., Chen, Z.: Multi-view meets average linkage: exploring the role of metadata in document clustering. Int. J. Inf. Retr. Res. **5**(2), 26–42 (2015)
8. Shi, J., Malik, J.: Normalized cuts and image segmentation. IEEE Trans. Pattern Anal. Mach. Intell. **22**(8), 888–905 (2000)

9. Sokal, R.R., Michener, C.D.: A statistical method for evaluating systematic relationships. Univ. Kansas Sci. Bull. **38**, 1409–1438 (1958)
10. Tao, H., Hou, C., Liu, X., Liu, T., Yi, D., Zhu, J.: Reliable multi-view clustering. In: AAAI Conference on Artificial Intelligence (2018)
11. Yan, Y., Chen, L., Nguyen, D.T.: Semi-supervised clustering with multi-viewpoint based similarity measure. In: The 2012 International Joint Conference on Neural Networks (IJCNN) (2012)

Triplet-CSSVM: Integrating Triplet-Sampling CNN and Cost-Sensitive Classification for Imbalanced Image Detection

Jiefan Tan, Yan Zhu[✉], and Qiang Du

School of Information Science and Technology,
Southwest Jiaotong University, Chengdu 611756, China
tanjiefan@163.com, yzhu@swjtu.edu.cn,
duqiang_swjtu@163.com,

Abstract. In real-world applications, image classes are often imbalanced, which result in detection performance decline and quite different misclassification costs. In order to deal with these issues, cost-sensitive learning based on manually designed features has been studied for many years. With the rapid development of Deep Learning, more comprehensive methods, such as CNN and RNN, have proven their strength on feature extraction and classification. In this paper, we develop triplet-sampling CNN to automatically obtain a great many in-depth features from images. Cost-sensitive SVM (CSSVM) is applied to deal with the classification performance degradation caused by imbalanced image dataset. Furthermore, two techniques are integrated as Triplet-CSSVM for classifying images accurately even over imbalanced image set. This approach can overcome the disadvantages of the conventional features extraction and improve the overall classification performance comparing with several other related schemes.

Keywords: Imbalanced image data · Image classification · Triplet loss CNN · Cost-Sensitive SVM

1 Introduction

Image classification has become one of the research hotspots because of the increase of multimedia data. However, one big issue influences the performance of image classification, i.e., image data is imbalanced in many cases, where the size of majority class is much bigger than that of minority class, for example, spammed images detection, cancer patients identification based on MRI (Magnetic Resonance Imaging) images. In such cases, applying traditional classification algorithms will result in serious problems as follows.

Decision Performance Decline: Most of conventional classification methods like SVM perform well based on the balanced datasets. When the data set is imbalanced, their performance, especial on the minority class, may lower down greatly.

Misclassification Cost: In a scenario of medical diagnosis of a certain cancer, there are only 10 cancer patients as positive samples, and 90 healthy persons as negative

S. Hartmann et al. (Eds.): DEXA 2019, LNCS 11707, pp. 341–350, 2019.
https://doi.org/10.1007/978-3-030-27618-8_25

samples. Using a normal classification method to identify the cancer patients, the overall accuracy rate will be higher than 90% when 90 healthy persons are correctly identified, even though only one cancer patient out of 10 is detected. Such accuracy is illusory and the good chance for saving many cancer patients' life may be lost due to misclassification. Therefore, misclassification of the minority class usually leads to "small data, big trouble".

In traditional machine learning, there are two kinds of methods to solve the problems of imbalanced classification. One is rescaling methods, including threshold usage, re-sampling and weighting-based methods. Another is cost-sensitive algorithms, which directly assign the much higher cost as a punishment factor to the minority class than to the majority one, if the misclassification appears.

Threshold methods use a threshold value to classify samples into positive or negative if the cost-sensitive classifiers can produce probability estimations. Zhou and Liu studied the training effect of sampling and threshold-moving in cost-sensitive neural networks [1]. Their experiments showed that threshold-moving is a good way in training cost-sensitive neural networks. Re-sampling [2] is a common method to deal with imbalanced data classification at the dataset level. Wolf and Martin [3] proposed Feature KO and applied it to GentleBoost for solving the problem of imbalanced class size. Ting [4] introduced an instance-weighting method to induce cost-sensitive trees, which is simple and effective in implementation. Rescaling methods change the distribution of data set and can improve the classification.

Cost-sensitive classification algorithms can tackle performance degradation over imbalanced data set without rescaling, such as cost-sensitive support vector machine (CSSVM). C4.5CS and Metacost [5] are the earliest cost-sensitive learning methods. Ali et al. [6] developed an effective cost-sensitive classifier with Gentleboost ensemble (Can-CSC-GBE) for finding breast cancer using protein amino acid features. This method has effectively reduced the misclassification costs and thereby improved the overall classification performance. In traditional machine learning, e.g. [5, 6], data is usually represented as hand-craft features or shallow features, which are easily interfered by human factors. Deep learning can solve such a problem well.

Deep learning is an advanced area of machine learning and it has recently achieved great success due to its high learning capacity. It has been increasingly applied to multimedia classification, Natural Language Processing (NLP) and so on. Convolution Neural Networks (CNN) is a kind of deep learning methods based on multilayer neural network which is specially designed for image classification and recognition. There are many famous CNN models, such as LeNet, AlexNet, VGG-Net, and GoogLeNet.

As to the classification of imbalanced data, there are a few researches based on CNN. Huang et al. [7] extended triplet loss [8] to Quintuplet loss to learn the in-depth features. Their method is more conducive to the minority class and shows good performance on several imbalanced datasets. However, this method cannot directly extract in-depth features from images.

In this paper, we combine deep learning approach with cost-sensitive classification to obtain the in-depth features automatically and to deal with imbalanced image classification problem. The image features are extracted from the training model by triplet-sampling CNN, which is modified by integrating triplet-sampling technique and triplet loss from [8] with CNN. Based on the extracted rich features, cost-sensitive SVM (CSSVM) is applied to complete the imbalanced image classification.

The rest parts of the paper are organized as follows: in Sect. 2 the relevant techniques including triplet-sampling CNN and CSSVM are introduced. Section 3 discusses our approach. The experiments are conducted in Sect. 4 for comparing our work with other relevant methods. Section 5 summarizes the research and point out the future work.

2 Discussion on Relevant Techniques

Two key problems must be addressed as to the imbalanced image classification. One is how to obtain the rich and prominent features; another is how to reduce the impact of data imbalance on detection. Triplet loss CNN is developed in this paper to learn the in-depth features of image data. Cost-sensitive classification is applied to improve detection performance over imbalanced image set.

2.1 Triplet Loss CNN

Several CNN techniques can deal with classification problem well with tens of thousands of iterations. Taking VGG-16 as an example, it consists of five convolution groups for feature learning, and two fully-connected layers for classification. Behind the fully-connected layer is softmax layer which is the error function to optimize the CNN model. For example, the input picture is expressed as a vector V and v_i is the ith element. When it is classified as class k, the softmax value of this picture is as follows:

$$S_k = \frac{e^{v_k}}{\sum_i e^{v_i}} \tag{1}$$

The loss is: $Loss = -\ln S_k$ $\tag{2}$

Schroff et al. [8] proposed a novel error function, triplet loss, which is based on metric learning for CNN. In this method, images are mapped to the Euclidean Space. If the distance of two images is closed in the Euclidean Space, the two images are very similar. The intra-class distances are reduced and the inter-class distances are extended by the triplet loss. The other parts of triplet loss CNN are as same as the traditional CNNs.

Fig. 1. Triplet loss CNN structure [8].

The structure of triplet loss CNN is shown in Fig. 1. This network consists of batch input layer, deep CNN and L2 normalization. L2 normalization produces an embedding f(x). The network is optimized by triplet loss during training.

Fig. 2. Triplet sampling [8].

Figure 2 shows the triplet sampling which comes with triple loss. Each triplet consists of three samples: anchor, positive and negative. For standard triplet, the positive sample (x^p) and the anchor (x^a) are in the same class and the negative one (x^n) is in the different class of the anchor (x^a). The triplet loss minimizes the distance between the positive and the anchor and maximizes the distance between the anchor and the negative. The mathematic expression is shown in Eq. (3).

$$\left\|x_i^a - x_i^p\right\|_2^2 + \alpha < \left\|x_i^a - x_i^n\right\|_2^2, \forall \left(x_i^a, x_i^p, x_i^n\right) \in T \tag{3}$$

In Eq. (3), α is the margin between positive and negative pairs. T denotes all triplet tuples in the training data set. In the embedding space, the goal is to minimize the Euclidean distance. Although the aim of triplet loss is to get a better performance of image recognition, it also cannot avoid the negative effect from imbalanced dataset.

$$loss = \sum_i^N \left[\left\|f(x_i^a) - f(x_i^p)\right\|_2^2 - \left\|f(x_i^a) - f(x_i^n)\right\|_2^2 + \alpha\right]_+ \tag{4}$$

2.2 Quintuplet Loss CNN

To solve the problem of imbalanced data, Huang et al. [7] extended triplet loss to Quintuplet loss. Based on this structure, the in-depth features obtained are more conducive to the minority class and can help classifiers to perform very well on several imbalanced data sets.

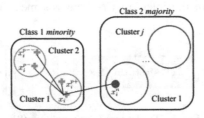

Fig. 3. Quintuplet sampling [7].

Compared with triplet loss, Quintuplet loss CNN needs some preparations, such as, clustering based on k-means using the learned features from the previous round of alternation [7], hand-crafted features in the first round. A quintuplet (See Fig. 3) consists of five samples: anchor (x^i), the most distant neighbor of anchor within-cluster (x_i^{p+}), the nearest neighbor of anchor within-class but between-clusters (x_i^{p-}), the most distant neighbor of the anchor within-class but between-clusters (x_i^{p--}) and the nearest neighbor of the anchor between-class (x_i^n). Quintuplet loss wants to obtain a relationship as follows.

$$
\begin{aligned}
\left\|f(x_i) - f(x_i^{p+})\right\|_2^2 &< \left\|f(x_i) - f(x_i^{p-})\right\|_2^2 \\
&< \left\|f(x_i) - f(x_i^{p--})\right\|_2^2 < \left\|f(x_i) - f(x_i^n)\right\|_2^2
\end{aligned}
\tag{5}
$$

Quintuplet loss CNN minimizes both distance within-class and within-cluster to build a clear demarcation between classes with most discriminative samples. Strictly speaking, Quintuplet loss CNN is a kind of re-sampling method. However, Quintuplet loss CNN cannot be used to train images directly, because it actually is applied on the clusters, which are built based on the feature set obtained from the image preprocessing.

2.3 Cost-Sensitive SVM

Support Vector Machine (SVM) is a popular machine learning method. Many studies focus on improving its performance, such as [9]. To deal with the imbalanced data classification, the most widely researched approach is to modify the SVM algorithm to be cost sensitive [10]. This consists of different penalty factors C_p and C_n for the misclassification cost of positive and negative samples. Assumed that the training set is as $R^N = \{(x_i, y_i)\}_{i=1,\dots,n}$ and the hyper-plane is as $(w^T.x) + b = 0, w \in R^N, b \in R$. The cost-sensitive SVM can be represented as,

$$
\underset{w,b,\delta}{argmin} \frac{1}{2}\|w\|^2 + C\left[C_p \sum_{\{i|y_i=p\}} \delta_i + C_n \sum_{\{i|y_i=n\}} \delta_i\right]
\tag{6}
$$
$$
(subject\ to\ y_i(w^T x_i + b) \geq 1 - \delta_i. \ i = 1, 2 \dots l)
$$

In Eq. (6) the misclassification cost C_p and C_n are given by domain knowledge and appear as precise values. While in many real-world applications, it is difficult to get the precise value of C_p and C_n.

3 Our Method Triplet-CSSVM

The proposed method is divided into two modules: Triplet-sampling CNN is used for learning feature, and CSSVM is for detecting images based on imbalanced data set.

3.1 Feature Extraction Based on Triplet-Sampling CNN

Because Quintuplet loss CNN is not suitable to produce features, we combined triplet loss CNN with re-sampling method to learn the in-depth features from images directly, which is named as **triplet-sampling** in this paper.

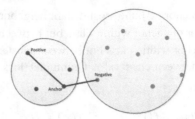

Fig. 4. Embed triplet sampling in terms of binary classification.

The triplet-sampling method is explained with Fig. 4. A triplet is chosen from the imbalanced dataset. If the selected triplets are not appropriate, it will result in a rather slow convergence. To solve this problem, re-sampling method is applied for selecting triplet, where the number of anchors selected from the majority-class must be equal to that selected from the minority-class in one batch. For example, if five triplets are built and their anchors are taken from the minority-class samples, the other five triplets should be built with the anchors taken from the majority-class samples in the same batch. Therefore the training data set of CNN could be balanced by building triplets.

The aim of triplet loss is to learn an embedding $f(x)$ to minimize the Euclidean distances between x^a and x^p and also to maximize the Euclidean distances between x^a and x^n over the iterations. We discard the margin α between positive and negative pairs to save the calculation time. So Eq. (4) is changed to Eq. (7).

$$loss = \sum_i^N \left[||f(x_i^a) - f(x_i^p)||_2^2 - ||f(x_i^a) - f(x_i^n)||_2^2 \right]_+ \qquad (7)$$

Given an imbalanced training dataset, the same number of anchors is selected from the minority and the majority class. The corresponding triplets are constructed in the batch according to the anchors. The selected triplets are fed into CNN to obtain an embedding $f(x)$, which then is normalized by L2. The triplet loss is computed by Eq. (7) afterwards. After hundreds of thousands of iterations, features are extracted from image training dataset.

The procedure is shown in Fig. 5 and the extraction method is described in Procedure 1. Firstly, the CNN is pre-trained to initialize parameters in the network. Then the softmax layer of CNN (error function) is replaced by triplet loss layer and the triplets are selected from the batches to fine-tune the triplet-sampling CNN. The feature vectors generated by using triplet-sampling CNN only have 512 dimensions.

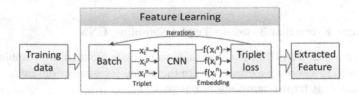

Fig. 5. The feature extraction module.

Procedure 1 Feature Extraction

Algorithm: Triplet-Sampling CNN
Input: Imbalanced image training dataset
  1: Pre-training model with softmax loss. Initialize the parameters of CNN
  2: Replace softmax layer with triplet loss layer
  3: **for** each iteration **do**
  4:   Generate triplets equally from majority and minority classes of the dataset
  5:   Encoding embedding f(x) from image x into Euclidean Space
  6:   L2 normalization for embedding
  7:   Optimize f(x) by minimizing the loss (Eq (7))
  8: **end for**
Output: The embedding model f(x)

3.2 Image Classification in Terms of CSSVM

Cost-sensitive support vector machine (CSSVM) algorithm is conducted to classify images based on the features extracted in Sect. 3.1. CSSVM aims to reduce the misclassification cost and improve the performance.

The cost model of Eq. (8) is adopted from paper [11]. The cost of correct classification is set to 0, i.e., $C(1, 1) = C(0, 0) = 0$. And the value of misclassification cost of the minority class depends on the imbalanced ratio.

$$Cp = \frac{number\ of\ y_i = n}{number\ of\ y_i = p}, \quad Cn = 1 \tag{8}$$

Where y_i represents the class label of a sample i. Cp and Cn denote the cost of the positive class (the minority) and negative class (the majority), respectively.

Procedure 2 shows the procedure of cost-sensitive classification based on CSSVM.

Procedure 2 Cost-Sensitive Classification

Algorithm: CSSVM
Input: Image features; misclassification cost: Cp, Cn; slack penalty C.
  1: **for** each iteration **do**
  2: train dataset to get hyper-plane $(w^T.x) + b = 0$
  3: Optimize $(w^T.x) + b = 0$ by minimizing Eq (6)
  4: **end for**
Output: Classification result

---

4 Experimental Results and Discussion

Two groups of experiments are conducted to comparing Triplet-CSSVM with the other different combinations of CNN and SVM, and with Quintuplet as well.

All experiments are carried out on Caffe [12] which is one of the most famous deep learning frameworks in recent years. Our experiments ran on GeForce GTX 960.

Evaluation Criteria: Precision and Recall based on Confusion Matrix are used as the criteria. Besides, the Receiver Operating Characteristic (ROC) curve and the Area Under the ROC Curve (AUC) are commonly used for assessing the performance of a classifier, as ROC curve and AUC can treat the minority class and the majority class fairly. We have to omit the results based on ROC and AUC due to the space limitation, but their results are similar to that based on Precision, Recall and F-Score.

Imbalanced Dataset: FaceScrub is used as the training dataset in our experiments, which is a large face dataset containing over 100,000 face images of 530 people, about 200 images per person. Two classes of images are chosen to conduct binary classification. For example, 141 images of "AaronEckhart" and 47 images of "AdamBrody" are used to build an imbalanced dataset. The cost of "AaronEckhart" is 3 and the cost of "AdamBrody" is 1 in terms of Eq. (8).

4.1 Comparison with the Different Combinations of CNN and SVM

Table 1. The comparison results.

Method	Imbalanced image dataset		
	Precision	Recall	F-Score
1. VGGNet-SVM	0.21	0.22	0.22
2. VGGNet-CSSVM	0.41	0.61	0.49
3. Triplet-SVM	0.47	0.57	0.52
4. Triplet-CSSVM (our approach)	0.52	0.93	0.67

The comparison results between our method and VGGNet-SVM, VGGNet-CSSVM and Triplet-SVM are shown in Table 1. When the classification method is the same (1 vs 3, 2 vs 4), the learning capability of triplet-sampling CNN is obviously better than that of VGGNet. Our approach shows a sharp contrast with conventional VGGNet-SVM, where the precision has increased by 31% and the recall has increased by 71% with 1:3 imbalanced rate. When investigating the effectiveness of cost-sensitive algorithm, our approach outperforms Triplet-SVM.

4.2 Comparison with Quintuplet

A comparison experiment between Quintuplet [7] and our approach is performed on MNIST-rot-back-image dataset [13]. The results (the mean of the per-class accuracy in percentage) are shown in Table 2. MNIST dataset is a balanced dataset with 10 digit classes. We construct a Gaussian-like imbalanced sample set by randomly removing data from the original data set using different percentages, such as 20% or 40% (ref. Data Remove in Table 2).

The results show that Quintuplet and Triplet-CSSVM have similar performance when the dataset is balanced (Data Remove = 0). If the dataset is imbalanced (Data Remove = 20% or 40%), Triplet-CSSVM outperforms Quintuplet. Besides, Triplet-CSSVM can perform similarly with it, if the combination of Quintuplet, resampling, and cost is applied [7]. The advantage of Triplet-CSSVM is that the rich features can be obtained directly from the original images. But Quintuplet should be applied only after features have been extracted and clusters have been built in other stages.

Table 2. Classification results of triplet-CSSVM and Quintuplet.

Data Remove (%)	0	20	40
Quintuplet [7]	77.62	72.26	65.27
Quintuplet + resample + cost [7]	77.64	75.58	70.13
Triplet-CSSVM (our approach)	77.65	75.55	70.23

5 Conclusion

In a lot of real world applications, the size of different data classes has clear distinction which is the problem of the imbalanced dataset. Many conventional classification algorithms perform well on the balanced image dataset, but decline greatly on the imbalanced one. One of main reasons is that they do not consider different costs (damages) when incorrectly classifying the images. Integrating the misclassification costs with the detection algorithms is a feasible way to improve the performance.

In this paper, we tackle two problems by using Triplet-CSSVM approach. On the one hand, in-depth image features are extracted from imbalanced dataset by integrating triplet-sampling and CNN. On the other hand, the detection performance in terms of imbalanced image set is improved clearly based on the learnt features with triplet-sampling CNN and by using Cost-Sensitive SVM. Our method outperforms several relevant techniques, such as VGGNet-CSSVM, Triplet-SVM, and also Quintuplet loss CNN.

In the future, an integrated deep learning approach for feature learning and cost-sensitive classification should be studied.

Acknowledgement. This work is supported by the Sichuan Science and Technology Program (No 2019YFSY0032) of China.

References

1. Zhou, Z.H., Liu, X.Y.: Training cost-sensitive neural networks with methods addressing the class imbalance problem. IEEE Trans. Knowl. Data Eng. **18**(1), 63–77 (2005)

2. Good, P.: Permutation Tests: A Practical Guide to Resampling Methods for Testing Hypotheses. Springer, New York (2013). https://doi.org/10.1007/978-1-4419-9863-7
3. Wolf, L., Martin, I.: Robust boosting for learning from few examples. In: Proceedings of the 2005 IEEE Computer Society Conference on Computer Vision and Pattern Recognition, pp. 359–364. IEEE Computer Society, San Diego (2005)
4. Ting, K.M.: An instance-weighting method to induce cost-sensitive trees. IEEE Trans. Knowl. Data Eng. **14**(3), 659–665 (2002)
5. Domingos, P.: MetaCost: a general method for making classifiers cost-sensitive. In: Proceedings of the fifth ACM SIGKDD International Conference on Knowledge Discovery and Data Mining, pp. 155–164. ACM, San Diego (1999)
6. Ali, S., Majid, A., Javed, S.G.: Can-CSC-GBE: developing cost-sensitive classifier with gentleboost ensemble for breast cancer classification using protein amino acids and imbalanced data. Comput. Biol. Med. **73**, 38–46 (2016)
7. Huang, C., Li, Y., Chen, C. L, et al.: Learning deep representation for imbalanced classification. In: Proceedings of the 29th IEEE Conference on Computer Vision and Pattern Recognition, pp. 5375–5384. IEEE Computer Society, Las Vegas (2016)
8. Schroff, F., Kalenichenko, D., Philbin, J.: FaceNet: a unified embedding for face recognition and clustering. In: Proceedings of the 2015 IEEE Conference on Computer Vision and Pattern Recognition, pp. 815–823. IEEE Computer Society, Boston (2015)
9. Shalev-Shwartz, S., Singer, Y., Srebro, N., et al.: Pegasos: Primal estimated sub-gradient solver for svm. Math. Program. **127**(1), 3–30 (2011)
10. Cao, P., Zhao, D., Zaiane, O.: An Optimized Cost-Sensitive SVM for Imbalanced Data Learning. In: Pei, J., Tseng, Vincent S., Cao, L., Motoda, H., Xu, G. (eds.) PAKDD 2013. LNCS (LNAI), vol. 7819, pp. 280–292. Springer, Heidelberg (2013). https://doi.org/10.1007/978-3-642-37456-2_24
11. Wang, K.J., Makond, B., Wang, K.M.: An improved survivability prognosis of breast cancer by using sampling and feature selection technique to solve imbalanced patient classification data. BMC Med. Inform. Decis. Mak. **13**, 124 (2013)
12. Deep learning framework by BAIR. http://caffe.berkeleyvision.org/. Accessed 12 Mar 2019
13. MNIST handwritten digit database. http://yann.lecun.com/exdb/mnist/. Accessed 14 June 2019

Discovering Partial Periodic High Utility Itemsets in Temporal Databases

T. Yashwanth Reddy[1], R. Uday Kiran[2,3(✉)], Masashi Toyoda[2],
P. Krishna Reddy[1], and Masaru Kitsuregawa[2,4]

[1] International Institute of Information Technology, Hyderabad, Hyderabad, India
yashwanth.t@research.iiit.ac.in, pkreddy@iiit.ac.in
[2] The University of Tokyo, Tokyo, Japan
{uday_rage,toyoda,kitsure}@tkl.iis.u-tokyo.ac.jp
[3] National Institute of Information and Communications Technology, Tokyo, Japan
[4] National Institute of Informatics, Tokyo, Japan

Abstract. High Utility Itemset Mining (HUIM) is an important model with many real-world applications. Given a (non-binary) transactional database and an external utility database, the aim of HUIM is to discover all itemsets within the data that satisfy the user-specified *minimum utility* (*minUtil*) constraint. The popular adoption and successful industrial application of HUIM has been hindered by the following two limitations: (*i*) HUIM does not allow external utilities of items to vary over time and (*ii*) HUIM algorithms are inadequate to find recurring customer purchase behavior. This paper introduces a flexible model of Partial Periodic High Utility Itemset Mining (PPHUIM) to address these two problems. The goal of PPHUIM is to discover only those interesting high utility itemsets that are occurring at regular intervals in a given temporal database. An efficient depth-first search algorithm, called PPHUI-Miner (Partial Periodic High Utility Itemset-Miner), has been proposed to enumerate all partial periodic high-utility itemsets in temporal databases. Experimental results show that the proposed algorithm is efficient.

Keywords: Data mining · Pattern mining · Utility Itemset Mining · Periodic itemsets

1 Introduction

High Utility Itemset Mining (HUIM) is an important model in data mining. HUIM algorithms discover all interesting itemsets whose *utility* (profit) in a transactional database is no less than the user-specified *minimum utility* (*minUtil*) constraint. The utility of an *itemset* is the summation of its utilities in all the transactions. The classic application of HUIM is market-basket analysis. HUIM has many other applications, such as website click stream analysis, cross-marketing and bio-medical applications [3]. HUIM has also inspired several other important data mining tasks such as high-utility occupancy pattern mining [4] and high-utility periodic pattern mining [1].

© Springer Nature Switzerland AG 2019
S. Hartmann et al. (Eds.): DEXA 2019, LNCS 11707, pp. 351–361, 2019.
https://doi.org/10.1007/978-3-030-27618-8_26

The popular adoption and successful industrial application of HUIM has been hindered by the following two obstacles: (*i*) Most previous studies on HUIM implicitly assume that the external utilities of the items do not change over time in the entire database. However, this is the seldom in real-world applications. In many applications, items' external utilities can vary with respect to time. For example, the prices of items in an eCommerce application can raise and/or fall depending on supply and demand. (*ii*) In many applications, high utility itemsets that are occurring at regular intervals can provide useful information to the users. For instance, in an eCommerce store, customers buy certain items (e.g. diapers and soaps) on a weekly or monthly basis. The knowledge pertaining to such periodically purchased high utility itemsets can facilitate an eCommerce application to improve its sales. Unfortunately, most studies on HUIM fail to discover such periodically occurring high utility itemsets in the data.

This paper makes an effort to address the above mentioned two issues. This paper introduces a novel model of Partial Periodic High Utility Itemset (PPHUI) in temporal databases. A temporal database not only facilitate multiple transactions to appear the same timestamp, but also facilitates irregular time gaps between the consecutive transactions. Partial Periodic High Utility Itemset Mining (PPHUIM) allows items' external utility values to vary overtime. Thus, addressing the first obstacle of HUIM. The PPHUIM tries to address the second obstacle of HUIM by finding partial periodically occurring high utility itemsets in temporal databases. A fast algorithm, called Partial Periodic High Utility Itemset-Miner (PPHUI-Miner), has been introduced to discover all PPHUIs by proposing new pruning techniques. Experimental results demonstrate that PPHUI-Miner is not only memory and runtime efficient, but also highly scalable as well.

The rest of the paper is organized as follows. Related work is presented in Sect. 2. Section 3 introduces the proposed model of PPHUIM. The proposed is presented in Sect. 4. Experimental results are reported in Sect. 5. Section 6 provides conclusions.

2   Related Work

High Utility Itemset Mining: Yao et al. [12] described HUIM by taking into account the importance of items and their occurrence *frequency* in every transaction. Since then, several algorithms have been proposed to discover high utility itemsets in transactional databases [2,7–9,11] and sequence databases [14]. To circumvent the fact that the utility is not anti-monotonic and to find all high utility itemsets, several HUIM algorithms (e.g. Two-Phase [9] and UP-Growth+) have employed *Transaction Weighted Utilization* (*TWU*) to reduce the search space. The *TWU* measure represents an upper bound on the *utility* of itemsets. Recently, algorithms like EFIM [13] introduced by proposing tighter measures to calculate upper bound on the utility of itemsets than *TWU*.

Periodic High Utility Itemset Mining: Tanbeer et al. [10] have introduced a model to find periodic-frequent itemsets in transactional databases. Philippe et al. [1] have extended the model [10] to discover full periodic high utility itemsets in a transactional database (i.e., a database in which transactions occur at a fixed time interval). This model discovers all periodic itemsets within the transactional database that satisfy the user specified minimum utility ($minUtil$), minimum average periodicity ($minAvgPer$), maximum average periodicity ($maxAvgPer$), minimum period ($minPer$) and maximum period ($maxPer$). This model suffers from the following limitations: (i) If an itemset has one instance where $period$ (or inter-arrival time) is more than the user-specified $maxPer$, the corresponding itemset is considered as an uninteresting itemset. (ii) This model assumes time gap between two consecutive transactions is constant, which is not the case in real-world databases and It requires too many input parameters from the user.

A model has been proposed in [6] to find partial periodic itemsets in temporal databases. It can overcome limitations of model proposed in [1]. However, the model [6] disregards the importance of the items and their occurrence $frequency$ in every transaction.

The proposed model of PPHUI mining does not suffer from any of the above mentioned limitations. A part from extracting PPHUI from a given transactional database, the proposed model is different from the model proposed in [1] as that model employs different measures to find periodic high utility itemsets.

3 Proposed Model

Let $I = \{i_1, i_2, \cdots, i_m\}$, $m \geq 1$, be a set of items. Let $X \subseteq I$ be an itemset. An itemset containing k items is known as k-itemset. A transaction $T_{tid} = (tid, ts, Y)$ is a triplet, where $tid \in R^+$ represents the transactional identifier, $ts \in R^+$ represents the timestamp of corresponding transaction and $Y \subseteq I$ is an itemset. A temporal database, denoted as TDB, represents a set of transactions. That is, $TDB = \{T_1, T_2, \cdots, T_n\}, 1 \leq n$. Let $p(i_j, tid)$ denote the **external utility** of an item $i_j \in I$ in a transaction whose transaction identifier is tid. Let $P(i_j) = \{p(i_j, 1), p(i_j, 2), \cdots, p(i_j, n)\}$ denote the set of all external utility values of i_j in the data. The (external) **utility database**, UD, is the set of external utility values of all items in I. That is, $UD = \bigcup_{i_j \in I} P(i_j)$. Every item $i_j \in T_{tid}$ has a positive number $q(i_j, tid)$, called its **internal utility**. The internal utility of an item generally represents its $frequency$ in a transaction and external utility represents cost/profit of item in a transaction.

Example 1. Let $I = \{a, b, c, d, e, f, g, h, i, j\}$ be the set of items. The set of items 'd' and 'f', i.e., $\{d, f\}$ (or df, in short) is an itemset. This itemset contains two items. Therefore, it is a 2-itemset. A temporal database generated from I is shown in Table 1. This database contains 8 transactions. The minimum and maximum timestamps of the transactions in this database are 1 and 12, respectively. It can be observed that temporal databases not only allow multiple transactions to share a common timestamp, but also encourage irregular time gaps between the consecutive transactions. Thus, a temporal database generalizes a transactional database by taking into account the temporal occurrence information of

Table 1. Temporal database

tid	ts	items
1	1	$(a,1),(b,2),(c,1)$
2	3	$(a,2),(b,2),(e,2),(h,1)$
3	4	$(c,1),(d,3),(f,2)$
4	6	$(b,1),(d,2),(e,3),(f,1),(g,2),(h,3)$
5	7	$(c,3),(f,1),(g,1)$
6	7	$(i,1),(j,3)$
7	9	$(a,1),(b,1),(d,2),(f,1),(g,2)$
8	12	$(c,3),(d,1),(e,1),(f,2),(g,2)$

Table 2. External utility database

tid	a	b	c	d	e	f	g	h	i	j
1	200	100	50	0	0	0	0	0	0	0
2	50	100	0	0	100	0	0	100	0	0
3	0	0	200	200	0	200	0	0	0	0
4	0	200	0	200	150	300	100	200	0	0
5	0	0	100	0	0	150	50	0	0	0
6	0	0	0	0	0	0	0	0	40	20
7	150	300	0	200	0	300	200	0	0	0
8	0	0	50	200	300	50	200	0	0	0

the transaction. Table 2 shows the external utilities (or prices/profit) of all items in every transaction. Let the currency of these prices be Japanese Yen (¥). The external utility of an item d in the third transaction, i.e., $p(d,3) = 200$¥. The internal utility of an item d in the third transaction T_3, i.e., $q(d,3) = 3$.

Definition 1 *(Utility of an item in a transaction). The utility of an item i_j in a transaction T_{tid} denoted as $u(i_j, T_{tid}) = p(i_j, T_{tid}) \times q(i_j, T_{tid})$.*

Definition 2 *(Utility of an itemset in a transaction). The utility of an itemset X in a transaction T_{tid}, denoted as $u(X, T_{tid}) = \Sigma_{i \in X} u(i, T_{tid})$.*

Definition 3 *(Utility of an itemset in a database). The utility of an itemset X in the database TDB, denoted as $u(X) = \Sigma_{T_{tid} \in g(X)} u(X, T_{tid})$, where $g(X)$ is the set of transactions containing X.*

Example 2. Continuing the previous example, the utility of 'd' in third transaction T_3, i.e., $u(d, T_3) = p(d, T_3) \times q(d, T_3) = 200 \times 3 = 600$¥. The utility of itemset df in T_3, $u(df, T_3) = u(d, T_3) + u(f, T_3) = 600$¥ $+ 400$¥ $= 1000$¥. In Table 1, the itemset df has appeared in the transactions T_3, T_4, T_7 and T_8. Therefore, $g(x) = \{T_3, T_4, T_7, T_8\}$. The *utility* of df in each of these three transactions: $u(df, T_3) = 1000$¥, $u(df, T_4) = 700$¥, $u(df, T_7) = 700$¥ and $u(df, T_8) = 300$¥. Therefore, the *utility* of df in the database, $u(df) = 2700$¥.

Definition 4 *(Periodic appearance of X). Let $TS^X = \{ts_a^X, ts_b^X, \cdots, ts_c^X\}$, $ts_{min} \leq ts_a^x \leq ts_b^x \leq ts_c^x \leq ts_{max}$, be an ordered list of timestamps in which X appeared in TDB. The terms ts_{min} and ts_{max} represent the minimal and maximal timestamps in TDB. Let $ts_j^X, ts_k^X \in TS^X$, $ts_{min} \leq ts_j^X \leq ts_k^X \leq ts_{max}$, denote any two consecutive timestamps in TS^X. The time difference between ts_k^X and ts_j^X is referred to an **inter-arrival time** of X, and denoted as iat_p^X, $p \geq 1$. That is, $iat_p^X = ts_k^X - ts_j^X$. Let $IAT^X = \{iat_1^X, iat_2^X, ..., iat_{|TS^X|-1}^X\}$, be the list of all inter-arrival times of X in TDB. An inter-arrival time of X is said to be **periodic** (or interesting) if it is no more than the user-specified maximum-inter arrival time (maxIAT). That is, an $iat_k^X \in IAT^X$ is said to be **periodic** if $iat_k^X \leq maxIAT$.*

Example 3. In Table 1, the itemset *df* has appeared in the transactions T_3, T_4, T_7 and T_8. Therefore, the set of timestamps of these four transactions, i.e., $TS^{df} = \{4, 6, 9, 12\}$. The inter-arrival times of '*df*' are: $iat_1^{df} = 6 - 4 = 2$, $iat_2^{df} = 9 - 6 = 3$, $iat_3^{df} = 12 - 9 = 3$. Thus, $IAT^{df} = \{iat_1^{df}, iat_2^{df}, iat_3^{df}\} = \{2, 3, 3\}$. If the user-specified $maxIAT = 3$, then iat_1^{df} is considered interesting (or periodic) occurrence of *df* within the database because $iat_1^{df} \leq maxIAT$. Similarly, iat_2^{df} and iat_3^{df} are also periodic occurrences of *df*.

Definition 5 (Periodic-Support of itemset). *Let* $\overline{IAT^X} \subseteq IAT^X$ *be the set of all inter-arrival times that have value no more than* $maxIAT$*. That is,* $\overline{IAT^X} \subseteq IAT^X$ *such that if* $\exists iat_k^X \in IAT^X : iat_k^X \leq maxIAT$*, then* $iat_k^X \in \overline{IAT^X}$*. The periodic-support of X, denoted as* $PS(X) = |\overline{IAT^X}|$*.*

Example 4. Continuing with the previous example, $\overline{IAT^{df}} = \{iat_1^{df}, iat_2^{df}, iat_3^{df}\}$. Therefore, the *periodic support* of '*df*', i.e., $PS(df) = |\overline{IAT^{df}}| = 3$. In other words, the itemset '*df*' has appeared 3 times periodically within the data.

The *periodic-support*, as defined above, determines the number of periodic occurrences of an *itemset* in the database. An inter-arrival time of an itemset can be expressed in percentage of $(ts_{max} - ts_{min})$. The *periodic-support* of an itemset also can be expressed in percentage of $|TDB| - 1$, where $|TDB| - 1$ represents the maximum *periodic-support* an itemset can have in the database.

Definition 6 (Partial Periodic High Utility Itemset X). *An itemset X is a Partial Periodic High Utility Itemset (PPHUI) if* $u(X) \geq minUtil$ *and* $PS(X) \geq minPS$*, where minUtil and minPS represent the user-specified minimum utility and minimum periodic-support, respectively.*

Example 5. If the user-specified $minUtil = 1500¥$, $maxIAT = 6$ and $minPS = 2$, then the itemset '*df*' is a PPHUI because $u(df) \geq minUtil$ and $PS(df) \geq minPS$. All PPHUIs generated from Table 1 are shown in Table 4.

Problem Statement: Given a temporal database (TDB), an external utility database (UD) and the user-specified $minUtil$, $maxIAT$ and $minPS$, the problem of finding PPHUIs involve discovering all itemsets in TDB whose *utility* and *periodic-support* is no less than the user-specified $minUtil$ and $minPS$, respectively.

4 Proposed Approach

The problem is to develop an efficient approach for discovering partial periodic high utility itemsets (PPHUIs) in Temporal Database subject to $minUtil$, $minPS$ and $maxIAT$ constraints. Given n data items, a naïve way to find PPHUIs is to mine set of all possible $2^n - 1$ combinations of items and test for $minUtil$, $minPS$ and $maxIAT$ constraints. Notably, such an approach suffers from exponential complexity. The basic idea is to define pruning techniques

Table 3. TWU values of items in Table 1

Item	f	d	g	b	e	c	a	h	i	j
TWU	6750	6250	5550	4950	3900	3300	2800	2750	100	100
PS	4	3	3	3	2	3	2	1	0	0

Table 4. PPHUIs generated from Table 1 at $minUtil = 2000, PS = 2, maxIAT = 6$

Itemset	d	f	g	ab	cf	dg	df	fg	dfg
Utility	1600	1250	1250	1150	1300	2200	2700	2100	2900
PS	3	4	3	2	2	2	3	3	3

based on Transaction Weighted Utilization (TWU), Periodic Support (PS) and Remaining Utility and proposed efficient approach to mine PPHUI. We briefly explain these techniques and discuss the proposed approach.

i. Pruning using TWU:
    We carry out the pruning based on TWU [9]. The notion of TWU is defined as follows.

Definition 7 (Transaction Weighted Utilization (TWU)). *The transaction utility (TU) of a transaction T_{tid} is the sum of the utility of all items in T_{tid}. i.e. $TU(T_{tid}) = \Sigma_{x \in T_{tid}} u(x, T_{tid})$. The transactional-weighted utilization (TWU) of an itemset X is defined as the sum of the transaction utility of transactions containing X, i.e. $TWU(X) = \Sigma_{T_c \in g(X)} TU(T_c)$.*

Example 6. Consider the first transaction in Table 1. The *transaction utility* of T_1, denoted as $TU(T_1)$, is the total revenue generated by all its items. That is, $TU(T_1) = u(a, 1) + u(b, 2) + u(c, 1) = 200 + 200 + 50 = 450 ¥$. In other words, the first transaction has generated the revenue of 450¥, Similarly, the *transaction utility* of T_2, T_3, T_4, T_5, T_6, T_7 and T_8 are 600, 1200, 2150, 500, 100, 1750 and 1150 respectively. Consider the item 'd,' which is appearing in the transactions T_3, T_4, T_7 and T_8. The *TWU* of d, i.e., $TWU(d) = TU(T_3) + TU(T_4) + TU(T_7) + TU(T_8) = 6250 ¥$.

    The pruning rule based on *TWU* is as follows. It can be observed that the *TWU* of item conveys the crucial information that it is equivalent to atmost utility that an item can generate by combining with other items in the database. *TWU* measure can be used to identify the items, whose supersets may generate PPHUIs. We ignore the extensions of items $i_j \in I$ whose $TWU(i_j) < minUtil$.

ii. Pruning using PS:
    We prune the itemsets based on the value of PS. *Periodic-Support* has **anti-monotonic property** that is an itemset cannot have PS greater than PS of its subsets. So, we can ignore extensions of those items/itemsets X, whose $PS(X) < minPS$.

iii. Pruning using Remaining Utility:

We carry out the pruning based on the notion of *Remaining Utility*. We define the notion of *Remaining Utility* and define the notion of *utility list*.

Definition 8 (Remaining utility). *Let $\succ$ be any total order on items from I and X be the itemset. The remaining utility of X in a transaction T_{tid} is defined as $rU(X, T_{tid}) = \Sigma_{i \in T_{tid} \wedge i \succ x \forall x \in X} u(i, T_{tid})$.*

Definition 9 (Utility-list). *Let $\succ$ be any total order on items from I. The utility-list of an itemset X in a database D is denoted as $UL(X)$ and defined as a set of tuples such that there is a tuple $< tid, ts, iutil, rutil >$ for each transaction T_{tid} containing X. The iutil element of a tuple is utility of X in T_{tid}. i.e., $u(X, T_{tid})$. The rutil element of a tuple is the remaining utility (see Definition 8).*

Example 7. For this example $\succ$ be lexicographical order i.e. ($i \succ h \succ g \succ f \succ e \succ d \succ c \succ b \succ a$) Remaining utility of df in T_4 is $rU(df, T_4) = u(g, T_4) + u(h, T_4) = 2 \times 100 + 3 \times 200 = 800$.

For pruning, we use **Remaining utility** measure to overestimate the utility value of *itemset*. Let X be an itemset. If $\Sigma iutil + \Sigma rutil < minUtil$, where $iutil, rutil \in ul(X)$, X and its extensions are low utility. So, such patterns can be pruned. The proof that the sum of $iutil$ and $rutil$ values of utility list an itemset X is an upper bound on the utility of X and its extensions is provided in [8].

The proposed **PPHUI-Miner** employs depth-first search of set enumeration tree and prunes patterns based on preceding pruning techniques.

Algorithm 1. PPHUI-Miner

1: **Input:** TDB: a temporal database; UD: a external utility database; $minUtil$: a user-specified minimum utility constraint; $maxIAT$: a user-specified period constraint; $minPS$: a user-specified periodic-support constraint.
2: **Output:** A set of partial periodic high-utility itemsets.
3: Let α denote the itemset that needs to be extended. Initially, set $\alpha = \emptyset$;
4: Scan TDB to compute $TWU(\{i_j\})$, $PS(\{i_j\})$ for each items $i_j \in I$;
5: $I^* = \{i_j | i_j \in I \wedge TWU(i_j) \geq minUtil \wedge PS(i_j) \geq minPS\}$;
6: Let us call I^* as candidate items;
7: Let $\succ$ be the total order of TWU descending values on candidate items;
8: Scan TDB to build the utility list(UL) of each item $i_j \in I^*$;
9: $Primary(\alpha) = \{i_j | i_j \in I^* \wedge \forall x \in \alpha, i_j \succ x\}$;
10: Search (UL, α, $Primary(\alpha)$, $minUtil$, $maxIAT$, $minPS$);

The Approach: PPHUI-Miner presented in Algorithms 1 and 2. We first scan the database to measure TWU and PS values for all items within the database. Table 3 shows the TWU amd PS values determined for all items after scanning the database. Next, we prune the items in the list that have PS value less than $minPS$ and/or TWU value less than $minUtil$. The remaining items in the list are considered as **candidate items** and sorted in TWU descending order of items. After finding **candidate items** and establishing $\succ$ total order (i.e., TWU descending order of items), the utility list (refer Definition 9) by scanning the database second time. The $Primary(\alpha)$ contains the candidate items, which are $\succ$ than every item in α.

After building the utility lists of candidate items, we call recursive search with α and UL utility list of candidate items. Next, we expand search by combining α with $Primary(\alpha)$ one by one using DFS technique. If $i_x \in Primary(\alpha)$, we build utility list of $\beta(\alpha \cup i_x)$. We check utility and periodic support of β from the above utility list. Then we have two cases: (i) if β is PPHUI, then $Primary(\beta)$ is generated and β is further extended by calling recursive search (ii) if β is not PPHUI, it may fail to satisfy either $minPS$ or $minUtil$ values. In the former case (i.e., when β fails to satisfy $minPS$), we stop performing depth-first search on α. In the latter case (i.e., when β fails to satisfy only $minUtil$), we calculate its remaining utility value. If this value is greater than $minUtil$, we continue exploring β same as in first case. If remaining utility of β is less than $minUtil$, then we stop exploring that branch in the DFS tree.

Algorithm 2. The search procedure

1: **Input**: α: an itemset; UL: utility lists of candidate items; $UL(\alpha)$: utility list of α;
  $Primary(\alpha)$: *Extension items of* α; $minUtil$; $maxIAT$; $minPS$.
2: **Output**: A set of periodic high-utility itemsets.
3: **for** $\forall$ itemsets $\beta = \alpha \cup i_j$, $i_j \in Primary(\alpha)$; **do**
4:    Calculate utility list of β from utility lists of α and i_j;
5:    Calculate utility and periodic support of β from utility list above;
6:    **if** $U(\beta) + rU(\beta) \geq minUtil \wedge PS(\beta) \geq minPS$ **then**
7:       **if** $U(\beta) \geq minUtil$ **then**
8:          Output β;
9:       **end if**
10:      generate itemset $Primary(\beta)$;
11:      Search(β, UL, $UL(\beta)$, $Primary(\beta)$, $minUtil$, $maxIAT$, $minPS$);
12:   **end if**
13: **end for**

---

5   Experimental Results

Since there exists no algorithm to find PPHUIs in temporal databases, we only evaluate the proposed PPHUI-Miner algorithm using both synthetic and real-world databases. Please note that we are not comparing the proposed PPHUI-Miner algorithm against the Periodic High Utility Mining (PHM) algorithm. It is because PHM employs different measures to find interesting itemsets.

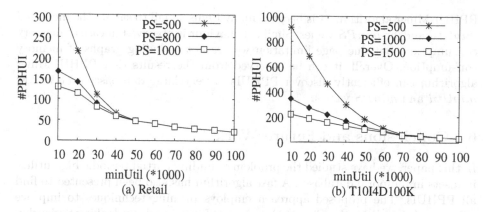

Fig. 1. PPHUI generated in Retail and T10I4D100k databases.

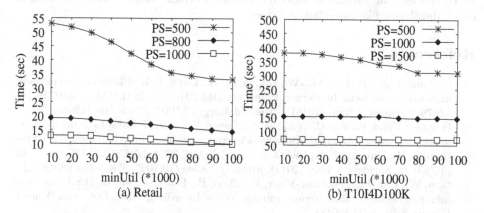

Fig. 2. Time taken by PPHUI-miner for Retail and T10I4D100k databases.

The algorithms, PHM and PPHUI-Miner, were written in C++ and executed on i5 1.5 GHz processor, with 16 GB ram. The experiments have been conducted using both synthetic (T10I4D100K) and real-world (Retail) databases. The Retail and T10I4D100K databases are available on SPMF toolkit.

The $maxIAT$ value for Retail database is fixed at 500 and for T10I4D100K database is fixed at 1000. We are not reporting results by varying $maxIAT$ value due page limitation. But in general, we observed that increase in $maxIAT$ increases number of PPHUIs generated [5].

Figure 1(a) and (b) show the number of PPHUIs generated by PPHUI-Miner in different databases at different $minUtil$ and $minPS$ values. It can be observed that increase in $minUtil$ and/or $minPS$ results in the decrease of PPHUIs as many itemsets fail to satisfy the increased $minUtil$ and/or $minPS$ values. Figure 2(a) and (b) show the runtime requirements of PPHUI-Miner in different databases at different $minUtil$ amd $minPS$ values. It can be observed that increase in $minUtil$ and/or $minPS$ results in the decrease of runtime for

PPHUI-Miner algorithm. It is because many itemsets fail to satisfy the increased $minUtil$ and/or $minPS$ values. Similar behaviour is observed in case of memory consumption, but due page limitation we are not including graphs of memory consumption. Overall, it can be observed from the results that PPHUI-Miner algorithm can efficiently discover PPHUIs in very large databases even at low $minUtil$ and $minPS$ values.

6   Conclusions and Future Work

In this paper, we have studied the problem of finding partial periodic high utility itemsets in temporal databases. A fast algorithm has also been presented to find all PPHUIs. The proposed approach employs pruning techniques to improve efficiency (or computational cost). As a part of future work, we looking to develop more efficient algorithms to discover Partial Periodic High Utility itemsets in other databases like uncertain database.

References

1. Fournier-Viger, P., Lin, J.C.-W., Duong, Q.-H., Dam, T.-L.: PHM: mining periodic high-utility itemsets. In: Perner, P. (ed.) ICDM 2016. LNCS (LNAI), vol. 9728, pp. 64–79. Springer, Cham (2016). https://doi.org/10.1007/978-3-319-41561-1_6
2. Fournier-Viger, P., Wu, C.-W., Zida, S., Tseng, V.S.: FHM: faster high-utility itemset mining using estimated utility co-occurrence pruning. In: Andreasen, T., Christiansen, H., Cubero, J.-C., Raś, Z.W. (eds.) ISMIS 2014. LNCS (LNAI), vol. 8502, pp. 83–92. Springer, Cham (2014). https://doi.org/10.1007/978-3-319-08326-1_9
3. Gan, W., Lin, J.C., Fournier-Viger, P., Chao, H., Hong, T., Fujita, H.: A survey of incremental high-utility itemset mining. Wiley Interdiscip. Rev. Data Min. Knowl. Discov. 8(2), 1–23 (2018)
4. Gan, W., Lin, J.C.W., Fournier-Viger, P., Chao, H.C., Philip, S.Y.: HUOPM: high-utility occupancy pattern mining. IEEE Trans. Cybern. (2019)
5. Uday Kiran, R., Yashwanth Reddy, T., Fournier-Viger, P., Toyoda, M., Krishna Reddy, P., Kitsuregawa, M.: Efficiently finding high utility-frequent itemsets using cutoff and suffix utility. In: Yang, Q., Zhou, Z.-H., Gong, Z., Zhang, M.-L., Huang, S.-J. (eds.) PAKDD 2019. LNCS (LNAI), vol. 11440, pp. 191–203. Springer, Cham (2019). https://doi.org/10.1007/978-3-030-16145-3_15
6. Kiran, R.U., Shang, H., Toyoda, M., Kitsuregawa, M.: Discovering partial periodic itemsets in temporal databases. In: Proceedings of the 29th SSDBM, p. 30. ACM (2017)
7. Lan, G.C., Hong, T.P., Tseng, V.S.: An efficient projection-based indexing approach for mining high utility itemsets. KAIS 38(1), 85–107 (2014)
8. Liu, M., Qu, J.: Mining high utility itemsets without candidate generation. In: Proceedings of the 21st ACM CIKM, pp. 55–64. ACM (2012)
9. Liu, Y., Liao, W.k., Choudhary, A.: A two-phase algorithm for fast discovery of high utility itemsets. In: PAKDD, pp. 689–695 (2005)
10. Tanbeer, S.K., Ahmed, C.F., Jeong, B.-S., Lee, Y.-K.: Discovering periodic-frequent patterns in transactional databases. In: Theeramunkong, T., Kijsirikul, B., Cercone, N., Ho, T.-B. (eds.) PAKDD 2009. LNCS (LNAI), vol. 5476, pp. 242–253. Springer, Heidelberg (2009). https://doi.org/10.1007/978-3-642-01307-2_24

11. Tseng, V.S., Shie, B.E., Wu, C.W., Yu, P.S.: Efficient algorithms for mining high utility itemsets from transactional databases. TKDE **25**(8), 1772–1786 (2013)

12. Yao, H., Hamilton, H.J., Butz, C.J.: A foundational approach to mining itemset utilities from databases. In: SIAM, pp. 482–486 (2004)

13. Zida, S., Fournier-Viger, P., Lin, J.C.W., Wu, C.W., Tseng, V.S.: EFIM: a fast and memory efficient algorithm for high-utility itemset mining. KAIS **51**(2), 595–625 (2017)

14. Zida, S., Fournier-Viger, P., Wu, C.-W., Lin, J.C.-W., Tseng, V.S.: Efficient mining of high-utility sequential rules. In: Perner, P. (ed.) MLDM 2015. LNCS (LNAI), vol. 9166, pp. 157–171. Springer, Cham (2015). https://doi.org/10.1007/978-3-319-21024-7_11

Using Mandatory Concepts
for Knowledge Discovery
and Data Structuring

Samir Elloumi[1]([⊠]), Sadok Ben Yahia[1,2] [iD], and Jihad Al Ja'am[3] [iD]

[1] Faculty of Sciences of Tunis, University of Tunis El Manar, LR11ES14,
Tunis, Tunisia
samir.elloumi@fst.utm.tn
[2] Department of Software Sciences, Tallinn University of Technology,
Akadeemia tee 15a, 12618 Tallinn, Estonia
[3] Qatar University, Doha, Qatar
jaam@qu.edu.qa

Abstract. A data scientist could apply several machine learning approaches in order to discover valuable knowledge from the data. While applying several techniques, he might discover that some pieces of knowledge are invariant, what ever the technique he used. We consider such knowledge as mandatory concepts, i.e., unavoidable knowledge to be discovered. As interesting property, a mandatory concept is characterized by a non-shared isolated point, that relates pieces of data, e.g., an object to a property, a document to specific words, an image to a specific topic, etc. Hence, the isolated points allow to make the distinction between the concepts. In this paper, we present a new approach for mandatory concepts extraction by making a level-based properties composition. Hence, the N-Composites isolated points are identified and constitute a key element for mandatory concept localization. We experiment our new algorithm by considering the coverage quality metrics.

Keywords: Mandatory formal concepts ·
N-Composites isolated points · Conceptual coverage

1 Introduction

Mandatory concepts (MC) play an important role in data mining as they allow to discover regular structures from data, based on Formal Concept Analysis (FCA). They are qualified as mandatory because they belong to any conceptual coverage of a formal context (FCT) [3]. From Relational Algebra (RA) perspective, a MC contains at least one isolated point as introduced by Riguet [17]. As a mathematical background, FCA and RA have been already combined and used to discover regularities in data [11]. In fact, a FC represents the atomic regular structure for decomposing a binary relation (BR). Riguet's difunctional relation [17], whose elements are defined as isolated points, describes invariant regular

S. Hartmann et al. (Eds.): DEXA 2019, LNCS 11707, pp. 362–375, 2019.
https://doi.org/10.1007/978-3-030-27618-8_27

structures that could be used for database decomposition and Textual Features Selection (TFS) [4]. Furthermore, an isolated point belongs to a unique formal concept (FC) that should exists in any conceptual coverage. Any FCA-based knowledge discovery process considers necessarily such concepts. As a matter of fact, several approaches have been proposed to locate the mandatory concepts in a formal context and have proposed strategies to build the conceptual coverage. In this paper we present alternatives for the conceptual coverage construction and we discuss their main characteristics and features. Nevertheless, finding the most efficient strategy remains a challenging perspective. As our main contribution, we introduce the mathematical properties related to isolated points and we propose a new approach for locating them while ensuring, level by level, the FCT coverage. The remainder of this paper is organized as follows: In Sect. 2, we present the mathematical background related to FCA and RA. In Sect. 3, we present the related work for conceptual coverage building, particularly based on isolated points. Our main contribution is thoroughly described in Sect. 4, in which we present the N-composites isolated points properties as well as the extraction algorithm. The results of the experimental evaluation are discussed in Sect. 5 before concluding and sketching some issues of future work in Sect. 6.

2 Mathematical Background

FCA is a mathematical tool for analyzing data and formally representing conceptual knowledge [6]. FCA helps forming conceptual structures from data. These structures such as: Closed itemset, Generic bases, Minimal Generators, etc., are very useful for data mining [2]. In the following, we recall the basic concepts of this theoretical framework.

2.1 FCA Background

Definition 1. Formal Context: *A formal context (FCT) is a triplet* $\mathcal{K} = (\mathcal{X}, \mathcal{Y}, \mathcal{R})$, *where* $\mathcal{X}$ *represents a finite set of objects or transactions,* $\mathcal{Y}$ *is a finite set of attributes (or items) and* $\mathcal{R}$ *is a binary (incidence) relation (i.e.,* $\mathcal{R} \subseteq \mathcal{X} \times \mathcal{Y}$). *Each couple* $(x, y) \in \mathcal{R}$ *expresses that the object, or transaction,* $x \in \mathcal{X}$ *contains the item* $y \in \mathcal{Y}$.

Example 1. Let's consider the FCT presented in Table 1 where $\mathcal{X} = \{o_0, o_1, o_2, o_3, o_4, o_5, o_6\}$ is a set of objects and $\mathcal{Y} = \{a, b, c, d, e, f, g, h\}$ is a set of properties.

Definition 2. Itemset: *An itemset is a set of items included in* $\mathcal{Y}$ *and representing an object or a transaction. In the sequel, we denote by* $\mathcal{I}$ *the power set of* $\mathcal{Y}$.

Definition 3. Galois Connection: *We define two functions, f and g, summarizing the links between subsets of objects and subsets of properties induced by* $\mathcal{R}$. *Thus,*

Table 1. A Binary relation illustrating a Formal Context

	a	b	c	d	e	f	g	h
o_0	1	1	1	1	0	0	0	0
o_1	0	0	0	1	1	0	0	0
o_2	0	0	1	0	0	0	0	0
o_3	1	1	0	0	0	1	1	0
o_4	0	0	1	0	0	1	1	0
o_5	0	0	0	0	0	1	1	0
o_6	0	0	0	0	0	0	0	1

- $f : \mathcal{P}(\mathcal{X}) \to (\mathcal{P}(\mathcal{Y}) = \mathcal{I})$, $f(X) = \{y \in \mathcal{Y} | \forall x \in X, (x,y) \in \mathcal{R}, X \subseteq \mathcal{X}\}$,
- $g : (\mathcal{P}(\mathcal{Y}) = \mathcal{I}) \to \mathcal{P}(\mathcal{X})$, $g(Y) = \{x \in \mathcal{X} | \forall y \in Y, (x,y) \in \mathcal{R}, Y \subseteq \mathcal{Y}\}$.

The operators f and g form a **Galois connection** between the sets $\mathcal{I}$ and $\mathcal{P}(\mathcal{X})$ [1]. Consequently, both compound operators $f \circ g$ and $g \circ f$ are closure operators defined respectively on $\mathcal{P}(\mathcal{X})$ and $\mathcal{I}$. The operator $f \circ g$ generates closed subsets of objects and $g \circ f$ generates closed subsets of properties or items called closed itemsets [14]. Operators f and g satisfy for any subsets $A, A_1, A_2 \subseteq \mathcal{X}$, and $B, B_1, B_2 \subseteq \mathcal{Y}$:

$$A_1 \subseteq A_2 \Rightarrow f(A_2) \subseteq f(A_1),$$
$$B_1 \subseteq B_2 \Rightarrow g(B_2) \subseteq g(B_1),$$
$$A \subseteq g \circ f(A),$$
$$B \subseteq f \circ g(B).$$

Definition 4. *Formal Concept:* *A formal concept $C = (A, B)$ is a pair where $A \subseteq \mathcal{X}, B \subseteq \mathcal{Y} / A = g(B)$ and $B = f(A)$. Set A is called the extent of C, and set B is called its intent [6]. A formal concept is also called* maximal rectangle *or a* non-enlargeable rectangle *[10, 11].*

Example 2. From the BR presented in Table 1, the following pair $(A, B) = (\{o_3, o_4, o_5\}, \{f, g\})$ is a FC since we have $f(A) = B$ and $g(B) = A$. The set B is a closed itemset.

2.2 Difunctionality

In the sequel, we recall some important definitions related to the difunctionality, as a closed notion to an uniformity aspect, that we may find or extract from binary relations. We focus on difunctional and fringe relations. Let us consider two sets $\mathcal{X}$ and $\mathcal{Y}$, a BR $\mathcal{R}$ as subset of the Cartesian product of $\mathcal{X}$ and $\mathcal{Y}$. We start by presenting the following definitions:

Definition 5. *Difunctional Relation [6]:* *Let $\mathcal{R}$ be a BR. $\mathcal{R}$ is said to be difunctional iff $\mathcal{R} \circ \mathcal{R}^{-1} \circ \mathcal{R} = \mathcal{R}$ where $\circ$ stands for the relative product and*

$\mathcal{R}^{-1}$ *is the inverse of* $\mathcal{R}$. $\mathcal{R}$ *can be written in block diagonal form by suitably rearranging rows and columns. This means that a difunctional relation presents uniformity of the association of objects and properties.*

Even though a BR is not always uniform as a difunctional, *Riguet* proved that there is an interesting difunctional embedded in it, which is generally not empty [17]. This relation is called a *fringe* relation defined as follows.

Definition 6. *Fringe Relation [17]: A fringe relation is a difunctional relation embedded in a BR* $\mathcal{R}$ *and computed by* $\mathcal{R} \circ \overline{\mathcal{R}^{-1}} \circ \mathcal{R} \cap \mathcal{R}$. *It is denoted* $\mathcal{R}_d$. *Moreover, a fringe relation is composed by a set of points contained in just one maximal rectangle inside the relation* $\mathcal{R}$ *which can play an important role in different applications.*

Example 3. In Table 2, we present the fringe relation corresponding to the BR $\mathcal{R}$ illustrated in Table 1. The difunctionality aspect is observed in the fringe relation as a sparse sub-relation.

Table 2. The Fringe relation corresponding to the BR presented in Table 1

	e	c	f	g	h	a	b	d
o_0	0	0	0	0	0	0	0	0
o_1	1	0	0	0	0	0	0	0
o_2	0	1	0	0	0	0	0	0
o_3	0	0	0	0	0	0	0	0
o_4	0	0	0	0	0	0	0	0
o_5	0	0	1	1	0	0	0	0
o_6	0	0	0	0	1	0	0	0

The elements belonging to a fringe relation are called *isolated points* and their usefulness are shown in extracting optimal coverage of formal contexts [4].

3 Related Work

The first attempt to define a lattice theory as a mathematical model was made by BIRKHOFF [8] in the 1940s. An underlying deep concept is the notion of the Galois connection that emerged in the early 40s after a long period of gestation that started at the beginning of the previous century. Later, the researches have demonstrated how concept lattices formalize conceptual structures by coding any kind of duality, such as the duality between the intent and the extent of a concept. Application in data analysis using this duality for analyzing questionnaire data was done by BARBUT and MONJARDET in the domain of social sciences [1].

The concept lattice, also named the "Galois lattice", was promoted by WILLE, and then extended to FCA [6]. FCA mathematical settings have recently been shown to provide a theoretical framework for the efficient resolution of many practical problems, e.g., data mining, conceptual reasoning, software engineering and information retrieval, to cite but a few. The reader is referred to [16] to a critical overview of the myriad of applicative cases of FCA. An interesting problem is to find a representation of a formal context by a minimal number of concepts. By considering only the isolated points, we may obtain an optimal, or a reduced formal context representation reflecting the potentially most relevant knowledge that might be discovered. In this respect, finding optimal coverage of a binary relation is known to be NP-hard problem [7]. Nevertheless, we witness a large number of work interested in tackling such a problem. BELKHITER et al. [13] introduced an optimal rectangular decomposition of a binary relation as well as an application to documentary databases. The introduced decomposition is based on the election of optimal maximal rectangles (or equivalently formal concepts) that achieve a maximal gain in storage space terms. Later Kcherif et al. [11] introduced a rectangular decomposition approach based on the RIGUET's difunctional relation [17]. The computation of this difunctional is reduced to the determination of a set of isolated points allowing the determination of the minimal set of rectangles covering a given binary relation. An extended isolated points based version, called *genCoverage*, was proposed in [4] by considering the extended context notion. BELOHLAVEK and VYCHODIL [3] tackled the same issue by attempting to solve the *Boolean factor analysis* problem by proposing a new method of decomposition an $n \times m$ binary matrix I into a Boolean product $A \circ B$ of an $n \times k$ binary matrix A and a $k \times m$ binary matrix B with k as small as possible. Mouakher and Ben Yahia [12] have introduced a new approach, based on a greedy algorithm, for the extraction of an optimal covering of a binary relation. The latter approach relies on the formal concept lattice representation. The guiding idea of MOUAKHER and BEN YAHIA's approach is that the coverage should not be extracted regardless of the quality of knowledge that may be drawn from it. That's why the authors introduced a gain function based on the assessment of the correlation of the intent part of pertinent formal concepts. Generally, the number of formal concepts grows exponentially with the size of the matrix (i.e. binary relation) [9]. Fortunately, the use of mandatory concepts, allows to select an information with minimal loss set of few relevant concepts in polynomial time. The extraction of the coverage of a binary relation $\mathcal{R}$ relies on the use of the Fringe relation. It has the advantage of helping discover *isolated properties*, e.g. those single properties belonging to only one concept. Moreover, from the given coverage, we may regenerate any concept belonging to the lattice of concepts. We may also generalize Galois Connection operators to be applied on any conceptual minimal coverage. In this way, for information retrieval purposes, it is possible to use parallel processing or cloud computing to make fast cooperative algorithms. In the following, we introduce our new approach for conceptual coverage construction based on N-composites isolated points.

4 Conceptual Coverage Based on N-Composites Isolated Points

As our main contribution, we propose to use the isolated points for producing progressively the conceptual coverage. More specifically, we propose to discover the FC for different levels $(l_1, l_2 \ldots, l_N)$ of i properties combinations, where $i \in [1, N]$ and N is the maximal number of properties to be combined together. In practice, $N = 2$ means that the properties in the formal context are combined 2 by 2 in order to locate the 2-composite isolated points. For each level l_i we show how the i-composite-isolated points are located and their corresponding FC is extracted. In the following, we present necessary definitions related to isolated points. We recall some of their useful properties and we present a novel approach for obtaining N-composite ones.

Definition 7. Isolated Point [11]: *Let $\mathcal{R}$ be a BR and $\mathcal{R}_d$ its associated fringe relation. If $p = (x, y) \in \mathcal{R}_d$ then p is called an isolated point. An isolated point belongs to only one formal concept and the latter is called an isolated concept.*

Example 4. With respect to the BR presented in Table 1, the FC $C = (A, B) = (\{o_3, o_4, o_5\}, \{f, g\})$ is an isolated one since (o_5, g) is an isolated point.

Proposition 1. *Let $p = (x, y)$ be an isolated point in a FC $C = (A, B)$. The following properties are satisfied [5]:*

$$
\begin{aligned}
g(\{y\}) &= A \\
f(\{x\}) &= B \\
f \circ g(\{y\}) &= B \\
g \circ f(\{x\}) &= A
\end{aligned}
$$

Remark 1. From Proposition 1, particularly the expression $f \circ g(\{y\}) = B$ means that if an object o in $\mathcal{X}$ contains the property y, then it contains necessarily all properties in B. Hence we have the implication $y \to B \backslash \{y\}$. The property y could be selected as an interesting candidate representing all properties in B since they are implied by y.

Definition 8. Composite Properties [5]: *Let $(\mathcal{X}, \mathcal{Y}, \mathcal{R})$ be a FCT, and let $SP \subseteq \mathcal{Y}$, i.e., a subset of properties such that $|SP| \geq 2$. A composition of the properties of SP defines a new composite property, denoted by $CP_{SP} = p_1.p_2 \ldots p_n$, $n = |SP|$, $p_i \in SP$, $i = 1 \ldots n$, which can be added to the set $\mathcal{Y}$. This new property CP_{SP} can be used to define an extended context as well as to select a composite label for a formal concept when a single label cannot be assigned.*

Definition 9. N-Composite Property: *Let N be a positive integer and $SP \subseteq \mathcal{Y}$. A N-Composite property denoted by NCP_{SP} is a CP_{SP} such that $|SP| = \mathcal{N}$.*

Table 3. Composite properties extracted for different levels

N-composites properties	Examples
1-composite	$\{a\}, \{b\}, \{f\}, \{g\}$
2-composite	$\{a, b\}, \{f, g\}\{d, e\}$
3-composite	$\{a, b, c\}, \{f, g, h\}$

Definition 10. *Expanded context:* *Let $FCT = (\mathcal{X}, \mathcal{Y}, \mathcal{R})$ be a Formal context, and let NCP_{SP} be a N-Composite property. NCP_{SP} can be added to the set $\mathcal{Y}$ to define an expanded context: $EXP_{(FC,SP)} = (\mathcal{X}, \mathcal{Y} \cup NCP_{SP}, \mathcal{R}')$ where $\mathcal{R}' = \mathcal{R} \cup \{(x, NCP_{SP}) | (x, y) \in R, \forall y \in SP\}$.*

Definition 11. *N-Composite property isolated point:* *Let $FC = (\mathcal{X}, \mathcal{Y}, \mathcal{R})$ and NCP_{SP} be a N-Composite property. NCP_{SP} is a N-composite Isolated point if there exists $x \in \mathcal{X} | (x, NCP_{SP})$ is an isolated point in the expanded context $EXP_{(FC,SP)}$.*

Remark 2. Why it is important to make properties composition?
There are two main reasons:

- Discover the hidden regularity in the data implied by the N-composite isolated points.
- Discover additional isolated concepts in order to cover more objects in a FCT. Note that for some applications, the coverage criteria is not the most important aspect. We are more focusing on extracting FC without having necessarily 100% of coverage percentage.

Remark 3. How could we make properties composition?
The isolated points have some important properties that would prevent any randomized properties combination. In particular, we would check whether some properties combinations are useless or not, mainly when isolated points are involved. As depicted in Fig. 1, we distinguish three cases for isolated points combination as illustrated by $y_1.y_2$:

- **Case 1:** The isolated points belong to two distinct concepts. Therefore the intersection between the concepts is empty and it is useless to make the isolated points composition. The Proposition 2 formalizes the case 1.
- **Case 2:** The composition $y_1.y_2$ is not empty and we would check if some additional isolated points might be derived.
- **Case 3:** Even if $y_1.y_2$ is not empty, the composition is useless. In fact, the isolated point (x_2, y_2) belongs to an already covered concept.

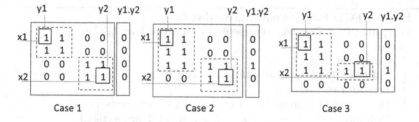

Fig. 1. Three cases for Isolated points compositions

Proposition 2. *Let $p_1 = (x_1, y_1)$ and $p_2 = (x_2, y_2)$ be two isolated points in two distinct concepts $C_1 = (A_1, B_1)$ and $C_2 = (A_2, B_2)$. Let $SP = \{y_1, y_2\}$ and $2CP_{SP}$ its corresponding 2-composite property. The following property is satisfied:*

- *If $A_1 \cap A_2 = \emptyset$ then $2CP_{SP}$ is empty, i.e., $g(\{SP\}) = \emptyset$. Roughly speaking, no 2-composite isolated point is derivable.*

Remark 4. Two points worth noting need to be outlined:

- The previous proposition is important in properties composition algorithm. Indeed, it shows that it is useless to make a composition between two isolated points when they belong to two distinct formal concepts.
- As long as we increase the value of N, for the N-Composite properties, we got, normally, new FC covering additional elements in the FCT. It depends on how the algorithm considers the properties combination. Furthermore, in practice, when the size of the data is high, it could be recommended to limit the value of N to a small value (e.g., 2 or 3) even in the case of non-covering the whole FCT.

4.1 Main Algorithm

The pseudo-code presented in Algorithm 1 reflects the steps we follow to compute the N-composite isolated properties and to derive the conceptual coverage. First, we identify the set of isolated points and their corresponding concepts. If the formal context is not covered and the composition level N is not reached, then we iterate on the remaining set of composite words and only keep those representing the isolated points.

- In lines 2–8: we prepare the closures sets related to objects and proportion respectively.
- In lines 9–13: we locate the current level-composite isolated point w.r.t the properties presented in Proposition 2. We update the coverage Cov and we remove the covered elements from ListNonCovered.
- In lines 16–19: we combine $(i+1)$ properties as new composite isolated points candidates. We relate this part to the Remark 3 and to Proposition 2.

Input: - FCT: Formal Context , N: Max Level
Output: - Cov: Coverage represented as a set of Formal Concepts

1 **begin**
2   Cov ← NULL
3   //**Phase 1:** *prepare the hMaps OL and PCLosure (or dictionaries as presented in Python)*
4   OL ← {objId : [properties] **for** objId **in** FCT.objects }
5   PClosure ← {p: [g(p), $f \circ g(p)$] **for** p **in** FCT.properties }
6   //**Phase 2:**
7   **level** ← 1
8   ListNonCovered ← OL **while** *(level ≤ N) and NotEmpty(ListNonCovered)* **do**
9   composites ← ExtractCompositeIsolated(PClosure,**level**)
10   //*For level 1, the composites correspond to the isolated points in the Fringe relation*
11   Cov ← Cov ∪ $f \circ g(composites)$
12   ListNonCovered.Remove(Concepts(composites))
13   **level** ← level+1
14   PClosure.addNewLevel(**level**)
15   i ← 0
16   **while** *(i ≤ PClosure.level)* **do**
17   ListComposites ← Combine(i,PClosure,**level**)
18   PClosure.add(ListComposites)

19   **return** *Cov*

Algorithm 1: N-Composites Coverage Algorithm

Complexity Analysis. Regarding the complexity analysis, let $n = |\mathcal{X}|$ and $m = |\mathcal{Y}|$ be respectively the number of objects and the number of properties in the FCT.

– In phase 1, the hMaps preparation require $n \times m$ operations in to build OL and $PClosure$.
– In phase 2, the algorithm performs the $PClosure$ traversal and locates the level-composite isolated points. In the worst case, we have $M = max(N,T)$ where N is the 'Max Level' specified by a user and T is the traversals number. Let us consider a composition level i for which $|PClosure| = m_i$. The $PClosure$ property composition requires $m_i + (m_i - 1) + (m_i - 2) + \ldots + 1 = \frac{m_i(m_i+1)}{2}$ operations. Hence, there are $\sum_{i=1}^{M} \frac{m_i(m_i+1)}{2}$ operations.

4.2   Illustrative Example

Let us consider the FCT given by Table 1. Figure 2 depicts the different phases of Algorithm 1. In phase 1, 'PClosure' and 'OL' were built. PClosure contains the properties, their closures and their associated objects obtained by Galois connection. OL indicates the properties for each object. In phase 2, we have two levels. At level 1, the "ExtractCompositeIsolated()" routine returns the set of 1-Composite isolated points. Since, with the obtained formal concepts from the isolated points, we do not cover the initial formal concepts, we move to combining at Level 2. At that level, the "combine()" routine combines 2 properties, w.r.t the Remark 3 and the Proposition 2, and returns the "ListComposites". The latter represents the new composite isolated points candidates. In the second iteration, the "ExtractCompositeIsolated()" returns the 2-composite isolated points and the context is entirely covered. The N-composite Algorithm output represents the set of isolated concepts. Hence, we have a partial-total FCT coverage, so it informs about the coverage percentage; As we can see in Fig. 2, at level 1, the 1-composite isolated points $\{(o_2, c), (o_1, e), \{(o_5, g), (o_5, f)\}, (o_6, h)\}$ are associated, respectively, to the formal concepts $\{C_1, C_2, C_4, C_6\}$. They cover 12/17 or (70.58%) elements in the FCT given by Table 1. At level 2, the 2-composite isolated points are associated to the additional concepts $\{C_0, C_3, C_5\}$. Hence, we achieve 100% coverage.

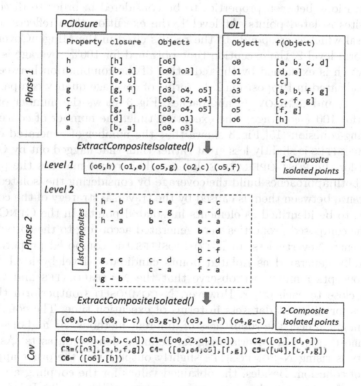

Fig. 2. Illustrative example

5   Experimental Results

Our results show the efficiency of the proposed algorithm. The implementation of the solution was realized and executed on a laptop i7-3610 QM 2.3 GHz, 8 GB of RAM, and Windows 7. The results concern some **Benchmark datasets** furnished by the UC Irvine Machine Learning Database Repository[1]. They are used extensively within the data mining and machine learning communities. Moreover, they are considered dense (i.e., yielding a large number of formal concepts even for a small number of objects and attributes). In Table 4, we present the number of N-composite isolated points, the execution time and the coverage percentage while increasing the composition level for different data sets. At level 1, the number of N-composite isolated points varies from 0 to 10. This reduced value reflects the absence of isolated points in the dense data sets. As long as we make the properties composition by 2 up to 5, the hidden data regularity is discovered and the concepts number increases. In most cases, starting from the level 3, more than 80% of the data is covered. In terms of temporal complexity, we remark that the time increases as the level does. This is due to the multiple combinations between properties to be considered in order to discover the N-composites isolated points. At level 1, the execution time reflects one data set traversal which corresponds to the results cited in the MingenCoverage [4]. Starting form level 2, the execution time obtained for 100% coverage is slightly reduced which is explained by the reduction of the combination between properties as indicated in Proposition 2. In terms of coverage quality compared with KCHERIF [11] and GENCOV [5] approaches, Fig. 3 shows the number of formal concepts (for 100% coverage), the execution time, the number of concepts and the coupling-cohesion [15] Fig. 3 shows that the number of generated concepts of N-COMPOSITES is slightly less great rather than that flagged out by GENCOV [5] except for the POST-OPERATIVE dataset. This result matches the predicted one since both approaches build the coverage by considering the isolated points. The difference between them is caused by the traversal strategy of the composed properties, to be identified as elements in isolated points. In the GENCOV procedure, the composed properties were generated according to the sorted matrix rows elements. Nevertheless, in N-COMPOSITES the composed properties were incrementally generated as isolated points candidates, level by level. Regarding the concepts number, we observe that the N-COMPOSITES and GENCOV are much close to each other. However, N-COMPOSITES outperform the other approaches for different datasets in terms of execution time. The coupling and cohesion might reflect the robustness of different methods. At a glance, we can see that most values are overlapping for reduced density datasets. As long as the density is rising N-COMPOSITES slightly outperforms the other approaches in terms of cohesion. Besides, the obtained values for the coupling are still the same except for the operative datasets known for its high density. It is important to mention that the obtained results may not outperform all those reported in the surveyed literature, yet they appear to be adhering to acceptable standards.

---

[1] http://archive.ics.uci.edu/ml/.

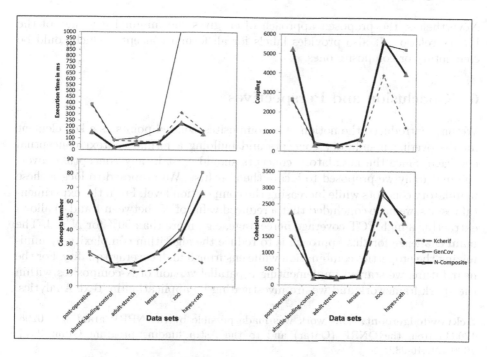

Fig. 3. Comparison between N-Composites and the other approaches

Table 4. Feature evolution w.r.t. the coverage percentage

	Level	#N-Composites	Time (ms)	Coverage percentage
Lenses	1	0	-	-
	2	4	29	16.66%
	3	11	34	45.83%
	4	24	33	100.00%
Car	1–3	0	-	00.00%
	4	1724	668	100.00%
Hayes-Roth	1	0	-	-
	2	52	70	50.75%
	3	78	79	100.00%
Lung-Cancer	1	10	160	31.00%
	2	32	1154	100.00%
Dermatology	1	0	-	-
	2	155	2470	42.62%
	3	332	26493	91.25%
	4	363	35268	99.72%
	5	364	36899	100.00%
Zoo	1	4	79	19.60%
	2	33	94	61.71%
	3	47	119	74.70%
	4	52	133	99.65%

Nevertheless, the proposed approach often gives the minimal coverage of the binary relation. It also provides labels for all formal concepts, which could be elementary or composite ones.

6   Conclusion and Perspectives

We have introduced the notion of N-composites isolated points as a key element for discovering mandatory concepts and building a formal context conceptual coverage. Since the mandatory concepts should exist in any conceptual coverage, naturally we proposed to locate them at first. We proposed to locate these mandatory concepts while increasing the composition level. From the experimental results, we have concluded that a reduced value of N (between 1 and 4) allows extracting a high FCT coverage percentage, e.g., more than 90% for $N = 4$. The main challenge for this approach is to reduce the algorithm complexity by minimizing the properties combination while ensuring a high coverage quality. For the near future, we started implementing a parallel version of N-composites within the spark framework and we are investigating its application in Text Analytics.

Acknowledgement. This work was made possible by the NPRP grant #10-0205-170346 from the QNRF (Qatar) and to the Astra funding program Grant 2014-2020.4.01.16-032.

References

1. Barbut, M., Monjardet, B.: Ordre et classification. Algèbre et Combinatoire. Hachette, Tome II (1970)
2. Bastide, Y., Pasquier, N., Taouil, R., Stumme, G., Lakhal, L.: Mining minimal non-redundant association rules using frequent closed itemsets. In: Lloyd, J., et al. (eds.) CL 2000. LNCS (LNAI), vol. 1861, pp. 972–986. Springer, Heidelberg (2000). https://doi.org/10.1007/3-540-44957-4_65
3. Belohlavek, R., Vychodil, V.: Discovery of optimal factors in binary data via a novel method of matrix decomposition. J. Comput. Syst. Sci. **76**(1), 3–10 (2010)
4. Elloumi, S., Ferjani, F., Jaoua, A.: Using minimal generators for composite isolated point extraction and conceptual binary relation coverage: application for extracting relevant textual features. Inform. Sci. **336**, 129–144 (2016)
5. Ferjani, F., Elloumi, S., Jaoua, A., Ben Yahia, S., Ismail, S., Ravan, S.: Formal context coverage based on isolated labels: an efficient solution for text feature extraction. Inform. Sci. **188**, 198–214 (2012)
6. Ganter, B., Wille, R.: Formal Concept Analysis. Springer, Heidelberg (1999). https://doi.org/10.1007/978-3-642-59830-2
7. Garey, M., Johnson, D.: Computers and Intractability: A Guide to the Theory of NP-Completness. W. H. Freeman & Co., New York (1979)
8. Birkhoff, G.: Lattice Theory, 1st edn. American Mathematical Society, Providence (1965)
9. Godin, R., Mineau, G.W., Missaoui, R., Mili, H.: Méthodes de Classification Conceptuelle Basées sur le Treillis de Galois et Applications. Revue d'intelligence Artificielle **9**(2), 105–137 (1995)

10. Jaoua, A., Elloumi, S., Hasnah, A., Jaam, J., Nafkha, I.: Discovering regularities in databases using canonical decomposition of binary relations. J. Relational Methods Comput. Sci. (JoRMiCS) **1**, 217–234 (2004)
11. Kcherif, R., Gammoudi, M., Jaoua, A.: Using difunctional relations in information organization. Inf. Sci. **125**, 153–166 (2000)
12. Mouakher, A., Ben Yahia, S.: Anthropocentric visualisation of optimal cover of association rules. In: Kryszkiewicz, M., Obiedkov, S. (eds.) Proceedings of the 7th International Conference on Concept Lattices and Their Applications (CLA 2010), Sevilla, Spain, pp. 211–222, October 2010
13. Bourhfir, N.B., Gammoudi, C., Jaoua, A., Thanh, N.L., Reguig, M.: Décomposition rectangulaire optimale d'une relation binaire: application aux bases de données documentaires. INFOR Inf. Syst. Oper. Res. **32**(1), 33–54 (1994)
14. Pasquier, N., Bastide, Y., Taouil, R., Lakhal, L.: Efficient mining of association rules using closed itemset lattices. Inform. Syst. J. **24**(1), 25–46 (1999)
15. Paul, G., Scott, D.: New coupling and cohesion metrics for evaluation of software component reusability. In: Proceedings of the 9th International Conference for Young Computer Scientists (ICYCS 2008), Hunan, China, pp. 1181–1186, 18–21 November 2008
16. Poelmans, J., Ignatov, D.I., Kuznetsov, S.O., Dedene, G.: Formal concept analysis in knowledge processing: a survey on applications. Expert Syst. Appl. **40**(16), 6538–6560 (2013). https://doi.org/10.1016/j.eswa.2013.05.009
17. Riguet, J.: Relations binaires, fermetures et correspondances de galois. Bull. Soc. Math. France **78**, 114–155 (1948)

Topological Data Analysis with ϵ-net Induced Lazy Witness Complex

Naheed Anjum Arafat[1(✉)], Debabrota Basu[2], and Stéphane Bressan[1]

[1] School of Computing, National University of Singapore, Singapore, Singapore
naheed_anjum@u.nus.edu
[2] Department of Computer Science and Engineering,
Chalmers University of Technology, Göteborg, Sweden

Abstract. Topological data analysis computes and analyses topological features of the point clouds by constructing and studying a simplicial representation of the underlying topological structure. The enthusiasm that followed the initial successes of topological data analysis was curbed by the computational cost of constructing such simplicial representations. The lazy witness complex is a computationally feasible approximation of the underlying topological structure of a point cloud. It is built in reference to a subset of points, called landmarks, rather than considering all the points as in the Čech and Vietoris-Rips complexes. The choice and the number of landmarks dictate the effectiveness and efficiency of the approximation.

We adopt the notion of ϵ-cover to define ϵ-net. We prove that ϵ-net, as a choice of landmarks, is an ϵ-approximate representation of the point cloud and the induced lazy witness complex is a 3-approximation of the induced Vietoris-Rips complex. Furthermore, we propose three algorithms to construct ϵ-net landmarks. We establish the relationship of these algorithms with the existing landmark selection algorithms. We empirically validate our theoretical claims. We empirically and comparatively evaluate the effectiveness, efficiency, and stability of the proposed algorithms on synthetic and real datasets.

1 Introduction

Topological data analysis computes and analyses topological features of generally high-dimensional and possibly noisy data sets. Topological data analysis is applied to eclectic domains namely shape analysis [6], images [2,22], sensor network analysis [8], social network analysis [3,24,25,27], computational neuroscience [21], and protein structure study [19,30].

The enthusiasm that followed the initial successes of topological data analysis was curbed by the computational challenges posed by the construction of an exact simplicial representation, the Čech complex, of the point cloud. A simplicial representation facilitates computation of basic topological objects such as simplicial complexes, filtrations, and persistent homologies. Thus, researchers

© Springer Nature Switzerland AG 2019
S. Hartmann et al. (Eds.): DEXA 2019, LNCS 11707, pp. 376–392, 2019.
https://doi.org/10.1007/978-3-030-27618-8_28

devised approximations of the Čech complex as well as its best possible approximation the Vietoris-Rips complex [7, 10, 26]. One of such computationally feasible approximate simplicial representation is the *lazy witness complex* [7]. The lazy witness complex is built in reference to a subset of points, called *landmarks*. The choice and the number of landmarks dictate the effectiveness and efficiency of the approximation.

We adopt the notion of ε-cover [18] from analysis to define and present the notions of ε-sample, ε-sparsity, and ε-net (Sect. 4) to capture bounds on the loss of topological features induced by the choice of landmarks. We prove that an ε-net is an ε-approximate representation of the point cloud with respect to the Hausdorff distance. We prove that the lazy witness complex induced by an ε-net, as a choice of landmarks, is a 3-approximation of the induced Vietoris-Rips complex. ε-net allows us to provide such approximation guarantees for lazy witness complex (Sect. 4.2) which was absent in the literature. Furthermore, we propose three algorithms to construct ε-net as landmarks for point clouds (Sect. 5). We establish their relationship with the existing landmark selection algorithms, namely *random* and *maxmin* [7]. We empirically and comparatively show that the size of the *ε-net landmarks* constructed by the algorithms varies inversely with ε and agrees with the known bound on the size of ε-net [20].

We empirically and comparatively validate our claim on the topological approximation quality of the lazy witness complex induced by the ε-net landmarks (Sect. 6). Furthermore, we empirically and comparatively validate the effectiveness, efficiency and stability of the proposed algorithm on representative synthetic point clouds as well as a real dataset. Experiments confirm our claims by showing equivalent effectiveness of the algorithms constructing ε-net landmarks with the existing maxmin algorithm. We also show the ε-net landmarks to be more stable than those selected by the algorithms maxmin and random as ε-net incurs narrower confidence band in the persistent landscape topological descriptor. We conclude (Sect. 7) with the theoretical and experimental pieces of evidence that validate the ε-nets constructed as a stable and effective way to construct landmarks and to induce lazy witness complexes.

2   Related Works

Applications of TDA. TDA is applied in different domains mostly on relatively small datasets and up to dimension 2 due to computational intractability of the popular Čech and Vietoris-Rips complexes. [6] computed homology classes at dimension 0 for their proposed tangential filtration of point clouds of handwritten digits for image classification (dataset size ∼69–294). [28] used the persistence pairs at dimension 0 for segmenting mesh on benchmark mesh segmentation datasets (size ∼50000). Researchers applying TDA to network analysis focus on characterising networks using features computed from persistence homology classes. [3] and [25] computed persistence homology at dimension 0,1 and 2 of the clique filtration to study weighted collaboration networks (size ∼36000) and weighted networks from different domains (size ∼54000) respectively. In biological networks, [11] clustered gene co-expression networks (size ∼400) based on distances between Vietoris-Rips persistence diagram computed on each network. In molecular biology, persistent homology reveals different conformations of proteins [19,30] based on the strength of the bonds of the molecules.

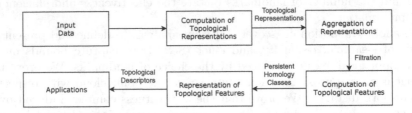

Fig. 1. Components of topological data analysis.

Topological Approximation. Computational infeasibility of the Čech complex and Vietoris-Rips complex motivates the development of approximate simplicial representations such as the lazy witness complexes, sparse-Rips complex [26] and graph induced complex (GIC) [10]. Sparse-Rips complex [26] perturbs the distance metric in such a way that when the regions covered by a point can be covered by its neighbouring points, that point can be deleted without changing the topology. Given a graph constructed on a point cloud as input, the graph induced complex is a simplicial complex built on a subset of vertices, where the vertices of a k-simplex are the nearest neighbours of a clique-vertices of a k-clique [10]. Due to their computational benefits, lazy witness complex and graph induced complexes have found applications in studying natural image statistics [7] and image classification [9].

Applications of ϵ-net. The concept of ϵ-net is a standard concept in analysis and topology [18] originating from the idea of (δ, ϵ)-limits formulated by Cauchy. Nets have been used in nearest-neighbor search [20]. [15] proposed the Net-tree data structure to represent ϵ-nets at all scales of ϵ. Net-tree is used to construct approximate well-separated pair decompositions [15] and approximate geometric spanners [15]. The simplicial complexes in the graph induced complex are nets. Sparse-Rips filtration constructs a net-tree on the point-cloud to decide which neighbouring points to delete. [14] used ϵ-net for manifold reconstruction.

3   Topological Data Analysis

Topological data analysis is the study of computational models for efficient and effective computation of topological features, such as persistent homology classes, from different datasets, and representation of the topological features using different topological descriptors, such as persistence barcodes, for further analysis and application [12,23]. In this section, we represent the computational blocks of topological data analysis in Fig. 1 and further describe each of the blocks.

Topological data analysis computes the topological features, such as persistent homology classes, by computing the topological objects called *simplicial complex* for a given dataset. A **simplicial complex** is constructed using simplices. Formally, a *k-simplex* is the convex-hull of $(k+1)$ data points. For instance, A 0-simplex $[v_0]$ is a single point, a 1-simplex $[v_0 v_1]$ an edge, and a 2-simplex $[v_0 v_1 v_2]$ a filled triangle. A *k*-**homology class** is an equivalent class of such *k*-simplicial complexes that cannot be reduced to a lower dimensional simplicial complex [12]. In order to compute the *k*-homology classes, a practitioner does not have direct access to the underlying space of the point cloud and it is combinatorially hard to compute the exact simplicial representation of Čech complex [31]. Thus, different approximations of the exact simplicial representation are proposed: *Vietoris-Rips* complex [17] and *lazy witness* complex [7].

The **Vietoris-Rips complex** $R_\alpha(D)$, for a given dataset D and real number $\alpha > 0$, is an abstract simplicial complex representation consisting of such *k*-simplices, where any two points v_i, v_j in any of these *k*-simplices are at distance at most α. Vietoris-Rips complex is the best possible ($\sqrt{2}$-)approximation of the Čech complex, computable with present computational resources, and is extensively used in topological data analysis literature [23]. Thus, we use the Vietoris-Rips complex as the baseline representation in this paper. In the worst case, the number of simplices in the Vietoris-Rips complex grows exponentially with the number of data points [23,31]. Lazy witness complex [7] approximates the Vietoris-Rips complex by constructing the simplicial complexes over a subset of data points L, referred to as the landmarks. Formally, given a positive integer ν and real number $\alpha > 0$, the **lazy witness complex** $LW_\alpha(D, L, \nu)$ of a dataset D is a simplicial complex over a landmark set L where for any two points v_i, v_j of a *k*-simplex $[v_0 v_1 \cdots v_k]$, there is a point w whose $(d^\nu(w) + \alpha)$-neighborhood[1] contains v_i, v_j. In the worst case, the size of the lazy witness complexes grows exponentially with the number of landmarks. Less number of landmarks facilitates computational acceleration while produces a bad approximation of Vietoris-Rips with loss of topological features. Thus, the trade-off between the approximation of topological features and available computational resources dictates the choice of landmarks.

As the value of filtration parameter α increases, new simplices arrive and the topological features, i.e. the homology classes, start to appear. Some of the homology classes merge with the existing classes in a subsequent simplicial complex, and some of them persist indefinitely [12]. In order to capture the evolution of topological structure with scale, topological data analysis techniques construct a sequence of simplicial complex representations, called a *filtration* [12], for an increasing sequence of α's. In a given filtration, the persistence interval of a homology class is denoted by $[\alpha_b, \alpha_d)$, where α_b and α_d are the filtration values of its appearance and merging respectively. The persistence interval of an indefinitely persisting homology class is denoted as $[\alpha_b, \infty)$.

Topological descriptors, such as persistence diagram [12] and persistence landscapes [1] represent persistence intervals as points and functions respectively

---

[1] $d^\nu(w)$ is the distance from point $w \in D$ to its ν-th nearest point in L.

in order to draw qualitative and quantitative inference about the topological features. Distance measures between persistent diagrams such as the bottleneck and Wasserstein distances [12] are often used to draw quantitative inference. The bottleneck distance between two diagrams is the smallest distance d for which there is a perfect matching between the points of the two diagrams such that any pair of matched points are at distance at most d [12]. The Wasserstein distance between two diagrams is the cost of the optimal matching between points of the two diagrams [12].

4 ϵ-net

As we discussed in Sect. 3, topological data analysis of a dataset begins with the computation of simplicial complex representations. Though Vietoris-Rips is the best possible approximation of the Čech complex, it incurs an exponential computational cost with respect to the size of the point cloud. Thus, lazy witness complex is often used as a practical solution for scalable topological data analysis. Computation of lazy witness complex is dependent on the selection of landmarks. Selection of landmarks dictates the trade-off between *effectiveness*, i.e. the quality of approximation of topological features, and *efficiency*, i.e. the computational cost of computing the lazy witness complex.

ϵ-cover is a construction used in topology to compute the inherent properties of a given space [18]. In this paper, we import the concept of ϵ-cover to define ϵ-net of a point cloud. We use the ϵ-net of the point cloud as the landmarks for constructing lazy witness complex. We show that ϵ-net, as a choice of landmarks, has guarantees such as being an ϵ-approximate representation of the point cloud, its induced lazy witness complex being a 3-approximation of its induced Vietoris-Rips complex, and also bounding the number of landmarks for a given ϵ. These guarantees are absent for the other existing landmark selection algorithms (Sect. 5) such as random and maxmin algorithms.

4.1 ϵ-net of a Point Cloud

ϵ-cover is a set of subsets of a point cloud in an Euclidean space such that these subsets together cover the point cloud, but none of the subsets has a diameter more than ϵ.

Definition 1 (ϵ-cover [18]). *An ϵ-cover of a point cloud P is the set of P_i's such that $P_i \subseteq P$, $P = \cup_i P_i$, and diameter^2 of P_i is at most $\epsilon \geq 0$ for all i.*

When the sets in the 2ϵ-cover of P are Euclidean balls of radius ϵ, the set of centres of the balls is termed as an ϵ-sample of set P.

Definition 2 (ϵ-sample [14]). *A set $L \subseteq P$ is an ϵ-sample of P if the collection $\{B_\epsilon(x) : x \in L\}$ of ϵ-balls of radius ϵ covers P, i.e. $P = \cup_{x \in L} B_\epsilon(x)$.*

---

^{2} The diameter $diam(P_i)$ of a set $P_i \subseteq P$ is defined as the largest distance $d(x,y)$ between any two points in $x, y \in P_i$.

According to the definition of ε-sample, P is an ε-sample of itself for $\epsilon = 0$. For decreasing further computational expense, it is desirable to have an ε-sample is sparse that means it contains as less number of points as possible. An ε-sparse subset of P is a subset where any two points are at least ϵ apart from each other.

Definition 3 (ε-sparse). *A set $L \subset P$ is ε-sparse if for all $x, y \in L$, $d(x, y) > \epsilon$.*

An ε-net of set P is an ε-sparse subset of P which is also an ε-sample of P.

Definition 4 (ε-net [18]). *Let (P, d) be a metric space and $\epsilon \geq 0$. A subset $L \subset P$ is called an ε-net of P if L is ε-sparse and an ε-sample of P.*

4.2 Properties of ε-nets

ε-net of a point cloud comes with approximation guarantees irrespective of its algorithmic construction. An ε-net of a point cloud of diameter Δ in Euclidean space $\mathbb{R}^D$ is an ε-approximation of the point cloud in Hausdorff distance. The lazy witness complex induced by an ε-net is a 3-approximation of the Vietoris-Rips complex on that ε-net. Furthermore, the size of an ε-net is at most $(\frac{\Delta}{\epsilon})^{\theta(D)}$ [20]. Here, we establish the first two approximation guarantees of ε-net theoretically.

Point-Cloud Approximation Guarantee of an ε-net. We use Lemma 1 to prove that the ε-net of a point cloud P is an ε-approximate representation of that point cloud in Hausdorff distance.

Lemma 1. *Let L be an ε-net of point cloud P. For any point $p \in P$, there exists a point $q \in L$ such that the distance $d(p, q) \leq \epsilon$.*

Theorem 1. *The Hausdorff distance between the point cloud P and its ε-net $L \subseteq P$ is at most ϵ.*

Proof. For any $l \in L$, there exists a point $p \in P$ such that $d(l, p) \leq \epsilon$, by definition of $B_\epsilon(l)$. Hence, $\min_{l \in L} d(l, p) \leq \epsilon$, and thus, $\max_{p \in P} \min_{l \in L} d(l, v) \leq \epsilon$. For any $p \in P$, there exists a landmark $l \in L$ such that $d(l, p) \leq \epsilon$, by Lemma 1. Thus, $\max_{l \in L} \min_{p \in P} d(l, p) \leq \epsilon$. Hence the Hausdorff distance $d_H(P, L)$ between P and L, defined as the maximum of $\max_{l \in L} \min_{p \in P} d(l, p)$ and $\max_{p \in P} \min_{l \in L} d(l, p)$ is bounded by ϵ.

Topological Approximation Guarantee of an ε-net Induced Lazy Witness Complex. In addition to an ϵ not being an ϵ approximation of the point-cloud, we prove that the lazy witness complex induced by the ε-net landmarks is a good approximation (Theorem 2) to the Vietoris-Rips complex on the landmarks. This approximation ratio is independent of the algorithm constructing the ε-net. As a step towards Theorem 2, we state Lemma 2 that follows from the definition of the lazy witness complex and ε-sample. Lemma 2 establishes the relation between 1-nearest neighbour of points in an ε-net.

Lemma 2. *If L is an ϵ-net landmark of point cloud P, then the distance $d(p, p')$ from any point $p \in P$ to its 1-nearest neighbour $p' \in P$ is at most ϵ.*

Theorem 2 shows that the lazy witness complex induced by the landmarks in an ϵ-net is a 3-approximation of the Vietoris-Rips complex on those landmarks above the value 2ϵ of filtration parameter.

Theorem 2. *If L is an ϵ-net of the point cloud P for $\epsilon \in \mathbb{R}^+$, $LW_\alpha(P, L, \nu = 1)$ is the lazy witness complex of L at filtration value α, and $R_\alpha(L)$ is the Vietoris-Rips complex of L at filtration α, then $R_{\alpha/3}(L) \subseteq LW_\alpha(P, L, 1) \subseteq R_{3\alpha}(L)$ for $\alpha \geq 2\epsilon$.*

Proof. In order to prove the first inclusion, consider a k-simplex $\sigma_k = [x_0 x_1 \cdots x_k] \in R_{\alpha/3}(L)$. For any edge $[x_i x_j] \in \sigma_k$, let w_t be the point in P that is nearest to the vertices of $[x_i x_j]$ and wlog, let the point corresponding to that vertex be x_j. Since w_t is the nearest neighbour of x_j, by Lemma 2, $d(w_t, x_j) \leq \epsilon \leq \frac{\alpha}{2}$. Since $[x_i x_j] \in R_{\alpha/3}$, $d(x_i, x_j) \leq \frac{\alpha}{3} < \frac{\alpha}{2}$. By triangle inequality, $d(w_t, x_i) \leq \frac{\alpha}{2} + \frac{\alpha}{2} \leq \alpha$. Hence, x_i is within distance α from w_t. The α-neighbourhood of point w_t contains both x_i and x_j. Since $d^1(w_t) > 0$, the $(d^1(w_t) + \alpha)$-neighbourhood of w_t also contains x_i, x_j. Therefore, $[x_i x_j]$ is an edge in $LW_\alpha(P, L, 1)$. Since the argument is true for any $x_i, x_j \in \sigma_k$, the k-simplex $\sigma_k \in LW_\alpha(P, L, 1)$.

In order to prove the second inclusion, consider a k-simplex $\sigma_k = [x_0 x_1 \cdots x_k] \in LW_\alpha(P, L, 1)$. Therefore, by definition of lazy witness complex, for any edge $[x_i x_j]$ of σ_k there is a witness $w \in P$ such that, the $(d^1(w) + \alpha)$-neighbourhood of w contains both x_i and x_j. Hence, $d(w, x_i) \leq d^1(w) + \alpha \leq \epsilon + \alpha$ (by Lemma 2) $\leq 3\alpha/2$. Similarly, $d(w, x_j) \leq 3\alpha/2$. By triangle inequality, $d(x_i, x_j) \leq 3\alpha$. Therefore, $[x_i x_j]$ is an edge in $R_{3\alpha}(L)$. Since the argument is true for any $x_i, x_j \in \sigma_k$, the k-simplex $\sigma_k \in R_{3\alpha}(L)$.

Discussion. Theorem 2 implies that the interleaving of lazy witness filtration $LW = LW_\alpha(L)$ and the Vietoris-Rips filtration $R = R_\alpha(L)$ occurs when $\alpha > 2\epsilon$. As a consequence, the weak-stability theorem [4] implies that the bottleneck distance between the partial persistence diagrams $Dgm_{>2\epsilon}(LW)$ and $Dgm_{>2\epsilon}(R)$ is upper bounded by $3 \log 3$. In Sect. 6, we empirically validate this bound.

Size of an ϵ-net. The size of an ϵ-net depends on ϵ, the diameter of the point-cloud and the dimension of the underlying space [15,20]. If a point cloud $P \subset \mathbb{R}^D$ has diameter Δ, the size of an ϵ-net of P is $(\frac{\Delta}{\epsilon})^{\theta(D)}$ [20]. The size of an ϵ-net does not depend on the size of the point cloud. In Sect. 6, we empirically validate this bound for the ϵ-net landmarks generated by the proposed algorithms. The framework of ϵ-net along with its approximation guarantees lead to the question of its algorithmic construction as landmarks.

5   Construction of an ϵ-net

The naïve algorithm [16] to construct an ϵ-net selects the first point l_1 uniformly at random. In i-th iteration, it marks the points at a distance less than ϵ from the previously selected landmark l_{i-1} as covered, and selects the new point l_i from the unmarked points arbitrarily until all points are marked [15]. The fundamental principle is to choose, at each iteration, a new landmark from the set of yet-to-cover points such that it retains the ϵ-net property. We propose three algorithms where this choice determines the algorithm.

5.1   Three Algorithms: ϵ-net-rand, ϵ-net-maxmin, and $(\epsilon, 2\epsilon)$-net

The algorithm ϵ-**net-rand**, at each iteration, marks the points at a distance less than ϵ from the previously chosen landmark as covered and chooses a new landmark uniformly at random from the unmarked points. The algorithm ϵ-**net-maxmin**, at each iteration, marks the points at a distance less than ϵ from the previously chosen landmark as covered and chooses the farthest unmarked point from the already chosen landmarks. It terminates when the distance to the farthest unmarked point is no more than ϵ. The algorithm $(\epsilon, 2\epsilon)$-**net**, at each iteration, marks the points at a distance less than ϵ from the previously chosen landmark as covered, and chooses a landmark uniformly at random from those unmarked points whose distance to the previously chosen landmark is at most 2ϵ. If there are no unmarked points at a distance in-between ϵ and 2ϵ from the previous landmark, it searches for unmarked points at a distance between 2ϵ and 4ϵ, 4ϵ and 8ϵ, and so on, until it either finds one to continue as before or all points are marked. The pseudo-code for $(\epsilon, 2\epsilon)$-net is in Algorithm 1.

$(\epsilon, 2\epsilon)$-net attempts to cover the point-cloud with intersecting balls of radius ϵ, whereas ϵ-net-maxmin attempts to cover the point-cloud with non-intersecting balls of radius ϵ. ϵ-net-rand does not maintain any invariant.

ϵ-net-rand and $(\epsilon, 2\epsilon)$-net have the time-complexity of $O(\frac{1}{\epsilon^D})$ and $O(\frac{1}{\epsilon^D} \log(\frac{1}{\epsilon}))$ respectively. Their run-time does not depend on the size of the input point cloud. On the other hand, the run-time of ϵ-net-maxmin depends on the size of the point-cloud as it has to search for the farthest point from the landmarks at each iteration. On a point cloud of sinze n, ϵ-net-maxmin has $O(\frac{n}{\epsilon^D})$ time-complexity.

5.2   Connecting ϵ-net to Random and Maxmin Algorithms

De Silva et al. [7] proposed random and maxmin algorithms for point clouds.

Random. Given a point cloud P, the algorithm *random* selects $|L|$ points uniformly at random from the set of points P. This algorithm is closely related to ϵ-nets. Given the number of landmarks $K > 1$, the set of landmarks selected by random is δ-sparse where δ is the minimum of the pairwise distances among the landmarks. However, the same choice of K may not necessarily make the landmarks a δ-sample of the point cloud.

The ϵ-net-rand algorithm is a modification of random that takes ϵ as a parameter instead of K and use ϵ to put a constraint on the domain of random choices. It continues to select landmarks until all points are marked to ensure the ϵ-sample property. The proof sketch of the fact that the constructed landmarks are ϵ-sparse and ϵ-sample is as follows:

Algorithm 1 Algorithm $(\epsilon, 2\epsilon)$-net

Input: Point cloud $P = \{p_1, p_2, \cdots, p_n\}$, $n \times n$ Distance matrix D, parameter ϵ.
Output: Set of Landmarks L.
 1: Select the initial landmark l_1 uniformly at random from P.
 2: Initialize $L = \{l_1\}$.
 3: Let $N^1_{(\epsilon, 2\epsilon)}$ be the set of points at a distance between ϵ and 2ϵ from l_1.
 4: Initialize candidate landmarks $C_1 = N^1_{(\epsilon, 2\epsilon)}$.
 5: $i = 1$.
 6: **repeat**
 7:    Let $N^i_{\leq \epsilon}$ be the set of points at a distance less than ϵ from l_i.
 8:    Mark all the points in $N^i_{\leq \epsilon}$ as covered.
 9:    Let C^u_i be the set of unmarked points in C_i.
10:    **if** C^u_i is empty **then**
11:       Find the first $\delta = [1, 2, \cdots, \log(\lceil \Delta/2\epsilon \rceil)]$ for which $N^i_{(2^\delta \epsilon, 2^{\delta+1}\epsilon)}$ contains any unmarked point.
12:       Set $C_i = C_i \cup N^i_{(2^\delta \epsilon, 2^{\delta+1}\epsilon)}$.
13:    **end if**
14:    Select l_{i+1} uniformly at random from C^u_i.
15:    Insert l_{i+1} to L.
16:    $C_{i+1} = C_i \cup N^{i+1}_{(\epsilon, 2\epsilon)}$.
17:    $i = i + 1$.
18: **until** all points are marked

Proof. The ϵ-net-rand algorithm does not terminate until all points are marked as covered. Hence the set of landmarks selected by ϵ-net-rand is an ϵ-sample, since otherwise, there would have been unmarked points. The pairwise distance between any two landmarks cannot be less than ϵ; otherwise, one of them would have been marked by the other and the marked point would not be a landmark. Hence the set landmarks selected by ϵ-net-rand is ϵ-sparse.

Maxmin. The *maxmin* algorithm selects the first landmark l_1 uniformly at random from the set of points, P. Following that; it selects the point which is furthest to the present set of landmarks at each step till a given number of landmarks, say $|L|$, are chosen. If $L_{i-1} = \{l_1, l_2, \ldots, l_{i-1}\}$ is the set of already chosen landmarks, it selects such a point $u \in P \backslash L_{i-1}$ as the ith landmark that maximises the minimum distance from the present set of landmarks L_{i-1}. Mathematically, $l_i \triangleq \arg\max_{u \in P \backslash L_{i-1}} \min_{v \in L_{i-1}} d(u, v)$. The maxmin algorithm selects landmarks such that the point cloud is covered as vastly as possible.

The maxmin algorithm is closely related to ϵ-net. Given the number of landmarks $K > 1$, the set of landmarks selected by maxmin is δ-sparse where δ is the minimum of the pairwise distances among the landmarks chosen. However, that choice of K may not necessarily make the landmarks a δ-sample of the point cloud. The ϵ-net-maxmin algorithm is a modification of maxmin that takes ϵ as a parameter instead of K and uses ϵ to control sparsity among the landmarks. It terminates when the minimum of the pairwise distances among the landmarks drops below ϵ to ensure the ϵ-sample property of the landmarks chosen. The proof sketch of the resulting landmarks being ϵ-sparse and ϵ-sample is as follows:

Proof. The ϵ-net-maxmin algorithm, at each iteration, selects only that point as a landmark whose minimum distance to the other landmark points is the largest among all unmarked points. If such a point's minimum distance to the other landmark points is no more than ϵ, the algorithm terminates. Hence the set of landmarks selected by ϵ-net-maxmin must be ϵ-sparse. A point that is not a landmark must be covered by some landmark point already. Otherwise, its minimum distance to the landmark set would have been at least ϵ, and hence would have been the only unmarked point available to be selected as a new landmark by ϵ-net-maxmin. Therefore the set of landmarks selected by ϵ-net-maxmin is also ϵ-sample of the point cloud.

6   Empirical Performance Evaluation

We implement the pipeline illustrated in Fig. 1 to empirically validate our theoretical claims and also the effectiveness, efficiency, and stability of the algorithms that construct ϵ-net landmarks compared to that of the random and maxmin algorithms. We test and evaluate these algorithms on two synthetic point cloud datasets, namely Torus and Tangled-torus, and a real-world point cloud dataset, namely 1grm. On each input point cloud, we compute the lazy witness filtration and Vietoris-Rips filtration induced by the landmarks, as well as the Vietoris-Rips filtration induced by the point cloud.

On each dataset, as we vary parameter ϵ of the algorithms constructing ϵ-nets, we study the relationship between ϵ to the number of landmarks, the quality of the topological features approximated by the lazy witness filtration induced by those landmarks, as well as the stability of those approximated features. As the algorithms maxmin and random require the number of landmarks a priori, for the sake of comparison, we use the same number of landmarks as that of the corresponding ϵ-net algorithm for a given ϵ.

We compute the quality of the features approximated by an algorithm in terms of the 1-Wasserstein distance between the lazy witness filtration induced by the landmarks selected by that algorithm to those of a Vietoris-Rips filtration on the same dataset. As there are elements of randomness in the algorithms, we run each experiment 10 times and compute distances averaged over the runs.

We compute the stability of the features approximated by the algorithms in terms of the 95% confidence band corresponding the rank 1 persistence landscape using bootstrap [5]. We use persistence landscape to validate the stability of the filtrations because unlike persistence diagrams and barcodes, two sets of persistence landscapes always have unique mean and by strong law of large numbers the empirical mean landscapes of sufficiently large collection converge to its expected landscapes [1].

6.1   Datasets and Experimental Setup

Datasets. We use the datasets illustrated in Fig. 2 for experimentation. The dataset **Torus** is a point cloud of size 500 sampled uniformly at random from the surface of a torus in $\mathbb{R}^3$. The torus has a major radius of 2.5 and minor radius of 0.5. The dataset **Tangled-torus** is a point cloud of size 1000 sampled uniformly at random from two tori tangled with each other in $\mathbf{R}^3$. Both tori has a major radius of 2.5 and minor radius of 0.5. The dataset **1grm** is the conformation of the gramicidin-A protein. It has a helical shape. Gramicidin-A has two disconnected chains of monomers consisting of 272 atoms.

Fig. 2. (left) Torus, (middle) Tangled-torus, and (right) 1grm Dataset

Experimental Setup. We implement the experimental workflow in Matlab 2018a (with 80 GB memory limit). All experiments are run on a machine with an Intel(R) Xeon(R)@2.20GHz CPU and 196 GB memory. We use the Javaplex library [29] to construct lazy witness filtrations and to compute their persistence intervals. We use the Ripser library to construct the Vietoris-Rips filtrations and to compute their persistence intervals. We use R-TDA package [13] to compute bottleneck and Wasserstein distances, and 95% confidence band for the landscapes. We set the lazy witness parameter $\nu = 1$ in all computations.

6.2   Validation of Theoretical Claims

Number of Landmarks Generated by the ϵ-net Algorithms. In Fig. 3, we illustrate the relation between the number of landmarks generated by the ϵ-net algorithms and ϵ on Torus dataset. Each algorithm is run 10 times for each ϵ, and the mean and standard deviation are plotted. We observe that the number of landmarks decreases as ϵ increases. We also observe that, for a fixed ϵ, the average number of landmarks selected by the ϵ-net algorithms is more or less stable across different algorithms. We use the number of landmarks of an ϵ-net-maxmin to fit a curve with values $\Delta = 5.9$ (the diameter of Torus) and coefficient $\theta(D) = 1.73$ (found from fitting with 95% confidence). This observation supports the theoretical bound of $(\frac{\Delta}{\epsilon})^{\theta(D)}$.

Topological Approximation Guarantee. In order to validate Theorem 2 on dataset Torus, we compute the bottleneck distance between the persistence diagram of the lazy witness filtration and that of the Vietoris-Rips filtration

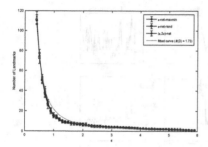

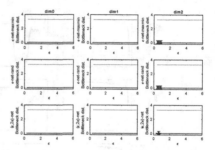

Fig. 3. Number of landmarks generated by the ϵ-net algorithms vs. ϵ on Torus dataset.

Fig. 4. Topological approximation guarantee of ϵ-net constructed by the algorithms on Torus dataset.

induced by the ϵ-net landmarks for different values of ϵ. For each ϵ and algorithm, we generate 10 sets of ϵ-net landmarks, compute their corresponding persistence diagrams and plot the mean and standard deviation of the bottleneck distances in Fig. 4. Since the theorem is valid for $\alpha \geq 2\epsilon$, we exclude the homology classes born below 2ϵ before the distance computation. The algorithms satisfy the bound as the distances are always less than the theoretical bound of $3 \log 3$. Since the plots on the other datasets support these claims, for the sake of brevity, we omit them.

6.3   Effectiveness and Efficiency of Algorithms Constructing ϵ-nets

For each ϵ, we compute the 1-Wasserstein distance between the persistent diagrams of the lazy witness filtration induced by each ϵ-net landmarks and that of the Vietoris-Rips filtration induced by the whole point cloud. We compute the mean distance and mean CPU-times across 10 runs. Unlike ϵ-net algorithms, the existing landmark selection algorithms take the number of landmarks as input. Since the average number of landmarks selected by the ϵ-net algorithms does not vary much across different algorithms (Fig. 3), for each ϵ, we take the same number of ϵ-net-maxmin landmarks as parameters to select the random and maxmin landmarks. Figure 5 illustrates result on Torus dataset.

We observe that maxmin performs well in dimensions 0 and 2 whereas $(\epsilon, 2\epsilon)$-net has competitive effectiveness. In dimension 1, we observe that ϵ-net-maxmin achieves the lowest minimum, whereas random achieves the highest minimum. All the ϵ-net algorithms has two local minima, the first of which at around $\epsilon = 0.5$ and the second in between $\epsilon = 2$ to $\epsilon = 4$. The first local minimum is due to the minor radius. As for the explanation of the second local minimum, it is sufficient to either cover the inner diameter of 5 or the outer diameter of 6 to capture the cycle. A 2.5- to 3-sparse sample suffices to do so. The performance of the maxmin and random landmarks is not as explainable as the ϵ-net landmarks. In terms of efficiency, we observe that that $(\epsilon, 2\epsilon)$-net algorithm has the lowest

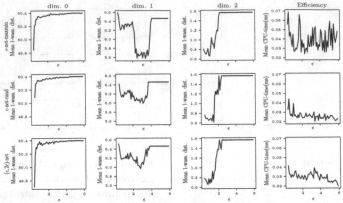

(a) Effectiveness and Efficiency of the ϵ-net algorithms on Torus dataset.

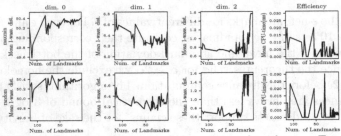

(b) Effectiveness and Efficiency of the existing algorithms on Torus dataset.

Fig. 5. Torus dataset.

run-time among all the ϵ-net algorithms. The $(\epsilon, 2\epsilon)$-net algorithm has competitive effectiveness and better efficiency among the proposed algorithms. Figure 6 illustrates the results for 1grm dataset. We observe that $(\epsilon, 2\epsilon)$-net achieves the smallest loss in dimensions 0 and 1. In dimension 2, maxmin achieves the smallest loss. ϵ-net-rand takes the smallest CPU-time among all the ϵ-net algorithms. We observe that the effectiveness of $(\epsilon, 2\epsilon)$-net and efficiency of ϵ-net-rand in the results on Tangled-torus dataset. We omit the plots due to space limitation.

Despite providing better efficiency and equivalent effectiveness on the datasets under study, the performance of the maxmin algorithm is less predictable and less explainable than the ϵ-net algorithms. Among the ϵ-net algorithms, $(\epsilon, 2\epsilon)$-net has better effectiveness at the expense of little loss in efficiency, whereas ϵ-net-rand has better efficiency than the others with effectiveness comparable to ϵ-net-maxmin.

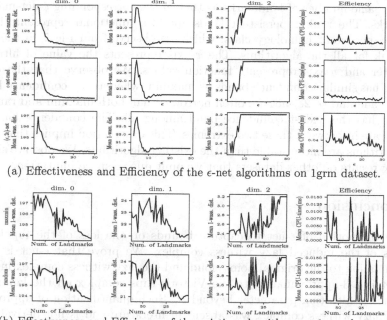

(a) Effectiveness and Efficiency of the ε-net algorithms on 1grm dataset.

(b) Effectiveness and Efficiency of the existing algorithms on 1grm dataset.

Fig. 6. 1grm dataset.

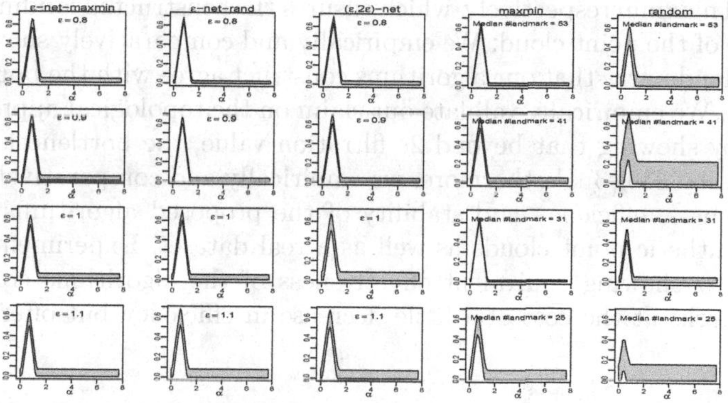

Fig. 7. 95% confidence band of the rank one persistence landscape at dimension 1 of the lazy witness filtration induced by the landmark selection algorithms on Tangled-torus dataset.

6.4 Stability of the ε-net Landmarks

In Fig. 7, we vary ϵ and plot the rank 1 persistence landscape at dimension 1 and its 95% confidence band corresponding to the lazy witness filtration induced by the different landmark selection algorithms. For maxmin and random, we take the same number of landmarks as that in the corresponding ϵ-net-maxmin landmarks. The rank 1 persistent landscape is a functional representation of the most persistent homology class, which we observe from Fig. 7 in the form of a peak for all the algorithms. The x-axis represents the value of filtration parameter and y-axis represents function values. We observe that the ϵ-net-maxmin has similar confidence bands as maxmin, whereas the confidence bands of ϵ-net-rand and $(\epsilon, 2\epsilon)$-net are often narrower than both maxmin and random. Random has the widest confidence band among all. The confidence bands of maxmin are in-between these two extremes. This observation implies that the ϵ-net algorithms are more stable than the existing algorithms. We observe similar stability results on other datasets that we omit due to space limitation.

7   Conclusion

We use the notion of ϵ-net to capture bounds on the loss of the topological features of the induced lazy witness complex. We prove that ϵ-net is an ϵ-approximation to the original point cloud and the lazy witness complex induced by ϵ-net is a 3-approximation to the Vietoris-Rips complex on the landmarks for values of filtration parameter beyond 2ϵ. Such quantification of approximation for lazy witness complex was absent in literature and is not derivable for algorithms limiting the number of landmarks.

We propose three algorithms to construct ϵ-net landmarks. We show that the proposed ϵ-net-rand and ϵ-net-maxmin algorithms are variants of the algorithm random and maxmin respectively, which ensures the constructed landmarks to be an ϵ-sample of the point cloud. We empirically and comparatively show that the sizes of the landmarks that our algorithms construct agree with the bound on the size of ϵ-net. We empirically validate our claim on the topological approximation guarantee by showing that beyond 2ϵ filtration value, the bottleneck distances are bounded by $3 \log 3$. Furthermore, we empirically and comparatively validate the effectiveness, efficiency and stability of the proposed algorithms on representative synthetic point clouds as well as a real dataset. Experiments confirm our claims by showing equivalent effectiveness of the algorithms constructing ϵ-net landmarks at the cost of a little decrease in efficiency but offering better stability.

Acknowledgement. This work is supported by the National University of Singapore Institute for Data Science project WATCHA: WATer CHallenges Analytics.

References

1. Bubenik, P.: Statistical topological data analysis using persistence landscapes. J. Mach. Learn. Res. **16**(1), 77–102 (2015)
2. Carlsson, G., Ishkhanov, T., de Silva, V., Zomorodian, A.: On the local behavior of spaces of natural images. Int. J. Comput. Vision **76**(1), 1–12 (2008)
3. Carstens, C.J., Horadam, K.J.: Persistent homology of collaboration networks. Math. Probl. Eng. **2013**(6), 1–7 (2013)

4. Chazal, F., Cohen-Steiner, D., Glisse, M., Guibas, L.J., Oudot, S.Y.: Proximity of persistence modules and their diagrams. In: Proceedings of the Twenty-Fifth Annual Symposium on Computational Geometry, pp. 237–246. ACM (2009)
5. Chazal, F., Fasy, B., Lecci, F., Michel, B., Rinaldo, A., Wasserman, L.: Subsampling methods for persistent homology. In: International Conference on Machine Learning, pp. 2143–2151 (2015)
6. Collins, A., Zomorodian, A., Carlsson, G., Guibas, L.J.: A barcode shape descriptor for curve point cloud data. Comput. & Graph. **28**(6), 881–894 (2004)
7. De Silva, V., Carlsson, G.: Topological estimation using witness complexes. In: Proceedings of the First Eurographics Conference on Point-Based Graphics, pp. 157–166. Eurographics Association (2004)
8. De Silva, V., Ghrist, R., et al.: Coverage in sensor networks via persistent homology. Algebraic & Geom. Topology **7**(1), 339–358 (2007)
9. Dey, T., Mandal, S., Varcho, W.: Improved image classification using topological persistence. In: Proceedings of the Conference on Vision, Modeling and Visualization, pp. 161–168. Eurographics Association (2017)
10. Dey, T.K., Fan, F., Wang, Y.: Graph induced complex on point data. Comput. Geom. **48**(8), 575–588 (2015)
11. Duman, A.N., Pirim, H.: Gene coexpression network comparison via persistent homology. Int. J. Genomics **2018** (2018)
12. Edelsbrunner, H., Harer, J.: Computational topology: an introduction. Am. Math. Soc. (2010)
13. Fasy, B.T., Kim, J., Lecci, F., Maria, C.: Introduction to the R package TDA. arXiv preprint. arXiv:1411.1830 (2014)
14. Guibas, L.J., Oudot, S.Y.: Reconstruction using witness complexes. Discrete & Comput. Geom. **40**(3), 325–356 (2008)
15. Har-Peled, S., Mendel, M.: Fast construction of nets in low-dimensional metrics and their applications. SIAM J. Comput. **35**(5), 1148–1184 (2006)
16. Har-Peled, S., Raichel, B.: Net and prune: a linear time algorithm for euclidean distance problems. J. ACM **62**(6), 44 (2015)
17. Hausmann, J.C., et al.: On the vietoris-rips complexes and a cohomology theory for metric spaces. Ann. Math. Studies **138**, 175–188 (1995)
18. Heinonen, J.: Lectures on Analysis on Metric Spaces. Springer Science & Business Media, New York (2012)
19. Kovacev-Nikolic, V., Bubenik, P., Nikolić, D., Heo, G.: Using persistent homology and dynamical distances to analyze protein binding. Stat. Appl. Genet. Mol. Biol. **15**(1), 19–38 (2016)
20. Krauthgamer, R., Lee, J.R.: Navigating nets: simple algorithms for proximity search. In: Proceedings of the Fifteenth Annual ACM-SIAM Symposium on Discrete algorithms, pp. 798–807 (2004)
21. Lee, H., Chung, M.K., Kang, H., Kim, B.N., Lee, D.S.: Discriminative persistent homology of brain networks. In: 2011 IEEE International Symposium on Biomedical Imaging: From Nano to Macro, pp. 841–844. IEEE (2011)
22. Letscher, D., Fritts, J.: Image segmentation using topological persistence. In: Kropatsch, W.G., Kampel, M., Hanbury, A. (eds.) CAIP 2007. LNCS, vol. 4673, pp. 587–595. Springer, Heidelberg (2007). https://doi.org/10.1007/978-3-540-74272-2_73
23. Otter, N., Porter, M.A., Tillmann, U., Grindrod, P., Harrington, H.A.: A roadmap for the computation of persistent homology. EPJ Data Sci. **6**(1), 17 (2017)
24. Patania, A., Petri, G., Vaccarino, F.: The shape of collaborations. EPJ Data Sci. **6**(1), 18 (2017)

25. Petri, G., Scolamiero, M., Donato, I., Vaccarino, F.: Topological strata of weighted complex networks. PloS One **8**(6), e66506 (2013)
26. Sheehy, D.R.: Linear-size approximations to the vietoris-rips filtration. Discrete & Comput. Geom. **49**(4), 778–796 (2013)
27. Sizemore, A., Giusti, C., Bassett, D.S.: Classification of weighted networks through mesoscale homological features. J. Complex Netw. **5**(2), 245–273 (2017)
28. Skraba, P., Ovsjanikov, M., Chazal, F., Guibas, L.: Persistence-based segmentation of deformable shapes. In: 2010 IEEE Computer Society Conference on Computer Vision and Pattern Recognition-Workshops, pp. 45–52. IEEE (2010)
29. Adams, H., Tausz, A., Vejdemo-Johansson, M.: javaPlex: a research software package for persistent (Co)homology. In: Hong, H., Yap, C. (eds.) ICMS 2014. LNCS, vol. 8592, pp. 129–136. Springer, Heidelberg (2014). https://doi.org/10.1007/978-3-662-44199-2_23
30. Xia, K., Wei, G.W.: Persistent homology analysis of protein structure, flexibility, and folding. Int. J. Numer. Meth. Biomed. Eng. **30**(8), 814–844 (2014)
31. Zomorodian, A.: Fast construction of the vietoris-rips complex. Comput. & Graph. **34**(3), 263–271 (2010)

Analyzing Sequence Pattern Variants in Sequential Pattern Mining and Its Application to Electronic Medical Record Systems

Hieu Hanh Le[1(✉)], Tatsuhiro Yamada[1], Yuichi Honda[1], Masaaki Kayahara[1], Muneo Kushima[2], Kenji Araki[2], and Haruo Yokota[1]

[1] Tokyo Institute of Technology, Tokyo, Japan
{hanhlh,yamada,honda,kayahara}@de.cs.titech.ac.jp,
yokota@cs.titech.ac.jp
[2] University of Miyazaki Hospital, Miyazaki, Japan
{muneo_kushima,taichan}@med.miyazaki-u.ac.jp

Abstract. Sequential pattern mining (SPM) is widely used for data mining and knowledge discovery in various application domains, such as medicine, e-commerce, and the World Wide Web. There has been much work on improving the execution time of SPM or enriching it via considering the time interval between items in sequences. However, no study has evaluated the sequence pattern variant (SPV) that is the original sequence containing frequent patterns including variants, and studied the factors that lead to the variants. Such a study is meaningful for medical tasks such as improving the quality of a disease's treatment method. This paper proposes methods for evaluating SPVs and understanding variant factors based on a statistical approach while considering the safety and efficiency of sequences and the relating static and dynamic information of the variants. Our proposal is confirmed to be effective by experimentally evaluating the electronic medical record system's real dataset and feedback from medical workers.

Keywords: Sequential pattern mining · Sequence pattern variant · Electronic medical record system

1 Introduction

Sequential pattern mining (SPM), which discovers frequent patterns in the sequence database is an important data-mining algorithm with various application domains, such as medicine, e-commerce, and the World Wide Web [3, 7,8,10,12,15,16]. Given a database based on a set of sequences, the problem is extracting frequent patterns in which the percentage of sequences containing them is greater than the predefined minimum support value, i.e., *MinSup*. For example, from the electronic medical record system that contains a set of

© Springer Nature Switzerland AG 2019
S. Hartmann et al. (Eds.): DEXA 2019, LNCS 11707, pp. 393–408, 2019.
https://doi.org/10.1007/978-3-030-27618-8_29

medical treatment order sequences, SPM can extract the frequent patterns automatically to generate clinical pathways, which can be used to review and improve medical tasks at hospitals. SPM can also be used to generate frequent flows of buying books and accessing websites; hence, it is useful for the recommended applications.

There have been many works on improving the speed of discovering frequent patterns [7,11] or considering the time interval between each item [4,5,14]. However, to our knowledge, studies on the sequence pattern variants (SPVs) have not been performed yet. The SPV is the sequence that contains the frequent patterns, and the frequent patterns include a number of variants. For example, in sequences of medical treatment orders, the variant indicates branched medical treatment in the clinical pathway. When there are two sequences, $(EnterHospital \rightarrow BodyTest \rightarrow Surgery \rightarrow Injection \rightarrow LeaveHospital)$ and $(EnterHospital \rightarrow NurseTask \rightarrow Surgery \rightarrow Prescription \rightarrow LeaveHospital)$, $(Surgery \rightarrow Injection \rightarrow LeaveHospital)$ and $(Surgery \rightarrow Prescription \rightarrow LeaveHospital)$ are two frequent patterns, $Injection$ and $Prescription$ are called the variants and the above two sequences are considered as SPVs. Studying the two SPVs greatly benefits the medical tasks because SPVs can be quantitatively compared between each other from the perspectives of safety and efficiency. Moreover, understanding the reason that leads to the variants also helps improve the quality of medical treatment, such as in determining the optimal treatment to a specific group of patients.

As a result, this paper aims to propose methods for quickly evaluating SPVs and understanding the factors of variants in SPM and verifying their effectiveness by experiments on an actual electronic medical record system dataset. To reach the research goal, we chose a statistical analysis approach to quantify the indicator score of each sequence based on safety and efficiency perspectives. Specifically, we consider the complication and severity of the disease for safety and the cost and average length of stay for efficiency. Moreover, to understand the factors of the variants, we extract relating static and dynamic information of the items in sequences and identify variants with significant differences by multivariate analysis. Additionally, because the task of mapping the frequent pattern to its original sequences is very inefficient in current SPM, we also propose a new SPM algorithm that mines the frequent patterns while retaining the sequence's identifier (SID). Last but not least, in the conventional visualization method [6], there is a problem that SPVs which do not exist also were generated as results of SPM. In this research, we propose a new visualization method to solve this problem and enrich the information for better grasp the analyzed results.

The contributions of this paper are as follows.

– We proposed a new SPM that mines frequent patterns while maintaining the sequence SID. It is confirmed through an experiment on a real dataset that using the new SPM decreased the execution time compared with using a method based on T-PrefixSpan.

- We proposed methods for evaluating SPVs by defining and calculating SPV's indicators, for understanding factors leading to variants by multivariate analysis.
- When applying our proposed methods on real medical dataset, it is statistically confirmed that complication risk, severity risk, length of stay, and cost can be used to quantitatively evaluate the safety and efficiency of SPVs. Moreover, the patient's age and hospitalized period are two important factors of variants. We gained positive feedback from medical workers regarding these results.
- Visualization showing the results obtained positive feedback from medical workers.

The remainder of this paper is organized as follows. Background knowledge and related works are summarized in Sect. 2. The proposed methods and experimental evaluation are described in Sects. 3 and 4. Conclusion and future works are discussed in Sect. 5.

2   Background Knowledge and Related Works

This section gives a brief review of background knowledge about sequential pattern mining, related works to improve the performance, and detect the variants of extracted sequential patterns.

2.1   Sequential Pattern Mining (SPM)

A well-known SPM algorithm is a Priori-based frequent pattern-mining algorithm [12]. However, it is very time-consuming with large data sets and generates many irrelevant patterns among its results. To exclude irrelevant patterns, PrefixSpan [7] was proposed to mine the complete set of patterns while reducing the effort of candidate pattern generation by exploring prefix projection. To improve efficiency further, CSpan [11] was proposed for mining closed sequential patterns. This algorithm uses a pruning method called occurrence checking that allows the early detection of closed sequential patterns during the mining.

2.2   Time Interval Sequential Pattern Mining

Initially, the proposed method of Agrawal et al. [12] did not consider the time interval between items. For example, the injection was performed on January 1, 2019, the sequence for performing surgery the next day, and the sequence for performing surgery three days after the injection was regarded as the same sequence. Chen et al. proposed a mining method called TI-SPM for sequences where the time interval is important, such as medical instructions, which should treat the above two sequences as different things [16]. TI-SPM outputs TI-frequent sequences by using sequence database, including a time interval, minimum support $MinSup$ ($0 \le MinSup \le 1$), and a preset time interval TI-set.

T-PrefixSpan [14] is a method to extract frequent sequential patterns from EMR logs that considers time intervals and the efficacy of medicines. T-CSpan [8] further improves the speed performance by applying the idea of mining only closed patterns. Based on T-CSpan, a study on the privacy protection during mining has been performed [9].

2.3 Sequential Variant Extraction Visualization

As a method of detecting SPV, Honda et al. detected the common part of the closed frequent pattern with the same number of items for each relative treatment day [6]. For example, A, B, C, D, E are items then $e = [[A], [B], [C, D], [E]]$ is the set of frequent patterns with variants C and D (Fig. 1).

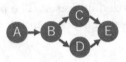

Fig. 1. Set e of frequent patterns that consider variants

3 Proposal

In order to evaluate the Sequential Pattern Variants (SPVs) and to discover the factors of the variants, we propose several methods summarized in Fig. 2. At first, from sequence database, we apply a SPM with retaining sequence ID (SID) information for better analyzing time. Then, from the variants detected by the existing method in [6], we generate SPVs and then calculate their indicator values using the information from the original sequences and other open national databases. Next, we evaluate SPVs by using statistical approach to enable comparing the SPVs. Moreover, we also utilize statistical methods to detect the factors of the variants. Finally, we visualize the evaluation for better grasp the results.

3.1 SPM Retaining SID Information

It is necessary to obtain from the original sequence database (SDB) the original information of the extracted frequent sequences for analyzing them using statistical methods by comparing each sequence. However, the calculated cost to map the extracted sequence to the original ones in SDB is very high. If the number of sequences in SDB is N, the number of extracted frequent sequences is M, and the sequence's average length is L, then the cost becomes $N \times M \times L$. Hence, to decrease the calculation cost, we propose a method to mine sequential sequences while retaining the SID information in the mining sequences. Hence, we can get the mapping information from the extracted frequent sequences. The costs of the proposed method are just the space cost for M sequences and the cost for

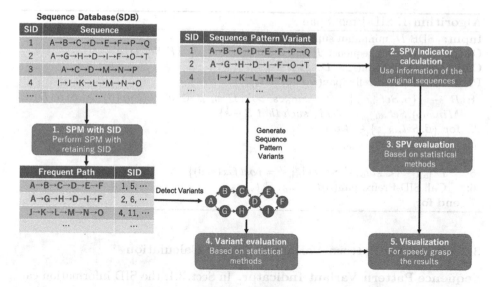

Fig. 2. An overview of our proposed methods

Table 1. Sequence Database D with time information

SID	T-sequence
1	$<(A,1),(B,3),(C,7),(E,10)>$
2	$<(A,1),(B,4),(E,7)>$
3	$<(A,2),(B,6),(B,9)>$
4	$<(A,2),(B,5),(F,10)>$
5	$<(A,2),(B,7)>$

adding SIDs to the mining sequences. The proposed method becomes more effective when the SDB or the number of frequent sequences is large. Algorithm 1 shows the detailed algorithm that applies the idea of retaining SID during mining in PrefixSpan. $postfix(s,\beta)$ is the postfix of β in sequence s.

Consider an example of mining an SDB with time information as shown in Table 1 with minimum support $MinSup = 0.4$. The number after the item is the time that the item occurs in the database. According to T-PrefixSpan, the extracted closed frequent sequences are $<A>$, $<B>$, $<A,(2,3,3,3,5),B>$, and $<A,(2,2,2,2,3),B,(3,5,5,5,7),E>$. The five values between each item are the minimum, maximum, average, median, and most frequent value of time intervals between successive items. However the outputs of Algorithm 1 are $\{<A>,<1,2,3,4,5>\}, \{<B>,<1,2,3,4,5>\}, \{<A,(2,3,3,3,5),B>,<1,2,3,4,5>\}$, and $\{<A,(2,2,2,2,3),B,(3,5,5,5,7),E>,<1,2>\}$. The last set of numbers including in $<>$ of the outputs are the SIDs of the original sequences that contain the patterns. E.g., Sequence 1 and 2 contain the pattern of $\{<A,B,E>\}$.

Algorithm 1. SID-PrefixSpan

Input: SDB D, minimum support $MinSup$

Output: Frequent sequence P and a union of Set_{sidP} $\{P, Set_{sidP}\}$

Call: SID-PrefixSpan($<>, D, <>$)

Procedure: SID-PrefixSpan($\alpha, D \mid_\alpha, Set_{sid_\alpha}$)

1: $B \leftarrow \{\{\beta, Set_{sid_{\alpha\beta}}\} \mid (< sid_\alpha, s >\subseteq D \mid_\alpha, \beta \in s) \land (Sup(\beta) \geq Size(D) \times Minsup), Set_{sid_{\alpha\beta}} \leftarrow \forall sid_\alpha$ such that $\beta \in s\}$

2: **for** $\{\beta, Set_{sid_{\alpha\beta}}\} \in B$ **do**

3:     $R \leftarrow \{\alpha\beta, Set_{sid_{\alpha\beta}}\}$

4:     **return** R

5:     $D \mid_{\alpha\beta} \leftarrow \{< sid_\alpha, s' >\in D \mid_\alpha \mid s' = postfix(s, \beta)\}$

6:     Call SID-PrefixSpan($\alpha\beta, D \mid_{\alpha\beta}, Set_{sid_{\alpha\beta}}$)

7: **end for**

3.2   Sequence Pattern Variant Indicator Calculation

Sequence Pattern Variant Indicator. In Sect. 3.1, the SID information can be obtained; however, the derived SIDs from the original frequent sequences may be duplicated. We remove such duplicated SIDs for precise statistical analysis. We propose the Indicator calculation of a frequent sequence shown in Eq. (1). Here, the number of original sequences of the analyzing frequent sequence s is N, and $indicator(k)$ is the aggregated value of information of an original sequence k.

$$Indicator(s) = \frac{\sum_{k=1}^{N} indicator(k)}{N} \tag{1}$$

Calculating Indicators in Electronic Medical Record System. In this paper, we focus on safety and efficiency which are medically required when analyzing SPVs. As safety indicators, the complication risk (CR) that considers the possibility of concurrently having more than one disease during the treatment. Moreover, the severity risk (SR) is the safety indicator that discriminates the seriousness of diseases. For the cost evaluation, we calculate the average cost for gaining the treatment from the sequent variant and the average length of stay. Therefore, $indicator$ may be one of CR, SR, cost or length of stay.

Complication Risk (CR). Equation (2) calculates the CR indicator by using the number of patients that received the treatment from analyzing the frequent pathway N and the number of patients that had the complicated diseases N_C.

$$CR = \frac{N_C}{N} \tag{2}$$

To calculate CR, we need SID information Set_{sid} and the patient classification open database called Diagnosis Procedure Combination (DPC). Here, DPC is the Japanese comprehensive method based on diagnosis group classification, contains the main disease and co-morbid complicated diseases.

Severity Risk (SR). SR can be calculated using Eq. (3). Here, N is the number of patients receiving the treatment of the frequent path, and $Sev(X_i)$ is the aggregation of the severity of all the complicated diseases of a patient X_i. Severity is considered by the average length of stay needed to treat the complicated disease.

$$SR = \frac{\sum_{i=1}^{N} Sev(X_i)}{N} \tag{3}$$

The method calculates SR from DPC and the electronic score sheet of DPC. The electronic score sheet of DPC defines the severity of each diseases by the standard length of stay in the hospital for treatment, which is managed by Japanese Ministry of Health, Labour and Welfare.

Average Cost (Cost$_{ave}$). Assume that a patient $X_i(i = 1, ..., N)$ receives a medical order $O_j(j = 1, ..., k)$ with the cost of $cost(O_j)$. Then, $Cost(i) = \sum_{j=1}^{k} cost(j)$ and the average cost that the patient X_i that receives the k medical orders is defined in Eq. (4).

$$Cost_{ave} = \frac{\sum_{i=1}^{N} Cost(i)}{N} \tag{4}$$

To calculate $Cost_{ave}$, the necessary information can be obtained from the medical order database of the hospital and the basic master database [2]. The basic master database contains the unit cost for medical order and medicine managed by national electronic receipt systems.

Average Length of Stay (Los$_{ave}$). The average length of stay (Los_{ave}) can be calculated by using the dates of entering $(InDate)$ and leaving $(OutDate)$ the hospital. Such information can easily be extracted from the medical order database. Assuming that there are N patients of $X_i(i = 1, ..., N)$, the average length of stay of the sequent variant is derived from Eq. (5).

$$Los_{ave} = \frac{\sum_{i=1}^{N} (OutDate_{X_i} - InDate_{X_i} + 1)}{N} \tag{5}$$

3.3 Sequential Pattern Variant Evaluation

In this research, to evaluate the SPV in relation to safety and efficiency, we applied the statistical method to perform the test of significant difference. CR is a name variable; SR, $Cost_{ave}$, and Los_{ave} are continuous variables; and the distribution of every indicator is non-parametric. The detailed methods used for the evaluation of each indicator in two-group and multigroup comparisons are summarized in Table 2. In the multigroup comparison, if the test is performed n times with a significance level of 0.05, then a multiplicity issue with $\alpha = 1 - (1 - 0.05)^n$ would occur. Thus, only indicators that have a significance level $(p - value)$ lower than 0.05 is considered meaningful in comparisons.

Table 2. Statistical method using indicator evaluation

Indicator	Two-group comparison	Multigroup comparison
CR	Fisher exact test	Fisher exact test and Holm test
SR	Mann–Whitney test	Steel–Dwass test
$Cost_{ave}$	Mann–Whitney test	Steel–Dwass test
Los_{ave}	Mann–Whitney test	Steel–Dwass test

3.4 Variant Factor Inference

At first, we extract the static and dynamic information of sequences as candidate factors of the variants. Then, we identify which information is the factor by multivariate analysis.

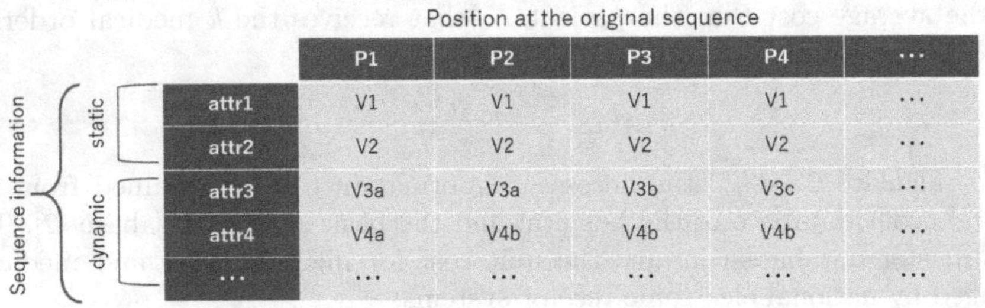

Fig. 3. Sequence information table

Extraction of Static and Dynamic Information Necessary for Multivariate Analysis.

To infer the factors of variants, the static and dynamic information of the sequences are considered. The static information is the information in the sequence that always refers to the same value regardless of which attribute is referenced. In contrast, the dynamic information is the information that refers to different values when different attributes are referenced. We performed multivariate analysis on both static and dynamic information to identify the background factors of variants.

In this section, the method of representing sequence information and the extract method are described. Sequence information is presented in a table where the attributes are present in rows and attribute classifications are presented in columns. Figure 3 shows an example of a sequence information table. The sequence information classification is the division when the elements of the sequence are divided so the values of all the sequence information of all the elements belonging to the information division become equal. The value of static information is constant regardless of the sequence information division; therefore, the same value is shown in each row (attr1 and attr2 in Fig. 3). In contrast, the value of the dynamic information varies depending on the information division, as shown in attr3 and attr4 in Fig. 3.

To use sequence information (especially dynamic information) as explanatory variables for multivariate analysis, it is necessary to determine the information classification of the original sequence at the time of branching. Such information classification can be derived as follows. First, for all sequences, a common reference attribute is established. Next, the distance from the information classification of the reference attribute and the variant position on the frequent sequence is calculated. Lastly, the information classification in the sequence that is branched from the original sequence is inferred by using the information classification of the reference attribute on the original sequence and the calculated distance (Fig. 3).

SID	Location	SPV	attr 1	attr 2	attr 3	attr 4	⋯
SID(n)	P1	SPV1	V1	V2	V3a	V4a	⋯
SID(n)	P2	SPV1	V1	V2	V3a	V4b	⋯
SID(n)	P3	SPV1	V1	V2	V3b	V4c	⋯
⋯	⋯	⋯	⋯	⋯	⋯	⋯	⋯
SID(m)	P1	SPV2	V'1	V'2	V'3a	V'4a	⋯
SID(m)	P2	SPV2	V'1	V'2	V'3b	V'4b	⋯
⋯	⋯	⋯	⋯	⋯	⋯	⋯	⋯

Explanatory variable

Objective variable

Fig. 4. Analyzed information table

Identification of Variants with Significant Differences by Multivariate Analysis. By performing multivariate analysis using the sequence information obtained in the previous section, it is possible to identify branches with significant differences. First, an analysis information table is created from the sequence information table of all original sequences, the information of the original sequences, and their corresponding frequent sequences (Fig. 4). By selecting the objective variable and explanatory variable set for each branch from the analysis information table, multivariate analysis is performed.

In multivariate analysis, a frequent sequence corresponding to the original sequence is used as an explanatory variable. This intuitively is an indicator representing the direction of the variant. The sequence information is obtained by using the information classification at the time of branching obtained from the previous part. By performing this multivariate analysis on each variant in the sequence, variants with significant differences and their factors are identified. In Fig. 4, the red box is the objective variable, and the green box is the set of explanatory variables.

For inferring the variant factor as a multivariate analyzing method, we adapted the logistic regression analysis method, which has been widely used for analyzing medical information. The significance level was also set to 0.05. Patient's age and hospitalized period were used as static information. The weight, body temperature, and systolic blood pressure values of patients were chosen to be dynamic information. We considered age as static information because it mostly remained constant while the patients received treatment in the hospital.

3.5 Visualization

In the conventional method [6], there is a problem that SPVs which do not exist in the database also generated as results by SPM. For example, SPV is detected and extracted from $<A, B, C, D, E>$ and $<A, F, C, G, E>$ and visualized as in Fig. 5. However, $<A, B, C, G, E>$ and $<A, F, C, D, E>$ are also visualized even they do not exist. Although it is meaningful from the viewpoint of discovering new sequences, they cannot be considered in evident-based evaluation.

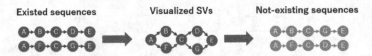

Fig. 5. Non-existing sequences also are visualized in traditional method

Therefore, in addition to this method, we introduce the concept of applying Sanky diagram, which is an expression method for visualizing the flow of things along with the progress of the process and time together. Thus, not only being able to distinguish between frequent patterns extracted by SPM and those that are not, but also only information of priority between items of the SPVs can be represented by the size of the directed path. Moreover, accurate numerical values of priorities, indicator values can be displayed through mouse operation, by which enabling accurate and smooth grasp of SPV evaluation results.

Examples of our visualization method can be viewed at Experimental Evaluation section.

3.6 Handling of Medical Treatment Data

We represent an item in the form of a set of four elements of text (*Type, Description, Code*, and *Name*). *Type* is the type of medical treatment, *Description* is the detailed record of the treatment, *Code* is a medicinal code representing the unique efficacy of the medicine used, and *Name* is the name of the medicine. For the treatment without using medicine, *Code* and *Name* are *null*. Moreover, we delete the sequences that are not medically meaningful because they do not have the *surgeon* treatment, which is considered an important order in our cases.

4 Experimental Evaluation

The effectiveness of our proposal described in Sect. 3 is verified by experiments using real datasets from the University of Miyazaki Hospital. First, we check whether performing SPM while retaining SID can reduce the execution time of mining the sequential patterns and looking for the SIDs. Then, we check that the proposed indicators are meaningful regarding the evaluation of an SPV from the perspectives of safety and efficiency. Lastly, we verify whether the proposed method can infer the background factor of an SPV.

4.1 Experimental Method and Environment

From the medical order database, we applied the SPM to obtain frequent patterns. Then, we extracted the original SPVs, the variants existing at each patterns and calculated their four indicators as described in Sect. 3.2. Next, we applied statistical methods to conclude which indicators are meaningful when evaluating the SPVs with regard to safety and efficiency. Lastly, we discuss the analyzed results of inferring background factors for each variant. We only took the original SPVs which consider the longest length because they contain the maximum number of medical orders that are medically meaningful.

We used an Intel Core i7-7700 3.60 GHz CPU, 8 GB Memory, Windows 10, Java 1.8 machine for measuring the execution time of extracting SPVs with SID.

4.2 Dataset

Our target data were medical treatment data coded as clinical pathways and recorded from November 19, 1991, to October 4, 2015, in the EMR system *WATATUMI* [13] in the Faculty of Medicine, University of Miyazaki Hospital. The data do not include information that could uniquely identify a patient to ensure patient privacy. When we extracted medical treatment data from patient records, we used anonymous patient IDs that do not allow the recovery of patient identification. In our research, the use of data from this EMR system for the support of medical treatments is described in [1], which is the website of the University of Miyazaki. Our research was approved by the Ethics Review Board of the University of Miyazaki and the Research Ethics Review Committee of the Tokyo Institute of Technology.

The target data for our experiments involved medical treatment pathways for *Transurethral Resection of a Bladder tumor (TUR-Bt)*. We chose *TUR-Bt* because it is a clinical pathway for which the flow is not well defined. Table 3 shows the characteristics of the dataset.

Table 3. Characteristics of TUR-Bt datasets

The number of patients	394
The maximum clinical orders for each patient	179
The average clinical orders for each patient	49.79
The maximum length of stay (days)	20
The average length of stay (days)	7.40

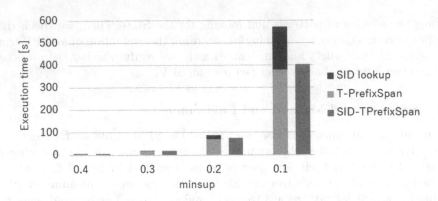

Fig. 6. Execution time of extracting SPVs with SID

4.3 Experimental Results

SID-Retaining SPM. Figure 6 shows the execution time for extracting SPVs with their SIDs with variable minimum support values ($MinSup$). The base method is performing T-PrefixSpan to extract frequent clinical paths and then looking for the SIDs of the original sequences that contain the extracted frequent clinical paths. In contrast, the proposal performs T-PrefixSpan while retaining SID information.

These results show that our proposal always reduces execution time. Especially, when $MinSup$ was 0.1, the proposal significantly reduced by approximately 170 s (30%). With smaller $MinSup$ values, a higher number of frequent pathways are extracted, so the computation cost becomes higher.

SPV Safety and Efficiency Evaluation with Visualization. Figure 7 shows an example of extracted SPVs with useful information such as medical orders, type of order, and explanation of order. Because the compared SPVs are too long to be fit in a figure, we focused only on the important parts. Table 4 summarizes the statistical analysis results of the two SPVs. The statistically significant results show that Seq2 has the lower average cost, compared with Seq1, because the prescription of *Cefazolin Na* in Seq2 occurs one time higher than in Seq1 (3 vs. 2 times). *Cefazolin Na* is an antibiotic and more expensive than other medicines. It was also confirmed in other results that CR, SR, and Los_{ave} also had statistically significant results when comparing other SPVs. Due to the page limitation, we do not include them it this paper.

The visualization of the results as shown in Fig. 7 has received positive feedback from medical workers that it can be used for quickly understanding the differences of the compared SPVs.

Variant Factor Inference. There were totally six variants, and the background factors of two of them were identified through our proposed method.

Figures 8 shows the result relating to *urgent test* and *verification test*. Relating to *urgent test*, 75 patients received the *urgent test*, while seven patients did not receive such orders during the treatment. The results show that because of medical reasons, the doctors might discontinue such tests during treatment. From this result, it is interesting that the proposal has a high potential for extracting the timing when the clinical pathway has been changed in the past.

Relating to *validation test*, ten patients got the *Cefazolin Na* without receiving the *body test*. Typically, old patients should take this test from this result, which agrees with the results, as the average age of the patients who received this test was 58 years old.

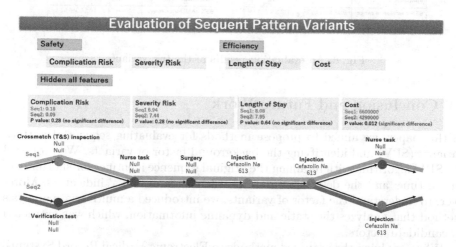

Fig. 7. An example of extracted SPVs to be compared

Table 4. Evaluation of the two SPVs Fig. 7

Indicator	Seq1 (#patients = 63)	Seq2 (#patients = 55)	p-value
CR	0.18	0.09	0.28
SR	6.94	7.44	0.25
Los_{ave} (Los_{sd})	8.08 (2.46)	7.95 (2.80)	0.64
$Cost_{ave}$ ($Cost_{sd}$)	6.60×10^6 (3.00×10^7)	4.29×10^6 (2.55×10^7)	0.0012

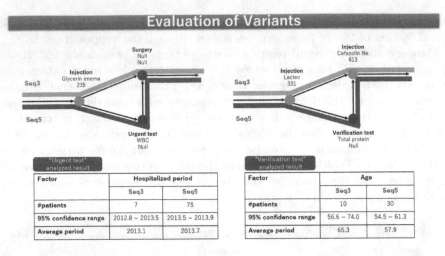

Fig. 8. An evaluation results of the two variants

Factor	Hospitalized period	
	Seq3	Seq5
#patients	7	75
95% confidence range	2012.8 ~ 2013.5	2013.5 ~ 2013.9
Average period	2013.1	2013.7

Factor	Age	
	Seq3	Seq5
#patients	10	30
95% confidence range	56.6 ~ 74.0	54.5 ~ 61.3
Average period	65.3	57.9

5   Conclusion and Future Work

In this paper, we aimed to propose methods for evaluating sequential pattern variants (SPV) and identifying the background factor of variants. We proposed the SPM algorithm with retaining the original sequence identifier for shorter execution time; and the determination and calculation of SPVs' indicators. Moreover, to understand the factor of variants, we introduced a multivariate analysis method that analyzes the static and dynamic information, which was extracted as candidate factors.

When applying the proposed methods to Electronic Medical Record Systems, we take into consideration the complications, severity of diseases, cost of medical orders, and length of stay of patients as indicators for SPVs from the perspectives of safety and efficiency. We also chose patient's age, hospitalized period, weight, body, systolic blood pressure values as information of variants.

We used real datasets from the University of Miyazaki Hospital to verify the effectiveness of our proposed methods. The experimental results show that our methods were effective in evaluating the safety and efficiency of the SPVs because all the proposed indicators were found to be meaningful. By analyzing the background factors leading to the variants, the patient's age and hospitalized period were important to understand the SPVs. The results show a high potential that our method can be used to discover the changes in past typical pathways in treating diseases over multiple hospitals.

In future work, we will apply our proposed methods to larger datasets, such as those from other hospitals, to increase the number of patients. Moreover, other static and dynamic information such as gender and disease history will be studied. We also want to verify the proposed effectiveness with other diseases.

Acknowledgement. This research has been supported by Health Labour Sciences Research Grant (Ministry of Health, Labour and Welfare, Japan) and the Kayamori Foundation of Information Science Advancement.

References

1. The Section of Medical Information at Faculty of Medicine University of Miyazaki Hospital. http://www.med.miyazaki-u.ac.jp/home/jyoho/. Accessed 19 June 2019
2. Various Information of Medical Fee, Ministry of Health, Labour and Welfare (Japan). http://www.iryohoken.go.jp/shinryohoshu/kaitei/. Accessed 19 June 2019
3. Fournier Viger, P., Lin, C.W., Rage, U., Koh, Y.S., Thomas, R.: A survey of sequential pattern mining. Data Sci. Pattern Recogn. **1**, 54–77 (2017)
4. Garg, N., Agarwal, S.: Process mining for clinical workflows. In: Proceedings of the International Conference on Advances in Information Communication Technology and Computing. pp. 5:1–5:5 (2016)
5. Hirate, Y., Yamana, H.: Sequential pattern mining with time intervals. In: Ng, W.-K., Kitsuregawa, M., Li, J., Chang, K. (eds.) PAKDD 2006. LNCS (LNAI), vol. 3918, pp. 775–779. Springer, Heidelberg (2006). https://doi.org/10.1007/11731139_90
6. Honda, Y., Kushima, M., Yamazaki, T., Araki, K., Yokota, H.: Detection and visualization of variants in typical medical treatment sequences. In: Begoli, E., Wang, F., Luo, G. (eds.) DMAH 2017. LNCS, vol. 10494, pp. 88–101. Springer, Cham (2017). https://doi.org/10.1007/978-3-319-67186-4_8
7. Han, J., et al.: PrefixSpan: mining sequential patterns efficiently by prefix-projected pattern growth. In: Proceedings of the 17th International Conference on Data Engineering, pp. 215–224 (2001)
8. Le, H.H., et al.: Fast generation of clinical pathways including time intervals in sequential pattern mining on electronic medical record systems. In: Proceedings the 4th International Conference on Computer Science and Computational Intelligent, pp. 1726–1731 (2017)
9. Le, H.H., Kushima, M., Araki, K., Yokota, H.: Differentially private sequential pattern mining considering time interval for electronic medical record systems. In: Proceedings of the 23rd International Database Engineering and Applications Symposium, pp. 95–103 (2019)
10. Mooney, C., Roddick, J.: Sequential pattern mining: approaches and algorithms. ACM Comput. Surv. **45**, 19:1–19:39 (2013)
11. Raju, V.P., Varma, G.S.: Mining closed sequential patterns in large sequence databases. Int. J. Database Manage. Syst. **7**(1), 29–39 (2015)
12. Agrawal, R., Srikant, R.: Fast algorithms for mining association rules in large databases. In: Proceedings of the 20th International Conference on Very Large Data Bases, pp. 487–499 (1994)
13. System, C.C.: Denshi Karte System WATATUMI (Electronic Medical Records WATATUMI. http://www.corecreate.com/02_01_izanami.html. Accessed 19 June 2019
14. Uragaki, K., et al.: Sequential pattern mining on electronic medical records with handling time intervals and the efficacy of medicines. In: Proceedings of the 2016 IEEE Symposium on Computers and Communication, pp. 20–25. IEEE (2016)

15. Yan, X., Han, J., Afshar, R.: CloSpan: mining closed sequential patterns in large datasets. In: Proceedings of the International Conference on Data Mining, pp. 166–177 (2003)
16. Chen, Y.-L., Chiang, M.-C., Ko, M.-T.: Discovering time-interval sequential patterns in sequence databases. Expert Syst. Appl. **25**(3), 343–354 (2003)

Web Services

Composing Distributed Data-Intensive Web Services Using Distance-Guided Memetic Algorithm

Soheila Sadeghiram[✉], Hui Ma, and Gang Chen

School of Engineering and Computer Science, Victoria University of Wellington,
Wellington, New Zealand
{Soheila.Sadeghiram,Hui.Ma,Aaron.Chen}@ecs.vuw.ac.nz

Abstract. Web services are fundamental elements of distributed computing and allow rapid development of distributed applications. Data-intensive Web services handle an enormous amount of data created by different companies. *Data-intensive Web service compositions (DWSC)* must fulfil functional requirements and optimise *Quality of Service (QoS)* attributes, simultaneously. *Evolutionary Computing (EC)* techniques allow for the creation of compositions that meets both requirements. However, current approaches to Web service composition have overlooked the impact of data transmission and the distribution of services, rendering them ineffective when applied to distributed *data-intensive Web service composition DWSC*. Especially, those approaches failed to consider important information from the problem that enables us to quickly determine the suitability of any solution. In this paper, we propose an EC-based algorithm with novel crossover operators to effectively address the above challenges. An evaluation is carried out and the results show that our proposed method is more effective than the existing methods.

Keywords: Web Service Composition (WSC) · Distribution · Data-intensive · Problem-specific crossover

1 Introduction

Various service providers in different parts of the world prepare Web services, i.e. software modules accessible by other programs over the Web to accomplish a task [5]. Web services require some inputs and subsequently generate a set of outputs after execution. In most cases, individual Web services are further composed together through a *Web Service Composition (WSC)* process to create new composite Web services, which consequently provide some new and complex functionality [7]. Although many Web services deliver the same functionality, non-functional properties, i.e. *Quality of Service (QoS)*, such as response time and cost (including communication time and cost), are discriminating factors which must be considered explicitly for effective composition.

© Springer Nature Switzerland AG 2019
S. Hartmann et al. (Eds.): DEXA 2019, LNCS 11707, pp. 411–422, 2019.
https://doi.org/10.1007/978-3-030-27618-8_30

Regarding the increasing quantities of available data on the Web, *data-intensive Web services* are crucial requirements for many companies to facilitate performing large-scale data analysis. They generate a high volume of new data as output. Therefore, *Data-intensive Web Service Composition (DWSC)* will be heavily influenced by massive data transmission and present new challenges since the quality of a composite service is affected not only by the QoS of component services but also by the locations of these Web services and the quantity of data transferred among them. However, existing composition approaches neglect the distribution of services over the network, i.e. they ignore the communication among services and just consider moving data from a data centre to a Web service, assuming that services are located in close proximity to each other [6,7,20]. Such assumptions have been frequently and significantly violated in practice. Specifically, it is highly desirable for a service composition system to automatically compose selected services from a large service repository and establish various service-oriented workflows. Unfortunately, this automated version of the DWSC problem is NP-hard in nature [18]. It is therefore very hard (or impossible) to solve large-scale DWSC problems by optimising the corresponding QoS metrics within a limited time frame [9].

Evolutionary Computing (EC) is widely demonstrated to be capable of automatically composing services with high quality, [2,6,7], to efficiently find "good enough" composite services that meet users' requirements [8]. Many EC algorithms, such as *Genetic Algorithm (GA)* [10], Genetic Programming (GP) [13] and Particle Swarm Optimisation (PSO) [11] have been used for WSC [6,7,15,16]. In GA, operators can be applied without restrictions since the functional correctness of the composition will be ensured through a decoding process. Additionally, GA is more capable of maintaining a desirable balance between solution quality and algorithm efficiency than PSO and integer linear programming methods in [7] and [4], respectively. To further enhance the effectiveness of EC algorithms, researchers hybridise EC with local search techniques [14]. This idea is called the *Memetic Algorithm (MA)* [14] and have been successfully applied to finding high-quality solutions for WSC [7,19]. MAs were also shown to clearly outperform simple EC on DWSC problems [15,17,20], where GA locates the regions where the global optimum exists, and local search enables the population to converge quickly to the optimum [17]. Therefore, in this paper, we will utilise a GA-based memetic algorithm to solve the distributed DWSC problem.

In spite of the recent success in some of the GA-based memetic algorithms for WSC and DWSC, only primitive forms of crossover, such as single-point crossover, have been exploited [15,20]. Apparently, such crossover operators may fail to effectively utilise solution structures in order to build high-quality solutions, making it challenging for GAs to evolve composite services with consistently increasing QoS. Therefore, new approaches, in particular, new crossover operators must be developed to successfully tackle large-scale DWSC problems. To fulfil this goal, in this paper, we will design crossover operators to generate new offspring composite services that eliminate the *bottleneck* communication links, i.e. the longest link between two adjacent services in a composite service.

In DWSC, bottleneck links determine to a large extent the QoS of a composite service. We will further develop an MA-based service composition algorithm that not only resolves bottleneck links but also preserves promising common sub-components in existing solutions while using them to build new composite services. Driven by those two ideas, the contributions of this paper are as follow:

(1) We will develop a new memetic algorithm armed with carefully designed crossover operators which will help to build new and high-quality composite services; (2) We will investigate the effectiveness of utilising domain knowledge, in the MA through crossover operators; (3) We will investigate the effectiveness of preserving promising sub-solutions among existing composite services in order to pass valuable information through generations by crossover operators; (4) We will conduct empirical comparisons between our proposed MA and several state-of-the-art MA methods proposed for composition problems.

A summary of this work has been previously accepted as a poster paper.

2   Related Work

Some of the existing approaches have focused on fully-automated DWSC [15,20]. However, those approaches have utilised blind recombination methods. For instance, [6,7,15,20] assign crossover points randomly and do not define which part of the solution should be maintained unchanged and passed to the next generation. A crossover where a random crossover point for each parent is determined (i.e., indices are independently chosen for the two parents) in [7]. Afterwards, each parent is split from the index point into a *prefix* and a *suffix*. In order to generate offspring, a parent is embedded within the *prefix* and the *suffix* of the other parent. Figure 3 illustrates an example of this crossover operator.

Different from existing WSC problems, the data size and the location of services are of great importance to DWSC due to their strong influence on the communication cost and time. As far as we know, the only existing approach which has considered communication characteristics for fully-automated DWSC is a clustering-based GA algorithm [15], where the information regarding the distribution of services is exploited to cluster existing services in a given repository. The clustered services are further used to generate the initial population of candidate composite services. In fact, this paper only used service location information when building the initial GA population. In this paper, however, we will explicitly and continuously use service location information to build new and better composite services with the help of newly designed crossover operators.

3   Problem Definition and Objective Function

In this section, first, we present the definition of the DWSC problem including basic concepts and terminology. Afterwards, we will present the objective function of the DWSC problem. First, we define the basic concepts involved in understanding the DWSC problem. These concepts have been introduced previously in [15], and described in more details in [17].

A *data-intensive Web service* is a tuple $S_i = (I_i, O_i, QoS_i, D_i, l_i)$, where S_i is the ith service in a repository $\mathcal{R}$. I_i is a set of inputs, and O_i is a set of outputs of service S_i. QoS_i is the set of quality attributes of the service which describes non-functional properties that are important to the DWSC problem. In this paper, for each Web service, we consider T_i and C_i, which refer to the total time and cost required for executing service S_i. D_i is the set of m data items d_j, $j \in \{1, ..., m\}$ required by service S_i and l_i is the location of S_i.

A *service repository* $\mathcal{R}$ consists of a finite collection of Web services S_i, $i \in \{1, ..., n\}$. A *service request* (also called a *composition task*) is a tuple $\mathcal{T} = (I_\mathcal{T}, O_\mathcal{T})$ where $I_\mathcal{T}$ is a set of inputs provided by a user, and $O_\mathcal{T}$ is a set of task outputs expected by the user to be produced by the composite service.

For a given *task* $\mathcal{T}$, we need to find a *composition* that fulfils $I_\mathcal{T}$ and produces $O_\mathcal{T}$. A composite service is often represented as a Directed Acyclic Graph (DAG) which includes a set of n services that could jointly accomplish the required task, where two special services can be used to represent the overall composition's inputs and outputs: a start service S_0 with $I(S_0) = \varnothing$ and $O(S_0) = I_\mathcal{T}$, and an end service S_{n+1} with $I(S_{n+1}) = O_\mathcal{T}$ and $O(S_{n+1}) = \varnothing$. In a composite Web service, there is a communication link between S and S' if there is a direct edge in the DAG that connects S and S' together. In this paper, composite services can support both parallel and sequential constructs.

Transferring and accessing data during the execution of a composite data-intensive Web service requires a significant amount of time, which, along with the quality of single Web services, affect the performance of the composite service. The following components will be utilised in the definition of the objective function. To further understand those components, Fig. 1 illustrates an example of a composite service and the time and cost involved in executing it, where for simplicity all associated time and cost components are shown only for one Web service, one connection link and one data item. In the following, we list the variables that contribute to the total cost and total execution of a composite service. For more information about the definition of these components refer to [17]: *Server access latency* ($Tsal$), *Data execution time* ($Tproc$), *Service cost* (Cs), *Data cost*($Cprov$), *Data transfer time* (Tt), *Propagation delay* (Tp) (including Tpd and Tps), and *Communication cost* (Cc) (including Ccs and Ccd).

Correspondingly, the total execution time and cost of a Web service S_i, i.e., T_i and C_i, (including data-related time and cost) are calculated in Eqs. (1) and (2), respectively.

$$T_i = \sum_{j=1}^{m} (Tpd_{d_j} + Tsal_{d_j} + Tproc_{d_j} + Tt) \tag{1}$$

$$C_i = \sum_{j=1}^{m} (Ccd_{d_j} + Cd_{d_j} + C_s) \tag{2}$$

In the above functions, m is the total number of data items in D_i.

The overall cost is obtained by summing up the costs of all services in the composition, i.e., nodes (services) and associated costs for edges (communication links) in the graph, as shown in Eq. (3):

$$C_{total} = \sum_{i=1}^{NODE} C_i + \sum_{i=1}^{EDGE} Ccs_i \tag{3}$$

Ccs is the communication cost. $NODE$ and $EDGE$ are the total numbers of nodes (Web services) and edges (links between services) included in that composition, respectively.

Response time T_{total} is the time of the most time-consuming path in the composition. Assuming h is the number of paths in a composite service, Np and Ep are the number of nodes and edges in a path p, respectively. The overall time is defined as in Eq. (4):

$$T_{total} = \max_{p=1}^{h} (\sum_{i=1}^{Np} T_i + \sum_{i=1}^{Ep} Tps_i) \tag{4}$$

The goal is to minimise the objective function in Eq. (5) by producing a suitable composite service constructed from the repository $\mathcal{R}$. Accordingly, the best solution will be a composition with a minimum value of F. for a detailed description of computing the objective function refer to [17].

$$F = w\hat{T}_{total} + (1 - w)\hat{C}_{total} \tag{5}$$

where $\hat{T}_{total}$ and $\hat{C}_{total}$ are normalised values of T_{total} and C_{total}, respectively. w is a positive weight to be determined by users who requested for a service composition to be performed. Therefore, the DWSC problem formulated in this section explicitly considers both the distribution of data and services over distant locations. During the process of building high-quality composite services, we must carefully manage the cost and delay caused by massive data communication among data centres and services, which is the central focus of this paper.

4   Representation of Solutions and the Decoding

We will utilise indirect representation, i.e. sequences, for representing chromosomes of GA (each individual solution in GA is called a chromosome). It is shown that the indirect representation outperformes other representations, such as graph and tree, both in efficiency and effectiveness for WSC [7], because it allows the optimisation to be carried out without any restrictions and functional constraints are enforced easily through a separate decoding step to transform sequences into an executable composite service in form of DAGs, i.e., a feasible workflow [7,15]. An example of the backward decoding of a service sequence (where the solution is built gradually from the end service, S_{n+1} to the start service S_0) is illustrated in Fig. 2. Redundant services, which have not been used in the solution, will be removed from the sequence after the decoding. Additionally, duplicated services added to the sequence through our EC operators which will be removed during the decoding process.

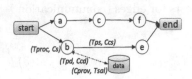

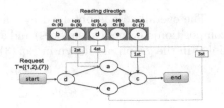

Fig. 1. A composite service and its components.

Fig. 2. Backward decoding (note that the sequence is traversed as many times as possible).

5   Distance-Guided Memetic Algorithm for DWSC

GAs employ the current population to locate a promising region [10], where the function F calculated in Sect. 3 will be used as the fitness measure. The initial population is created by randomly ordering all the services in the repository to form sequences. GAs rely heavily on crossover operators to derive an offspring population by combining parts of existing solutions. New solutions may also be modified by other operators, such as mutation, before being added to the new population. Algorithm 1 presents the pseudocode of our new algorithm. To investigate the potential of crossover operators, we will design three crossover operators. The relative distance of services is clearly important information to DWSC that we will on a continual basis throughout the full evolutionary process.

Algorithm 1: Memetic Algorithm for DWSC

    Input : *Service Repository* ($\mathcal{R}$), *Task* ($\mathcal{T}$), *Number of Generations* ($\mathcal{G}$)
    Output: *A Service Composition Solution*
1: Generate sequences with randomly ordered services in $\mathcal{R}$;
2: Apply *decoding* to sequences to create a composite service for each sequence, and calculate their fitness;
3: Update sequences by removing redundant services not used during the *decoding*;
4: **while** *number of iterations* < $\mathcal{G}$ **do**
5:     | Use tournament selection to select individuals based on their fitness values;
6:     | ***Apply a distance-guided crossover operator to the tournament winners;***
7:     | Apply mutation operator to the tournament winners;
8:     | Apply local search operator to the tournament winners;
9: **end**
10: **return** *SequenceWithBestFitness*;

In this paper, efforts will be put into the development of new crossover operators to improve the performance of MAs. In fact, three distance-guided crossover operators will be developed to utilise domain knowledge in the form of bottleneck, i.e. the longest communication, links. It is expected that the integration of MA and our crossover operators will enable us to build new algorithms that will significantly outperform several state-of-the-art algorithms in terms of both solution quality and efficiency [7,16].

The new crossover operators will be created from a baseline operator, (B-MA), which has been widely used in recent MAs for WSC [7], which has been explained in Sect. 2. To design three new crossover operators, we focus on the bottleneck link. Since the distance of every communication link is the deciding factor in obtaining the crossover point, we call it *distance-guided crossover*. We, therefore, set the crossover point based on the longest distance between any two consecutive services in the service sequence. Following this idea, we first introduce *distance-guided single-point* crossover operators. We will then enhance it either with the LCS heuristic so as to preserve good sub-solutions (or building blocks) in the offspring solutions or with *distance-guided two-point crossover*.

Distance-Guided Single-Point Crossover: this crossover uses the location of services as the key decision factor. As illustrated in Fig. 4, this crossover is very similar to *B-MA*; however, distance-guided single-point crossover picks the crossover point based on the distance of services to each other. For example, in Fig. 4, the largest distance in *Parent1* is 150 km which is between service *b* and service *c*, and 140 km for *Parent2* between service *e* and service *g*. The aim of this crossover is to enhance the fitness of the offspring by eliminating the bottleneck links of parents. To achieve this goal, parents are broken apart from the longest communication distance point.

Distance-Guided Two-Point Crossover: different from the crossover operator above, for this two-point crossover operator, the crossover points in each parent are chosen based on the first and second longest distances between any pair of consecutive services. Using these two crossover points, each parent is split into three parts. In order to produce offspring, portions of the first parent are combined with those of the second parent, one in between. An example of distance-guided two-point crossover is illustrated in Fig. 5. Two offspring differ from each other in the order of combination. The diversity between children and parents is expected to be increased through a three-part combination mechanism.

Distance-Guided LCS Crossover: as illustrated in Fig. 6, a heuristic is incorporated in this crossover operator to preserve the promising part of each parent, which will be inherited directly by their children without any change. This new heuristic is called the longest common sub-sequence (LCS), i.e. the longest sequence of services which appears in both parents. To avoid any change to LCS, the crossover point is selected in the same way as the *distance-guided single-point crossover* operator except that this point cannot be inside the LCS. In that way, children can easily preserve good sub-solutions embedded in the LCS.

Fig. 3. Index crossover

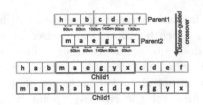

Fig. 4. Distance-guided single-point crossover

6   Evaluation and Experiment Design

In this section, we will conduct experimental evaluations to examine the effectiveness of all the three newly developed crossover operators in the context of an MA-based algorithm for DWSC. A set of experiments is carried out using WSC-2008 [3] and WSC-2009 [12] benchmark datasets. WSC-2008 contains eight service repositories of varying sizes, while WSC-2009 has five repositories with the number of services in a repository up to 15211. The total number of services for each dataset is presented in Table 1. One associated composition task per repository of services are also given in advance [3,12]. These datasets were chosen because they are the largest benchmarks that have been broadly exploited in the WSC literature; however, the original WSC datasets did not include all required information. Therefore, we further obtain QoS settings from the QWS dataset [1], and the location information of the servers hosting Web services from WS-Dream open dataset [21].

The data-intensive information such as $Tsal$ and Ccd are obtained using the distance between two Web services which is estimated by the same method as proposed in [15] based on the location information in the WS-Dream open dataset [21]. Specifically, the network bandwidth used in our experiments is randomly sampled from a normal distribution. Network bandwidth and data size are utilised to calculate Tt, for each connection link in the interval (0,1]. Additionally, each data item has its own $Tsal$ and $Cprov$, which are both generated randomly in the interval (0,1]. Since the data size has been considered the same (i.e. 3) for all methods, the values of Cc, including Ccs and Ccd, and Tp only depend on the distance between relevant services which is calculated from the WS-Dream dataset. Values of $Tproc$ and Cs are obtained from datasets WSC-2008 and WSC-2009, and then normalised within the range 0 and 1. Two recent approaches, i.e. B-MA [7] and Cluster-guided MA [15], will be evaluated as baselines. Each method will be run 30 independent times. For the algorithm, number of generations and population size are 100 and 30, respectively. Local search, crossover and mutation operators probability are 0.05, 0.95 and 0.05, respectively. Tournament selection with size 2 is used to select individuals for the operators. Therefore, all methods share the same parameter set which follow the common practice in the literature [13]. The user can set the weight

(w in Eq. (5)) to specify the relative importance of total time and cost according to their preference. Since we do not have access to the real preferences from any users in our experiments, following other WSC research works [6,7], we set $w = 0.5$, which means that the time and cost have same importance to the fitness (Note that since the value of $w = 0.5$ is normally provided by service users, it will not affect the generality of the algorithm.).

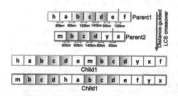

Fig. 5. Distance-guided two-point crossover

Fig. 6. Distance-guided LCS crossover (the longest common sub-sequence is identified with grey).

7 Results and Discussions

Table 1 shows the average solution fitness and standard deviation for 30 independent runs of each approach. Distance-MA-I and distance-MA-II represent the MA algorithms that use respectively the distance-guided single-point and distance-guided two-point crossover operators. We performed ANOVA statistical analysis on the average of these values at 0.05 significance level. For dataset 08-2 the fitness achieved by distance-MA-II was significantly better in all comparisons against other approaches. On the other hand, for dataset 08-1 there was no significant difference between distance-MA-LCS and distance-MA-II. Results show that utilising distance-guided crossover operators in general performed better than B-MA and Cluster-MA. Table 1, therefore, clearly demonstrates the importance of using domain knowledge and preserving good sub-solutions in crossover operators. Distance-MA-LCS method has performed the best out of the five methods, thanks to its capability of maintaining valuable sub-solutions embedded in the LCS. Additionally, distance-MA-II performed mostly better than distance-MA-I. This implies that increasing the diversity between offspring chromosomes and their parents can improve the effectiveness of MAs.

Figure 7 illustrates the mean fitness values over 30 runs for the tasks 09-4 and 08-5, where, for both tasks, Cluster MA outperformed B-MA even in the early stages of the evolutionary process. Additionally, Cluster-MA marginally outperformed distance-MA-I on task 09-4, which might be due to the substantial number of local search evaluations in it. Task 09-4 includes more services resulting in longer service sequences to be evaluated by GAs which need more evaluations during the local search process. According to Table 1, the fitness can be improved much faster by using our algorithms with newly designed crossover

Table 1. Mean fitness values and standard deviations per 30 runs. Significantly better values in all four comparisons are highlighted. (Note: the lower the fitness the better)

Task (size)	B-MA [7]	Cluster-MA [15]	Distance-MA-I	Distance-MA-II	Distance-MA-LCS
WSC08-1 (158)	0.54 ± 0.04	0.46 ± 0.04	0.45 ± 0.01	**0.42 ± 0.04**	**0.41 ± 0.12**
WSC08-2 (558)	0.51 ± 0.09	0.44 ± 0.14	0.46 ± 0.03	**0.42 ± 0.05**	0.42 ± 0.02
WSC08-3 (604)	0.55 ± 0.18	0.53 ± 0.03	0.52 ± 0.04	0.48 ± 0.01	**0.44 ± 0.02**
WSC08-4 (1041)	0.52 ± 0.03	0.5 ± 0.09	0.49 ± 0.02	0.45 ± 0.06	**0.4 ± 0.01**
WSC08-5 (1090)	0.55 ± 0.09	0.53 ± 0.041	0.51 ± 0.01	**0.5 ± 0.17**	**0.47 ± 0.09**
WSC08-6 (2198)	0.58 ± 0.13	0.56 ± 0.08	0.57 ± 0.01	0.55 ± 0.02	**0.46 ± 0.2**
WSC08-7 (4113)	0.57 ± 0.01	0.55 ± 0.236	0.59 ± 0.01	0.58 ± 0.04	**0.53 ± 0.02**
WSC08-8 (8119)	0.54 ± 0.08	0.49 ± 0.04	0.53 ± 0.02	**0.44 ± 0.09**	0.45 ± 0.05
WSC09-1 (572)	0.59 ± 0.03	0.54 ± 0.23	0.55 ± 0.03	0.57 ± 0.07	**0.53 ± 0.02**
WSC09-2 (4129)	0.56 ± 0.01	0.51 ± 0.04	0.5 ± 0.002	0.46 ± 0.1	**0.47 ± 0.02**
WSC09-3 (8138)	0.55 ± 0.09	0.54 ± 0.04	0.52 ± 0.06	0.52 ± 0.09	**0.49 ± 0.06**
WSC09-4 (8301)	0.539 ± 0.08	0.515 ± 0.04	0.525 ± 0.01	**0.48 ± 0.01**	**0.47 ± 0.16**
WSC09-5 (15211)	0.58 ± 0.09	0.46 ± 0.04	0.49 ± 0.02	0.51 ± 0.02	**0.42 ± 0.03**

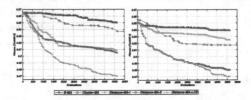

Fig. 7. Mean fitness values over 30 runs. (Cluster-MA evaluation is truncated at 5000 instead of displaying up to 10,000 as the variation in values is minimal after this point.)

operators. Particularly, distance-MA-LCS finds high-quality composite services much faster than all other algorithms. Based on the results of the experiments we can finally conclude that our new crossover operators open a promising avenue of research with the potential to generalise to similar problems in other domains.

8   Conclusions

In this paper, we proposed an MA-based approach with novel crossover operators designed to utilise information from the problem and preserve good sub-solutions from the parents. We applied an appropriate fitness function to the distributed DWSC to include quality of services, properties of data items and communication attributes between services. We consequently implemented our method on DWSC and made comparisons with other existing works. Our experimental evaluations using various benchmark datasets confirmed that our proposed method were able to effectively produce better quality results. In the future, our proposed method should be applied to other problems in this domain. Appropriate attention should be given to user preferences to specify the value of parameters.

References

1. Al-Masri, E., Mahmoud, Q.H.: Investigating Web services on the world wide web. In: Proceedings of the 17th International Conference on World Wide Web, pp. 795–804. ACM (2008)
2. Aversano, L., Di Penta, M., Taneja, K.: A genetic programming approach to support the design of service compositions. Int. J. Comput. Syst. Sci. Eng. **21**(4), 247–254 (2006)
3. Bansal, A., Blake, M.B., Kona, S., Bleul, S., Weise, T., Jaeger, M.C.: WSC-08: continuing the Web services challenge. In: 2008 10th IEEE Conference on E-Commerce Technology and the Fifth IEEE Conference on Enterprise Computing, E-Commerce and E-Services, pp. 351–354. IEEE (2008)
4. Canfora, G., Di Penta, M., Esposito, R., Villani, M.L.: An approach for QoS-aware service composition based on genetic algorithms. In: Proceedings of the 7th Annual Conference on Genetic and Evolutionary Computation, pp. 1069–1075. ACM (2005)
5. Channabasavaiah, K., Holley, K., Tuggle, E.: Migrating to a service-oriented architecture. IBM DeveloperWorks **16**, 727–728 (2003)
6. da Silva, A.S., Mei, Y., Ma, H., Zhang, M.: A memetic algorithm-based indirect approach to web service composition. In: IEEE Congress on Evolutionary Computation (CEC) (2016)
7. da Silva, A.S., Mei, Y., Ma, H., Zhang, M.: Evolutionary computation for automatic Web service composition: an indirect representation approach. J. Heuristics **24**(3), 425–456 (2018)
8. Fogel, D.B.: What is evolutionary computation? IEEE Spectr. **37**(2), 26–32 (2000)
9. Gabrel, V., Manouvrier, M., Murat, C.: Web services composition: complexity and models. Discrete Appl. Math. **196**, 100–114 (2015)
10. Holland, J.H.: Genetic algorithms. Sci. Am. **267**(1), 66–73 (1992)
11. Kennedy, J.: Particle swarm optimization. In: Sammut, C., Webb, G.I. (eds.) Encyclopedia of Machine Learning, pp. 760–766. Springer, Boston (2011). https://doi.org/10.1007/978-0-387-30164-8
12. Kona, S., Bansal, A., Blake, M.B., Bleul, S., Weise, T.: WSC-2009: a quality of service-oriented Web services challenge. In: 2009 IEEE Conference on Commerce and Enterprise Computing, CEC 2009, pp. 487–490. IEEE (2009)
13. Koza, J.R.: Genetic Programming: On the Programming of Computers by Means of Natural Selection, vol. 1. MIT Press, Cambridge (1992)
14. Moscato, P., et al.: On evolution, search, optimization, genetic algorithms and martial arts: towards memetic algorithms. Caltech concurrent computation program, C3P Report, 826 (1989)
15. Sadeghiram, S., Ma, H., Chen, G.: Cluster-guided genetic algorithm for distributed data-intensive Web service composition. In: 2018 IEEE Congress on Evolutionary Computation (CEC) (2018)
16. Sadeghiram, S., Ma, H., Chen, G.: Distance-guided GA-based approach to distributed data-intensive Web service composition. arXiv preprint. arXiv:1901.05564 (2019)
17. Sadeghiram, S., Ma, H., Chen, G.: Composing distributed data-intensive Web services using a flexible memetic algorithm. In: IEEE Congress on Evolutionary Computation (CEC) (2019, in press)
18. Strunk, A.: QoS-aware service composition: a survey. In: 2010 Eighth IEEE European Conference on Web Services, pp. 67–74. IEEE (2010)

19. Yan, L., Mei, Y., Ma, H., Zhang, M.: Evolutionary Web service composition: a graph-based memetic algorithm. In CEC, pp. 201–208 (2016)
20. Yu, Y., Ma, H., Zhang, M.: A hybrid GP-Tabu approach to QoS-aware data intensive Web service composition. In: Dick, G., et al. (eds.) SEAL 2014. LNCS, vol. 8886, pp. 106–118. Springer, Cham (2014). https://doi.org/10.1007/978-3-319-13563-2_10
21. Zheng, Z., Lyu, M.R.: WS-dream: a distributed reliability assessment mechanism for Web services. In: 2008 IEEE International Conference on Dependable Systems and Networks with FTCS and DCC, DSN 2008, pp. 392–397. IEEE (2008)

Keyword Search Based Mashup Construction with Guaranteed Diversity

Huanyu Cheng, Ming Zhong(✉), Jian Wang, and Tieyun Qian

School of Computer Science, Wuhan University, Wuhan 430072, China
{chy,clock,jianwang,qty}@whu.edu.cn

Abstract. To assist system engineers in efficiently constructing mashups, the keyword search based approach is proposed recently, which finds the optimal mashup of services with respect to QoS. However, we claim that the diversity of mashups should be taken into account due to the ambiguity of input keywords, so that the returned diverse set of mashups can improve user satisfaction by covering various possible user demands behind the keywords. For that, we present a novel keyword search based service composition approach that finds the top-k mashups dissimilar to each other. Specifically, our approach firstly searches for specific subtrees that contain all given keywords in a service connection graph as candidate mashups, and then uses an efficient graph-based algorithm to select the final diverse top-k set without evaluating the similarity between each pair of mashups. We conduct the evaluations of our approach on a real world data set crawled from the ProgramableWeb.com. The experimental results demonstrate that our approach outperforms the previous work on two selected metrics.

Keywords: Diversification · Service composition · Keyword search · Web service · Quality of service

1 Introduction

Web services are stand-alone, modular applications described by standardized web protocols and provide publishing and discovery in a standardized way. They can be composed loosely for building complex distributed software systems under a framework called Service Oriented Architecture (SOA). Due to the advantage of SOA, there has been a rapid growth of demands of constructing mashups by composing web services. A web mashup is a programming environment which allows end-users to integrate information and web services. In mashups, a service is regarded as a black-box component. The traditional service composition approaches consist of three key tasks, namely, system planning, service discovery and service selection. These traditional approaches (e.g. [2,6,8]) are too complicated and thereby not efficient enough for ordinary system engineers without comprehensive and in-depth knowledge of the three tasks.

The original version of this chapter was revised: the acknowledgement section was updated. The correction to this chapter is available at https://doi.org/10.1007/978-3-030-27618-8_34

© Springer Nature Switzerland AG 2019
S. Hartmann et al. (Eds.): DEXA 2019, LNCS 11707, pp. 423–433, 2019.
https://doi.org/10.1007/978-3-030-27618-8_31

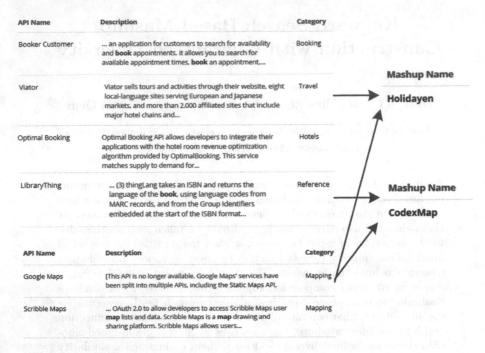

API Name	Description	Category
Booker Customer	... an application for customers to search for availability and **book** appointments. It allows you to search for available appointment times, **book** an appointment,....	Booking
Viator	Viator sells tours and activities through their website, eight local-language sites serving European and Japanese markets, and more than 2,000 affiliated sites that include major hotel chains and...	Travel
Optimal Booking	Optimal Booking API allows developers to integrate their applications with the hotel room revenue optimization algorithm provided by OptimalBooking. This service matches supply to demand for...	Hotels
LibraryThing	... (3) thingLang takes an ISBN and returns the language of the **book**, using language codes from MARC records, and from the Group Identifiers embedded at the start of the ISBN format...	Reference

Mashup Name

Holidayen

Mashup Name

CodexMap

API Name	Description	Category
Google Maps	[This API is no longer available. Google Maps' services have been split into multiple APIs, including the Static Maps API,	Mapping
Scribble Maps	... OAuth 2.0 to allow developers to access Scribble Maps user **map** lists and data. Scribble Maps is a **map** drawing and sharing platform. Scribble Maps allows users...	Mapping

Fig. 1. An example from ProgrammableWeb.com.

To help system engineers construct mashups efficiently, He et al. [5] propose KS3, a keyword search based service composition approach, which integrates and automates the procedures of system planning, service discovery and service selection. By leveraging the studies of keyword search over graphs (e.g. [4]), the keyword search based service composition approach can automatically produce prototype-like solution with respect to a given set of keywords describing the mashup.

However, only the optimal solution may not satisfy the real user requirements underlying the given keywords, because a same set of keywords may represent different user requirements due to the inherent ambiguity of keywords. For example, a user inputs two keywords "book" and "map" into ProgrammableWeb.com (abbr. PW)[1] for a system that interacts with a map to find books in nearby public libraries as shown in Fig. 1. However, most of the search results offer the functionality of booking something like hotels. In contrast, only several search results in the category "Book" are really used for deriving the information of books. As a result, we only get a few of compositions that help travelers plan and book personalized holiday experiences, but not the mashup "CodexMap" that lets the user find books graphically on a map.

Therefore, the diversity should be taken into account in keyword search based service composition. In contrast, the existing service diversification works just exploit the users' query history for disambiguation of keywords and are not devised to find a set of diverse service compositions. Mei et al. [9] proposed DivRank based on a reinforced random walk, which automatically balanced the

---

[1] http://www.programmableweb.com/.

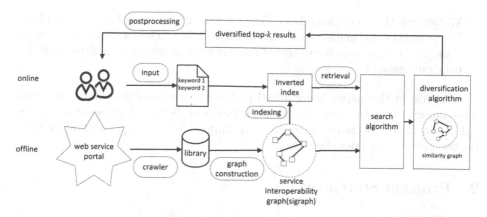

Fig. 2. An overview of our proposed system

prestige and the diversity. Kang et al. [7] used expansion ratio to evaluate the diversity of web services. However, they can only diversify individual services but not compositions of services. Naim et al. [10] first achieved the diverse functional web services and then leveraged web service dependency network to diversify the compositions. Although they try to diversify service compositions, they actually have a totally different definition of "diversity". Their diversity is the scalability of compositions, which means measuring how many dependent services of its services exist for each single composition. In contrast, we focus on find a set of service compositions that are dissimilar enough to each other.

In this paper, we propose a novel keyword search based approach that finds the diversified top-k mashups with different optimization goals of QoS, such as reliability, throughput and cost. As shown in Fig. 2, our approach has two phases, namely, offline and online. In the offline phase, we build a service connection graph from a library of web services crawled from a given web service portal. Moreover, we build an inverted index from keywords to the nodes on the graph by parsing the service descriptions. In the online phase, the system engineers can input a few keywords describing their demands or the main functionality of expected mashups. The nodes/services matched by the keywords will be retrieved from the index and be passed to the search algorithm. The search algorithm will return the results of composition on the service connection with QoS better than a given threshold. The diversification algorithm will gradually build a *similarity graph* by adding the results as nodes in the descending order of QoS, and meanwhile, incrementally compute the *maximal independent sets* of the similarity graph until convergence, namely, a maximal independent set of k results have been found. Lastly, the diversified top-k results will be postprocessed as in [5] to generate the final prototype-like service compositions.

The main contributions of this paper are as follows.

- We formalize a diversified keyword search based service composition problem, which is to find the top-k mashups that are dissimilar to each other with respect to QoS.

- We present the search and diversification algorithms to address the problem.
- We perform experiments on a real world data set. The experimental results show that our approach can reveal more diverse mashups compared with the previous method.

The rest of this paper is organized as follows. Section 2 describes our graph data model and formalizes the research problem. Section 3 presents our algorithms in detail. Section 4 evaluates our approach with experimental results. Lastly, Sect. 5 concludes the paper.

2   Problem Statement

2.1   Preliminaries

In the scenario of service composition, we can generally model a library of web services as a service connection graph (scgraph). A scgraph is a node-labeled directed simple graph, where the nodes represent the web services and the edges between them represent the service connection, namely, whether the two corresponding web services can be composed in the order specified by the direction of the edge. Due to limited space, it is not specifically introduced here. More details can be found in e.g., [3].

2.2   Problems

Following the definition of [5], we define a mashup as a result tree generated by keyword search on a scgraph. Given a scgraph G and a keyword query $Q = \{t_1, ..., t_l\}$ containing l ($l \geq 2$) keywords, we denote the set of result trees of Q on G as $Q(G)$, which are subtrees of G and contain all keywords in Q. To formalize this concept, let us consider the following definitions.

Definition 1 (Search Path). *Given a graph $G = (V, E)$ and a keyword t, a search path P in G is a sequence of nodes $v_1/v_2/\cdots$, where (1) v_1 contains t, (2) $v_i \in V$, (3) $e(v_i, v_{i+1}) \in E$, with $i \geq 1$.*

Definition 2 (Result Tree). *Given a graph $G = (V, E)$ and a query Q, a result tree T is comprised of a set of search paths from each keyword in Q on G, whose last nodes are a same node, namely, the root of T.*

In order to rank the result trees, we need to measure the goodness of mashups represented by them. There could be various measurements like QoS, relevance to queries, etc. In this paper, we consider reliability, throughput and cost as the metrics of result trees respectively. These *structure-independent system qualities* [5] are calculated independently of the structure and dynamics of mashups, and only based on the quality of their component services. Since the topology of the result trees will change in the postprocessing [5], it is difficult to consider the structure-dependent system qualities.

Let $q_{rb}(T)$, $q_{tp}(T)$, $q_c(T)$ be the reliability, throughput and cost of the result tree T. They can be calculated by the following equations: $q_{rb}(T) = \prod_{v \in V^T} q_{rb}(v)$, $q_{tp}(T) = \min_{v \in V^T} q_{tp}(v)$, $q_c(T) = \sum_{v \in V^T} q_c(v)$. For conciseness, we assume the cost of all services is equal to 1.

In order to identify the diverse sets of the result trees, we need to be able to compute the similarity between each pair of result trees. For that, let the services be categorized. We assume the more common service categories two result trees have, the more similar they are. Formally, given two result trees T and T', we compute their Jaccard similarity as

$$sim_{jsc}(T, T') = \frac{|C(T) \bigcap C(T')|}{|C(T) \bigcup C(T')|} \tag{1}$$

where $C(T)$ is the set of all distinct categories of services contained by T.

Different from the previous work [5] that finds only the optimal (i.e., top-1) connected tree in scgraphs, we aim to find the top-k diverse result trees, so that the result set could satisfy different user demands. Formally, the problem to be addressed in this paper is as follows.

Definition 3 (Diversified Top-k Mashups). *Given a scgraph G, a keyword query Q, an objective function F (e.g., q_c, q_{rb} or q_{tp}), a diversity function D (e.g. similarity), a positive real number $\alpha \in [0, 1]$ as diversity threshold and a positive real number β as quality constraint, find a set of result trees $R \subseteq Q(G)$ such that $|R| = k$, $\min_{T,T' \in R} D(T, T') \geqslant \alpha$ and $\min_{T \in R} F(T) \geqslant \beta$ is maximized.*

Note that α is a parameter for adjusting the diversity of results, and β is a threshold for excluding the results with poor QoS.

3 Diversified Search

In this section, we address the problem of finding the diversified top-k result trees on a scgraph. Our diversified search approach is divided into two phases. The first phase is to search the results with objective values greater than β, and the second phase is to derive the diversified top-k result set.

3.1 Result Generation

Following the line of *backward search* [1], our result generation algorithm decomposes the problem of finding result trees of a keyword query Q into $|Q|$ independent subproblems, each of which is to iteratively enumerate search paths from one of the query keywords by heuristics. Once a set of search paths from all keywords meet at a same root node, a result tree is generated by joining the paths.

Algorithm 1 presents the pseudo-code of the function result generation algorithm. Let R be a priority queue of intermediate result trees in descending order of their objective values, PQ_t be a priority queue of search paths from $t \in Q$ that

have not been traversed yet, and $H_{t,v}$ be a container that records the traversed
search paths from t to a specific node $v \in V$. Firstly, all PQ_t for each keyword
in query will be initialized (line 1–3). This algorithm traverses the scgraph G
iteratively, until the upper bounds are lower than β or all search path queues
are empty (line 4–20). At each iteration, a nonempty queue PQ_t is chosen in a
round-robin way (for avoiding search stagnation), and a search path P_t whose
end node is v is dequeued from PQ_t and is added into $H_{t,v}$ (line 5–8). Then,
the newly found search paths from t through appending a neighbor edge of v
at the end of P_t will be generated and enqueued into PQ_t (line 9–12). Each
combination of traversed search paths from other keywords $t' \in Q$ in $H_{t',v}$ will
be joined with P_t to generate a new result tree, and all new result trees will be
enqueued into R if their objective values are higher than β (line 13–18). At the
end of iteration, the upper bounds of objective values of results that have not
been generated yet are estimated (line 19). Lastly, the result trees in R will be
returned (line 21).

In addition, we explain the sorting of search paths in PQ_t and the estimation
of upper bounds $\eta(v)$ as follows.

A search path can be considered as a special result tree with only one branch.
Thus, we can use the equations in Sect. 2 to calculate the objective value $F(P_t)$
of a search path P_t respectively. Heuristically, a search path P_t takes precedence
over another search path P_t' in PQ_t if and only if $F(P_t)$ is better than $F(P_t')$.

For each node $v \in V$, we denote by $\eta(v)$ the upper bound of objective values of
result trees rooted at v (including unknown results). Let $\eta(v)_c$, $\eta(v)_{rb}$ and $\eta(v)_{tp}$
be the upper bound of correlation, reliability and throughput respectively. They
can be estimated by using the following equations.

$$\eta(v)_c = \min_{t \in Q}(\sum_{t' \in Q, t' \neq t} \min_{P_{t'} \in H_{t',v}} q_c(P_{t'}) + PQ_t.peek) \tag{2}$$

$$\eta(v)_{rb} = \max_{t \in Q}(\prod_{t' \in Q, t' \neq t} \max_{P_{t'} \in H_{t',v}} q_{rb}(P_{t'}) \cdot PQ_t.peek) \tag{3}$$

$$\eta(v)_{tp} = \max_{t \in Q} \min_{t' \in Q, t' \neq t} \{ \max_{P_{t'} \in H_{t',v}} q_{tp}(P_{t'}), PQ_t.peek \} \tag{4}$$

where $PQ_t.peek$ is the peek value of $q_c(P_t)$, $q_{rb}(P_t)$ or $q_{tp}(P_t)$ for $P_t \in PQ_t$.

3.2   Result Diversification

In order to find the sets of diverse top results, we find maximal independent sets
(MISs) in a specific similarity graph like [11].

Algorithm 2 presents the pseudo-code of the result diversification algorithm,
which can incrementally find the new MISs on a similarity graph that is updated
constantly by adding a new node representing the next top result into it. As a
result, our algorithm can eliminate the redundant computation that occurs while
finding the MISs on a constantly evolving graph, thereby reducing the overall
overheads.

Algorithm 1: Result Generation

    Input: G, Q, β;
    Output: the result trees with objective values higher than β;
1 **foreach** *keyword $t \in Q$* **do**
2   put the neighbor edges of each node containing t into PQ_t;
3 **end**
4 **while** $\exists v \in V$, $\eta(v) \geqslant \beta$ *and* $\exists t \in Q$, PQ_t *is not empty* **do**
5   choose a nonempty PQ_t in a round-robin way;
6 $P_t \leftarrow PQ_t$.dequeue();
7 $v \leftarrow$ the last node on P_t;
8 $H_{t,v}$.add(P_t);
9   **foreach** *neighbor edge $e \in E$ of v* **do**
10   generate a new search path P_t' by appending e at the end of P_t;
11 PQ_t.enqueue(P_t');
12   **end**
13   **foreach** *combination of paths in $H_{t',v}$ with $t' \neq t \in Q$* **do**
14   generate a result tree T comprised of the combination and P_t;
15   **if** $F(T) \geqslant \beta$ **then**
16 R.insert(T);
17   **end**
18   **end**
19   update $\eta(v)$ for each $v \in V$;
20 **end**
21 **return** R;

Our algorithm is to call function findMIS() iteratively. The details of function findMIS() is as follows. Let $MISs$ be the set of MISs on the previous similarity graph G_n^s, and $MISs'$ be the set of new MISs on the new similarity graph G_{n+1}^s. There are two steps to find $MISs'$. Firstly, for each MIS $MIS \in MISs$, create a new set $MIS' = MIS \cup \{v\}$ (line 15). If there are nodes in MIS' adjacent to v, remove them from MIS' so that MIS' is an independent set (line 17). Then, put MIS' into $MISs'$ (line 19). Secondly, remove the independent sets in the $MISs'$ that are non-maximal, namely, are the subsets of some other sets (line 21), and merge the previous set of MISs (with non-maximal sets removed) $MISs$ and the new set of MISs $MISs'$ (line 22). Lastly, we can find all MISs on the new similarity graph. The proof of correctness of our algorithm can be found in [12].

4   Experimental Evaluation

4.1   Experiment Setup

The experiments are performed on a Windows 7 PC with 3.20 GHz CPU and 16 GB memory. We have conducted a series experiments on the Programmable Web (PW) dataset, which contains 1104 web services and 2429 mashups. We manually generate 20 queries which contain keywords obtained from the description of services used by the corresponding mashup.

Compared Algorithm. We compare our diversification method (MIS) with the method (EXP) that uses expansion ratio to diversify the compositions in [7]. The default value of β, k and α is 4, 3 and 0.3 respectively.

4.2   Evaluation

We employ the metric proposed in [9] to measure diversity. The metric makes use of density of the induced sub-graph. The density of a graph is defined as the number of edges presenting in the graph divided by the maximal possible number of edges in the graph. Given a sub-graph S, the density is as followed:

$$Density(S) = \frac{|\{(u,v)|u \in S, v \in S, (u,v) \in E\}|}{|S| \cdot (|S| - 1)} \tag{5}$$

Algorithm 2: Result Diversification

    Input: k, a set of results R and a similarity threshold α;
    Output: the diversified top-k results;
1 **while** $\forall MIS \in MISs$, $|MIS| < k$ **do**
2      $T \leftarrow R.\text{pop}()$;
3      $MISs \leftarrow \text{findMIS}(MISs, T, \alpha)$;
4 **end**
5 **return** $MIS \in MISs$ with $|MIS| = k$;
6 **Function** $\text{findMIS}(MISs, T, \alpha)$
7      $G_{n+1}^s \leftarrow G_n^s \cup v$;
8   **foreach** $v' \in V_n^s$ **do**
9   **if** $sim(v, v') \geqslant 1 - \alpha$ **then**
10   add an edge $e(v, v')$;
11   **end**
12   **end**
13      $MISs' \leftarrow \emptyset$;
14   **foreach** $MIS \in MISs$ **do**
15         $MIS' = MIS \cup \{v\}$;
16   **while** $\exists u \in MIS$, u is adjacent to v **do**
17   remove u from MIS';
18   **end**
19   add MIS' to $MISs'$;
20   **end**
21   remove the non-maximal independent sets in $MISs'$;
22      $MISs \leftarrow MISs \cup MISs'$;
23   **return** $MISs$;
24 **end**

In addition, to evaluate the redundancy of returned results, we use the following equation:

$$Redundancy(R) = \frac{\sum_{T \in R} |C(T)| - |\bigcup_{T \in R} C(T)|}{\sum_{T \in R} |C(T)|} \tag{6}$$

where R is the set of returned results and $C(T)$ is the set of categories of services in T.

We conduct experiments to study the performance of our approach with the diversity metric and make comparison with its competitors.

As shown in Fig. 3(a), our approach always achieves zero density since our approach is to find the maximal independent set which contains no edges between results. In contrast, the density of results generated by EXP is always more than zero, even though it decreases with the increase of k. Thus, we can conclude that our approach must outperform EXP in terms of density.

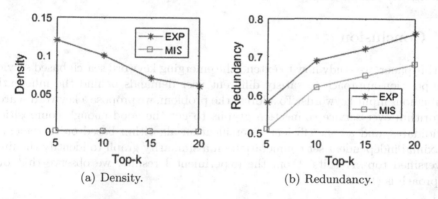

(a) Density.

(b) Redundancy.

Fig. 3. The effectiveness comparision

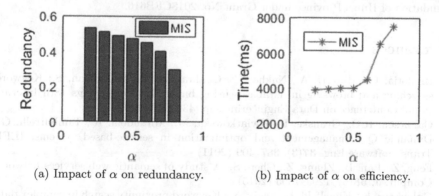

(a) Impact of α on redundancy.

(b) Impact of α on efficiency.

Fig. 4. Impact of α.

As shown in Fig. 3(b), the redundancy of MIS is always lower than EXP no matter the vary of k. Different from the approach using expansion ratio, our approach ensures that the minimum dissimilarity between mashups is more than a threshold. In this way, our method can apparently reduce the redundancy. Therefore, we can conclude that our approach also outperforms EXP in terms of redundancy.

Next, we study the impacts of α for MIS while keeping other parameters unchanged. From Fig. 4(a), we can see that the redundancy decreases with the increase of α. Obviously, the higher the dissimilarity between the results, the fewer the overlapped categories, and the lesser the redundancy. From the observation, we can conclude that larger α offers lower redundancy, namely, better diversity. From Fig. 4(b), we can see that the time cost of our approach increases with the increase of α. In particular, the time cost starts to increase dramatically when $\alpha > 0.5$. Therefore, the evaluation of α should achieve a balance between diversity and efficiency.

Overall, our approach outperforms the EXP approach in spite of redundancy and density comparison.

5   Conclusion

In this paper, we study how to extend the emerging keyword search based service composition approach to satisfy different user demands behind the inherently ambiguous input keywords. To address the problem, we propose a keyword search algorithm over service connection graphs to get the good enough composition candidates, and a very efficient diversification algorithm based on incremental maximal independent set enumeration on a similarity graph to identify the final diversified top-k results. From the experimental results, we observe that our approach is effective.

Acknowledgement. This paper was supported by National Natural Science Foundation of China under Grant No. 61202036, 61502349 and 61572376 and Natural Science Foundation of Hubei Province under Grant No. 2018CFB616.

References

1. Bhalotia, G., Hulgeri, A., Nakhe, C., Chakrabarti, S., Sudarshan, S.: Keyword searching and browsing in databases using banks. In: Proceedings 18th International Conference on Data Engineering, pp. 431–440 (2002)
2. Calinescu, R.C., Grunske, L., Kwiatkowska, M., Mirandola, R., Tamburrelli, G.: Dynamic QoS management and optimization in service-based systems. IEEE Trans. Software Eng. **37**(3), 387–409 (2011)
3. Feng, Z., Lan, B., Zhang, Z., Chen, S.: A study of semantic web services network. Comput. J. **58**(6), 1293–1305 (2015)
4. Golenberg, K., Kimelfeld, B., Sagiv, Y.: Keyword proximity search in complex data graphs. In: Proceedings of the 2008 ACM SIGMOD International Conference on Management of Data, pp. 927–940 (2008)
5. He, Q., et al.: Keyword search for building service-based systems. IEEE Trans. Software Eng. **43**(7), 658–674 (2017)

6. He, Q., Yan, J., Jin, H., Yang, Y.: Quality-aware service selection for service-based systems based on iterative multi-attribute combinatorial auction. IEEE Trans. Software Eng. **40**(2), 192–215 (2014)
7. Kang, G., Tang, M., Liu, J., Liu, X.F., Cao, B.: Diversifying web service recommendation results via exploring service usage history. IEEE Trans. Serv. Comput. **9**(4), 566–579 (2016)
8. Klusch, M., Fries, B., Sycara, K.P.: OWLS-MX: a hybrid semantic web service matchmaker for OWL-S services. J. Web Semant. **7**(2), 121–133 (2009)
9. Mei, Q., Guo, J., Radev, D.R.: DivRank: the interplay of prestige and diversity in information networks. In: Proceedings of the 16th ACM SIGKDD International Conference on Knowledge Discovery and Data Mining, Washington, DC, USA, 25–28 July 2010, pp. 1009–1018 (2010)
10. Naim, H., Aznag, M., Quafafou, M., Durand, N.: Probabilistic approach for diversifying web services discovery and composition. In: 2016 IEEE International Conference on Web Services (ICWS), pp. 73–80, June 2016
11. Qin, L., Yu, J.X., Chang, L.: Diversifying top-k results. Proc. VLDB Endow. **5**(11), 1124–1135 (2012)
12. Zhong, M., Wang, Y., Zhu, Y.: Coverage-oriented diversification of keyword search results on graphs. In: Pei, J., Manolopoulos, Y., Sadiq, S., Li, J. (eds.) DASFAA 2018. LNCS, vol. 10828, pp. 166–183. Springer, Cham (2018). https://doi.org/10.1007/978-3-319-91458-9_10

Using EDA-Based Local Search to Improve the Performance of NSGA-II for Multiobjective Semantic Web Service Composition

Chen Wang[✉], Hui Ma, and Gang Chen

School of Engineering and Computer Science, Victoria University of Wellington,
Wellington, New Zealand
{chen.wang,hui.ma,aaron.chen}@ecs.vuw.ac.nz

Abstract. Service-oriented computing is a computing paradigm that creates reusable modules over the Internet, often known as Web services. Web service composition aims to accomplish more complex functions by loosely coupling web services. Researchers have been proposing evolutionary computation (EC) techniques for efficiently building up composite services with optimized non-functional quality (i.e., QoS). Some of these techniques employ multi-objective EC algorithms to handle conflict qualities in QoS for fully automated service composition. One recent state-of-art work hybridizes NSGA-II and MOEA/D, which allows the multi-objective service composition problem to be decomposed into many scalar optimization subproblems, where a simple form of local search can be easily applied. However, their local search is considered to be less effective and efficient because it is randomly applied to a predefined large number of subproblems without focusing on the most suitable candidate solutions. In this paper, we propose a memetic NSGA-II with probabilistic model-based local search based on Estimation of Distribution Algorithm (EDA). In particular, a clustering technique is employed to select suitable Pareto solutions for local search. Each selected solution and its belonged cluster members are used to learn a distribution model that samples new solutions for local improvements. Besides that, a more challenging service composition problem that optimizes both functional and non-functional quality is considered. Experiments have shown that our method can effectively and efficiently produce better Pareto optimal solutions compared to other state-of-art methods in the literature.

Keywords: Web service composition · QoS optimisation · EDA

1 Introduction

Service-oriented computing (SOC) is a computing paradigm that creates reusable modules to achieve cost-efficient and integrable enterprise applications [5]. These modules are known as *Web services*, which are self-describing and self-containing

© Springer Nature Switzerland AG 2019
S. Hartmann et al. (Eds.): DEXA 2019, LNCS 11707, pp. 434–451, 2019.
https://doi.org/10.1007/978-3-030-27618-8_32

applications that can be deployed, discovered and invoked over the Internet. Often, web services are loosely coupled into an execution workflow to build up an entirely new service. This idea is known as *Web service composition* [15]. Many researchers have been working on *fully automated service composition* to automatically create execution workflows with required functionalities while optimizing the overall non-functional quality of composite services (i.e., Quality of Service (QoS)) [15]. Due to the complexity of the fully automated service composition problem, finding optimal solutions in polynomial time is impossible [11]. Evolutionary computation (EC) approaches [8,16] are proposed to efficiently find "good enough" composite services that meet users' QoS requirements reasonably well [10]. Recently, comprehensive quality-aware semantic web service composition has gained increasing interests, where both functional and non-functional quality criteria, i.e., quality of semantic matchmaking (QoSM) and QoS are simultaneously optimized as a single objective [18–20,22,23].

EC-based fully automated service composition approaches are mainly classified into two groups based on the number of objectives to be optimized: single-objective or multi-objective approaches. Many single-objective EC algorithms, such as Genetic Algorithm (GA), Genetic Programming (GP), Particle Swarm Optimization (PSO), Estimation of Distribution Algorithm (EDA), have been used for fully automated service composition, achieving promising results [8,16,18–20,23]. On the other hand, users often do not have clear preferences on trade-off solutions before they see the trade-offs of the solutions. For example, some users are willing to trade QoS for QoSM. Multi-objective algorithms can address these issues, and provide a set of trade-off solutions. Some recent works [6,7] investigated multi-objective optimization techniques, such as NSGA-II [9], for QoS-aware fully automated service composition, tackling conflicting QoS attributes (i.e., one objective combines time and cost, another objective combines availability and reliability).

To further enhance the effectiveness of NSGA-II, memetic algorithms have been successfully utilized in many applications for finding higher quality solutions using local search [25]. A recently published memetic approach to multi-objective fully automated service composition problem (henceforth referred to as Hybrid [6]) effectively combines the use of two optimization algorithms, i.e., NSGA-II and MOEA/D. This approach takes advantage of the "divide and conquer" strategy supported by MOEA/D, allowing the local search to be performed on numerous decomposed single-objective scalar optimization subproblems.

Despite this recent success, the number of decomposed subproblems is predefined (e.g., 500 subproblems in Hybrid [6]), and a simple form local search (i.e., so-called one-point "swap") is less effective and efficient to make local improvements because it is randomly applied to every subproblem without focusing on the best candidate solutions in each generation. Meanwhile, each one-point "swap" local search searches solutions in the space of candidate solutions based on only one solution (i.e., subproblem representative), ignoring any information of other promising candidate solutions that could be jointly used for guiding the local search. Therefore, new memetic approaches must be developed to address these two limitations. Besides that, to the best of our knowledge, existing EC-based multi-objective fully automated approaches only focus on QoS and overlook QoSM of composition solutions. In practice, some customers often demand highly accurate and reliable outputs of composite services (i.e., high QoSM), therefore, are willing to trade QoS for QoSM. However, a portion of customers may prefer (demand) a more highly responsible composite service at an affordable cost (i.e., high QoS). In this paper, we *propose a memetic NSGA-II with EDA-based local search (henceforth referred to as MNSGA2-EDA) for multi-objective fully automated semantic service composition*, where EDA can effectively handle the two limitations of the local search in Hybrid [6]. Besides that, MNSGA2-EDA tackles two practical objectives, i.e., two objective functions in Eqs. (2) and (3), with respect to the functional and non-functional quality criteria, achieving substantially high performances in effectiveness and efficiency. The contributions of this paper are listed below, and some initial ideas have been recently accepted in a poster [21].

1. To avoid pre-determining a large number of single-objective subproblems in advance, we propose a new clustering technique to select candidate Pareto-optimal solutions for local search, which is performed separately and concurrently in different regions of the Pareto front that contributes to wide and uniformly distributed near-optimal Pareto solutions produced by our MNSGA2-EDA.
2. To perform effective local search using the useful information of good candidate solutions in each generation. We propose a model-guided local search, which first constructs distribution models from suitable Pareto front solutions and other good candidate solutions selected by our proposed clustering technique, and then samples effective solutions from the distribution models.
3. To generate a set of trade-off solutions regarding both QoSM and QoS, NHSGA2-EDA is effectively utilized in this paper to solve challenging multi-objective service composition problems with requirements for both QoSM and QoS. Empirical comparison with NSGA-II and Hybrid [6] shows that NHSGA2-EDA is much more effective and efficient. To explore the scalability of multi-objective approaches we propose a new benchmark dataset. Experiments conducted with this dataset show that NSGA2-EDA can maintain high performance on problems with significantly larger sizes.

2   Related Work

EC techniques have been widely used to automatically find optimal or near-optimal composite service solutions efficiently, and the optimization target can be either or both of QoSM and QoS. [4,6–8,16,18–20,22–24]. These works can be mainly divided into two groups: single-objective or multi-objectives web service composition.

EC-based single-objective fully automated service composition approaches are well studied, resulting in many new designs of effective solution representations and problem-specific genetic operators [8]. Specifically, there are two categories of solution representations—*direct representations* and *indirect representations*. The *direct single-objective approaches* employ GP variants to evolve tree and graph-based composite solutions [16,19]. For example, [19] proposes a tree-like representation to eliminate the replicas of subtrees and specific genetic operators to generate offsprings. [20] uses EDA to learn one Edge Histogram Matrix (EHM) of service dependencies in every generation, and samples valid promising DAG-based solutions from the EHM. However, this approach suffers from a scalability issue when size of service repository is double of the reported size in [20].

The *indirect single-objective approaches* often employ vector-based representations to find an optimized queue of services, which will be decoded into an interpretable solution in the form of a direct representation with the help of a decoding method. As suggested in [8], utilizing the indirect representation often contributes to more effective performance, compared to direct representation, because the search space is not unwittingly restricted by unconstrained random initialization of solutions and operators. PSO, GA, and EDA have been employed for this purpose [8,18,20,22,23]. For example, [22] learns one Node Histogram Matrix (NHM) for the current population. This learned NHM will be used to sample new candidate solutions for the next population through the use of EDA. Empirical experiment are later conducted in [23]. In this paper, we also employ an indirect representation, which also simplifies the use of EDA for local search.

Very limited works have ever proposed EC-based multi-objective fully automated service composition approaches, although many works on multi-objective semi-automated service composition have been reported [4,24]. To the best of our knowledge, [6,7] are the two recent attempts on fully automated service composition with the aim of handling trade-offs in QoS alone. [7] develop a multi-objective method using NSGA-II and a fragmented tree-based representation. However, this fragmented tree-based representation does not show its effectiveness for finding better Pareto solutions in their experiment, comparing to an indirect representation. The same authors later proposed Hybrid [6] with the indirect representation. Hybrid [6] decomposes the multi-objective problem into single-objective subproblems, where local search can be applied based on Tchebycheff scores on each subproblem. The limitations of this work have already addressed in Sect. 1, e.g., a large number of decomposed subproblems is pre-defined. Despite some promising results have been achieved, opportunities still exist to address these limitations.

3 The Multiobjective Semantic Web Service Composition Problem

In this paper, we study *comprehensive quality-aware semantic web service composition* problem that concerns the quality of composite solutions in both functional (i.e., QoSM) and non-functional (i.e., QoS) aspects. This problem has been well approached in the literature using EC-based single-objective techniques, where QoSM and QoS are combined to be one globally optimized objective

[18–20,22,23]. Some concepts related to this web service composition problem, such as *semantic web service, service repository ($\mathcal{SR}$), service request ($\mathcal{T}$), composite service* are not demonstrated in this paper due the page limit, please refer to [18–20,23]. However, according to our knowledge, no attempts have ever been reported in literature to address this problem in a multi-objective setting where QoSM and QoS must be optimized separately. Such a new problem will be referred to as **M**ulti-objective **C**omprehensive **Q**uality-aware semantic web service composition **P**roblem (MOCQP, for short) in this paper.

Here we formulate MOCQP based on two objectives that reflect the functional (i.e., QoSM) and non-functional quality criteria (i.e., QoS) as follows:

$$\text{Minimize } \boldsymbol{f}(C) = (f_1(C), f_2(C)) \tag{1}$$
$$\text{subject to } C \in \mathcal{Z}$$

$$f_1(C) = w_1(1 - \hat{MT}) + w_2(1 - \hat{SIM}) \tag{2}$$

$$f_2(C) = w_3(1 - \hat{A}) + w_4(1 - \hat{R}) + w_5\hat{T} + w_6\hat{CT} \tag{3}$$

where $\mathcal{Z}$ denotes the set of all composite services over a given repository of atomic services, and f_1, f_2 are two objective functions that capture the QoSM and QoS, respectively, for every service C in $\mathcal{Z}$. In particular, QoSM is calculated based on the normalized semantic matching type $\hat{MT}$ and the semantic similarity $\hat{SIM}$ while QoS is calculated based on the normalized availability $\hat{A}$, reliability $\hat{R}$, response time $\hat{T}$, and execution cost $\hat{CT}$, see calculations in [18–20,23]. $\hat{MT}$, $\hat{SIM}$, $\hat{A}$ and $\hat{R}$ are offset by 1, so that lower scores correspond to better quality.

The goal of MOCQP is to find the set of Pareto optimal composite services $PF^\star = \{C^\star \in \mathcal{Z}\}$, where $C^\star$ is Pareto optimal if $\nexists C' \in \mathbf{C}$, such that $C^\star \prec C'$. Note that $C^\star \prec C'$ means C' dominates $C^\star$ if $f_1(C^\star) \geq f_1(C')$ and $f_2(C^\star) > f_2(C')$ or if $f_1(C^\star) > f_1(C')$ and $f_2(C^\star) \geq f_2(C')$.

4   Our New Method MNSGA2-EDA

In this section, we present our new method for solving MOCQP, starting with an overview of MNSGA2-EDA, which enables EDA to be employed in NSGA-II as an effective local search component. Subsequently, we discuss MNSGA2-EDA in detail.

4.1   An Overview of MNSGA2-EDA

MNSGA2-EDA enhances NSGA-II by EDA-based local search, where EDA is exploited to discover better solutions based on some non-dominated solutions in each generation generated by NSGA-II. These solutions are determined separately and concurrently in different regions of the Pareto front for each generation. These regions are created by grouping the current Pareto front into multiple clusters, see details in Sect. 4.4.

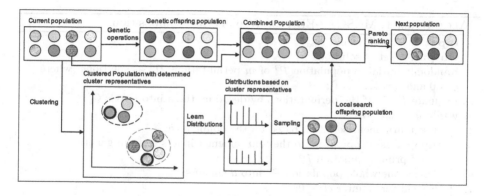

Fig. 1. Generation updates in MNSGA2-EDA

The generation updates in MNSGA2-EDA is illustrated in Fig. 1. From the current population in Fig. 1, two offspring populations are produced: genetic offspring population is produced by genetic operators, including both crossover and mutation (see details in Sect. 4.3); local search offspring population is produced by sampling from the distribution models constructed from the most suitable cluster representatives of the Pareto front (see details in Sect. 4.5).

4.2 Outline of MNSGA2-EDA

MNSGA2-EDA is outlined in Algorithm 1. Initially, we generate m permutations Π_k^g as composition solutions for population $\mathcal{P}^g$ of generation g, where $0 \leq k < m$ and $g = 0$. Each permutation is a randomly ordered sequence of task-related service indexes. For example, Let $\Pi = (\pi_0, \ldots, \pi_t, \ldots, \pi_{n-1})$ be a permutation-based composite solution of service indexes $\{0, \ldots, t, \ldots, n-1\}$ such that $\pi_i \neq \pi_j$ for all $i \neq j$. f_1, f_2 in Eqs. (2) and (3) of any newly produced permutations will be evaluated by decoding each permutation into a DAG-based solution, $\mathcal{G}_k^g$. Subsequently, the following steps (Step 3 to 15) are repeated until a maximum number of generation g_{max} is reached. Particularly, the production of the first offspring population starts with tournament selection in favor of winners with higher dominance regarding ranks and sparsity suggested in NSGA-II. The tournament winners will be processed by genetic operators (see details in Sect. 4.3) to produce genetic offspring population $\mathcal{P}_a^g$ based on a predefined probability. Afterwards, offspring $\mathcal{P}^g$ will be clustered into d clusters based on the values of f_1, f_2. For each cluster, we start by identifying its cluster representative Rep_{cl}^g, and then transform each cluster member $\mathcal{G}_k^q$ into a different permutation Π'_k^g, element of which are ordered based on $\mathcal{G}_k^g$, see details in Sect. 4.5. As suggested in [23], this transformation process allows more reliable and accurate learning of the distribution models in the form of Node Histogram Matrix $\mathcal{NHM}_{cl}^g$ (NHM). The contribution of each cluster member to NHM is adjusted decreasingly according to the Euclidean distance in the objective space between the cluster member and Rep_{cl}^g. Subsequently NHM is used to sample

---

ALGORITHM 1. MNSGA2-EDA for Web Service Composition.

Input : T, $\mathcal{SR}$, d and g_{max}
Output: A set of solutions
1: Randomly initialize population $\mathcal{P}^g$ of m permutations Π_k^g as solutions (where $g = 0$ and $k = 1, \ldots, m$);
2: Evaluate f_1, f_2 of the permutations by decoding them into DAGs $\mathcal{G}_k^g$;
3: **while** $g < g_{max}$ **do**
4:     Use tournament selection based on the dominance;
5:     Apply genetic operators to the tournament winners to form genetic offspring population $\mathcal{P}_a^g$;
6:     Divide the whole population $\mathcal{P}^g$ into d clusters;
7:     Set cluster counter $cl \leftarrow 0$;
8:     **while** $cl < d$ **do**
9:        Identify the cl^{th} cluster representative Rep_{cl}^g;
10:        Update each $\mathcal{G}_k^g$ in the cl^{th} cluster into a different permutation Π'^g_k;
11:        Learn $\mathcal{NHM}_{cl}^g$ over the cl^{th} cluster based on the representative Rep_{cl}^g to form sampling local search offspring population $\mathcal{P}_b^g$;
12:     $\mathcal{P}^{g+1} = \mathcal{P}^g \cup \mathcal{P}_a^g \cup \mathcal{P}_b^g$;
13:     Evaluate f_1, f_2 of each permutation in $\mathcal{P}^{g+1}$ by decoding it into $\mathcal{G}_k^g$;
14:     Perform a fast non-dominated sorting on $\mathcal{P}^{g+1}$;
15:     Keep top m solutions in $\mathcal{P}^{g+1}$;
16: Return non-dominated solutions in $\mathcal{P}^{g_{max}}$;

---

new local search offspring population $\mathcal{P}_b^g$, see details in Sect. 4.5. Consequently, we produce the next population $\mathcal{P}^{g+1}$ by combining the current population $\mathcal{P}^g$, genetic offspring population $\mathcal{P}_a^g$ and local search offspring population $\mathcal{P}_b^g$. After evaluating newly generated solutions and performing the fast non-dominated sorting in NSGA-II, the top m individuals are chosen to form the next generation $\mathcal{P}^{g+1}$. When the stopping criterion is finally met, the non-dominated solutions in $\mathcal{P}^{g_{max}}$ are returned as the output of NSGA2-EDA.

4.3 Genetic Operators

One two-point crossover and one one-point swap mutation [6,14] are employed to produce the genetic offspring population. An example of this crossover and mutation operator is illustrated in Fig. 2. The crossover operator produces two children. Each child preserves part of the elements of one parent, while elements of another parent (excluding those preserved elements by the child) fill the remaining parts of this child from left to right. The mutation randomly swaps two elements of one parent to produce a new permutation.

We produce genetic offspring population $\mathcal{P}_a^g$ more efficiently than that in Hybrid [6]. Although two children are produced by one crossover in Hybrid [6], only one child associated with a higher Tchebycheff score will be added to the offspring population $\mathcal{P}_a^g$. Compared to [6], we put both children in population

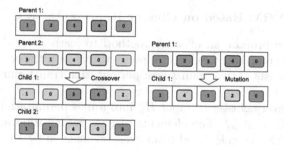

Fig. 2. Examples of crossover and mutation for parents

$\mathcal{P}_a^g$. Therefore, to produce an offspring population with equal sizes, we only need to evaluate half the number of offspring solutions as required by Hybrid [6].

4.4   Identify a Cluster Representative of Each Cluster

Unlike single-objective optimization problems in [22, 23], it is not straightforward to determine promising solutions for learning NHM in EDA under the multi-objective optimization setting since they often have two objectives. To address this issue, we propose to define one cluster representative as a promising solution based on the dominance relationships among all solutions in the cluster it belongs to. In particular, we cluster d groups of close individuals in one generation using some existing clustering techniques, such as K-means++ [2]. The sensitivity of parameter d is studied in Sect. 5.1 for the effectiveness of our NSGA2-EDA. We infer a group of individuals that represents close similarities measured by fitness values, f_1 and f_2 in Eqs. (2) and (3). We choose with equal probability one solution that is not dominated by any other solutions of the same cluster as the representative of the cluster. Consequently, we can learn an NHM based on the cluster representatives, see details in Sect. 4.5.

An example of identifying promising solutions of clusters is illustrated in Fig. 3. In Fig. 3, we consider one population of 8 individuals, which are clustered into two groups of individuals based on their fitness values using K-mean++. Subsequently, we can randomly pick up one non-dominant solution of its related cluster as the cluster representatives, see the labels in Fig. 3.

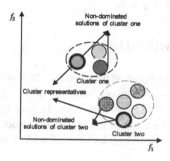

Fig. 3. Examples of identifying two cluster representatives

4.5   Learn a NHM Based on Cluster Representatives

In this paper, we propose an effective method to learn a suitable distribution model (i.e. NHM) based on the cluster representative with respect to each cluster. This method consists of two main steps: permutation transformation and NHM learning.

We transfer every cluster member Π_k^g into a new permutation Π'^g_k based on its decoded DAG form $\mathcal{G}_k^g$. The elements of new permutation are sorted based on the longest distance calculated from every element in $\mathcal{G}_k^g$ to the $Start$ node, see details in [23]. We can now learn a NHM based on the cluster representative formed in this permutation. Based on [23], we propose a different way of learning NMH, which is more likely to make local improvements on the cluster representatives through sampling. In particular, we use the Euclidean distances between the cluster representative and other members of this cluster to weight the influences of every cluster members on NHM because cluster members far from cluster representative contribute less to the distribution model that we aim to learn.

The *node histogram matrix* (NHM) for the cl^{th} cluster with the cluster representative Rep^{cl} in generation g is denoted as $\mathcal{NHM}_{cl}^g$, which is an $n \times n$-matrix with entries $e_{i,j}$ as follows:

$$e_{i,j} = \sum_{k=0}^{m-1} \delta_{i,j}(\Pi'^g_k) + \varepsilon \tag{4}$$

$$\delta_{i,j}(\Pi'^g_k) = \begin{cases} w(\Pi'^g_k) & \text{if } \pi_i = j \\ 0 & \text{otherwise} \end{cases} \tag{5}$$

$$w(\Pi'^g_k) = 1 - ||\boldsymbol{f}(\Pi'^g_k) - \boldsymbol{f}(Rep_{cl}^g)||_2 \tag{6}$$

where $i, j = 0, 1, \ldots, n-1$, $\varepsilon = \frac{m}{n-1}b_{ratio}$ is a predetermined bias, and $||\boldsymbol{f}(\Pi'^g_k) - \boldsymbol{f}(Rep_{cl}^g)||_2$ measures a Euclidean distance between one cluster member and the cluster representative. This distance value is offset by 1, so the higher values correspond to less weights in learning an NHM. Roughly speaking, entry $e_{i,j}$ counts how often service π_i appears in position j of all the permutations in the cl^{th} cluster, and the weight of the frequency is penalized by Eq. (6). Afterwards, we can use node histogram-based sampling [17] to sample local search offspring population $\mathcal{P}_b^g$ for generation $g + 1$.

5   Experimental Evaluation

We conduct two experiments for studying the performance of our MNSGA2-EDA approach using two augmented benchmarks in [6] that originally comes from WSC-08 [3] and WSC-09 [13] extended with real QoS attributes in [1]. Both WSC-08 and WSC-09, define a set of composition tasks. However, the number of web services in augmented benchmarks [6] is doubled as a new benchmark (with much bigger searching space) to demonstrate that NSGA2-EDA can maintain high performance on our problem with significantly larger sizes. In particular, each service in WSC-08 and WSC-09 is duplicated with the same functionality (i.e., inputs and outputs) but different QoS attributes extended from QWS [1]. The first experiment investigates the sensitivity of parameters on EDA based on task WSC08-03. In particular, we investigate three groups of EDA settings with increasing size of $\mathcal{P}_b^g$ (see details in Sect. 5.1). The following experiment further investigates the effectiveness and efficiency of MNSGA2-EDA in comparison to the baseline method NSGA-II and to Hybrid [6]. These two approaches have recently been proposed to solve a similar service composition problem for the fully automated and multi-objective purpose. Note that in [6] a further method (called Hybrid-L) has been proposed, that uses a so-called swap operator as a local search to Hybrid. However, Hybrid-L observes very bad convergence rates. We use two tasks WSC09-3 and WSC09-5 to exemplify the very bad performance of Hybrid-L in Figs. 4 and 5. Therefore, we do not further report on the performance of Hybrid-L for the remaining tasks, when compared to our MNSGA2-EDA method.

We follow the settings in [6] for all approaches, where the size of both $\mathcal{P}^g$ and $\mathcal{P}_a^g$ are set to 500. The maximum generation g is 51, and the probability rates of crossover, mutation, and reproduction are 0.8, 0.1 and 0.1. For EDA settings in MNSGA2-EDA, b_{ratio} of ε is set to 0.0002 according to [23]. The weights in the fitness function Eqs. (2) and (3) are set to balance quality criteria in both QoSM and QoS, i.e., w_1 and w_2 are set to 0.5, and w_3, w_4, w_5 and w_6 to 0.25. We have also conducted tests with other weights and parameters and generally observed the same behavior.

5.1   Parameters Sensitivity

To determine suitable parameters of EDA-based local search in MNSGA2-EDA, we use task WSC08-3 to perform parameters sensitivity tests over a set of parameters with an increasing size of $\mathcal{P}_b^g$ in MNSGA2-EDA.

We use Wilcoxon rank-sum testing with a significance level of 5% to verify the observed differences in IGD and $hypervolume$ over 30 runs. This test method is used consistently to detect any noticeable differences in the experiment results in Sects. 5.2 and 5.3.

IGD and $hypervolume$ are commonly used performance evaluation metrics for multi-objective optimization [12]. IGD measures the distance from the nearest point of the non-dominated set produced by an approach to an approximated true Pareto front obtained by using all approaches. Hypervolume measures the dominated volume covered by a reference point (e.g., a point (1,1) is chosen in our case) and the front evolved by each algorithm. In particular, we highlight IGD and hypervolume values of all the top performances for all approaches.

The first column of Tables 1 and 2 show the size of $\mathcal{P}_b^g$. The second and third column of Tables 1 and 2 show a pair of parameters used in EDA, which are the number of clusters d and their sampling size. The fourth column of Tables 1 and 2 show the mean values of IGD and hypervolume and the standard deviation over 30 repetitions.

Table 1. Mean IGD of MNSGA2-EDA with three groups of parameter settings over WSC08-3 (Note: the lower the IGD the better)

Size of $\mathcal{P}_b^g$	d	Sampling size	MNSGA2-EDA
160	2	80	$6e - 04 \pm 3e - 04$
	4	40	$2e - 04 \pm 0$
	6	27	$2e - 04 \pm 1e - 04$
200	2	100	$4e - 04 \pm 2e - 04$
	4	50	$1e - 04 \pm 0$
	6	34	$2e - 04 \pm 1e - 04$
240	2	120	$4e - 04 \pm 3e - 04$
	4	60	$1e - 04 \pm 1e - 04$
	6	40	$1e - 04 \pm 0$

Table 2. Mean hypervolume of MNSGA2-EDA with three groups of parameter settings over WSC08-03 (Note: the higher the hypervolume the better)

Size of $\mathcal{P}_b^g$	d	Sampling size	MNSGA2-EDA
160	2	80	$0.2302 \pm 1e - 04$
	4	40	$0.2304 \pm 1e - 04$
	6	27	$0.2304 \pm 1e - 04$
200	2	100	$0.2303 \pm 1e - 04$
	4	50	0.2305 ± 0
	6	34	$0.2305 \pm 1e - 04$
240	2	120	$0.2303 \pm 1e - 04$
	4	60	$0.2305 \pm 1e - 04$
	6	40	0.2305 ± 0

Tables 1 and 2 show that 200 local search offspring population size based on 4 clusters with 50 sampling size is the best-found parameter setting over all designed parameter settings for task WSC08-3. As shown in Tables 1 and 2, MNSGA2-EDA with this setting is highlighted as one top performance regarding mean IGD and hypervolume, but with the smallest size of $\mathcal{P}_b^g$. We will use this setting in our second experiment.

5.2   Comparison of the Execution Time

Table 3 shows the mean execution times (in seconds) and the standard deviation observed for the three methods MNSGA2-EDA, NSGA-II and Hybrid over 30 repetitions. More specifically, Table 4 summarizes the results of pairwise comparisons of the three methods without Bonferroni correction. The table displays win/draw/loss of one method compared to all other methods. That is, it is reported how often one method outperforms, equals or is outperformed by the competing method.

Table 3. Mean execution time (in s) for our method in comparison to the baseline NSGA-II, and to Hybrid (Note: the shorter the time the better)

Task	MNSGA2-EDA	NSGA-II	Hybrid [6]
WSC08-1	224 ± 12	190 ± 48	418 ± 65
WSC08-2	81 ± 17	58 ± 14	139 ± 32
WSC08-3	5539 ± 464	8095 ± 1437	20793 ± 4149
WSC08-4	210 ± 17	317 ± 58	805 ± 147
WSC08-5	4242 ± 562	6090 ± 1704	14735 ± 5166
WSC08-6	62966 ± 10943	65051 ± 8592	158737 ± 27171
WSC08-7	5489 ± 814	9132 ± 2578	23074 ± 6030
WSC08-8	9917 ± 3788	12443 ± 1818	33077 ± 6164
WSC09-1	198 ± 67	155 ± 76	327 ± 90
WSC09-2	5634 ± 679	6139 ± 1678	14634 ± 2816
WSC09-3	2968 ± 301	2820 ± 714	6527 ± 2403
WSC09-4	269207 ± 23542	255195 ± 28813	646897 ± 117538
WSC09-5	39370 ± 5125	35338 ± 8350	86281 ± 19944

Table 4. Summary of statistical significance tests for the execution time, where each column shows the win/draw/loss score of one method against a competing one for all tasks of WSC08 and WSC09.

Dataset	Method	MNSGA2-EDA	NSGA-II	Hybrid [6]
WSC08 (8 tasks)	MNSGA2-EDA	-	2/1/5	0/0/8
	NSGA-II	5/1/2	-	0/0/8
	Hybrid [6]	8/0/0	8/0/0	-
WSC09 (5 tasks)	MNSGA2-EDA	-	3/2/0	0/0/5
	NSGA-II	0/2/3	-	0/0/5
	Hybrid [6]	5/0/0	5/0/0	-

The mean execution time for MNSGA2-EDA and NSGA-II are very comparable (but not equal) to each other for tasks in WSC08 and WSC09. In comparison, Hybrid consistently takes twice the execution time for each task. This observation does not agree with the findings in [6] that Hybrid and NSGA-II achieve competitive execution time. This is because they do not point out one assumption that evaluation time of every candidate solution is indistinct. Here in this paper, a more challenging benchmark is utilized for testing, and a larger number of evaluations is required for computing QoSM of each solution. In Hybrid, every crossover operator requires two evaluations of two produced children in order to keep a child with a higher Tchebycheff score, while MNSGA2-EDA and NSGA-II both keep two children. For example, let us say that 500 children are kept for the next generation from the crossover, then Hybrid requires 1000 evaluations while MNSGA2-EDA and NSGA-II only require 500 evaluations. Therefore, Hybrid consumes much more execution time than both MNSGA2-EDA and NSGA-II.

5.3   Comparison of the IGD and Hypervolume

Tables 5, 6, 7 and 8 show the mean IGD and the mean hypervolume, respectively, observed for MNSGA2-EDA, NSGA-II, and Hybrid with the standard deviation over 30 repetitions. We note that MNSGA2-EDA achieves significantly better values of IGD for all tasks except for one task (i.e., WSC09-1) and significantly better values of hypervolume for all tasks. On the other hand, NSGA-II only achieves significantly better values of both IGD and hypervolume for 2 of the 13 tasks, and Hybrid only obtained significantly better values of IGD and hypervolume for 4 out of the 13 tasks and 3 out of the 13 tasks respectively.

Table 5. Mean IGD for our method in comparison to the baseline NSGA-II, and to Hybrid (Note: the lower the IGD the better)

Task	MNSGA2-EDA	NSGA-II	Hybrid [6]
WSC08-1	0 ± 0	$1e - 04 \pm 7e - 04$	$1e - 04 \pm 5e - 04$
WSC08-2	0 ± 0	0 ± 0	0 ± 0
WSC08-3	$1e - 04 \pm 0$	$0.001 \pm 4e - 04$	$0.001 \pm 3e - 04$
WSC08-4	0 ± 0	$3e - 04 \pm 3e - 04$	$1e - 04 \pm 1e - 04$
WSC08-5	0.0029 ± 0.0014	0.0043 ± 0.0015	0.0027 ± 0.0011
WSC08-6	$7e - 04 \pm 3e - 04$	$0.0014 \pm 3e - 04$	$0.0012 \pm 3e - 04$
WSC08-7	$1e - 04 \pm 2e - 04$	$0.002 \pm 9e - 04$	0.0015 ± 0.001
WSC08-8	$0 \pm 1e - 04$	$9e - 04 \pm 5e - 04$	$6e - 04 \pm 3e - 04$
WSC09-1	0.0701 ± 0.0132	$0.0731 \pm 6e - 04$	0.0654 ± 0.0199
WSC09-2	0.0055 ± 0.001	0.0065 ± 0.0011	$0.0061 \pm 9e - 04$
WSC09-3	$0.002 \pm 9e - 04$	0.0126 ± 0.0085	0.0107 ± 0.0076
WSC09-4	0.0025 ± 0.001	$0.0061 \pm 7e - 04$	0.0056 ± 0.0012
WSC09-5	0.0025 ± 0.0014	0.0052 ± 0.0011	$0.0045 \pm 7e - 04$

Table 6. Summary of statistical significance tests for IGD, where each column shows win/draw/loss scores of one method against a competing one for all tasks of WSC08 and WSC09.

Dataset	Method	MNSGA2-EDA	NSGA-II	Hybrid [6]
WSC08 (8 tasks)	MNSGA2-EDA	-	0/2/6	0/3/5
	NSGA-II	6/2/0	-	4/4/0
	Hybrid [6]	5/3/0	0/4/4	-
WSC09 (5 tasks)	MNSGA2-EDA	-	0/0/5	1/0/4
	NSGA-II	5/0/0	-	2/3/0
	Hybrid [6]	4/0/1	0/3/2	-

Table 7. Mean Hypervolume for our method in comparison to the baseline NSGA-II, and to Hybrid (Note: the higher the hypervolume the better)

Task	MNSGA2-EDA	NSGA-II	Hybrid [6]
WSC08-1	0.3825 ± 0	$0.3824 \pm 4e - 04$	$0.3825 \pm 1e - 04$
WSC08-2	0.5798 ± 0	0.5798 ± 0	0.5798 ± 0
WSC08-3	0.2305 ± 0	$0.2298 \pm 2e - 04$	$0.23 \pm 1e - 04$
WSC08-4	0.3217 ± 0	$0.3213 \pm 7e - 04$	$0.3215 \pm 5e - 04$
WSC08-5	$0.278 \pm 7e - 04$	0.2752 ± 0.0022	0.2767 ± 0.0014
WSC08-6	$0.2341 \pm 1e - 04$	$0.2338 \pm 2e - 04$	$0.2341 \pm 2e - 04$
WSC08-7	$0.2808 \pm 2e - 04$	0.278 ± 0.0014	0.2788 ± 0.0014
WSC08-8	$0.2475 \pm 1e - 04$	$0.2465 \pm 7e - 04$	$0.2471 \pm 4e - 04$
WSC09-1	0.4435 ± 0.0028	$0.4424 \pm 9e - 04$	0.4434 ± 0.0031
WSC09-2	$0.2751 \pm 1e - 04$	0.2742 ± 0.0016	$0.2747 \pm 7e - 04$
WSC09-3	$0.3693 \pm 1e - 04$	0.361 ± 0.0064	0.3618 ± 0.0054
WSC09-4	0.239 ± 0.0014	$0.2346 \pm 9e - 04$	0.2355 ± 0.0017
WSC09-5	0.2376 ± 0.001	$0.235 \pm 5e - 04$	$0.2353 \pm 5e - 04$

5.4 Comparison of the Convergence Rate

To investigate the effectiveness and scalability of the three methods, we further investigate the convergence rates for IDG and hypervolume over 30 repetitions using WSC09-3 and WSC09-5 as two examples.

Figures 5 and 4 depict the evolution of the mean values of the IGD and hypervolume over mean execution time for MNSGA2-EDA, NSGA-II, Hybrid, and Hybrid-L. We cut mean execution time to fit the maximal required time of Hybrid because Hybrid-L results in a much higher order of magnitude in execution time, and it also never gets a chance to catch up with MNSGA2-EDA. For Hybrid, it converges much better than Hybrid-L, but the scalability of Hybrid still suffers when competing with the baseline NSGA-II. In contrast,

Table 8. Summary of the statistical significance tests for hypervolume, where each column shows win/draw/loss scores of one method against a competing one for all tasks of WSC08 and WSC09.

Dataset	Method	MNSGA2-EDA	NSGA-II	Hybrid [6]
WSC08 (8 tasks)	MNSGA2-EDA	-	0/2/6	0/3/5
	NSGA-II	6/2/0	-	5/3/0
	Hybrid [6]	5/3/0	0/3/5	-
WSC09 (5 tasks)	MNSGA2-EDA	-	0/0/5	0/0/5
	NSGA-II	5/0/0	-	2/3/0
	Hybrid [6]	5/0/0	0/3/2	-

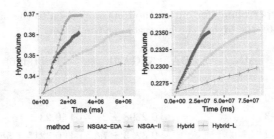

Fig. 4. Mean hypervolume over time for non-dominated solutions, for WSC09-3 (left) and WSC09-5 (right) (Note: the larger the hypervolume the better)

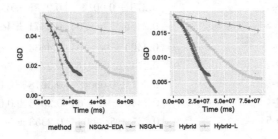

Fig. 5. Mean IGD over time for non-dominated solutions, for WSC09-3 (left) and WSC09-5 (right) (Note: the smaller the IGD the better)

our MNSGA2-EDA approach achieves significantly better IGD and hypervolume values with the fastest convergence rate.

5.5   Comparison of the Pareto Optimal Solutions

We present a plot of the Pareto optimal solutions of WSC09-3 and WSC09-5 obtained by the three methods over 30 independent runs in Fig. 6. The best Pareto optimal solutions are identified based on the combined results of all 30 runs of each method. It is easy to observe that the Pareto front generated by

MNSGA2-EDA is much more widely distributed. In other words, extreme solutions are more likely to be found by MNSGA2-EDA. For task WSC09-3, a trade-off solution at the knee point of the Pareto front is found by MNSGA2-EDA. We hasten to point out that it is highly important and desirable to discover a solution like this. The other two methods (NSGA-II and Hybrid) fail to discover this solution, which may be regarded as a weakness. For task WSC09-05, much better Pareto optimal solutions are obtained by MNSGA2-EDA, and these solutions consistently dominate all solutions obtained by other methods.

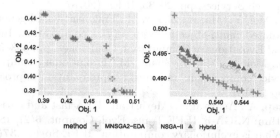

Fig. 6. Pareto optimal solutions obtained for tasks WSC09-3 (left) and WSC09-5 (right)

6 Conclusion

In this paper, we proposed a novel memetic NSGA-II with an EDA-based local search for fully automated multi-objective web service composition, where two objectives related to the functional and non-functional quality of composite services are optimized, i.e., QoSM and QoS. Our experimental evaluation demonstrates that our proposed approach can effectively and efficiently produce better Pareto optimal solutions, thus, outperforming two recently proposed approaches in the literature. Future work in this field demands more research in EC techniques that can be applied to service composition, achieving better results that benefit the application side. For example, we can investigate sampling techniques to design problem-specific templates that can be used to sample solutions with good quality effectively.

References

1. Al-Masri, E., Mahmoud, Q.H.: Qos-based discovery and ranking of web services. In: Proceedings of 16th International Conference on Computer Communications and Networks, ICCCN 2007, pp. 529–534. IEEE (2007)
2. Arthur, D., Vassilvitskii, S.: K-means++: the advantages of careful seeding. In: Proceedings of the Eighteenth Annual ACM-SIAM Symposium on Discrete Algorithms, pp. 1027–1035. Society for Industrial and Applied Mathematics (2007)

3. Bansal, A., Blake, M.B., Kona, S., Bleul, S., Weise, T., Jaeger, M.C.: WSC-08: continuing the web services challenge. In: 2008 10th IEEE Conference on E-Commerce Technology and the Fifth IEEE Conference on Enterprise Computing, E-Commerce and E-Services, pp. 351–354. IEEE (2008)

4. Chen, Y., Huang, J., Lin, C.: Partial selection: an efficient approach for QoS-aware web service composition. In: IEEE ICWS, pp. 1–8. IEEE (2014)

5. Curbera, F., Nagy, W., Weerawarana, S.: Web services: why and how. In: Workshop on Object-Oriented Web Services-OOPSLA (2001)

6. Da Silva, A.S., Ma, H., Mei, Y., Zhang, M.: A hybrid memetic approach for fully automated multi-objective web service composition. In: 2018 IEEE International Conference on Web Services, pp. 26–33. IEEE (2018)

7. Da Silva, A.S., Mei, Y., Ma, H., Zhang, M.: Fragment-based genetic programming for fully automated multi-objective web service composition. In: Proceedings of the Genetic and Evolutionary Computation Conference, pp. 353–360. ACM (2017)

8. Da Silva, A.S., Mei, Y., Ma, H., Zhang, M.: Evolutionary computation for automatic web service composition: an indirect representation approach. J. Heuristics **24**(3), 425–456 (2018)

9. Deb, K., Pratap, A., Agarwal, S., Meyarivan, T.: A fast and elitist multiobjective genetic algorithm: NSGA-ii. IEEE Trans. Evol. Comput. **6**(2), 182–197 (2002)

10. Fogel, D.B.: What is evolutionary computation? IEEE Spectr. **37**(2), 26–32 (2000)

11. Gabrel, V., Manouvrier, M., Murat, C.: Web services composition: complexity and models. Discrete Appl. Math. **196**, 100–114 (2015)

12. Jiang, S., Ong, Y.S., Zhang, J., Feng, L.: Consistencies and contradictions of performance metrics in multiobjective optimization. IEEE Trans. Cybern. **44**(12), 2391–2404 (2014)

13. Kona, S., Bansal, A., Blake, M.B., Bleul, S., Weise, T.: WSC-2009: a quality of service-oriented web services challenge. In: 2009 IEEE Conference on Commerce and Enterprise Computing, pp. 487–490. IEEE (2009)

14. Lacomme, P., Prins, C., Ramdane-Cherif, W.: Competitive memetic algorithms for arc routing problems. Ann. Oper. Res. **131**(1–4), 159–185 (2004)

15. Rao, J., Su, X.: A survey of automated web service composition methods. In: Cardoso, J., Sheth, A. (eds.) SWSWPC 2004. LNCS, vol. 3387, pp. 43–54. Springer, Heidelberg (2005). https://doi.org/10.1007/978-3-540-30581-1_5

16. Rodriguez-Mier, P., Mucientes, M., Lama, M., Couto, M.I.: Composition of web services through genetic programming. Evol. Intel. **3**(3–4), 171–186 (2010)

17. Tsutsui, S.: A comparative study of sampling methods in node histogram models with probabilistic model-building genetic algorithms. In: IEEE International Conference on Systems, Man and Cybernetics, SMC 2006, vol. 4, pp. 3132–3137. IEEE (2006)

18. Wang, C., Ma, H., Chen, A., Hartmann, S.: Comprehensive quality-aware automated semantic web service composition. In: Peng, W., Alahakoon, D., Li, X. (eds.) AI 2017. LNCS (LNAI), vol. 10400, pp. 195–207. Springer, Cham (2017). https://doi.org/10.1007/978-3-319-63004-5_16

19. Wang, C., Ma, H., Chen, A., Hartmann, S.: GP-based approach to comprehensive quality-aware automated semantic web service composition. In: Shi, Y., et al. (eds.) SEAL 2017. LNCS, vol. 10593, pp. 170–183. Springer, Cham (2017). https://doi.org/10.1007/978-3-319-68759-9_15

20. Wang, C., Ma, H., Chen, G., Hartmann, S.: Towards fully automated semantic web service composition based on estimation of distribution algorithm. In: Mitrovic, T., Xue, B., Li, X. (eds.) AI 2018. LNCS (LNAI), vol. 11320, pp. 458–471. Springer, Cham (2018). https://doi.org/10.1007/978-3-030-03991-2_42

21. Wang, C., Ma, H., Chen, A., Hartmann, S.: A memetic NSGA-II with EDA-based local search for fully automated multiobjective web service composition. In: Genetic and Evolutionary Computation Conference Companion. ACM (2019), (To appear)
22. Wang, C., Ma, H., Chen, G.: EDA-based approach to comprehensive quality-aware automated semantic web service composition. In: Proceedings of the Genetic and Evolutionary Computation Conference Companion, pp. 147–148. ACM (2018)
23. Wang, C., Ma, H., Chen, A., Hartmann, S.: Knowledge-driven automated web service composition—an EDA-based approach. In: Hacid, H., Cellary, W., Wang, H., Paik, H.-Y., Zhou, R. (eds.) WISE 2018. LNCS, vol. 11234, pp. 135–150. Springer, Cham (2018). https://doi.org/10.1007/978-3-030-02925-8_10
24. Yin, H., Zhang, C., Zhang, B., Guo, Y., Liu, T.: A hybrid multiobjective discrete particle swarm optimization algorithm for a SLA-aware service composition problem. Math. Probl. Eng. **2014**, 14 (2014)
25. Zhou, A., Qu, B.Y., Li, H., Zhao, S.Z., Suganthan, P.N., Zhang, Q.: Multiobjective evolutionary algorithms: a survey of the state of the art. Swarm Evol. Comput. **1**, 32–49 (2011)

Adaptive Caching for Data-Intensive Scientific Workflows in the Cloud

Gaëtan Heidsieck[1]([✉]) [iD], Daniel de Oliveira[4] [iD], Esther Pacitti[1] [iD], Christophe Pradal[1,2] [iD], François Tardieu[3] [iD], and Patrick Valduriez[1] [iD]

[1] Inria & LIRMM, Univ. Montpellier, Montpellier, France
gaetan.heidsieck@inria.fr
[2] CIRAD & AGAP, Montpellier SupAgro, Montpellier, France
[3] INRA & LEPSE, Montpellier SupAgro, Montpellier, France
[4] Institute of Computing, UFF, Niterói, Brazil

Abstract. Many scientific experiments are now carried on using scientific workflows, which are becoming more and more data-intensive and complex. We consider the efficient execution of such workflows in the cloud. Since it is common for workflow users to reuse other workflows or data generated by other workflows, a promising approach for efficient workflow execution is to cache intermediate data and exploit it to avoid task re-execution. In this paper, we propose an adaptive caching solution for data-intensive workflows in the cloud. Our solution is based on a new scientific workflow management architecture that automatically manages the storage and reuse of intermediate data and adapts to the variations in task execution times and output data size. We evaluated our solution by implementing it in the OpenAlea system and performing extensive experiments on real data with a data-intensive application in plant phenotyping. The results show that adaptive caching can yield major performance gains, *e.g.*, up to 120.16% with 6 workflow re-executions.

Keywords: Adaptive caching · Scientific workflow · Cloud · Workflow execution

1 Introduction

In many scientific domains, *e.g.*, bio-science [8], complex experiments typically require many processing or analysis steps over huge quantities of data. They can be represented as scientific workflows (SWfs), which facilitate the modeling, management and execution of computational activities linked by data dependencies. As the size of the data processed and the complexity of the computation keep increasing, these SWfs become data-intensive [8], thus requiring execution in a high-performance distributed and parallel environment, *e.g.* a large-scale virtual cluster in the cloud.

Most Scientific Workflow Management Systems (SWfMSs) can now execute SWfs in the cloud [12]. Some examples of such SWfMS are Swift/T, Pegasus,

S. Hartmann et al. (Eds.): DEXA 2019, LNCS 11707, pp. 452–466, 2019.
https://doi.org/10.1007/978-3-030-27618-8_33

SciCumulus, Kepler and OpenAlea, the latter being widely used in plant science for simulation and analysis.

It is common for workflow users to reuse other workflows or data generated by other workflows. Reusing and re-purposing workflows allow for the user to develop new analyses faster [7]. Furthermore, a user may need to execute a workflow many times with different sets of parameters and input data to analyze the impact of some experimental step, represented as a workflow fragment, *i.e.* a subset of the workflow activities and dependencies. In both cases, some fragments of the workflow will be executed many times, which can be highly resource consuming and unnecessary long. Workflow re-execution can be avoided by storing the intermediate results of these workflow fragments and reuse them in later executions.

In OpenAlea, this is provided by a lazy evaluation technique, *i.e.* the intermediate data is simply kept in memory after the execution of a workflow. This allows for a user to visualize and analyze all the activities of a workflow without any re-computation, even with some parameter changes. Although lazy evaluation represents a step forward, it has some limitations, *e.g.* it does not scale in distributed environments and requires much memory if the workflow is data-intensive.

In a single user perspective, the reuse of the previous results can be done by storing the relevant outputs of intermediate activities (intermediate data) within the workflow. This requires the user to manually manage the caching of the results that she wants to reuse, which can be difficult as she needs to be aware of the data size, execution time of each task, *i.e.* the instantiation of an activity during the execution of a workflow, or other factors that could allow deciding which data is the best to store.

A complementary, promising approach is to reuse intermediate data produced by multiple executions of the same or different workflows. Some SWfMSs support the reuse of intermediate data, yet with some limitations. VisTrails [4] automatically makes the intermediate data persistent with the workflow definition. With a plugin [20], VisTrails allows SWf execution in HPC environments, but does not benefit from reusing intermediate data. Kepler [2] manages a persistent cache of intermediate data in the cloud, but does not take data transfers from remote servers into account. There is also a trade-off between the cost of re-executing tasks versus storing intermediate data that is not trivial [1,6]. Yuan *et al.* [18] propose an algorithm based on the ratio between re-computation cost and storage cost at the task level. The algorithm is improved in [19] to take into account workflow fragments. Both algorithms are used before the execution of the workflow, using the provenance data of the intermediate datasets. However, this approach is static and cannot deal with variations in tasks' execution times. In data intensive SWf, such variations can be very important depending on the input data, *e.g.*, data compression tasks can be short or long depending on the data itself, regardless of size.

In this paper, we propose an adaptive caching solution for efficient execution of data-intensive workflows in the cloud. By adapting to the variations in tasks' execution times, our solution can maximize the reuse of intermediate data produced by workflows from multiple users. Our solution is based on a new SWfMS architecture that automatically manages the storage and reuse of intermediate data. Cache management is involved during two main steps: SWf preprocessing, to remove all fragments of the workflow that do not need to be executed; and cache provisioning, to decide at runtime which intermediate data should be cached. We propose an adaptive cache provisioning algorithm that deals with the variations in task execution times and output data. We evaluated our solution by implementing it in OpenAlea and performing extensive experiments on real data with a complex data-intensive application in plant phenotyping.

This paper is organized as follows. Section 2 presents our real use case in plant phenotyping. Section 3 introduces our SWfMS architecture in the cloud. Section 4 describes our cache algorithm. Section 5 gives our experimental evaluation. Finally, Sect. 6 concludes.

2   Use Case in Plant Phenotyping

In this section, we introduce in more details a real SWf use case in plant phenotyping that will serve as motivation for the work and basis for the experimental evaluation.

In the last decade, high-throughput phenotyping platforms have emerged to allow for the acquisition of quantitative data on thousands of plants in well-controlled environmental conditions. These platforms produce huge quantities of heterogeneous data (images, environmental conditions and sensor outputs) and generate complex variables with *in-silico* data analyses. For instance, the seven facilities of the French Phenome project (https://www.phenome-emphasis.fr/phenome_eng/) produce each year 200 Terabytes of data, which are heterogeneous, multiscale and originate from different sites. Analyzing automatically and efficiently such massive datasets is an open, yet important, problem for biologists [17].

Computational infrastructures have been developed for processing plant phenotyping datasets in distributed environments [14], where complex phenotyping analyses are expressed as SWfs. Such analyses can be represented, managed and shared in an efficient way, where compute- and data-based activities are linked by dependencies [5].

One scientific challenge in phenomics, *i.e.*, the systematic study of phenotypes, is to analyze and reconstruct automatically the geometry and topology of thousands of plants in various conditions observed from various sensors [16]. For this purpose, we developed the OpenAlea Phenomenal software package [3]. Phenomenal provides fully automatic workflows dedicated to 3D reconstruction, segmentation and tracking of plant organs, and light interception to estimate plant biomass in various scenarios of climatic change [15].

Phenomenal is continuously evolving with new state-of-the-art methods that are added, thus yielding new biological insights (see Fig. 1). A typical workflow

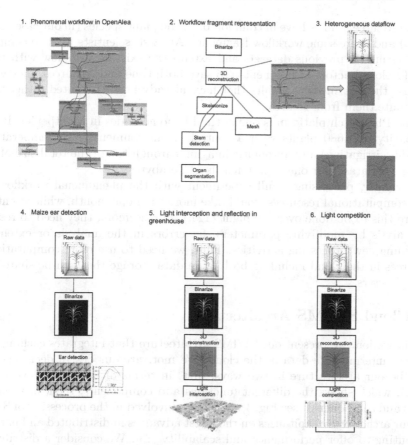

Fig. 1. Use Cases in Plant Phenotyping. (1) The Phenomenal workflow in OpenAlea's visual programming environment. The different colors represent different workflow fragments. (2) A conceptual view of the same workflow. (3) Raw and intermediate data such as RGB images, 3D plant volumes, skeleton, and mesh. (4–6) Three different SWfs that reuse the same workflow fragments to address different scientific questions.

is shown in Fig. 1(1). It is composed of different fragments, *i.e.*, reusable sub-workflows. In Fig. 1(2), the different fragments are for binarization, 3D reconstruction, skeletonization, stem detection, organ segmentation and mesh generation. Other fragments such as greenhouse or field reconstruction, or simulation of light interception, can be reused.

Based on these different workflow fragments, different users can conduct different biological analyses using the same datasets. The SWf shown in Fig. 1(4) reuses the *Binarize* fragment to predict the flowering time in maize. In Fig. 1(5), the same *Binarize* fragment is reused and the *3D reconstruction* fragment is added to reconstruct the volume of the 1,680 plants in 3D. Finally, in the SWf shown in Fig. 1(6), the previous SWf is reused, but with different parameters to study the environmental versus genetic influence of biomass accumulation.

These three studies have in common both the plant species (in our case maize plants) and share some workflow fragments. At least, scientists want to compare their results on previous datasets and extend the existing workflow with their own developed actors or fragments. To save both time and resources, they want to reuse the intermediate results that have already been computed rather than recompute them from scratch.

The Phenoarch platform is one of the Phenome nodes in Montpellier. It has a capacity of 1,680 plants with a controlled environment (*e.g.*, temperature, humidity, irrigation) and automatic imaging through time. The total size of the raw image dataset for one experiment is 11 Terabytes.

Currently, processing a full experiment with the phenomenal workflow on local computational resources would take more than one month, while scientists require this to be done over the night (12 h). Furthermore, they need to restart an analysis by modifying parameters, fix errors in the analysis or extend it by adding new processing activities. Thus, we need to use more computational resources in the cloud including both large data storage that can be shared by multiple users.

3   Cloud SWfMS Architecture

In this section, we present our SWfMS architecture that integrates caching and reuse of intermediate data in the cloud. We motivate our design decisions and describe our architecture in two ways: first, in terms of functional layers (see Fig. 2), which shows the different functions and components; then, in terms of nodes and components (see Fig. 3), which are involved in the processing of SWfs.

Our architecture capitalizes on the latest advances in distributed and parallel computing to offer performance and scalability [13]. We consider a distributed architecture with on premise servers, where raw data is produced (*e.g.*, by a phenotyping experimental platform in our use case), and a cloud site, where the SWf is executed. The cloud site (data center) is a shared-nothing cluster, *i.e.* a cluster of server machines, each with processor, memory and disk. We choose shared-nothing as it is the most scalable architecture for big data analysis.

In the cloud, metadata management has a critical impact on the efficiency of SWf scheduling as it provides a global view of data location, *e.g.* at which nodes some raw data is stored, and enables task tracking during execution [9]. We organize the metadata in three repositories: catalog, provenance database and cache index. The catalog contains all information about users (access rights, etc.), raw data location and SWfs (code libraries, application code). The provenance database captures all information about SWf execution. The cache index contains information about tasks and intermediate data produced, as well as the location of files that store the intermediate data. Thus, the cache index itself is small (only file references) and the cached data can be managed using the underlying file system. A good solution for implementing these metadata repositories is a modern key-value store, such as Cassandra (https://cassandra.apache.org), which provides efficient key-based access, scalability and fault-tolerance through replication in a shared-nothing cluster.

The raw data (files) are initially produced at some servers, *e.g.* in our use case, at the phenotyping platform and get transferred to the cloud site. The cache data (files) are produced at the cloud site after SWf execution. A good solution to store these files in a cluster is a distributed file system like Lustre (http://lustre.org) which is used a lot in HPC as it scales to high numbers of files.

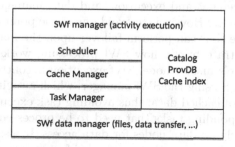

Fig. 2. SWfMS functional architecture

Figure 2 extends the SWfMS architecture proposed in [10], which distinguishes various layers, to support intermediate data caching. The SWf manager is the component that the user clients interact with to develop, share and execute workflows, using the metadata (catalog, provenance database and cache index). It determines the workflow activities that need to be executed, and generates the associated tasks for the scheduler. It also uses the cache index for SWf preprocessing to identify the intermediate data to reuse and the tasks that need not be re-executed.

The scheduler exploits the catalog and provenance database to decide which tasks should be scheduled to cloud sites. The task manager controls task execution and uses the cache manager to decide whether the task's output data should be placed in the cache. The cache manager implements the adaptive cache provisioning algorithm described in Sect. 4. The SWf data manager deals with data storage, using a distributed file system.

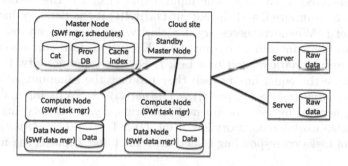

Fig. 3. SWfMS technical architecture

Figure 3 shows how these components are involved in SWf processing, using the traditional master-worker model. There are three kinds of nodes, master, compute and data nodes, which are all mapped to cluster nodes at configuration time, *e.g.* using a cluster manager like Yarn (http://hadoop.apache.org). The master node includes the SWf manager, scheduler and cache manager, and deals with the metadata. The master node is lightly loaded as most of the work of serving clients is done by the compute and data nodes (or worker nodes), which perform task management and execution, and data management, respectively. So, it is not a bottleneck. However, to avoid any single point of failure, there is a standby master node that can perform failover upon the master node's failure.

Let us now illustrate briefly how SWf processing works. User clients connect to the cloud site's master node. SWf execution is controlled by the master node, which identifies, using the SWf manager, which activities in the fragment can take advantage of cached data, thus avoiding task execution. The scheduler schedules the corresponding tasks that need to be processed on compute nodes which in turn will rely on data nodes for data access. It also adds the transfers of raw data from remote servers that are needed for executing the SWf. For each task, the task manager decides whether the task's output data should be placed in the cache taking into account storage costs, data size, network costs. When a task terminates, the compute node sends to its master the task's execution information to be added in the provenance database. Then, the master node updates the provenance database and may trigger subsequent tasks.

4   Cache Management

This section presents in details our techniques for cache management.

We start by introducing some terms and concepts. A SWf $W(A, D)$ is the abstract representation of a directed acyclic graph (DAG) of computational activities A and their data dependencies D. There is a dependency between two activities if one consumes the data produced by the other. An activity is a description of a piece of work and can be a computational script (computational activity), some data (data activity) or some set-oriented algebraic operator like map or filter [11]. The parents of an activity are all activities directly connected to its inputs. A task t is the instantiation of an activity during execution with specific associated input data. The input $Input(t)$ of t is the data needed for the task to be computed, and the output $Output(t)$ is the data produced by the execution of t. Whenever necessary, for clarity, we alternatively use the term intermediate data instead of output data. Execution data corresponds to the input and output data related to a task t. For the same activity, if two tasks t_i and t_j have the equal inputs then they produce the same output data, *i.e.*, $Input(t_i) = Input(t_j) \Rightarrow Output(t_i) = Output(t_j)$. A SWf's input data is the raw data generated by the experimental platforms, *e.g.*, a phenotyping platform. An executable workflow for workflow $W(A, D)$ is $W_{ex}(A, D, T, Input)$, where T is a DAG of tasks corresponding to activities in A and $Input$ is the input data.

In our solution, cache management is involved during two main steps: SWf preprocessing and cache provisioning. SWf preprocessing occurs just before execution and is done by the SWf manager using the cache index. The goal is to transform an executable workflow $W_{ex}(A, D, T, Input)$ into an equivalent, simpler subworkflow $W'_{ex}(A', D', T', Input')$, where A' is a subgraph of A with dependencies D', T' is a subgraph of T corresponding to A' and $Input' \in Input$. This is done by removing from the executable workflow all tasks and corresponding input data for which the output data is in the cache. The preprocessing step uses a recursive algorithm that traverses the DAG starting from the leaf nodes (corresponding to tasks). For each task t, if $Output(t)$ is already in the cache, this means that the entire subgraph of T whose leaf is t can be removed.

Figure 4 illustrates the preprocessing step on the Phenomenal SWf. The yellow tasks have their output data stored in the cache. They are replaced by the corresponding data as input for the subgraphs of tasks that need to be executed.

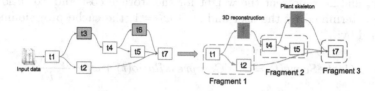

Fig. 4. DAG of tasks before pre-processing (left) and the selected fragments that need to be executed (right). (Color figure online)

The second step, cache provisioning, is performed during workflow execution. Traditional (in memory) caching involves deciding, as data is read from disk into memory, which data to replace to make room, using a cache replacement algorithm, *e.g.* LRU. In our context, using a disk-based cache, the question is different, *i.e.* to decide which task output data to place in the cache using a cache provisioning algorithm. This algorithm is implemented by the cache manager and used by the task manager when executing a task.

A simple cache provisioning algorithm, which we will use as baseline, is to use a *greedy* method that simply stores all tasks' output data in the cache. However, since SWf executions produce huge quantities of output data, this approach would incur high storage costs. Worse, for some short duration tasks, accessing cache data from disk may take much more time than re-executing the corresponding task subgraph from the input data in memory.

Thus, we propose a cache provisioning algorithm with an adaptive method that deals with the variations in task execution times and output data. The principle is to compute, for each task t, a score based on the sizes of the input and output data it consumes and produces, and the execution time of t. During workflow execution, the execution time of each task t, denoted by $ExTime(t)$, is stored in the provenance database. If t has already been executed, $ExTime(t)$ already exists in the provenance database. When t is re-executed, its execution time is recomputed and $ExTime(t)$ is updated as the average between the new and old execution times.

The adaptive aspect of our solution is to take into account task compression behavior. With a high compression ratio, it may be efficient to store the output data rather than the input data and recomputing it. In contrary, with a high expansion ratio, storing the input data rather than the output may save disk space.

Let $size(Input(t))$ and $size(Output(t))$ denote the input and output data size of a task t, respectively. The data compression ratio of a task quantifies the reduction of the input data processed by the task, $i.e.$,

$$CompressRatio(t) = \frac{size(Input(t))}{size(Output(t))} \tag{1}$$

Based on this data compression ratio, a cache provisioning score, denoted by $CacheScore$, is defined. For a task t, let F be a constant to normalize the time factor, ω_s and ω_t represent the weight for the storage cost and execution time, they are determined by the user and $\omega_s + \omega_t = 1$, the cache provisioning score is obtained by:

$$CacheScore(t) = \omega_s * CompressRatio(t) + \omega_t * \frac{T_{exec}(t)}{F} \tag{2}$$

The cache score reveals the relevancy of caching the output data of t and takes into account the compression metric and execution time. According to the weights provided by the user, she may prefer to give more importance to the compression ratio or executions time, depending on the storage capacity and available computational resources.

Then, during each task t execution, the task manager calls the cache manager to compute $CacheScore(t)$. If the computed value is bigger than the threshold provided by the user, then t's output data will be cached. This threshold is chosen based on the overhead of cache provisioning ($i.e.$, the time spent to store t's output data) and the cache size.

5   Experimental Evaluation

In this section, we first present our experimental setup. Then, we present our experiments and comparisons of different caching methods in terms of speedup and monetary cost in single user and multiuser modes. Finally, we give concluding remarks.

5.1   Experimental Setup

Our experimental setup includes the cloud infrastructure, SWf implementation and experimental dataset.

The cloud infrastructure is composed of one site with one data node (N1) and two identical compute nodes (N2, N3). The raw data is originally stored in an external server. During computation, raw data is transferred to N1, which contains Terabytes of persistent storage capacities. Each compute node has much computing power, with 80 vCPUs (virtual CPUs, equivalent to one core each of a 2.2 GHz Intel Xeon E7-8860v3) and 3 Terabytes of RAM, but less persistent storage (20 Gigabytes).

We implemented the Phenomenal workflow (see Sect. 2) using OpenAlea and deployed it on the different nodes using the Conda multi-OS package manager. The master node is hosted on one of the compute node (N2). The metadata repositories are stored on the same node (N2) using the Cassandra key-value store. Files for raw and cached data are shared between the different nodes using the Lustre file system. File transfer between nodes is implemented with ssh.

The Phenoarch platform has a capacity of 1,680 plants with 13 images per plant per day. The size of an image is 10 Megabytes and the duration of an experiment is around 50 days. The total size of the raw image dataset represents 11 Terabytes for one experiment. The dataset is structured as 1,680 time series, composed of 50 time points (one per plant and per day).

We use a version of the Phenomenal workflow composed of 9 main activities. We execute it on a subset of the use case dataset, that is $\frac{1}{25}$ of the size of the full dataset, or 440 Gigabytes of raw data, which represents the execution of 30,240 tasks.

5.2   Experiments

We execute the workflow on the subset dataset with different number of vCPUs and different caching methods. We consider workflow executions from a single user or multiple users to test the re-execution of the same workflow several times.

We compare three different caching methods: (1) no cache, (2) greedy, and (3) adaptive. Greedy and adaptive are described in Sect. 4.

In the single user scenario, the execution time corresponds to the transfer time of the raw data from the remote servers, the time to run the workflow and the time for cache provisioning, if any. In the multiuser scenario, the same workflow is executed on the same data several times (up to 6 times).

The raw data is fetched on the data node as follows: a first chunk is fetched from the remote data servers, then the remaining chunks are fetched while the execution starts on the first chunk. As the execution takes longer than transferring the raw data, we only count the time of transferring the first chunk in the execution time.

For the adaptive method, the coefficients ω_d and ω_t defined by the user are set to 0.5 each. The threshold is set to 0.4.

In the rest of this section, we compare the three methods in terms of speedup and monetary cost.

Speedup. We compare the speedup of the three caching methods. We define speedup as $speedup(n) = \frac{T_n}{T_{10}}$ where T_n is the execution time on n vCPUs and T_{10} is the execution time of the no cache method on 10 vCPUs.

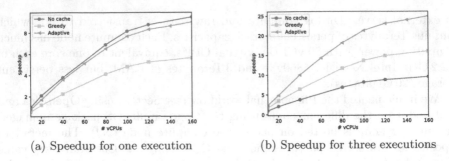

(a) Speedup for one execution   (b) Speedup for three executions

Fig. 5. Speedup versus number of vCPUs: without cache (red), greedy caching (blue), and adaptive caching (green). (Color figure online)

The workflow execution is distributed on nodes N2 and N3, for different numbers of vCPUs. For one execution, Fig. 5(a) shows that the fastest method is no cache (red curve). This is normal because there is no additional time to make data persistent and provision the cache. However, the overhead of cache provisioning with the adaptive method is very small (green curve in Fig. 5(a) compared with the greedy method (blue curve in Fig. 5(a) where all the output data are saved in the cache.

The speedup with adaptive goes up to 94.4% of that with no cache, while the speedup with greedy goes up to 59.9%. For instance, with 80 vCPUs, the execution time of the adaptive method (*i.e.*, 3,714 s) is only 5.8% higher than that of the no cache method (*i.e.* 3,510 s). This is much faster than the greedy method, which adds 68.2% of computation time in comparison with the no cache method. Re-execution with the greedy and adaptive methods have much smaller execution time than the first execution. The greedy method re-execution time is the fastest, with only 2.3% (*i.e.*, 129 s) of the no cache method execution time, because all the output data is already cached. Furthermore, as only the master node is working although no computation is done, the re-execution time is independent of the number of vCPUs and can be computed from a personal computer with limited vCPUs. The adaptive method re-execution time is a bit higher as 16.3% (*i.e.*, 572 s) of the no cache method execution time for a gain of 513%. With the adaptive method, some computation still needs to be done when the workflow is re-executed, but such re-execution on the whole dataset can be done in less than a day (*i.e.*, 19.4 h) on a 10 vCPUs machine, compared with 6.9 days with the no cache method. For three executions, starting without cache, Fig. 5(b) shows that the adaptive method is much faster than the other methods. The greedy method is faster than the no cache method in this case, because the additional time for the cache provisioning is compensated by the very short re-execution times of the greedy method. With 80 vCPUs, the speedup of the adaptive method (*i.e.*, 18.1) is 54.70% better than that of the greedy method (*i.e.*, 11.7) and 162.31% better than that of the no cache method (*i.e.*, 6.9). The adaptive method is faster on three executions than the other methods, despite having re-execution time higher than the greedy method, because the overhead of the cache provisioning is 57% smaller.

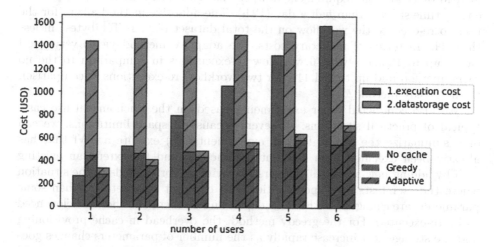

Fig. 6. Monetary cost depending on the number of workflow executions.

Monetary Cost. To compare the monetary of the three caching methods, we first define execution cost in US$ as follows:

$$Cost = Cost_{cpu} * ExTime + Cost_{disk} * TotalCache$$

where $ExTime$ is the total time of one or multiple executions in seconds, $TotalCache$ represents the size of the data in the cache in Gigabytes. $Cost_{cpu}$ and $Cost_{disk}$ are the pricing coefficients determined in $ per cpu per hour and $ per Gigabyte per month, respectively.

To set the price parameters, we use Amazon's cost model, *i.e.*, $Cost_{disk}$ is $0.1 per Gigabyte per month for storage and two instances at $5.424 per hour for computation, *i.e.*, $Cost_{cpu}$ is $10.848 per hour. As we can see from Fig. 6, the monetary cost of the adaptive method is much smaller than the greedy method due to the amount of cached data produced by the adaptive method (*i.e.*, 390 Gigabytes for the whole experimentation), which is much smaller than for the greedy method (*i.e.*, 3.9% of the total output data). In terms of monetary cost, the greedy method becomes more efficient than the no cache method at the sixth user in the month. The adaptive method is 28.40% less costly than the no cache method and 254.44% less costly than the greedy method for two executions. For six executions, the adaptive method is still 120.16% less costly than no cache method and 114.38% less costly than the greedy method.

5.3   Discussion

The adaptive method has better speedup compared to the no cache and greedy methods, with performance gains up to 162.31% and 54.70% respectively for three executions. The direct execution time gain for each re-execution is 344.9% for the adaptive method in comparison with the no cache method (*i.e.*, 3.9 h instead of 17.7). One requirement from the use case was to make workflow execution time shorter than half a day (12 h). The adaptive method allows for the user to re-execute the workflow on the total dataset (*i.e.*, 11 Terabytes) in less than 4 h. In terms of monetary costs, the adaptive method yields very good gains, up to 120.16% with 6 workflow re-executions in comparison to the no cache method and up to 254.44% for two workflow re-executions in comparison to the greedy method.

We also conducted other experiments based on the Phenomenal use case, typical of practical situations. However, because of space limitations, we can only summarize the results for two experiments: (1) execute a SWf that has already been executed with different parameters, and (2) extend an existing SWf by adding new activities. The first experiment corresponds to the situation where the user tests other possibilities with different parameters. When some parameters are changed, all the tasks depending on them and the one below need to be re-executed. For the greedy method, the overhead in cache provisioning and the storage cost increase rapidly as the number of parameters changes goes up. But the adaptive method has small overhead due to less data storage, and thus the increase of the storage cost is an order of magnitude smaller than that with greedy.

In the second experiment, the structure of the workflow is modified by adding new activities as discussed in Sect. 2. Similar to what happens with re-execution of a single SWf, the monetary cost of the greedy method is higher than the no cache method for up to 6 executions with different fragments or different parameters. And the execution time of greedy is always better than no cache. The adaptive method is both faster and cheaper than both no cache and greedy.

6   Conclusion

In this paper, we proposed an adaptive caching solution for efficient execution of data-intensive workflows in the cloud. Our solution automatically manages the storage and reuse of intermediate data and adapts to the variations in task execution times and output data size. The adaptive aspect our solution is to take into account task compression behavior.

We implemented our solution in the OpenAlea system and performed extensive experiments on real data with the Phenomenal workflow, with 11 Terabytes of raw data. We compared three methods: no cache, greedy, and adaptive. Our experimental validation shows that the adaptive method allows caching only the relevant output data for subsequent re-executions by other users, without incurring a high storage cost for the cache. The results show that adaptive caching can yield major performance gains, *e.g.*, up to 120.16% with 6 workflow re-executions.

This work solves an important issue in experimental science like biology, where scientists extend existing workflows with new methods or new parameters to test their hypotheses on datasets that have been previously analyzed.

Acknowledgments. This work was supported by the #DigitAg French Convergence Lab. on Digital Agriculture (http://www.hdigitag.fr/com), the SciDISC Inria associated team with Brazil, the Phenome-Emphasis project (ANR-11-INBS-0012) and IFB (ANR-11-INBS-0013) from the Agence Nationale de la Recherche and the France Grille Scientific Interest Group.

References

1. Adams, I.F., Long, D.D., Miller, E.L., Pasupathy, S., Storer, M.W.: Maximizing efficiency by trading storage for computation. In: HotCloud (2009)
2. Altintas, I., Barney, O., Jaeger-Frank, E.: Provenance collection support in the Kepler scientific workflow system. In: Moreau, L., Foster, I. (eds.) IPAW 2006. LNCS, vol. 4145, pp. 118–132. Springer, Heidelberg (2006). https://doi.org/10.1007/11890850_14
3. Artzet, S., Brichet, N., Chopard, J., Mielewczik, M., Fournier, C., Pradal, C.: OpenAlea.Phenomenal: a workflow for plant phenotyping, September 2018. https://doi.org/10.5281/zenodo.1436634
4. Callahan, S.P., Freire, J., Santos, E., Scheidegger, C.E., Silva, C.T., Vo, H.T.: VisTrails: visualization meets data management. In: ACM SIGMOD International Conference on Management of Data (SIGMOD), pp. 745–747 (2006)
5. Cohen-Boulakia, S., et al.: Scientific workflows for computational reproducibility in the life sciences: status, challenges and opportunities. Future Gener. Comput. Syst. (FGCS) **75**, 284–298 (2017)
6. Deelman, E., Singh, G., Livny, M., Berriman, B., Good, J.: The cost of doing science on the cloud: the montage example. In: International Conference for High Performance Computing, Networking, Storage and Analysis, pp. 1–12 (2008)
7. Garijo, D., Alper, P., Belhajjame, K., Corcho, O., Gil, Y., Goble, C.: Common motifs in scientific workflows: an empirical analysis. Future Gener. Comput. Syst. (FGCS) **36**, 338–351 (2014)
8. Kelling, S., et al.: Data-intensive science: a new paradigm for biodiversity studies. BioScience **59**(7), 613–620 (2009)
9. Liu, J., et al.: Efficient scheduling of scientific workflows using hot metadata in a multisite cloud. IEEE Trans. Knowl. Data Eng. 1–20 (2018)
10. Liu, J., Pacitti, E., Valduriez, P., Mattoso, M.: A survey of data-intensive scientific workflow management. J. Grid Comput. **13**(4), 457–493 (2015)
11. Ogasawara, E., Dias, J., Oliveira, D., Porto, F., Valduriez, P., Mattoso, M.: An algebraic approach for data-centric scientific workflows. Proc. VLDB Endow. (PVLDB) **4**(12), 1328–1339 (2011)
12. de Oliveira, D., Baião, F.A., Mattoso, M.: Towards a taxonomy for cloud computing from an e-Science perspective. In: Antonopoulos, N., Gillam, L. (eds.) Cloud Computing. Computer Communications and Networks, pp. 47–62. Springer, London (2010). https://doi.org/10.1007/978-1-84996-241-4_3
13. Özsu, M.T., Valduriez, P.: Principles of Distributed Database Systems, 3rd edn. Springer, New York (2011). https://doi.org/10.1007/978-1-4419-8834-8
14. Pradal, C., et al.: InfraPhenoGrid: a scientific workflow infrastructure for plant phenomics on the grid. Future Gener. Comput. Syst. (FGCS) **67**, 341–353 (2017)
15. Pradal, C., Cohen-Boulakia, S., Heidsieck, G., Pacitti, E., Tardieu, F., Valduriez, P.: Distributed management of scientific workflows for high-throughput plant phenotyping. ERCIM News **113**, 36–37 (2018)

16. Roitsch, T., et al.: Review: new sensors and data-driven approaches–a path to next generation phenomics. Plant Sci. **282**, 2–10 (2019)
17. Tardieu, F., Cabrera-Bosquet, L., Pridmore, T., Bennett, M.: Plant phenomics, from sensors to knowledge. Curr. Biol. **27**(15), R770–R783 (2017)
18. Yuan, D., Yang, Y., Liu, X., Chen, J.: A cost-effective strategy for intermediate data storage in scientific cloud workflow systems. In: IEEE International Symposium on Parallel and Distributed Processing (IPDPS), pp. 1–12 (2010)
19. Yuan, D., et al.: A highly practical approach toward achieving minimum data sets storage cost in the cloud. IEEE Trans. Parallel Distrib. Syst. **24**(6), 1234–1244 (2013)
20. Zhang, J., et al.: Bridging VisTrails scientific workflow management system to high performance computing. In: 2013 IEEE Ninth World Congress on Services, pp. 29–36. IEEE (2013)

Correction to: Keyword Search Based Mashup Construction with Guaranteed Diversity

Huanyu Cheng, Ming Zhong, Jian Wang, and Tieyun Qian

Correction to:
Chapter "Keyword Search Based Mashup Construction
with Guaranteed Diversity" in: S. Hartmann et al. (Eds.):
Database and Expert Systems Applications, **LNCS 11707,**
https://doi.org/10.1007/978-3-030-27618-8_31

In the originally published version of chapter 31 the funding information in the acknowledgement section was incomplete. This has now been corrected.

The updated version of this chapter can be found at
https://doi.org/10.1007/978-3-030-27618-8_31

© Springer Nature Switzerland AG 2020
S. Hartmann et al. (Eds.): DEXA 2019, LNCS 11707, p. C1, 2020.
https://doi.org/10.1007/978-3-030-27618-8_34

Author Index

Printed in the United States
By Bookmasters